Electromagnetic Waves and Signal Processing

Electromagnetic Waves and Signal Processing

Edited by **George Pilato**

Willford Press

New York

Published by Willford Press,
118-35 Queens Blvd., Suite 400,
Forest Hills, NY 11375, USA
www.willfordpress.com

Electromagnetic Waves and Signal Processing
Edited by George Pilato

International Standard Book Number: 978-1-68285-283-5 (Hardback)

Printed in China.

Contents

	Preface	IX
Chapter 1	**Investigations for broadband internet within high-speed trains** I. Gaspard and G. Zimmermann	1
Chapter 2	**Cordic based algorithms for software defined radio (SDR) baseband processing** B. Heyne and J. Götze	7
Chapter 3	**A hybrid equalizer/rake receiver for the wideband CDMA downlink in large delay spread channels** C. Buchacher, M. Zimmermann, S. Paul and M. Huemer	13
Chapter 4	**Multicarrier systems and filter banks** D. S. Waldhauser and J. A. Nossek	19
Chapter 5	**Improving the performance of BICM-ID and MLC systems with different FEC codes** T. Arafa, W. Sauer-Greff and R. Urbansky	24
Chapter 6	**Design space exploration of high throughput finite field multipliers for channel coding on Xilinx FPGAs** C. de Schryver, S. Weithoffer, U. Wasenmüller and N. Wehn	31
Chapter 7	**On parallel random number generation for accelerating simulations of communication systems** C. Brugger, S. Weithoffer, C. de Schryver, U. Wasenmüller and N. Wehn	38
Chapter 8	**Analytical calculation of intrinsic shielding effectiveness for isotropic and anisotropic materials based on measured electrical parameters** M. Kühn, W. John and R. Weigel	45
Chapter 9	**A relation between algebraic and transform-based reconstruction technique in computed tomography** S. Kiefhaber, M. Rosenbaum, W. Sauer-Greff and R. Urbansky	52
Chapter 10	**The road surveying system of the federal highway research institute – a performance evaluation of road segmentation algorithms** R. Streiter and G. Wanielik	58
Chapter 11	**A wireless MIMO CPM system with blind signal separation for incoherent demodulation** O. Weikert and U. Zölzer	63

Chapter 12 **Validation of a measurement method for magnetic shielding effectiveness of a wire mesh enclosure with comparison to an analytical model** **68**
M. Kühn, W. John and R. Weigel

Chapter 13 **Reproducing system-level bulk current injection test in direct power injection setup for multiple-port DUTs** **75**
S. Miropolsky and S. Frei

Chapter 14 **Analytic investigations on the susceptibility of nonlinear analog circuits to substrate noise** **81**
C. Widemann, S. Stegemann, W. John and W. Mathis

Chapter 15 **Substituting EMC emission measurement by field and cable scan method using measured transfer function** **86**
D. Rinas, J. Jia, A. Zeichner and S. Frei

Chapter 16 **DMT transmission in the context of industrial telecontrol applications** **92**
S. Edinger, C. Bauer and N. J. Fliege

Chapter 17 **Adaptive decoding of convolutional codes** **98**
K. Hueske, J. Geldmacher and J. Götze

Chapter 18 **Optimized power allocation for iterative multiuser detection for a SC-CDMA uplink** **104**
P. Weitkemper, K. Zielinski, K.-D. Kammeyer and R. Laur

Chapter 19 **Calibration method for direct conversion receiver front-ends** **110**
R. Müller and H. -J. Jentschel

Chapter 20 **Considering even-order terms in stochastic nonlinear system modeling with respect to broadband data communication** **116**
O. Schmitz, S. K. Hampel, H. Eul and D. Schwingshackl

Chapter 21 **Quantitative design space exploration of routing-switches for Network-on-Chip** **123**
M. C. Neuenhahn, H. Blume and T. G. Noll

Chapter 22 **Simulating circuits with impasse points** **129**
T. Thiessen and W. Mathis

Chapter 23 **Automatic focus algorithms for TDI X-Ray image reconstruction** **135**
J. Dörr, M. Rosenbaum, W. Sauer-Greff and R. Urbansky

Chapter 24 **Auxiliary branch method and modified nodal voltage equations** **142**
A. Reibiger

Chapter 25 **An energy efficient weakly programmable MIMO detector architecture** **149**
C. Gimmler-Dumont and N. Wehn

Chapter 26 **Best usage of free-space capacitors in ASIC regulators** **155**
T. Maier, D. Droste and M. Siegel

Chapter 27 **Comparison of reconfigurable structures for flexible word-length multiplication** **161**
O. A. Pfänder, R. Nopper, H. -J. Pfleiderer, S. Zhou and A. Bermak

Chapter 28 **FPGA implementation of trellis decoders for linear block codes** **167**
S. Scholl, E. Leonardi and N. Wehn

Chapter 29 **Systematic design of output filters for audio class-D amplifiers via simplified real frequency technique** **174**
E. Hintzen, T. Vennemann and W. Mathis

Chapter 30 **Parameter estimation under non-stationary circumstances using extended signal model** **178**
M. Lechtenberg, J. Götze, K. Görner and C. Rehtanz

Chapter 31 **Indoor radio channel modeling and mitigation of fading effects using linear and circular polarized antennas in combination for smart home system at 868 MHz** **184**
S. Wunderlich, M. Welpot and I. Gaspard

Chapter 32 **Investigation of 3G radio network topologies considering performance and exposure aspects** **191**
H. Buddendick, P. Wertz, T. F. Eibert, F. M. Landstorfer and G. Wölfle

Chapter 33 **High-end silicon PDICs** **197**
H. Zimmermann

Chapter 34 **Analysis of communications and implementation performance of FFT based carrier synchronization of BPSK/QPSK bursts** **203**
U. Wasenmüller, T. Brack and N. Wehn

Chapter 35 **The influence of different PAST-based subspace trackers on DaPT parameter estimation** **209**
M. Lechtenberg and J. Götze

Chapter 36 **Matrix-vector based fast fourier transformations on SDR architectures** **215**
Y. He, K. Hueske, J. Götze and E. Coersmeier

Chapter 37 **A first approach to the distortion analysis of nonlinear analog circuits utilizing X-parameters** **221**
H. Weber, C. Widemann and W. Mathis

Chapter 38 **A code-aided synchronization IP core for iterative channel decoders** **226**
I. Ali, U. Wasenmüller and N. Wehn

Chapter 39 **A design study on complexity reduced multipath mitigation** **232**
U. Wasenmüller, T. Brack, I. Groh, E. Staudinger, S. Sand and N. Wehn

Chapter 40 **Application of syndrome based Turbo decoding with adaptive computational complexity in LTE downlink** **239**
J. Geldmacher, K. Hueske, M. Kosakowski and J. Götze

Chapter 41 **An identification procedure of multi-input Wiener models for the distortion analysis of nonlinear circuits** **246**
T. Weng, S. Stegemann, W. John and W. Mathis

Chapter 42 **Design of RF MEMS based switch matrix for space applications** **252**
S. Di Nardo, P. Farinelli, T. Kim, R. Marcelli, B. Margesin, E. Paola, D. Pochesci, L. Vietzorreck and F. Vitulli

Permissions

List of Contributors

Preface

The development of communications and information technology could not have been as rapid without the progress in electromagnetic waves and signal processing. This book provides comprehensive insights into this field. It is a complete source of knowledge on its present status as it presents various important topics like wave particles, propagation speed, types of waves, types of signal processing, etc. In this book, using case studies and examples, constant effort has been made to make the understanding of the difficult concepts of signal processing and theories associated with electromagnetic waves as easy and informative as possible.

All of the data presented henceforth, was collaborated in the wake of recent advancements in the field. The aim of this book is to present the diversified developments from across the globe in a comprehensible manner. The opinions expressed in each chapter belong solely to the contributing authors. Their interpretations of the topics are the integral part of this book, which I have carefully compiled for a better understanding of the readers.

At the end, I would like to thank all those who dedicated their time and efforts for the successful completion of this book. I also wish to convey my gratitude towards my friends and family who supported me at every step.

Editor

1

Investigations for broadband internet within high-speed trains

I. Gaspard and G. Zimmermann

T-Systems, BU Technologiezentrum, Am Kavalleriesand 3, 64295 Darmstadt, Germany

Abstract. Broadband IP-based multimedia services are rapidly growing and there is the trend to use these services everywhere. Due to the fact that passengers in long distance high speed trains have a very comfortable environment to use their own terminal equipment, e.g. laptops with IEEE802.11×WLAN capability, it seems to be very promising to deliver broadband internet into trains. The challenge is the land-train connection where a high bit data rate link, e.g. 1 Mbps in average in up- and downlink, has to be provided to enable several users within the train broadband internet access at the same time even at high speeds of the train. This paper describes a systematic approach starting at basic propagation measurements in a typical train environment and ending up with lab investigations by means of hardware channel simulation for different existing radio technologies which are candidates for the land-train connection.

1 Introduction

Market studies show that WLAN in trains has an even bigger potential than airport hotspots (BWCS report, 2003) for the coming years. A first key finding is that the market for train-based wireless internet services is six times bigger than that for airport hotspots. The second main finding is that laptop users spend more than 12.6 billion h on trains per year but just under two billion hours at airports. Thus it seems very attractive to provide broadband internet to the passengers of long distance high speed trains. In principle the solution is as shown in Fig. 1.

The passenger uses his or her own terminal with WLAN capability and is thus connected within a WLAN inside the train. The access points of the train WLAN infrastructure are connected to an universal train communication gateway which provides a wireless broadband link to one or more landside access networks. The challenging requirement for the train-land connection is twofold: first it has to be broadband, e.g. 1 Mbps each up- and downlink, to be able to carry the traffic generated of several passengers at the same time. Second the performance of this link has to be of reasonable performance even at very high speeds, e.g. 300 km/h.

Correspondence to: I. Gaspard
(ingo.gaspard@t-systems.com)

2 Channel sounder measurements

A versatile representation of the real characteristics of the mobile radio channel is given by the measured broadband channel impulse responses. These channel sounder measurements measured in typical reception scenarios can be used as a basis for radio system link performance investigations as well as for coverage planning investigations as given in Fig. 2.

The time-consuming and costly field measurements have to be done only once. Variation of parameters, like investigations at different velocities of the mobile terminal, can be considered by post-processing of the measured impulse responses.

2.1 Field measurement setup

To provide a realistic basis for further investigations channel sounder measurements were done with a RUSK WLL channel sounder (see: http://www.medav.de). Measurements were taken for two different transmitter positions A and B, see Fig. 3. In both experiments the mobile receiver was moved along a track in parallel to the railway track. For transmitter position A metal structures like poles and electrical power lines of the train infrastructure may have an impact onto wave propagation whereas for position B always a line of sight condition was given.

The main parameters of the channel sounder measurement experiments are given in Table 1. The distance between successive impulse responses was chosen to fulfil the spatial sampling theorem and thus allowing Doppler spectra derivation.

Figure 4 shows two photographs of the field measurement environment. The metallic train infrastructure towering into the line of sight for transmitter position A can clearly be seen.

2.2 Evaluation of field measurements

To verify the assumption that the metal infrastructure of the train track has an impact onto radio wave propagation the median values of path loss out of sections of 1 m length corresponding to ≈40 impulse responses along the measurement track were analyzed for transmitter position A and B. Figure 5 shows the distribution of path loss for the two cases

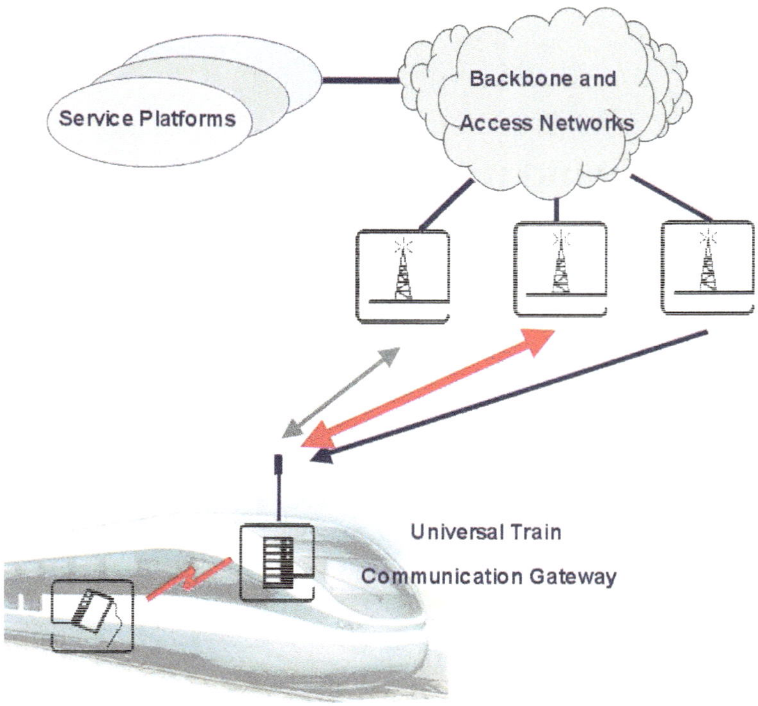

Fig. 1. Broadband internet in train – principle solution.

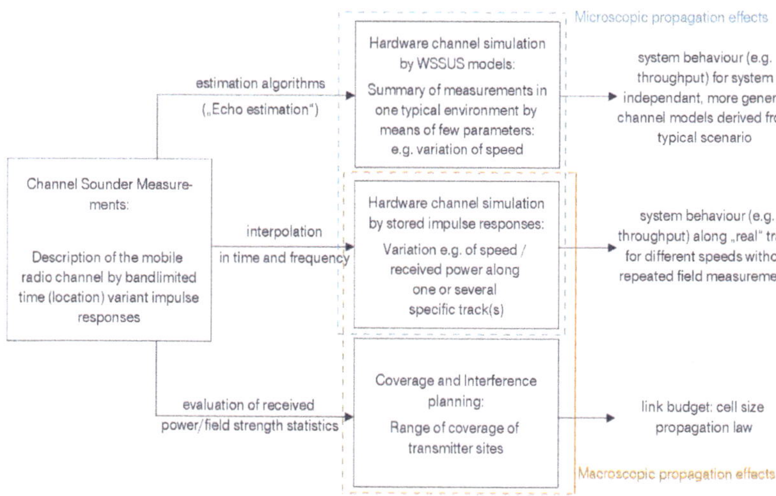

Fig. 2. Channel sounder measurements as the basis of further investigations.

Table 1. Main parameters of channel sounder experiments.

Carrier frequencies	1.8 GHz / 5.2 GHz
Measurement bandwidth	22 MHz
Max. length / resolution of impulse response	25.6 μs / 25 ns
Distance between successive impulse responses	2.46 cm
TX antenna height /gain	9.7 m / 7.5 dBi
RX antenna height /gain	1.7 m / 2.15 dBi (1.8 GHz), 0 dBi (5.2 GHz)
Distances between RX and TX	40 m . . . 400 m

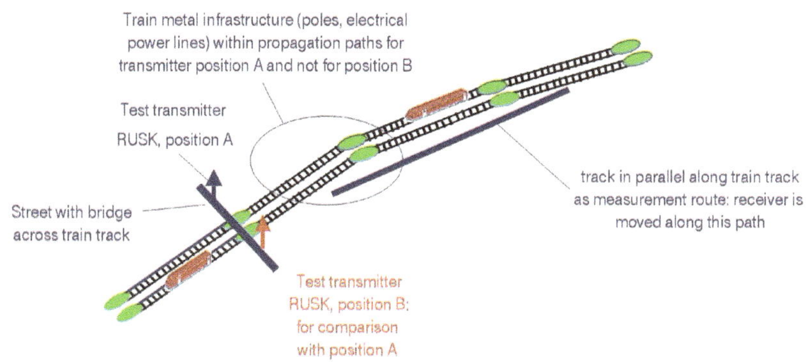

Fig. 3. Field measurement setup with different transmitter positions A and B.

Fig. 4. Field measurement environment: view from RX towards TX antennas positions (left), view from TX antenna position B towards receiver (right).

with and without metal structures reaching into the line of sight between RX and TX. It can clearly be conducted that the presence of the metal structures for transmitter position B causes an increased path loss.

For further analysis WSSUS models were derived for successive blocks consisting of 64 impulse responses each by the methods developed in U. Martin (1994). The evaluation of estimated far field paths given in Table 2 shows that by the presence of metal structures in the line of sight between TX and RX more paths are estimated due to the more complicated propagation by additional diffraction and scattering.

Out of all WSSUS models estimated along the measurement run for 1.8 GHz carrier frequency and TX position A three new simple two-path WSSUS models were derived as typical representative models and are given in Table 3. These models are only differing in the power ratio of the first to second path and are thus differing in frequency selectivity.

3 Measurement based lab investigations

Several different existing radio technologies were investigated by means of hardware simulation of the radio channel. The main important performance measures were the throughput as a function of maximum Doppler and thus speed and the sensitivity which determines at the end the density of base stations along the track to have coverage.

In Fig. 6 a generic lab setup including the channel simulator is depicted.

In the following some measurement examples are given.

Figure 7 shows the throughput performance for UMTS FDD in downlink in case of TCP/IP traffic over 384 kbps bearer for different simulated radio channels at constant speeds as a function of receiver input power. The "constant" channel corresponds to the AWGN channel whereas the two "2 path models" are derived from the measurements, see Table 3. The channel for the "Rice/Ratio 1 dB" is a frequency flat Rice channel with a power ratio of direct to

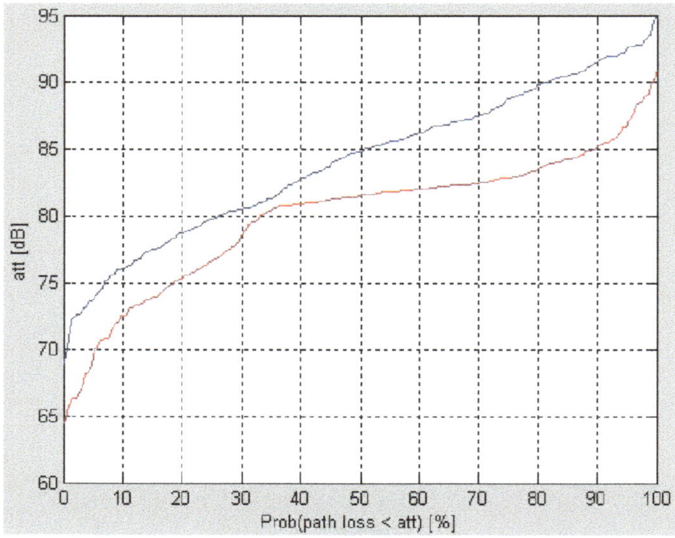

Fig. 5. Distribution of path loss (or attenuation) for TX position A/with metal structures (blue) and B/without metal structures (red).

Table 2. Evaluation of estimated WSSUS models.

Number of paths in WSSUS models (@1.8 GHz)	1	2	3	>=4
Distance RX–TX:216 m . . . 400, without metal structures (%)	17.7	82.3	0	0
Distance RX–TX:216 m . . . 400, with metal structures (%)	12.4	38.9	41.2	7.5

Table 3. Two-path WSSUS models derived from measurement campaign @1.8 GHz.

Model #	1	2	3
Rice factor of first path		10 dB	
Rice factor of second path		−5 dB	
Power ratio first to second path	5 dB	10 dB	15 dB
Delay between first and second path		0.2 μs	
Doppler spectra for first and second path	Classical Jakes, in addition: Dirac at $+f_{Dmax}$		

scattered components of 1 dB. The "Rural Area" channel is a 6 path COST207 channel which is the most frequency selective channel in comparison to the aforementioned channels. As can be seen from the curves throughput is nearly constant with decreasing input power down to –105 dBm. For the frequency flat channel "Rice/Ratio 1 dB" as well as for the frequency selective channel "Rural Area" only 64 kbps are possible at the given speeds whereas for the two channels derived from measurements more than 300 kbps are reached.

A similar measurement result of an IEEE802.11b link where data rate against input power was measured for different radio channels is given in Fig. 8. Throughput drops down already at moderate receiver input power values, e.g. for the "Rice/Ratio 1 dB" channel an input power of at least –85 dBm is necessary to allow at throughput of 1 Mbps. This input power value is a very sensitive to the applied channel. Thus and because of EIRP limitation in the 2.4 GHz band a very high density of access points along the track would be necessary for full coverage.

As a further example throughput measurement results for a preliminary IEEE802.16 (WiMAX) system are given in Fig. 9. In this figure the TCP/IP throughput as function of speed in downlink is given for the system operated at 3.5 GHz carrier frequency occupying a bandwidth of 3.5 MHz for different modulation schemes of the OFDM subcarriers as well as for different radio channels at a constant receiver input power of –65 dBm. For most of the channels and most of the modulation schemes a fast decreasing throughput with increasing speed can be observed. Only for the most robust modes and for the frequency flat channels there is less dependence on speed. Especially channel estimation seems to be a critical item due to the fact that IEEE802.16 focused on fixed and not on mobile wireless access. It has to be noticed that IEEE802.16e enhancements will allow also enhanced channel estimation as well as it will include handover mechanisms.

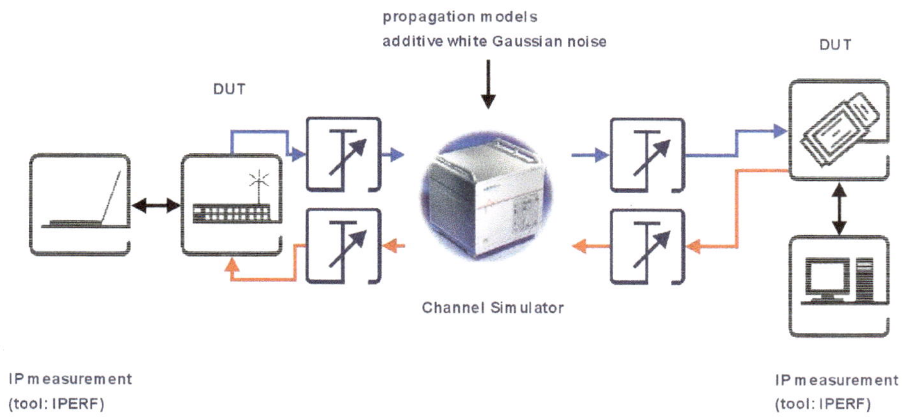

Fig. 6. Generic lab setup for channel simulator measurements. DUT= Device Under Test.

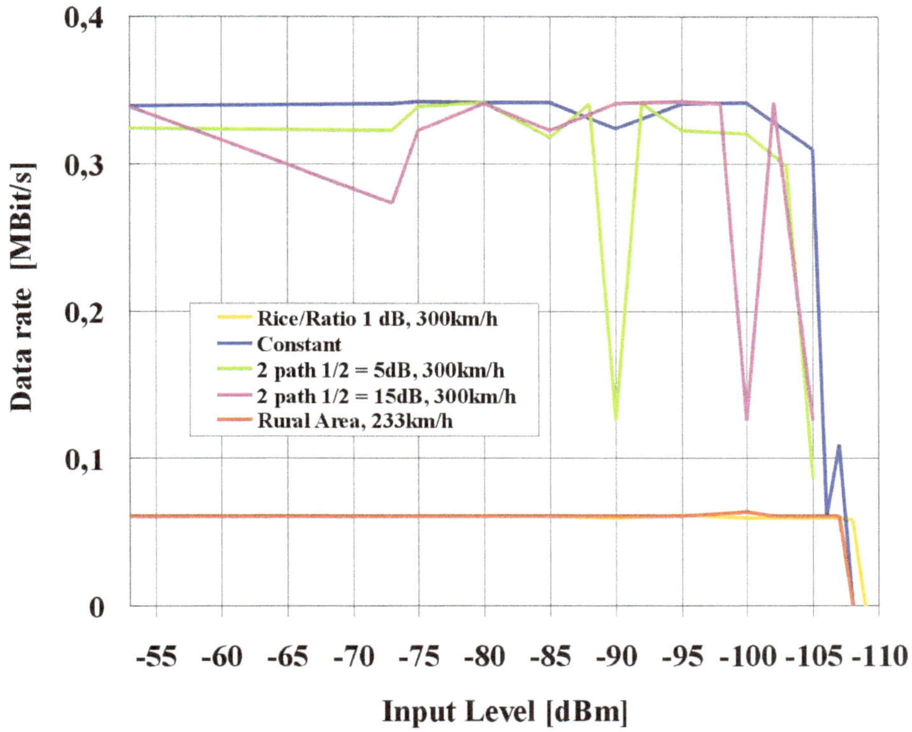

Fig. 7. Measured TCP/IP throughput in UMTS FDD downlink as a function of receiver input power for different channels.

4 Conclusion

In this paper we showed a systematic approach to investigate existing radio systems in a high speed train application which was not their primary application area. Starting with realistic channel impulse response measurements which allow a general description of the radio channel properties WSSUS models were derived. It could be shown measurement evaluation that typical infrastructure like poles and electrical power lines along the tracks have an impact onto the propagation characteristics of the electromagnetic waves between transmitter and receiver. By means of hardware simulation measurements were performed regarding IP throughput on the basis of the derived WSSUS models. For comparison reasons the performance for 'Rural Area' and frequency flat channel models were also investigated. It turns out that the UMTS system is performing well for various channels but delivers only relatively low data rates. The IEEE802.11b system is able to deliver high data rates in principle but these are fast decreasing for realistic multipath channels at high velocities. This holds even for high S/N values. Also the preliminary IEEE802.16 system which was investigated is not well suited for high speed applications. Enhancements are expected for the near future because mobility aspects like enhanced channel estimation capabilities and handover mechanisms are currently in the IEEE standardization process.

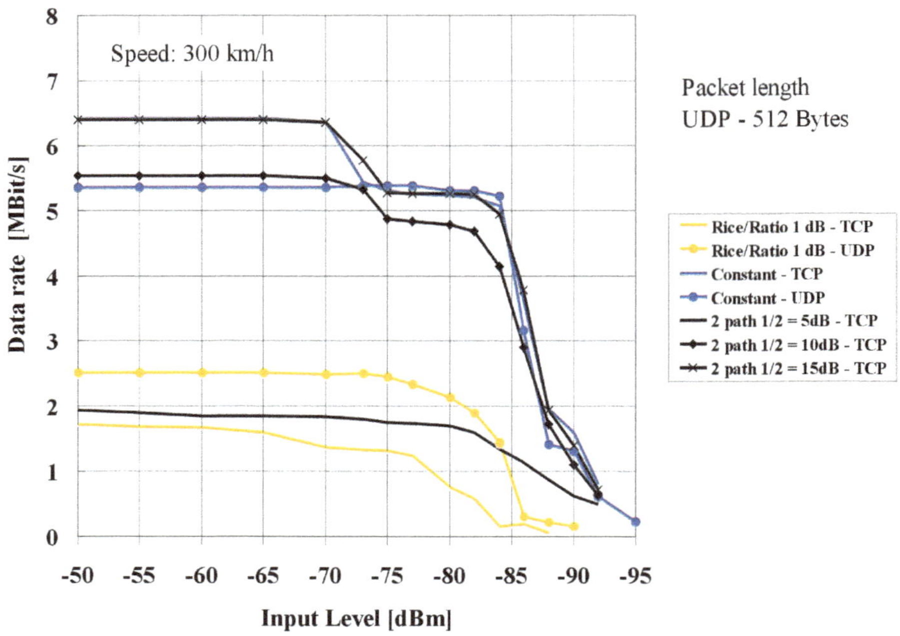

Fig. 8. Measured TCP/IP and UDP throughput in IEEE802.11b as a function of receiver power for different channels.

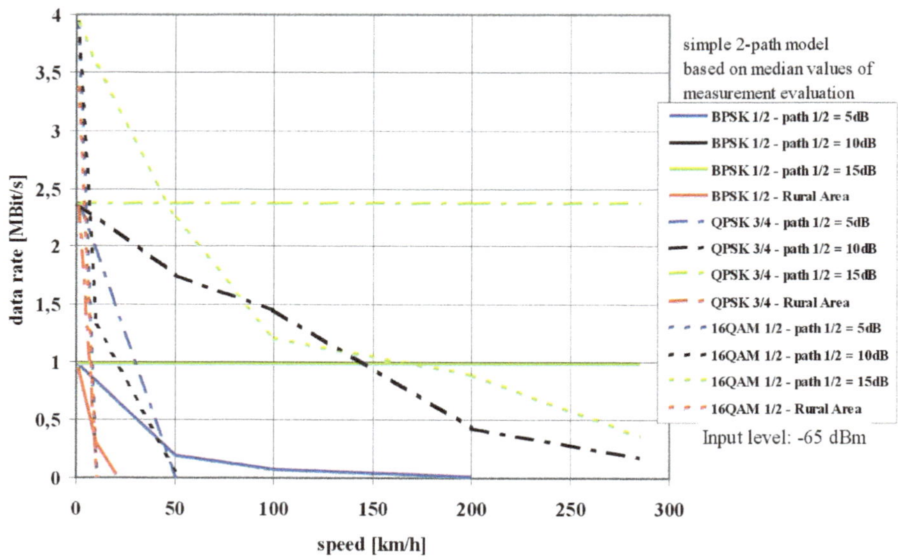

Fig. 9. Measured TCP/IP throughput as function of speed for a preliminary IEEE802.16 system operated at 3.5 GHz with 3.5 MHz bandwidth for different modulation schemes and different radio channels.

References

BWCS report: Railway Wireless LAN Services, BWCS limited, 2003.

http://www.medav.de

Martin, U.: Ausbreitung in Mobilfunkkanälen: Beiträge zum Entwurf von Messgeräten und zur Echoschätzung, Dissertation Universität Erlangen, 1994.

2

Cordic based algorithms for software defined radio (SDR) baseband processing

B. Heyne and J. Götze

University of Dortmund, Information Processing Lab, Otto-Hahn-Str. 4, 44227 Dortmund, Germany

Abstract. This paper presents two Cordic based algorithms which may be used for digital baseband processing in OFDM and/or CDMA based communication systems. The first one is a linear least squares based multiuser detector for CDMA incorporating descrambling and despreading. The second algorithm is a pure Cordic based FFT implementation. Both algorithms can be implemented using solely Cordic based architectures (e.g. coprocessors or ASIPs). The algorithms exactly fit the needs of a multistandard terminal as they both are freely parameterizable. This regards to the accuracy of the results as well as to the parameters of the performed function (e.g. size of the FFT).

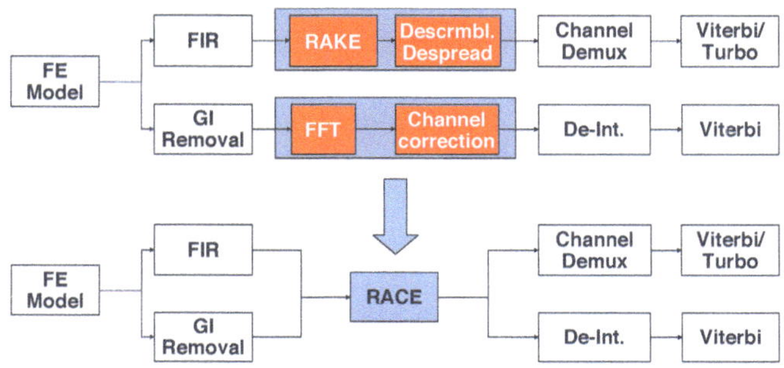

Fig. 1. Software defined baseband processing for WLAN and UMTS using the RACE accelerator.

1 Introduction

A lot of modern signal processing applications require such a high computational power that only ASICs can fulfill the technical demands. Unfortunately, ASICs are inflexible, costly (development and debugging) and only economical for mass-products. As a consequence, system designers are striving to replace specialized hardware solutions with software based solutions as developments in the field of software radio demonstrate. Due to the fact that even the most commonly used programmable devices, i.e. DSPs, often lack the required processing power, one tries to develop a solution that lays somewhere in between the two extrema programmable signal processing and dedicated hardware. The efforts in this area are summarized with the term reconfigurable computing.

We are using this approach to create a common software defined baseband implementation for UTRA/UMTS-FDD and WLAN as shown in Fig. 1. The most computational intensive tasks are performed on a dedicated hardware accelerator called RACE using Cordic processing elements. The implementation of a Cordic based the FFT and a Cordic based linear equalizer/multiuser detector for CDMA, including descrambling and despreading, is presented in this paper.

As the equalizer and the FFT can now be build upon solely Cordics, we have derived a software defined architecture for a mobile multi-standard terminal. The main processing blocks of the WLAN and UMTS baseband can be replaced by this programmable architecture.

The paper is organized as follows. At first the multiuser detector including the system model, algorithm and simulations is described in Sect. 2. Secondly the Cordic based FFT algorithm and simulation results are presented in Sect. 3, followed by an overview of the RACE coprocessor in Sect. 4. Finally the conclusions are given in Sect. 5.

2 CDMA multiuser detection

2.1 System model

Consider a CDMA downlink where all u users share the same channel. Thus the received signal also contains the signal of

Correspondence to: B. Heyne
(benjamin.heyne@uni-dortmund.de)

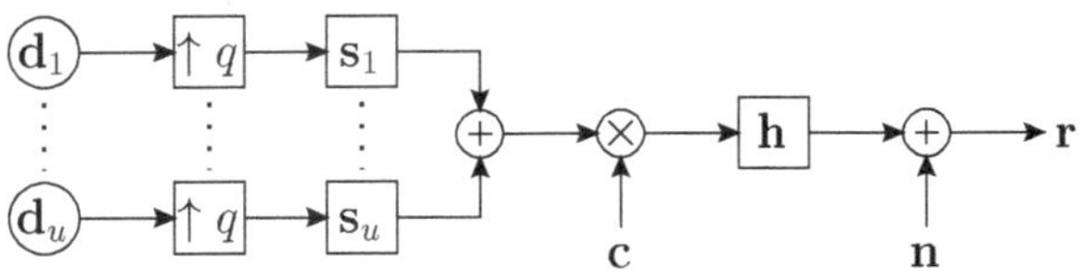

Fig. 2. Block diagram of the system model.

the other users. The system model used is shown in Fig. 2.

The incoming m complex data symbols of user i, collected in the vector $\mathbf{d}_i$, are first upsampled by the spreading factor q, so that one symbol now consists of q chips. Each upsampled symbol is now convolved with an OVSF code (3GPP, 2002) contained in the vector $\mathbf{s}_i$ of length q. Finally, the summed data streams are scrambled with the complex data sequence in vector $\mathbf{c}$ which is repeated for every data frame (38400 chips in UTRA/FDD).

The received chips are now obtained by propagating the signal through a channel, which is characterized by its complex valued channel impulse response vector $\mathbf{h}$ of length h_l, and adding an AWGN component $\mathbf{n}$.

Therefore the received data vector $\mathbf{r}$ is given by

$$\mathbf{r} = \mathbf{n} + \mathbf{HC} \sum_{i=1}^{u} \mathbf{S}_i \mathbf{d}_i \tag{1}$$

where $\mathbf{H}$ is a convolutional matrix describing the time variant complex channel, $\mathbf{C}$ is a complex valued diagonal matrix containing the scrambling code $\mathbf{c}$ on its main diagonal and $\mathbf{S}_i$ is a block Toeplitz spreading matrix. In this matrix each block is one column wide and contains the OVSF code for the i-th data stream.

For the proposed algorithm it is assumed that the received signal $\mathbf{r}$ has already passed the chip matched filter and has been sampled at chip rate. We also assume that the channel impulse response $\mathbf{h}$ (or the strongest taps of it) is known, as the channel estimation is not part of this paper.

2.2 Multiuser detection

The derivation of the single user detector/equalizer from this model is described in Heyne et al. (2003). For the multiuser detection we are interested in the first j users and will treat the other users as an additional noise component, so that $\mathbf{r}$ is changed to

$$\mathbf{r} = \mathbf{n}' + \mathbf{HC} \sum_{i=1}^{j} \mathbf{S}_i \mathbf{d}_i \tag{2}$$

with

$$\mathbf{n}' = \mathbf{n} + \mathbf{HC} \sum_{i=j+1}^{u} \mathbf{S}_i \mathbf{d}_i \, .$$

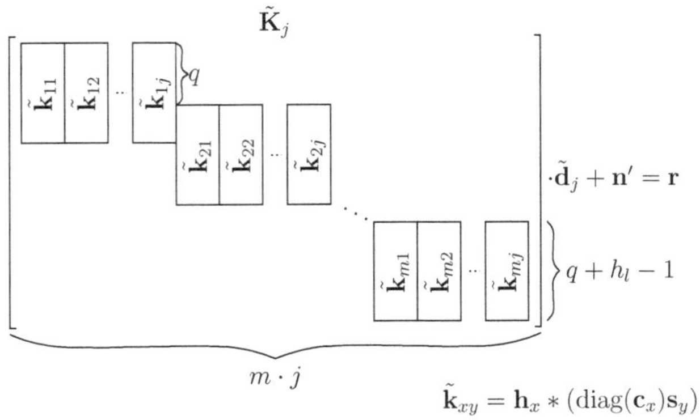

Fig. 3. Resulting structure of the system matrix $\tilde{\mathbf{K}}_j$. The $*$-operator indicates a convolution.

To describe $\mathbf{r}$ as a simple matrix-vector multiplication, we create a new symbol vector $\tilde{\mathbf{d}}_j$ which contains the time interleaved data symbols $\mathbf{d}_i$ of the first j users

$$\tilde{\mathbf{d}}_j = [d_{11} \; d_{12} \; \cdots \; d_{1j} \; d_{21} \; d_{22} \; \cdots \; d_{2j} \; \cdots \; d_{mj}]^T \, ,$$

and a new spreading matrix $\tilde{\mathbf{S}}_j$ containing the j corresponding spreading codes.

Now the vector $\mathbf{r}$ can be rewritten as:

$$\mathbf{r} = \mathbf{n}' + \mathbf{HC}\tilde{\mathbf{S}}_j \tilde{\mathbf{d}}_j \tag{3}$$

When we have a close look at the structure of the matrices involved in the computation, we will notice that there are several characteristic properties that can be exploited to simplify the calculation of the data symbols. In our approach the H, C and $\tilde{S}_j$ matrices are multiplied to get the system matrix $\tilde{K}_j$ of width $m \cdot j$:

$$\tilde{\mathbf{K}}_j = \mathbf{HC}\tilde{\mathbf{S}}_j \, . \tag{4}$$

This matrix will be used to calculate the desired data symbols. The structure of $\tilde{\mathbf{K}}_j$ is shown in Fig. 3.

The nonzero column vectors are calculated by convolving the channel impulse response with the scrambled spreading code. $\mathbf{c}_x$ denotes the x-th code block of length q in $\mathbf{c}$. As the scrambled spreading code just has got $(\pm 1 \pm i)$ entries the system matrix can be build without using multiplications.

It is obvious that $\tilde{\mathbf{K}}_j$ has got a very sparse structure which can be exploited as described in Sect. 2.3, to reduce the computational effort to solve the linear system.

2.3 Implementation

The detection of the estimated data symbols $\tilde{\mathbf{d}}'_j$ can now be performed by solving the overdetermined linear system

$$\tilde{\mathbf{K}}_j \tilde{\mathbf{d}}'_j = \mathbf{r} \, . \tag{5}$$

The equation is solved in the least squares sense by a QR-decomposition which can be implemented efficiently on a

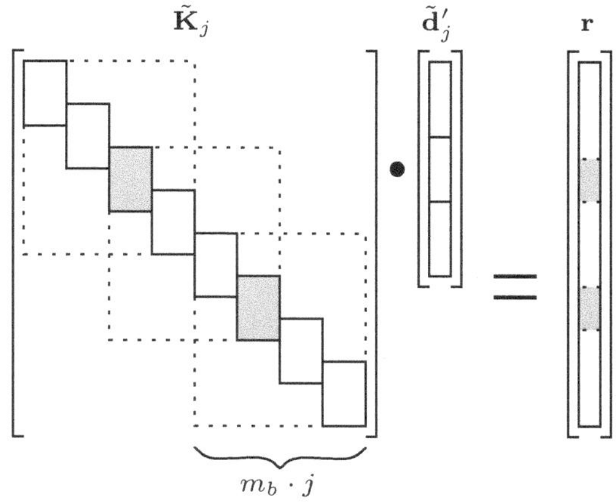

Fig. 4. Overlapping for easier calculation.

processor array (Otte et al., 2002) using Cordic processor elements (PE) performing Givens transformations. But, as the structure of $\tilde{\mathbf{K}}_j$ is known, a direct approach is used for the calculation of the QR decomposition. A new system matrix is build from $\tilde{\mathbf{K}}_j$ and $\mathbf{r}$. Then the required Givens transformations are applied only to the nonzero elements of $\tilde{\mathbf{k}}_{xy}$ in $\tilde{\mathbf{K}}_j$.

In each step one vector $\tilde{\mathbf{k}}_{xy}$ is annihilated. For each annihilation of one element in $\mathbf{k}_{xy}$ it is necessary to recompute two rows of the matrix composed of $\tilde{\mathbf{K}}_j$ and $\mathbf{r}$. The last step shows the matrix $\mathbf{R}$ and the vector $\mathbf{r}'$ which are used to perform the back-substitution.

2.4 Subdividing the calculation

It is obvious that $\tilde{\mathbf{K}}_j$ can grow to a very large matrix. A whole data frame of $m \cdot j = 2\,400 \cdot 16 = 38\,400$ symbols at spreading factor $q = 16$ and a channel of length $h_l = 10$ would require a $\tilde{\mathbf{K}}_j$ matrix of size $38\,409 \times 38\,400$.

To overcome this problem the system $\tilde{\mathbf{K}}_j \tilde{\mathbf{d}}'_j = \mathbf{r}$ is subdivided into overlapping subsystems of manageable size as shown in Fig. 4 and described in Vollmer et al. (2001).

The linear system is subdivided into blocks of size $m_b \cdot j$. Using this method it is possible to solve the complete system without the need to store the whole matrix which would also involve large latency and memory needs. Overlapping of the blocks is necessary, as the independent calculation of the subproblems leads to higher symbol errors at the block edges. These errors are nearly eliminated by using this overlapping method.

This method involves a certain amount of computational overhead as the grey blocks have to be calculated twice. The overhead can be reduced by choosing larger block sizes m_b. For good results the overlapping factor should be chosen at least as high as the block overlapping factor p of $\tilde{\mathbf{K}}_j$.

Table 1. Computational complexity comparison.

	Rake	**Cordic based LS**
$\frac{\textbf{Operations}}{\textbf{Symbol}}$	4352	6023

2.5 Computational complexity

If the overlap method is used for a data frame with $q = 16$, $m_b = 8$, $p = 2$, $j = 16$ and $m = 2\,400$ about 400 blocks have to be calculated for one frame. It is also assumed that the channel length h_l is 10.

In this case the decomposition/back-substitution for each block needs *approx* 179 000 (real valued) Cordic operations and $\approx$ 20 500 complex additions for the creation of the system matrix. Therefore the detection of a whole data-frame of $m \cdot j = 38400$ symbols uses

$$\approx 1865 \; \frac{\text{Cordic Operations}}{\text{Symbol}}$$

and

$$\approx 214 \; \frac{\text{Complex Additions}}{\text{Symbol}} .$$

Note that there is no further descrambling/despreading necessary. Furthermore the Cordic based QR decomposition can make nearly 100% use of two parallel Cordics.

The complexity comparison of our proposed algorithm to other implementations is based on the numbers given in Nahler et al. (2002) for Rake and PIC based receivers. An equivalent of three array multipliers for one Cordic operation is used to include some overhead. Therefore a Cordic operation would be equal to three operations, and a complex addition equals two operations. Then complexity of our approach for this example is on the same order than the conventional Rake receiver as shown in Table 1.

Of course this is only a rough estimate of the computational complexity. But it shows that it is about the same as for the conventional Rake receiver, while the performance is significantly increased as shown in Sect. 2.6. For a detailed description of the algorithm and the complexity analysis see Heyne and Goetze (2005).

2.6 Simulations

Figure 5 shows the frame error rate for a 16-QAM based system with $q = 16$, $j = 16$, $h_l = 10$, $m_b = 8$ and $p = 2$. The channel is assumed to be constant throughout the simulation and contains four, randomly distributed, strong taps. Hence the Rake receiver is using four fingers.

The Figure shows that the Rake has got no chance to detect the symbols in this case, and that the LS approach has got a large performance gain for rising SNR values.

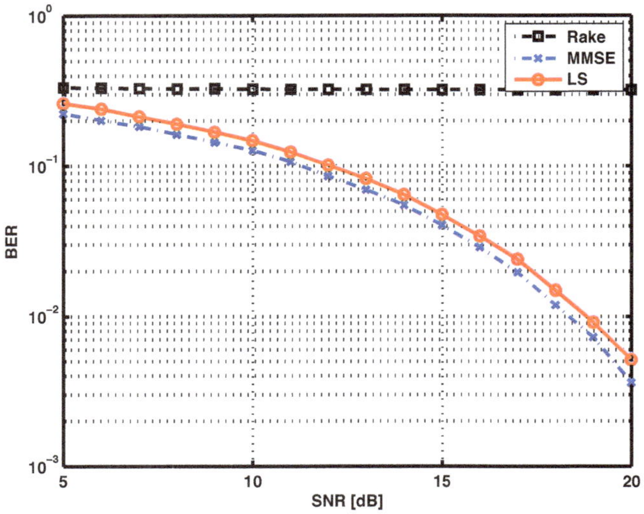

Fig. 5. Error rate for 16-QAM modulation.

3 FFT

3.1 MAC based FFT

A DFT with N input values **s** can be described as a matrix-vector multiplication. By exploiting the properties of the DFT matrix, operations can be greatly reduced and the well known Fast Fourier Transformation (FFT) (Oppenheim and Schafer, 1999) is derived.

The listing below shows a recursive implementation of a MAC based FFT in Matlab style.

```
1  function y=fft(x,n)
2    if n=1
3      y=x
4    else
5      m=n/2
6      w=exp(-2*pi*i/n)
7      om=diag(1,w,... ,w^(m-1))
8      zt=fft(x(0:2:n-1),m)
9      zb=om*fft(x(1:2:n-1),m)
10     I_m=eye(m)
11     y=[I_m I_m;I_m -I_m]*[zt;zb]
12   end
13 end
```

3.2 Algorithm

To derive a Cordic based FFT we will stop the recursion at $n=2$. In this case line 11 will look like:

$$y=\begin{bmatrix}1 & 1\\ 1 & -1\end{bmatrix}\cdot\begin{bmatrix}zt\\ zb\end{bmatrix} \tag{6}$$

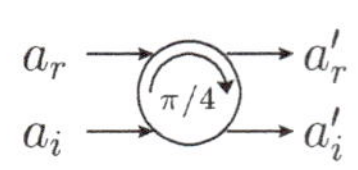

$$\hat{=} \quad \begin{bmatrix}a'_r\\ b'_r\end{bmatrix}=\begin{bmatrix}\frac{1}{\sqrt{2}} & \frac{1}{\sqrt{2}}\\ -\frac{1}{\sqrt{2}} & \frac{1}{\sqrt{2}}\end{bmatrix}\cdot\begin{bmatrix}a_r\\ b_r\end{bmatrix}$$

Fig. 6. Symbol and function of one real valued Cordic.

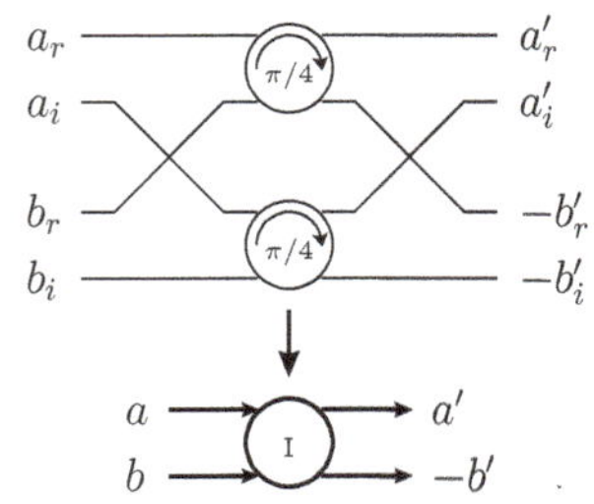

Fig. 7. Inner structure of complex Cordic type I.

The first matrix can then be decomposed to:

$$\begin{bmatrix}1 & 1\\ 1 & -1\end{bmatrix}=\begin{bmatrix}\sqrt{2} & \\ & -\sqrt{2}\end{bmatrix}\cdot\begin{bmatrix}\frac{1}{\sqrt{2}} & \frac{1}{\sqrt{2}}\\ -\frac{1}{\sqrt{2}} & \frac{1}{\sqrt{2}}\end{bmatrix} \tag{7}$$

This equals a Cordic rotating the input values by $\pi/4$, followed by a scaling of $\sqrt{2}/-\sqrt{2}$.

As the Cordic elements are real valued but the input values are complex valued, the complex Cordic operation has to be separated into real valued operations. Due to the special structure of the rotation matrix, this is quite easy to perform. If we assume two complex numbers $a, b\in\mathbb{C}$, the result of the complex rotation will be (with $t=1/\sqrt{2}$):

$$\underbrace{\begin{bmatrix}t & t\\ -t & t\end{bmatrix}}_{T}\cdot\begin{bmatrix}a_r+\mathrm{j}a_i\\ b_r+\mathrm{j}b_i\end{bmatrix}=\begin{bmatrix}t(a_r+b_r)+\mathrm{j}t(a_i+b_i)\\ t(b_r-a_r)+\mathrm{j}t(b_i-a_i)\end{bmatrix} \tag{8}$$

$$=T\cdot\begin{bmatrix}a_r\\ b_r\end{bmatrix}+\mathrm{j}T\cdot\begin{bmatrix}a_i\\ b_i\end{bmatrix} \tag{9}$$

Thus the operation can be applied to the real and the imaginary part of the input values independently, and the complex "butterfly" can be performed by using two of the real valued Cordics shown in Fig. 6. The scaling of the results can be performed at the end of the flow graph. The symbol for the resulting complex Cordic, called type I, rotating two complex valued numbers by $\pi/4$ is shown in Fig. 7.

As shown in Eq. 7, the second scaling factor has got a negative sign. Therefore all "lower" results of the complex Cordic operation have got a reversed sign. Fortunately, because of the structure of the FFT stages, this compensation does not result in additional computational overhead. In each stage of the FFT the sign reversed results are just combined with other sign reversed results. Therefore a $-3\pi/4$ rotation is applied to the results in these cases to project the result from the third quadrant back into the first one. This equals a multiplication of the complex input value by $-T$. Therefore

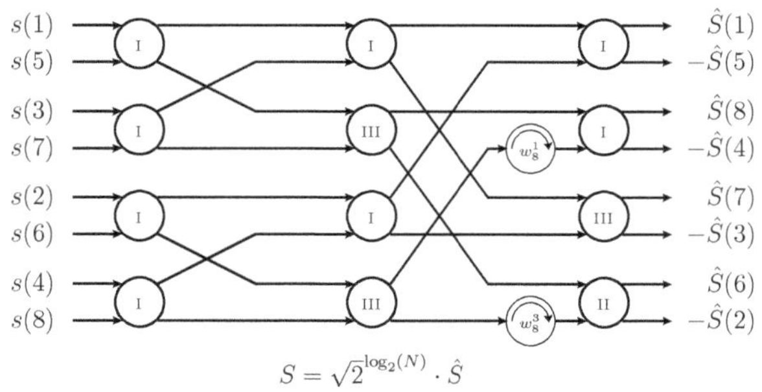

Fig. 8. Optimized Cordic based FFT.

this operation can be decomposed, too. The complex Cordic performing this operation is called type II.

The structure is the same compared to the type I Cordic, except that the real valued Cordics now perform a rotation by $-3\pi/4$.

Further optimizations are possible for $w_y^x = -\mathrm{j}$. The result of this operation can also be obtained by swapping the real and imaginary part of the input value, and then inverting the sign of the new imaginary part. A closer look at the algorithm reveals, that this operation is always performed on sign reversed results of the previous stage. This also implies that the result of the multiplication with $w = -\mathrm{j}$ is always provided to complex Cordics of type II.

Therefore the input value of the real valued Cordic is $-a$. Thus a $w = -\mathrm{j}$ multiplication followed by a type II complex Cordic can be replaced by another complex Cordic, called type III.

The final optimized version of the FFT is shown in Fig. 8. Note that there are no more twiddle factor multiplications after the first FFT stage, which saves one stage in a pipelined implementation.

The twiddle factors shown are the same as used for the standard FFT. As they are located on the unit circle, the multiplication with ω_y^x can be replaced by a Cordic rotation directly.

It is obvious that the FFT like butterfly structure is kept. The $\sqrt{2}^{\log_2(N)}$ scaling of the result can be performed after the computation of the three stages.

3.3 Complexity

If N is the size of the FFT, the number of Cordic operations is:

$$\mathrm{OP_{Cordic}} = \frac{3N}{2}(\log_2(N) - 1) + 2 \tag{10}$$

The same architecture used to implement the Cordic array also provides a MAC processing element (PE) (Lange et al., 2002). This PE needs

$$\mathrm{OP_{MAC}} = (N+2)\log_2(N) \tag{11}$$

activations for an FFT of size N.

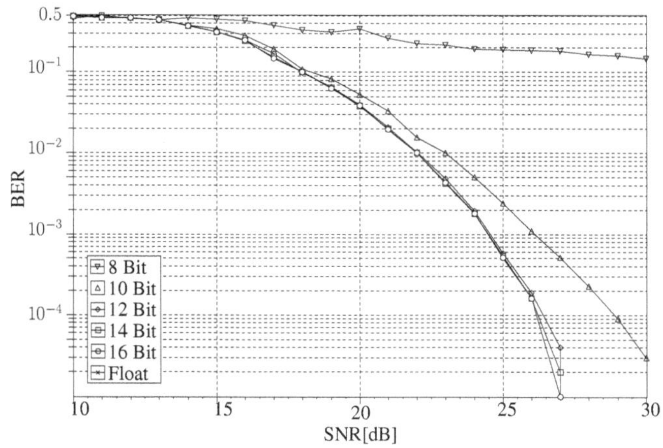

Fig. 9. BER for 54 MBit.

For small FFTs the operation count $\mathrm{OP_{Cordic}}$ is even lower than for $\mathrm{OP_{MAC}}$. For the 64-FFT used in a WLAN receiver $\mathrm{OP_{Cordic}}$ is 482 and $\mathrm{OP_{MAC}}$ is 396. As the hardware accelerator currently provides two parallel PEs these numbers can be halved to get the number of accelerator activations (198 for the MAC, 241 for the Cordic).

So the Cordic based FFT is slower than the MAC based implementation, but on the other hand one Cordic based reconfigurable hardware architecture can now be used to implement the FFT for WLAN and the Rake substitute for UMTS. Also the OFDM channel correction can be implemented very efficiently on a Cordic, as it supports division. Hence, in the case of a FFT followed by an OFDM channel correction, the computational overhead is considerably small. A more detailed description of the algorithm and it's properties can be found in Heyne and Goetze (2004).

3.4 Simulation results

Finally, the proposed FFT has been implemented in a WLAN transmitter/receiver simulation to replace the regular FFT. The simulation shown in Fig. 9 has been performed using an AWGN channel and the coding parameters defined in IEEE Std 802.11a-1999 (1999).

The results for the 54 MBit case show that a wordlength ≥ 12 Bit is enough to achieve the same BER performance than the floating point FFT implementation. So the Cordic based FFT has to use just 12 Bit arithmetics to replace the standard FFT in a WLAN environment.

4 Hardware coprocessor

Both algorithms have been implemented on the reconfigurable hardware accelerator (RACE) shown in Fig. 10. The RACE can be described as an algorithm specific instruction set processor (ASIP) with a limited instruction set that is optimized for different classes of algorithms.

Table 2. Basic operations of the Cordic PE.

Mode	Operation
Ortogonal Rotation	$x_{out} = x_{in}\cos(\phi_z) + y_{in}\sin(\phi_z))$ $y_{out} = -x_{in}\sin(\phi_z) + y_{in}\cos(\phi_z))$ $z_{out} = z_{in}$
Ortogonal Vector	$x_{out} = \sqrt{x_{in}^2 + y_{in}^2}$ $y_{out} = 0$ $z_{out} = \arctan_2(x_{in}, y_{in})$
Linear Rotation	$x_{out} = x_{in}$ $y_{out} = -x_{in}z_{in} + y_{in}$ $z_{out} = z_{in}$
Linear Vector	$x_{out} = x_{in}$ $y_{out} = 0$ $z_{out} = y_{in}/x_{in}$

In our case the class of algorithms is composed of matrix based algorithms that can be implemented by enhanced Cordic operations. For this purpose the accelerator contains Cordic processing elements which are based on simple shift-add operations, but there are other PE types such as e.g. MACs that can be used as well. The operations that can be performed, and which are used to implement the QR decomposition of the system matrix, are given in Table 2. For example in the "Orthogonal Rotation" mode the Cordic rotates a two dimensional input vector $a = (x, y)$ by an angle ϕ_z.

The accelerator contains several processing elements (PE) in parallel that perform the computations, a Data RAM to store values and a Configuration RAM in conjunction with a finite state-machine (FSM) to control the data flow.

Here, the RACE is embedded in a processor environment where it is connected to the system bus via a bus wrapper, which has direct memory access (DMA) capability and thereby controls the dataflow into and out of the RACE. The processor itself is freed from all data moving tasks and is just informed by an interrupt when the results of an operation are available.

The number and the interfaces of the PEs as well as the amount and structure of the memory inside the Data-RAM can be parameterized. Hence it exactly fits the needs of the multistandard terminal where it is used.

5 Conclusions

We have presented two Cordic based algorithms usable for communication systems based on OFDM and CDMA, namely a FFT and a linear least squares based CDMA multiuser detector/equalizer. The good performance and the low computational complexity of both algorithms make an implementation feasible.

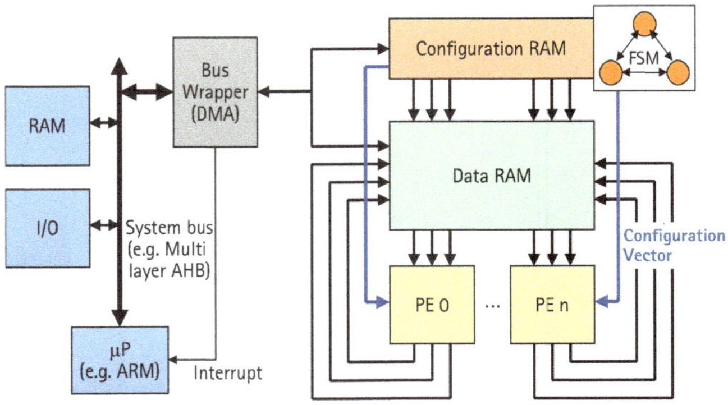

Fig. 10. Reconfigurable accelerator.

Hence, they can be used to implement a reconfigurable (software defined) architecture for multi standard terminal digital basebands, replacing dedicated hardware by using Cordic coprocessors like the RACE accelerator.

References

3GPP: TS25.213 V5.0.0 - Spreading and Modulation (FDD), Technical specification group radio access network, 3rd Generation Partnership Project (3GPP), 2002.

Heyne, B. and Goetze, J.: A Cordic Based Equalizer For Multiuser Detection in WCDMA Systems, in: IEEE Workshop on Signal Processing Systems (SiPS2005), Athens, Greece, 2005.

Heyne, B. and Goetze, J.: A Pure Cordic Based FFT For Reconfigurable Digital Signal Processing, in: 12th European Signal Processing Conference (EUSIPCO2004), Vienna, Austria, 2004.

Heyne, B., Otte, M., and Goetze, J.: A Performance Adjustable and Reconfigurable CDMA Receiver Concept for UMTS-FDD, in: 14th IEEE Intern. Symposium on Personal, Indoor and Mobile Radio Communications (PIMRC2003), Beijing, China, 2003.

IEEE Std 802.11a-1999: Wireless LAN Medium Access Control (MAC) and Physical Layer (PHY) specifications, High-speed Physical Layer in the 5 GHz Band, IEEE, 1999.

Lange, H., Franzen, O., Schröder, H., Bücker, M., and Oelkrug, B.: Reconfigurable Multiply-Accumulate-based Processing Element, in: IEEE Workshop on Heterogeneous Reconfigurable Systems on Chip, Hamburg, Germany, 2002.

Nahler, A., Irmer, R., and Fettweis, G.: Reduced And Differential Parallel Interference Cancellation For CDMA Systems, IEEE J. on Select. Areas in Commun., 20, 237–247, 2002.

Oppenheim, A. V. and Schafer, R. W.: Discrete Time Signal Processing, Prentice-Hall, Upper Saddle River, New Jersey, third edn., 1999.

Otte, M., Goetze, J., and Buecker, M.: Matrix Based Signal Processing on a Reconfigurable Hardware Accelerator, in: 10th Digital Signal Processing Workshop, Pine Moutain, Georgia, USA, 2002.

Vollmer, M., Haardt, M., and Götze, J.: Comparative Study of Joint-Detection Techniques for TD-CDMA Based Mobile Radio Systems, IEEE J. on Selected Areas in Communications, 19, 1461–1475, 2001.

3

A hybrid equalizer/Rake receiver for the Wideband CDMA Downlink in large delay spread channels

C. Buchacher[1], M. Zimmermann[1], S. Paul[2], and M. Huemer[3]

[1]Infineon Technologies, Am Campeon 1–12, 85579 Neubiberg, Germany
[2]Institut für Theoretische Elektrotechnik und Mikroelektronik, Univ. of Bremen, Otto-Hahn Allee, 28359 Bremen, Germany
[3]Embedded Systems and Signal Processing Group, University of Klagenfurt, Lakeside B02b, 9020 Klagenfurt, Austria

Abstract. Signals transmitted over a multi-path propagation channel exhibit Inter-Path Interference (IPI) and fading. The receiver has to employ measures to mitigate these effects or it will incur severe performance degradation. A classic approach in Code Division Multiple Access (CDMA) communications is the Rake receiver. Alternatively, a Linear Minimum Mean Square Error (LMMSE) equalizer can further improve performance. This paper compares performance and computational complexity of these two algorithms. A hybrid solution will be introduced in order to decrease computational complexity while retaining most of the interference suppression capabilities of the LMMSE equalizer.

1 Introduction

Spread-spectrum signals inherently exhibit frequency diversity. Due to the increased bandwidth, a spectral null is less likely to affect the entire signal spectrum. For the same reason, however, the radio channel is likely dispersive. In order to exploit frequency diversity, the receiver has to collect signal energy from several multi-paths. For this purpose, a Rake receiver allocates so-called *fingers* to multi-paths. Each finger despreads the receive signal synchronized to the corresponding path delay. The Rake receiver subsequently computes a weighted sum of the Rake fingers' output.

It is important to note, however, that the Rake receiver does not take into account IPI caused by correlation of delayed versions of the spreading codes. Paths separated by less than one chip duration significantly degrade Rake receiver performance.

Multi-path interference caused by paths separated by less than one chip duration is called Inter-Chip Interference (ICI). The despreader does not suppress this interference, because the corresponding signals' pulse shapes overlap. The multipath delay profiles used in High Speed Downlink Packet Access (HSDPA) test scenarios – HSDPA is a high data rate extension of the Universal Mobile Telecommunications System (UMTS) standard – exhibit ICI. In order to achieve higher data rates it is necessary to introduce a more sophisticated algorithm which is capable of suppressing multi-path interference. The chip-level LMMSE equalizer, which maximizes the Signal-to-Interference and Noise Ratio (SINR), is such an algorithm.

Typical channel delay spreads are in the range of $0\,\mu$s to $5\,\mu$s. At the UMTS chip rate of 3.84 MHz and an oversampling factor of two, $5\,\mu$s correspond to roughly $n{=}40$ receive samples. An LMMSE equalizer of this length exceeds the Rake receiver's computational complexity by far. The LMMSE equalizer coefficient computation involves inversion of a channel autocorrelation matrix of size $n \times n$. Approximating the inverse of this matrix has runtime complexity of at least $\mathbb{O}(n^2)$. Thus, the equalizer's complexity increases significantly with increasing channel length. The complexity of the Rake receiver, on the other hand, is independent of the magnitude of the path delays, because the computation of Rake finger weights does not involve inversion of the channel autocorrelation matrix.

Even though in theory the LMMSE equalizer always performs better than the Rake receiver, this is not true in practice, because the length of an equalizer is limited by the available computing power. In the case of a propagation channel with large delay spread a limited-length equalizer will therefore suffer from interference caused by multi-paths exceeding its observation window.

As a compromise to this steep performance versus complexity tradeoff we will propose a hybrid algorithm. This algorithm will not achieve the performance of an equalizer of arbitrary length, but due to the shorter equalizer length, computational complexity is comparatively small and it is independent of the channel delay spread.

Correspondence to: C. Buchacher
(clemens.buchacher@infineon.com)

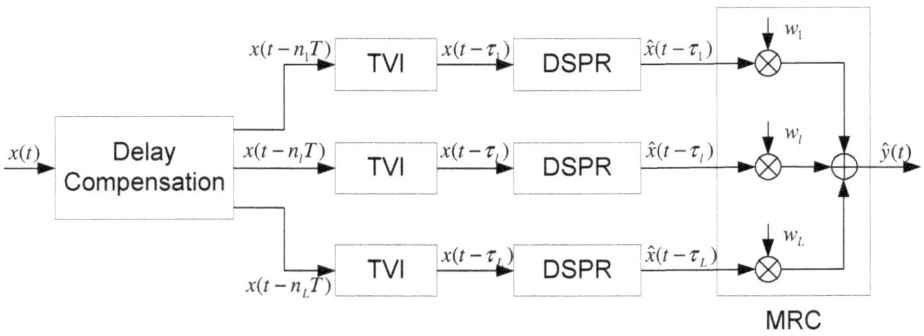

Fig. 1. Symbol-level Rake.

2 System model

We consider the CDMA signal

$$y(t) = \sum_{k\in\mathbb{Z}_K} A_k \sum_{m\in\mathbb{Z}_M} b_m^{(k)} \sum_{n\in\mathbb{Z}_N} c_{m,n}^{(k)}\, \delta(t-(m\,N+n)\,T_c)\ ,$$

where $\mathbb{Z}_r=\{0,\ldots,r-1\}$, M is the observation block length, A_k is the signal amplitude of user k and $(b_0^{(k)},\ldots,b_{M-1}^{(k)})$ is the symbol sequence of user k. N is the spreading factor and $(c_{m,0}^{(k)},\ldots,c_{m,(N-1)}^{(k)})$ is the pseudo-random spreading code sequence for symbol m of user k. T_c is the chip duration and $\delta(t)$ is the Dirac delta.

Transmitted over a multi-path channel with additive white Gaussian noise, after matched filtering (matched to the transmit pulse shape) the receiver obtains the signal

$$\begin{aligned} x(t) &= y(t) * h(t) + \nu(t) \\ &= \sum_{\substack{k\in\mathbb{Z}_K\\ m\in\mathbb{Z}_M\\ n\in\mathbb{Z}_N}} A_k\, b_m^{(k)}\, c_{m,n}^{(k)}\, h(t-(m\,N+n)\,T_c) + \nu(t)\ , \end{aligned}$$

where $h(t)$ is the channel impulse response convolved with transmit and receive pulse shaping filters $\psi(t)$ and $\psi^*(-t)$, respectively. $\psi(t)$ is non-zero only in the interval $[0, T_c)$. It is sufficient to sample $h(t)$ at twice the chip rate, because we will assume the pulse shape to be band limited to $1/T_c$[1]. $\nu(t)$ is a white Gaussian random process of variance σ^2, filtered by the receive pulse shaping filter.

Without loss of generality the user of interest shall be user number 0. Furthermore, we will assume the users' power and the channel impulse response to be normalized to unit energy, i.e., $\sum_{k=0}^{K-1} A_k^2=1$ and $\sum_{l=0}^{L-1} |h_l|^2=1$, where $h_l=h(l\,T_c/2)$ and L is chosen such that $h(l\,T_c/2)=0$ for all $l\geq L$.

[1]Even though it is impossible for a waveform to be confined to one chip interval *and* band limited at the same time, the pulse shapes used in Wideband CDMA communications systems approximate these properties reasonably well.

3 Rake receiver

The Rake Receiver was named after the ordinary garden rake because it collects the different paths by setting a (rake) *finger* at each corresponding delay. Each finger represents a Time-Variant Interpolator (TVI) used for fine-synchronization of the path delay followed by a despreader. Figure 1 depicts the architecture of a *Symbol-Level Rake*.

For each Rake finger j with the corresponding path delay τ_j, the time-variant interpolator eliminates the delay and the despreader reverses the spreading operation. This yields a processing gain equal to the spreading factor N. Neglecting multi-path interference, the output Signal-to-Interference and Noise Ratio (SINR) of Rake finger j is

$$N\,A_0^2\,\frac{|h(\tau_j)|^2}{\rho_j+\sigma^2}\ ,$$

where

$$\rho_j := \sum_{\substack{i\in\mathbb{Z}\\ i\neq 0}} \phi_h(\tau_j - i\,T_c)$$

is the cross-correlation induced by other paths and

$$\phi_h(t) := \left|\sum_{l=0}^{L-1} h(t-l\,T_c/2)\right|^2$$

is the power delay profile.

The Rake fingers' output is now combined in order to maximize the Signal-to-Noise ratio (SNR). Provided the radio channel adds stationary noise to the receive signal, the output Signal-to-Noise Ratio (SNR) of each finger is proportional to the corresponding path's power. Including phase compensation in the combining operation, the combining weights thus amount to the complex conjugate of the path phasors. This combining scheme is known as Maximum Ratio Combining (MRC).

If we disregard multi-path interference, the output Signal-to-Noise Ratio (SNR) after MRC evaluates to

$$\mathrm{SNR} = N\ \|\boldsymbol{h}\|_2^2\ A_0^2/\sigma^2 = N\,A_0^2/\sigma^2\ ,$$

where $\boldsymbol{h}$=$(h_0, \ldots, h_{L-1})^T$. Note, however, that this measure can be used as an indicator for the signal quality only if the noise power is dominant over multi-path interference, because MRC is sub-optimum with respect to maximizing the SINR, cf. Burke (2005).

4 LMMSE equalizer

Figure 2 shows a block diagram of the discrete random vector model of a linear equalizer. $\boldsymbol{y}, \boldsymbol{n}$ and $\mathbf{H}$=$[h((j-i)\,T_c/2)]_{i,j}$ denote the discrete time equivalents of the transmit signal, noise signal and channel impulse reponse, respectively. $\boldsymbol{x}$ denotes the random vector obtained by sampling $x(-t)$ at twice the chip rate[2]. $\mathbf{R}$ and $\boldsymbol{r}$ denote the autocorrelation matrix of $\boldsymbol{x}$ and the cross-correlation vector of $\boldsymbol{x}$ with a transmit chip d, respectively. In terms of the channel convolution matrix $\mathbf{H}$, noise autocorrelation matrix $\sigma^2\hat{\mathbf{I}}$ and equalizer filter delay R, they can be expressed as

$$\mathbf{R} = E(\boldsymbol{x}\boldsymbol{x}^H) = \mathbf{H}\mathbf{H}^H + \sigma^2\hat{\mathbf{I}} \quad \text{and}$$
$$\boldsymbol{r} = E(\boldsymbol{x}d) = \mathbf{H}\boldsymbol{e}_R\ ,$$

where $\boldsymbol{e}_R$ denotes the R-th unit vector.

Given the observation vector $\boldsymbol{x}$ one desires an estimate $\hat{d}$ of the transmit chip d which minimizes the mean square error

$$J_{\hat{d}} = E(|d-\hat{d}|^2)\ ,$$

where $E(\cdot)$ denotes the expected value. The estimate is also required to be a linear combination of the receive samples $\boldsymbol{x}$, that is

$$\hat{d}(\boldsymbol{w}) = \boldsymbol{w}^H\boldsymbol{x}\ ,$$

where the weight vector $\boldsymbol{w}$ is to be determined such that $J_{\hat{d}}$ is minimum. In other words:

$$\boldsymbol{w}_\text{o} = \underset{\mathbf{w}}{\text{argmin}}\ J_{\hat{d}}(\boldsymbol{w})$$

This estimator is known as the LMMSE estimator (or equalizer). The solution to this minimization problem is obtained by solving the Wiener-Hopf equation

$$\mathbf{R}\boldsymbol{w}_\text{o} = \boldsymbol{r}\ . \tag{1}$$

If the noise power is greater than zero $\mathbf{R}$ is guaranteed to be non-singular.

Furthermore, $\mathbf{R}$ is Hermitian and block Toeplitz. These properties are exploited by efficient matrix inversion algorithms such as the conjugate gradient method or the Levinson-Durbin algorithm. Nevertheless, approximating the inverse of an $n \times n$ matrix has runtime complexity of at least $\mathbb{O}(n^2)$. The equalizer length is therefore limited by the available computing power.

[2]The discrete time signal vectors are reversed in time to simplify notation of the channel convolution matrix $\mathbf{H}$.

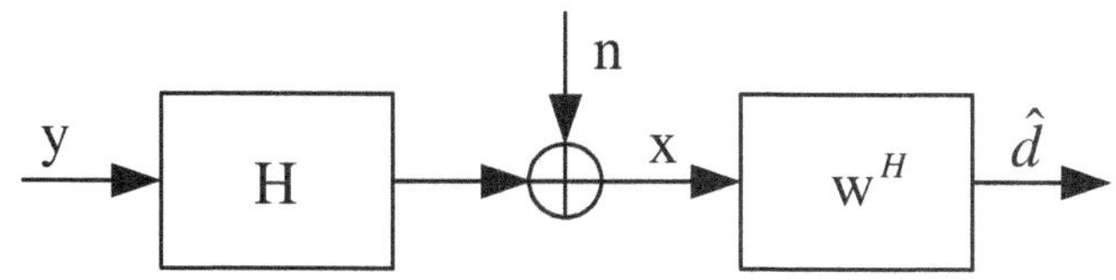

Fig. 2. Discrete random vector model of channel and equalizer.

Provided the observation window is large enough, the equalizer will perform better than the Rake receiver, because in contrast to the Rake receiver the equalizer does not only collect signal energy from the individual paths, but it also reduces multi-path interference. As the channel delay spread gets larger, exceeding the equalizer length, however, the equalizer performance will gradually deteriorate.

5 Hybrid solution

Since both of the solutions described in the previous sections are inadequate for high data rate services in multi-path propagation environments with large delay spread, we propose a hybrid approach. Usually, the paths at the beginning of the channel impulse response bear most of the total receive signal energy, because the path power typically decreases exponentially with increasing delay. The channel's strongest source of signal energy and interference can be equalized by a short LMMSE equalizer. The remaining paths not covered by the equalizer are assigned a conventional Rake finger each. This is illustrated in Fig. 3.

In order to motivate this approach we consider the unbiased LMMSE solution. Given an arbitrary transmit vector $\boldsymbol{y}$ and the desired response d=$\boldsymbol{e}_R^T\boldsymbol{y}$, the estimate's expected value is

$$E(\hat{d}) = E(\boldsymbol{w}_\text{o}^H\boldsymbol{x}) = \boldsymbol{w}_\text{o}^H E(\mathbf{H}\boldsymbol{y}+\boldsymbol{n}) = \boldsymbol{w}_\text{o}^H\mathbf{H}\boldsymbol{e}_R d$$
$$= \boldsymbol{w}_\text{o}^H\boldsymbol{r}d\ .$$

The unbiased LMMSE solution is therefore $\boldsymbol{w}'_\text{o}$=$(\boldsymbol{w}_\text{o}^H\boldsymbol{r})^{-1}\boldsymbol{w}_\text{o}$. In general, the unbiased estimate increases the MSE. The likelihood ratios generated from the symbol estimates for the channel decoder, however, take signal and noise power into account. Introducing an arbitrary scaling factor at this point is therefore of no consequence.

Provided $\sigma^2 \gg 1$, the signal autocorrelation matrix can be approximated as

$$\mathbf{R} = \mathbf{H}\mathbf{H}^H + \sigma^2\hat{\mathbf{I}} \approx \sigma^2\mathbf{I}\ , \tag{2}$$

so that the unbiased LMMSE solution reduces to

$$\boldsymbol{w}'_\text{o} = \frac{\mathbf{R}^{-1}\boldsymbol{r}}{\boldsymbol{r}^H\mathbf{R}^{-1}\boldsymbol{r}} \approx \frac{\boldsymbol{r}}{\boldsymbol{r}^H\boldsymbol{r}} = \boldsymbol{r} = \mathbf{H}\boldsymbol{e}_R\ .$$

As the SNR approaches zero, the unbiased LMMSE equalizer thus degenerates to the Rake receiver. This asymptotic

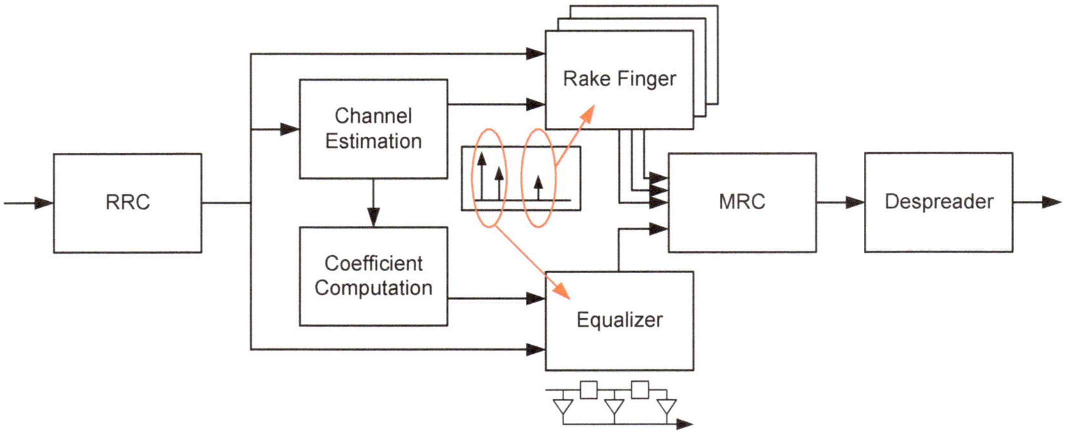

Fig. 3. Combined equalizer/Rake receiver structure.

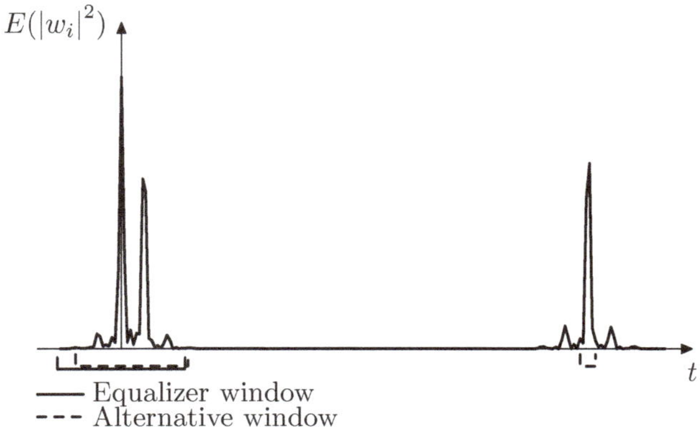

Fig. 4. Case 2 weight statistics for 200-tap equalizer.

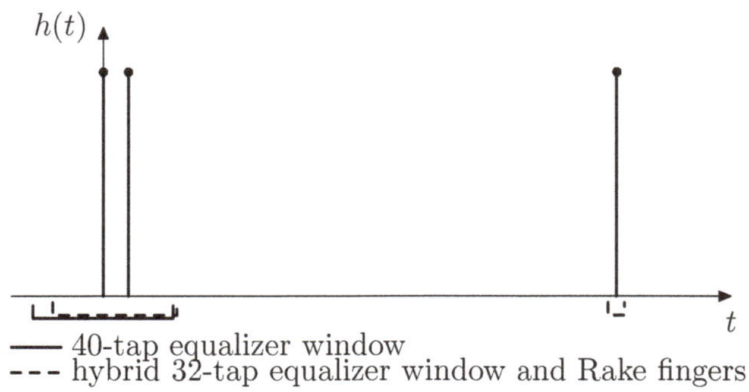

Fig. 5a. Case 2 path delay profile.

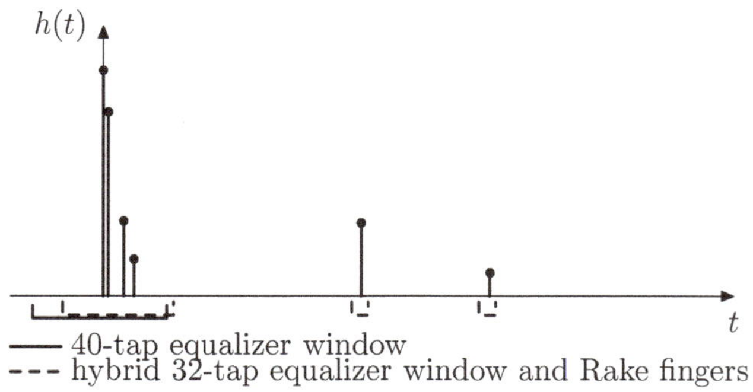

Fig. 5b. PB SP2 5 dB path delay profile.

behavior can be observed in typical UMTS simulation scenarios as well. Figure 4 shows the average equalizer weight power $E(|w_i|^2)$ simulated for a 200-tap LMMSE equalizer. The weight energy is concentrated around strong path energy. Using Rake fingers as approximation of parts of the equalizer solution is therefore a reasonable approach.

In order to perform MRC of equalizer output and Rake fingers, the equalizer gain has to be estimated. The SINR at the filter output is, cf. Haykin (1996),

$$\mathrm{SINR} = \frac{E(|\hat{d}|^2)}{J_{\hat{d}}} = \frac{E(|d|^2) - J_{\hat{d}}}{J_{\hat{d}}} = \frac{\boldsymbol{w}^H \mathbf{R} \boldsymbol{w}}{J_{\hat{d}}(\boldsymbol{w})} .$$

If $\boldsymbol{w}$ is the optimum weight vector this relation can be simplified to

$$\mathrm{SINR} = \frac{\boldsymbol{w}_{\mathrm{o}}^H \boldsymbol{r}}{1 - \boldsymbol{w}_{\mathrm{o}}^H \boldsymbol{r}} .$$

We thus have the means to estimate the equalizer gain. However, while the equalizer also reduces interference, Rake fingers only minimize noise. Using the equalizer output SINR and the SNR of the Rake fingers to estimate the individual receiver path's gain wrongly favors the Rake fingers.

We therefore similarly ignore the interference term, using the output SNR as equalizer gain estimate:

$$\mathrm{SNR} = \frac{E(|\hat{d}|^2)}{E(|\boldsymbol{w}_{\mathrm{o}}^H \boldsymbol{n}|^2)} - 1 = \frac{\boldsymbol{w}_{\mathrm{o}}^H \boldsymbol{r}}{\|\boldsymbol{w}_{\mathrm{o}}\|_2^2 \, \sigma^2} - 1 \qquad (3)$$

Table 1. Case 2 path delay profile.

Delay [μs]	Attenuation [dB]
0.0	0.0
0.976	0.0
20.0	0.0

Table 2. PB SP2 5 dB path delay profile.

Delay [μs]	Attenuation [dB]
0.0	0.0
0.2	0.9
0.8	4.9
1.2	8.0
10.0	5.0
15.0	10.0

6 Simulation results

We have implemented an HSDPA simulation chain to simulate performance of the Rake receiver and LMMSE equalizer, as well as the hybrid solution described in the previous section. All simulations use floating-point arithmetic.

The 200-tap equalizer is provided as a reference because it approximates the performance of an infinite-length equalizer. The immense computational complexity of this equalizer makes it infeasible for implementation in a mobile receiver.

The simulation environment is equivalent to the Fixed Reference Channel (FRC) specifications for HSDPA described in 3rd Generation Partnership Project TS 25.101 (2002), which specifies UMTS User Equipment (UE) minimum requirements. Parameter variations are defined for a number of so-called "H-Sets". For example, they specify the number of multi-codes (i.e., the number of parallel HSDPA physical channel codes), which is denoted by p in the performance plots. The parameters E_c/I_{or} and $\hat{I}_{or}/I_{oc}$ correspond to the desired user's power A_0^2 and the SNR$=1/\sigma^2$ in our system model, respectively. Performance is measured in terms of HSDPA physical layer throughput. Channel coding and Hybrid Automatic Repeat Request (HARQ) are included in the simulation.

Because delay profiles with large delay spread are of particular interest, instead of the HSDPA minimum requirement test channels, the following power delay profiles are used for performance evaluation.

- The "Case 2" Rake test case, cf. 3rd Generation Partnership Project TS 25.101 (2002), as illustrated in Fig. 5a, is characterized by 3 widely spaced paths of equal power. In particular, this channel exhibits no ICI and is therefore best suited for the Rake receiver. A conventional equalizer with a manageable number of taps, however, can at best cover the first two taps in its observation window.

- The "PB SP2 5 dB" channel is a modified version of the International Telecommunication Union (ITU) Pedestrian B Reference Channel. It introduces two additional paths at 10 μs and 15 μs delay, attenuated by 5 dB and 10 dB, respectively. The delay profile is shown in Fig. 5b. It exhibits both ICI and a large delay spread.

The delay profile specifications are reproduced in Tables 1 and 2.

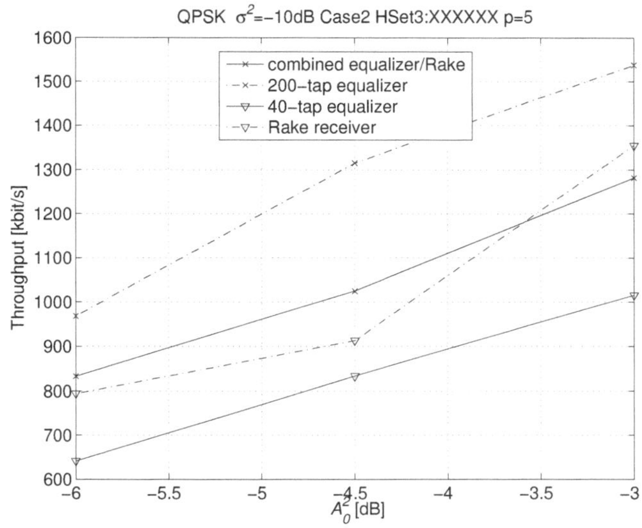

Fig. 6a. Performance results QPSK H-Set 3, σ^2=−10 dB, Case 2.

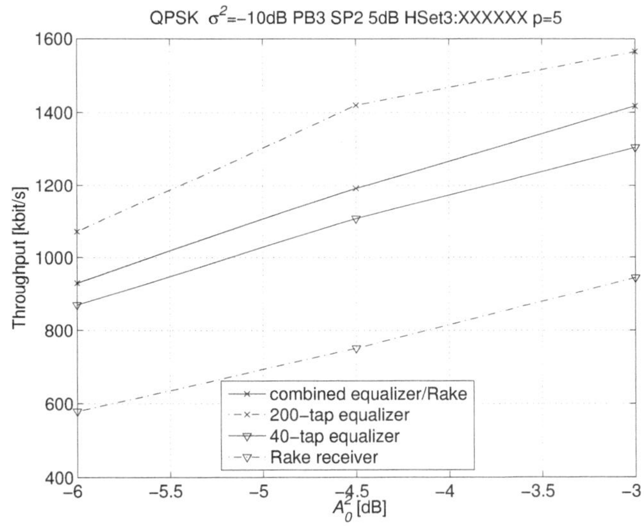

Fig. 6b. Performance results QPSK H-Set 3, σ^2=−10 dB, PB3 SP 5 dB.

6.1 Discussion of results

Performance results for each channel are now discussed individually:

Case 2. Clearly, the Rake receiver is expected to perform better than the 40-tap equalizer for the Case 2 test case. The 40-tap equalizer is not even "aware" of the third path, which simply remains as a significant amount of interference at the equalizer output. Furthermore, the Rake receiver does not have to deal with ICI. Accordingly, Fig. 6a shows that the Rake receiver performance exceeds the equalizer performance by 1.5 dB. The 200-tap equalizer reference shows that an infinite-length equalizer can still improve performance by approximately 1 dB.

The hybrid solution does not achieve a significant performance gain. The additional Rake finger compensates for the too-short equalizer, however, so its performance still competes with the Rake receiver.

PB SP2 5 dB. For the modified Pedestrian B channel, the situation is just the opposite. The two late paths are not strong enough to cause too much interference with the 40-tap equalizer. The Rake receiver, however, is impaired by strong ICI, caused by closely spaced paths at the beginning of the delay profile. It comes as no surprise that the plot in Fig. 6b shows a huge performance gap between equalizer and Rake receiver. The hybrid solution benefits from the advantages of both the equalizer and the Rake receiver. It achieves a small performance gain over the 40-tap equalizer.

7 Conclusions

We have shown that both the Rake receiver and the limited-length equalizer perform badly in certain scenarios. The Rake receiver performance is severely affected by multi-path interference. An equalizer can reduce multi-path interference, but, for the number of taps which are required to equalize a path with large delay spread, the LMMSE equalizer is computationally too complex.

The solution presented was a hybrid equalizer/Rake receiver. It uses a subset of the available observation window for equalization, allocating Rake fingers to the remaining multi-paths. The equalizer is treated as a single Rake finger, so that after appropriate gain estimation, Rake finger and equalizer outputs can be combined. Our simulation results show that the hybrid algorithm can exceed performance of both, a limited-length LMMSE equalizer and a Rake receiver, with comparative computational complexity in scenarios with strong multi-path interference and large delay spread.

References

3rd Generation Partnership Project: Technical Specification 25.101: User equipment (UE) radio transmission and reception (FDD), Release 5, 2002.

Burke, J. P., Zeidler, J. R., and Rao, B. D.: CINR difference analysis of optimal combining versus maximal ratio combining, IEEE Transactions on Wireless Communications, 4(1), 1–5, Jan., 2005, postprint available at: http://repositories.cdlib.org/postprints/596.

Haykin, S.: Adaptive Filter Theory, Prentice Hall, 3rd Edition, 1996.

4

Multicarrier systems and filter banks

D. S. Waldhauser and J. A. Nossek

Institute for Circuit Theory and Signal Processing, Munich University of Technology, Germany

Abstract. Multicarrier systems, especially orthogonal frequency division multiplex (OFDM), are already established in a lot of current communications standards. Moreover, they are very prominent candidates for future systems in mobile communications.

Filter bank based multicarrier systems can be designed to provide a better spectral shaping than OFDM systems. That way, they have some advantages against OFDM, which will be shown in this work and motivate further research in the direction of channel equalizers with low complexity. This leads to optimum bandwidth efficient multicarrier systems without guard interval, i. e. cyclic prefix.

1 Introduction

Optimum adaptability to the time and frequency selectivity of the propagation channel is crucial for mobile communication systems (Kammeyer, 1992). Multicarrier systems provide the possibility to find a trade-off between robustness against frequency selectivity and time selectivity by adjusting the number of subcarriers.

Especially orthogonal frequency division multiplex (OFDM) resp. discrete multitone modulation (DMT) are widely recognized for high data rate communications. They are already established in a lot of current standards like Wireless LAN, HiperLAN, WiMAX, DAB, DVB-T, and xDSL (Akansu et al., 1998). Moreover, they are very prominent candidates for the mobile communication systems of the 4th generation (4G) at least in the downlink, i. e. from the base station to the mobile stations. Even the extension of the system of the 3rd generation (3G+) is likely to be based on multicarrier systems.

Correspondence to: D. S. Waldhauser
(waldhauser@tum.de)

OFDM resp. DMT multicarrier systems used in current standards transmit some kind of redundancy — the cyclic prefix (CP) with a length of at least the equivalent discrete-time channel impulse response — for transforming the time-invariant multipath channel into a periodically time-varying additive white Gaussian noise (AWGN) channel (Ivrlac, 2005). The time-invariance is assumed for the length of at least one OFDM symbol. This represents an elegant way of avoiding channel equalizers, which remove the intersymbol interference (ISI) and adjacent channel interference (ACI), at the expense of reduced bandwidth efficiency up to 25%.

Linear time-domain equalizers are state of the art in OFDM / DMT systems in order to shorten the length of the equivalent discrete time channel impulse response and, therefore, the necessary length of the cylic prefix. In this way some bandwidth efficiency is saved (Melsa et al., 1996).

Moreover, some research has been done in the field of bandwidth efficient OFDM transmission without any cyclic prefix (Toeltsch and Molisch, 2000; Trautmann and Fliege, 2002; Sinn et al., 2002).

Generally, multicarrier systems can be realized by filter banks, called transmultiplexer filter banks (TMUX) or simply transmultiplexers (Tzannes et al., 1994).

The most important implementation uses one prototype filter $H_0(z)$, from which all other transmit and receive filters $H_k(z)$, $k = 0, 1, \dots, M-1$, are derived by complex modulation according to Eq. (1) , which represents a modulation with the equidistant carrier frequencies (c. f. Fig. 1).

Ideal bandwidth efficiency can be achieved by filters satisfying an overall transfer function according to the first Nyquist criterion. Offset quadrature amplitude modulation (OQAM) (Kammeyer, 1992) or staggered quadrature amplitude modulation (SQAM) (Bingham, 1990) have been proposed in the literature to avoid adjacent channel interference (ACI).

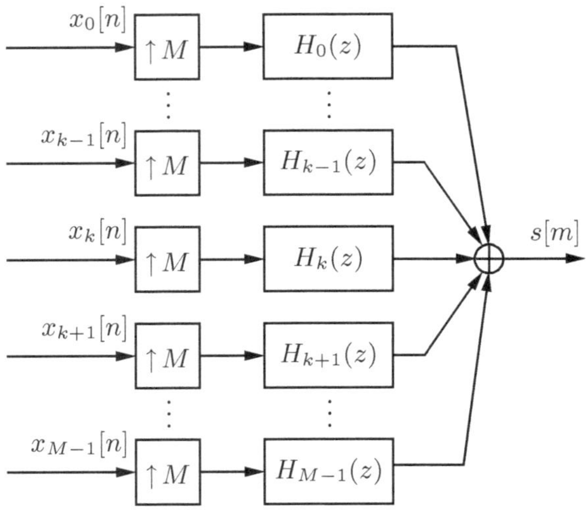

Fig. 1. Synthesis part of a transmultiplexer filter bank.

2 Complexity estimation

The subband filters $H_k(z)$, $k = 0, 1, \ldots, M-1$, of complex modulated transmultiplexer filter banks with M subcarriers are deduced from one prototype filter $H_0(z)$, which determines the spectral shaping of the subbands. Subband frequency responses can be designed which overlap only with directly adjacent subbands and provide separation from other subbands by a high attenuation characteristic. The complex modulation

$$H_k(z) = H_0(W_M^k z), \quad W_M = \exp(-\mathrm{j}\,2\pi/M) \tag{1}$$

in combination with a polyphase decomposition – if the prototype filter is designed as finite impulse response filter (FIR) or infinite impulse response filter (IIR) with special properties – leads to extremely efficient implementations of the transmultiplexers. Figure 2 shows the resulting structure of the the synthesis part of the transmultiplexer filter bank. The subband filters are realized as IDFT operation and subsequent polyphase component filters at the low symbol rate. Due to the evolving IDFT operation, which is advantageously calculated by an inverse fast Fourier transformation (IFFT), those transmultiplexer filter banks are denoted as DFT transmultiplexers (Fliege, 1993).

At this point it becomes obvious, that orthogonal frequency division multiplex (OFDM) is a special case of these general DFT-transmultiplexer filter banks with the following prototype filter

$$H_0^{\mathrm{OFDM}}(z) = 1 + z^{-1} + \ldots + z^{-(M-1)}. \tag{2}$$

The polyphase components $H_{0,0}(z)$, $H_{0,1}(z), \ldots$, $H_{0,M-1}(z)$ become unity in the case of OFDM.

In the following a square root raised cosine FIR filter (RRC) of length MK is assumed for the the prototype filter $H_0(z)$ in the general transmultiplexer case because of its beneficial properties as square root Nyquist filter. The number of subcarriers M for the multicarrier system is varied between $128, 256, 512$ and 1024.

Table 1. Real multiplications per symbol (synthesis part of FB).

M subcarriers:	128	256	512	1024
OFDM	1792	4096	9216	20480
RRC-TMUX	6912	14336	29696	61440

Both in the OFDM and the general DFT-transmultiplexer case with the RRC prototype filter (RRC-TMUX), the same number of real multiplications per symbol $2M\log_2(M)$ is needed for the IFFT operation, because it only depends on the number of subcarriers. Additionally, the RRC polyphase filters require $4K$ real multiplications per symbol. This is the computational overhead required for the improved spectral shaping.

Because of the duality of the synthesis part and the analysis part of the filter bank, the same observations are sincerely true for the receiver.

One additional issue should be taken into consideration: Especially in OFDM systems with cyclic prefix, a large number of subcarriers M is desired in order to keep the proportion of the cyclic prefix small. If no other reasons, e.g. a very strong frequency diversity, prevent the reduction of the number of subcarriers in the RRC-TMUX case, the number of subcarriers can be reduced compared to OFDM in order to compensate for the additional multiplications of the polyphase filters by saving multiplications of the IFFT operations. Table 1 shows the number of real multiplications per symbol for the OFDM and the RRC-TMUX case for different numbers of subcarriers and an RRC polyphase filter length of $K = 10$. The additional complexity for the RRC polyphase filters can be seen by the difference in each column. The reduction of the number of subcarriers in the RRC-TMUX case by a factor of 4 to $M/4$ leads to an overall complexity even smaller than that of an OFDM system with M subcarriers.

The main difference is that for OFDM with cyclic prefix no additional channel equalizer is needed as long as the cyclic prefix is sufficient in length. A comparison at the basis of the achievable data rates shows that OFDM is able to transmit $M\log_2(\mathcal{M}_\mathrm{O})$ bit per symbol, whereas the transmultiplexer transmits $(M + M_\mathrm{cp})\log_2(\mathcal{M}_\mathrm{T})$ bit per symbol, where $\mathcal{M}_\mathrm{O}$ resp. $\mathcal{M}_\mathrm{T}$ denote the modulation alphabets for OFDM resp. the RRC-TMUX. Moreover, the same modulation alphabets are assumed for all subcarriers. Equal rates are obtained by compensating the cyclic prefix length by the modulation scheme. If 25% cyclic prefix is assumed, i.e. $M_\mathrm{cp}/M = 1/3$, 256 QAM resp. 16 QAM is necessary for OFDM against 64 QAM resp. 8 QAM for a TMUX without CP. This means that more transmit power is necessary for OFDM in order to get the same (uncoded) bit error rates for the higher modulation schemes.

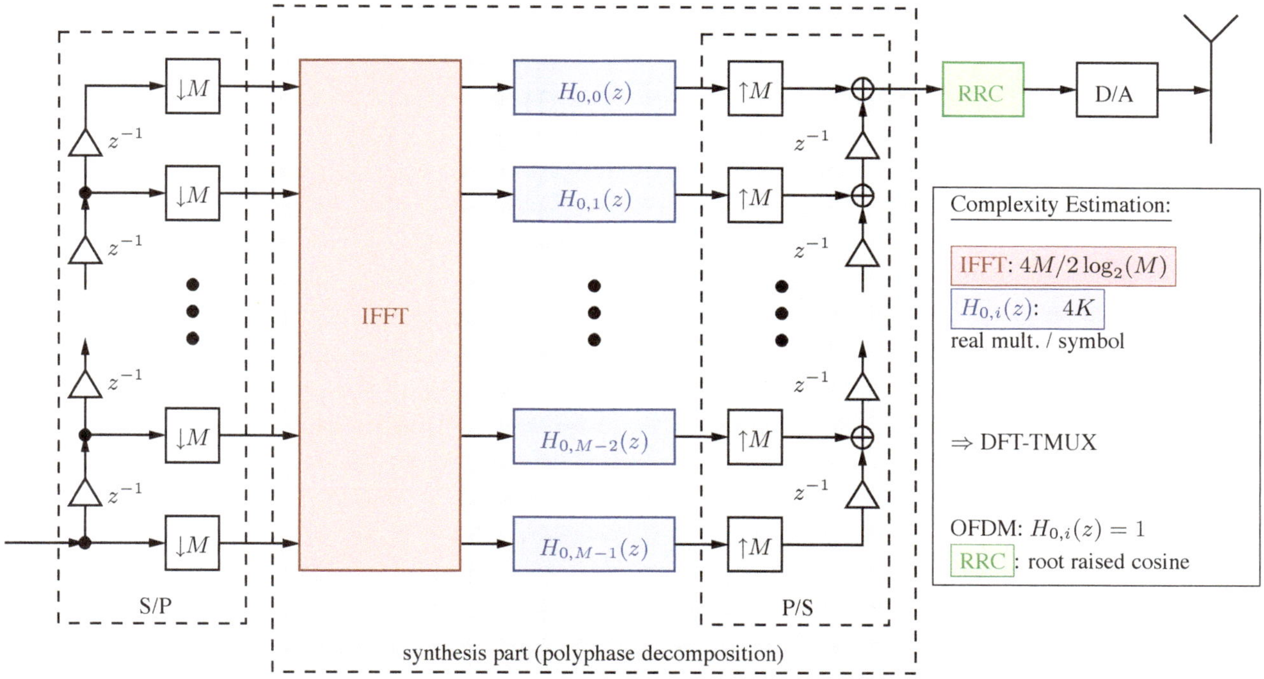

Fig. 2. DFT-transmultiplexer.

3 Power spectral density

The power spectral density of the input symbol vector of both the OFDM and the TMUX system is assumed to be white with unity power, i.e. $\mathrm{E}\left\{\boldsymbol{x}[n]\boldsymbol{x}[n]^{\mathrm{H}}\right\} = \mathbf{1}_M$. It is easily shown, that both for OFDM and the RRC-TMUX the transmitted symbol vector – disregarding the cyclic prefix for OFDM – keeps the white power spectral density. With OFDM the IDFT operation is a unitary matrix transformation which doesn't change the spectral properties of the vector it is applied to. The RRC filters of the TMUX are approximately square root M-th band filters which means that

$$\sum_{k=0}^{M-1} H^2(zW_M^k) = 1. \tag{3}$$

Therefore, the approximately M-th band spectral power densities of each subcarrier add up to a white power spectral density at the transmitter output.

In practice a certain number of subcarriers at the band edges is omitted for data transmission as frequency guard band in order to simplify the subsequent signal processing tasks concerning the band limitation and D/A conversion. Figures 3a and 3b show the power spectral densities both for an OFDM and an RRC-TMUX multicarrier system. The number of total subcarriers is $M = 32$, and $M_g = 8$ subcarriers are omitted at the band edges. Therefore, in both cases the total power spectral density is not white any more. The power spectral densities of the output signals of each subchannel of the TMUX have only overlap with the directly adjacent subchannels and separate the other subchannels by the large attenuation characteristic, whereas the power spectral densities for OFDM show large overlap with all other subchannels.

The subsequent RRC pulse shaping filter – operating at the high, M-fold symbol rate – has to limit the bandwidth to the existing bandwidth restrictions while avoiding intersymbol interference. In Figs. 3a and 3b a rolloff factor $\rho = 0.20$ has been used. The resulting power spectral densities are given in Figs. 3c and 3d. Obviously, the high stopband attenuation of the RRC subchannel filters allows relatively relaxed requirements on the RRC pulse shaping filter. The resulting bandwidth for the OFDM system is about 1.5 times the bandwidth for the TMUX system, when an attenuation threshold of 40 dB is considered.

Therefore, a smaller rolloff factor for the pulse shaping filter is required for the same bandwidth demand in an OFDM system. This increases the complexity for the pulse shaping filter and deteriorates the peak-to-average ratio of the transmit signal.

4 Conclusions

Complex modulated transmultiplexer filter banks with RRC subband filters provide an improved spectral shaping compared to OFDM systems. This means additional costs for the

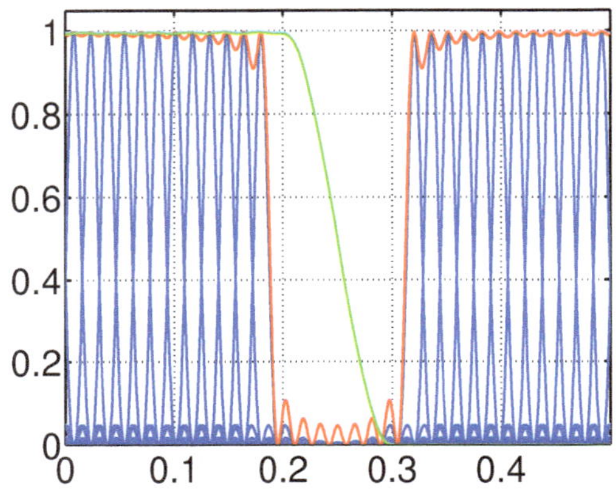

(**a**) PSD of OFDM: $M_{ch} = 32, M_g = 8$

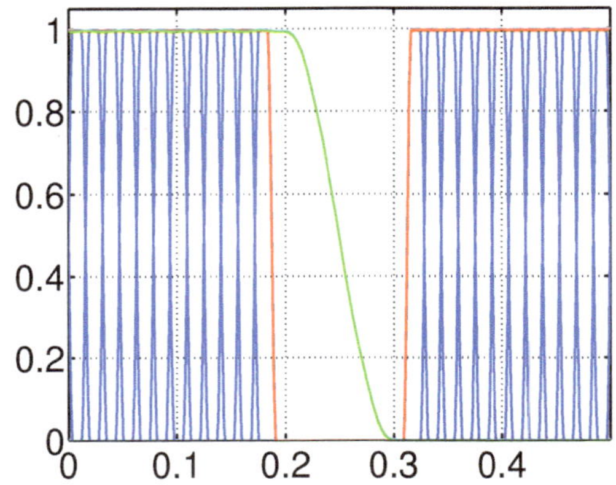

(**b**) PSD of RRC-TMUX: $M_{ch} = 32, M_g = 8$

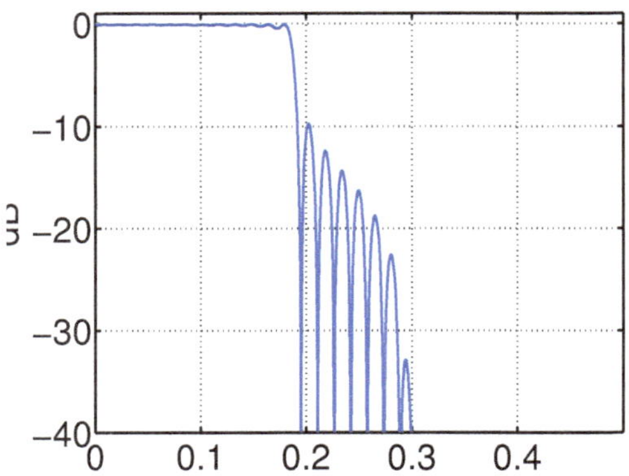

(**c**) PSD of OFDM: $M_{ch} = 32, M_g = 8$

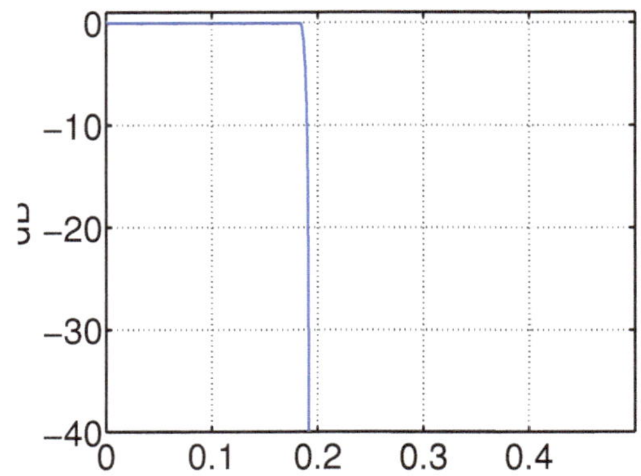

(**d**) PSD of RRC-TMUX: $M_{ch} = 32, M_g = 8$

Fig. 3. Comparison: Power spectral densities of OFDM vs. DFT.

pulse shaping filters, but the polyphase decomposition allows for a very efficient implementation.

The power spectral density of the transmit signal with a frequency guard band is nearly band-limited by the subband filters alone, the RRC pulse shaping filter is mainly needed to eliminate periodic repetitions. Therefore, the requirements on the RRC pulse shaping filter are relaxed and a relatively large rolloff factor can be applied. This leads to less complex pulse shaping filters and a lower peak-to-average ratio of the transmit signal because of the faster decaying impulse response of the pulse shaping filter.

The cyclic prefix of OFDM systems is an elegant solution to cope with the distortions of the multipath channel, but it reduces the bandwidth efficiency, because the number of subcarriers cannot be increased arbitrarily due to the time-selectivity of mobile communication channels. Therefore, the modulation alphabet has to be increased to achieve the same data rate as an RRC-TMUX. This requires more transmit power for the same (uncoded) bit error rate. The increase of the transmit power not only boosts the complexity of the transmit amplifier but also enhances the inter-cell interference in cellular systems like mobile communication systems.

The RRC-TMUX necessitates an additional channel equalizer to cancel intersymbol and adjacent channel interference caused by the frequency selective channel. The alleged complexity of this equalizer has been the main obstacle in the past to the application of DFT-transmultiplexer filter banks with RRC subband filters. The good attenuation properties of the subband filters, which only show overlap with directly adjacent subchannels, should reduce the complexity of the equalization considerably. This will be the subject of future work.

References

Akansu, A. N., Duhamel, P., Lin, X., and de Courville, M.: Orthogonal Transmultiplexers in Communication: A Review, IEEE Transactions on Signal Processing, 46, 979–995, 1998.

Bingham, J. A. C.: Multicarrier Modulation for Data Transmission: An Idea Whose Time Has Come, IEEE Communications Magazine, 5–14, 1990.

Fliege, N.: Multiraten-Signalverarbeitung, Informationstechnik, B. G. Teubner, Stuttgart, 1993.

Ivrlac, M. T.: Wireless MIMO Systems — Models, Performance, Optimization, Ph.D. thesis, Munich University of Technology, Shaker Verlag, 2005.

Kammeyer, K. D.: Nachrichtenübertragung, Informationstechnik, B. G. Teubner Stuttgart, 1992.

Melsa, P. J., Younce, R. C., and Rohrs, C. E.: Impulse response shortening for discrete multitone transceivers, IEEE Transactions on Communications, 44, 1662–1672, 1996.

Sinn, C. V., Götze, J., and Haardt, M.: Efficient Data Detection Algorithms in Single- and Multi-Carrier Systems without the Necessity of a Guard Period, in: IEEE ICASSP, 3, 2737–2740, 2002.

Toeltsch, M. and Molisch, A. F.: Efficient OFDM Transmission without Cyclic Prefix over Frequency-selective Channels, in: IEEE Int. Symp. on PIMRC, 2, 1363–1367, 2000.

Trautmann, S. and Fliege, N. J.: Perfect Equalization for DMT Systems Without Guard Interval, IEEE J. on Selected Areas in Communications, 20, 987–996, 2002.

Tzannes, M. A., Tzannes, M. C., Proakis, J., and Heller, P. N.: DMT Systems, DWMT Systems and Digital Filter Banks, in: IEEE Int. Conf. Comm., New Orleans, USA, 1, 311–315, 1994.

5

Improving the performance of BICM-ID and MLC systems with different FEC codes

T. Arafa, W. Sauer-Greff, and R. Urbansky

University of Kaiserslautern, Communications Engineering, Gottlieb-Daimler-Str., 67653 Kaiserslautern, Germany

Correspondence to: T. Arafa (arafa@eit.uni-kl.de)

Abstract. In bandwidth limited communication systems, the high data rate transmission with performance close to capacity limits is achieved by applying multilevel modulation schemes in association with powerful forward error correction (FEC) coding, i.e. coded modulation systems. The most important practical approaches to coded modulation systems are multilevel coding with multistage decoding (MLC/MSD) and bit interleaved coded modulation with iterative demapping and decoding (BICM-ID).

Multilevel modulation formats such as M-QAM, which can be used as a part of coded modulation systems, have the capability of multilevel protection. Based on this fact, we investigate the methods to improve the performance of BICM-ID using multiple interleavers with different binary channel coding schemes such as convolutional codes, turbo codes and low-density parity-check (LDPC) codes. Moreover, an MLC system with parallel decoding on levels (PDL) at the receiver is considered. In our contribution, we propose to design the individual coding schemes using the extrinsic information transfer (EXIT) charts for individual bit levels in the constellation. Our simulation results show that the BICM-ID systems, taking into account different bit-level protections, can provide an improvement of 0.65 dB, 1.2 dB and 1.5 dB for 256-QAM with turbo, LDPC and convolutional codes, respectively. On the other hand, MLC systems with PDL designed using EXIT charts for individual bit levels can slightly improve the performance and eliminate the error floor compared to the systems with MSD.

1 Introduction

The increasing demand to achieve higher data rates in digital communication systems with limited spectrum resources leads to employing multilevel modulation formats. In order to approach the channel capacity, an appropriate forward error correction (FEC) code is highly required to protect the multilevel signaling schemes. The combination of high level modulation schemes with FEC coding schemes is denoted as a coded modulation system.

The field of coded modulation was first introduced by (Massey, 1974) who proposed to jointly design coding and modulation. Ungerboeck presented a practical approach for coded modulation known as trellis coded modulation (TCM) (Ungerboeck and Csajka, 1976; Ungerboeck, 1982). In this system, Ungerboeck proposed the set partitioning strategy for mapping scheme in order to maximize the minimum intra-subset Euclidean distance. The key point of this system is to deal with the free Euclidean distance measure instead of the Hamming distance as in classical coding schemes. Therefore, the modulation and the channel coding based on convolutional code are combined in a single entity which enables to transmit only constrained sequences of symbols that maximize the minimum free Euclidean distance.

Another approach of constructing coded modulation system is multilevel coding (MLC) proposed by (Imai and Hirakawa, 1977). A more comprehensive study of this system with information-theoretic analysis including MLC capacity, error bounds and rate design rules was reported in (Wachsmann et al., 1999). The basic idea behind MLC scheme is to protect each bit level of the signal constellation by an individual binary code. The common key point of this system is to employ the Ungerboeck mapping by set partitioning and multistage decoding (MSD) at the receiver. Although both TCM and MLC have the same principle, which is optimizing the system in the Euclidean space, the MLC approach provides more flexibility regarding to implementation issues since mapping schemes and channel coding schemes are decoupled from each other. Moreover, different component codes such as convolutional codes, turbo codes and low-density parity-check (LDPC) codes can be used in MLC systems.

A bit interleaved coded modulation with iterative demapping and decoding (BICM-ID) was proposed by (ten Brink et al., 1998; ten Brink, 1999) as an attractive technique to achieve both high spectral efficiency and high performance. Unlike MLC systems, BICM-ID systems use only one channel encoder separated from modulation by an interleaver which achieve a reasonable complexity.

The task of mapping in high order modulation schemes such as M-QAM is to assign a number of bits to one symbol of the constellation points. The most important characteristic of these schemes is the constellation capacity. However, each bit position has a distinct capacity, i.e. a different amount of protection. In this paper, we investigate the methods to improve the performance of BICM-ID using the idea of protection matching with different binary channel coding schemes such as convolutional codes, turbo codes and LDPC codes. Moreover, an MLC system employing Gray mapping for modulation and parallel decoding on levels (PDL) at the receiver is considered. We proposed to design the proper individual coding schemes with code rates using the extrinsic information transfer (EXIT) charts for individual bit levels in the constellation.

The remainder of this paper is organized as follows. The basic BICM-ID systems and the new design based on multilevel protection is briefly introduced in Sect. 2. The concept of MLC/MSD systems with an idea of code design is presented in Sect. 3. The proposed code design based on EXIT charts for MLC/PDL systems is explained in Sect. 4. Simulation results and discussions are provided in Sect. 5. Finally, conclusions are drawn in Sect. 6.

2 BICM-ID systems

The BICM-ID system is a serial concatenation of channel encoding, bit-interleaving and multilevel modulations at the transmitter, while the turbo principle is adapted to iterative soft-demapping together with channel decoding at the receiver (ten Brink et al., 1998). The performance mainly depends on matching between mapping and coding schemes. Therefore, a Gray mapping scheme should be combined with a strong channel code such as a serial or parallel concatenated convolutional code (turbo code) and an LDPC code. On the other hand, a mapping different from Gray mapping should be combined with a weak channel code. This system achieves more flexibility and simplicity from an implementation point of view than the other coded modulation systems.

The basic structure of the BICM-ID system is depicted in Fig. 1. At the transmitter, the sequence u of k data bits is firstly encoded into a codeword c of n bits. The generated codeword c is randomly interleaved by a bit-wise interleaver. After interleaving, m consecutive bits are mapped onto one complex symbol of $M = 2^m$ according to a mapping

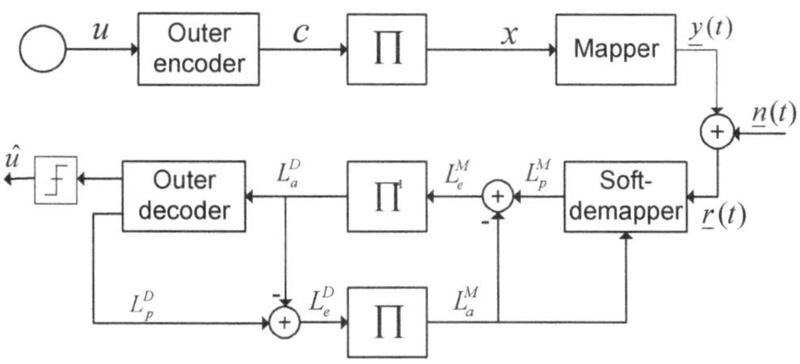

Fig. 1. Block diagram for conventional BICM-ID systems.

scheme. The originated complex symbols $\underline{y}$ are corrupted by noise through the AWGN channel, so the received symbols are $\underline{r} = \underline{y} + \underline{n}$ where $\underline{n}$ is a complex circular symmetric Gaussian noise with zero mean and variance $N_o/2$ in each dimension. At the receiver, and based on channel observations $\underline{r}$ and a-priori information of unmapped bits L_a^M (at first iteration $L_a^M = 0$), the soft-demapper calculates the a-posteriori log-likelihood ratio (LLR) for each of the m coded bits per symbol by using (ten Brink et al., 1998)

$$L_p^M(x_k|\underline{r}) = L_a^M(x_k) + \ln \frac{\sum_{\underline{y}\in\chi_k^1} \exp[-\frac{|\underline{r}-\underline{y}|^2}{N_o} + \sum_{j=1,j\neq k}^{m} \mu_j^{-1}(\underline{y}) L_a^M(x_j)]}{\sum_{\underline{y}\in\chi_k^0} \exp[-\frac{|\underline{r}-\underline{y}|^2}{N_o} + \sum_{j=1,j\neq k}^{m} \mu_j^{-1}(\underline{y}) L_a^M(x_j)]}, \quad (1)$$

where $\mu_j^{-1}(\underline{y})$ gives the j-th bit of the symbol $\underline{y}$ while χ_k^1, χ_k^0 are the sets of symbols having the k-th bit set to 1 and 0, respectively. Then, the extrinsic information of the soft-demapper $L_e^M = L_p^M - L_a^M$ is deinterleaved to become the a-priori input L_a^D to the outer soft-in/soft-out decoder, which gives the a-posteriori value L_p^D based on the BCJR algorithm (Hagenauer, 1997). The extrinsic information of the outer decoder $L_e^D = L_p^D - L_a^D$ is passed through the interleaver and fed back as a-priori information L_a^M to the soft-demapper for the next iteration.

Although the BICM-ID systems can provide a desirable BER performance by combining different channel coding and mapping schemes, further improvements can be achieved by considering multilevel protection of high order modulation schemes. The task of mapping in high order modulation schemes such as M-QAM is to assign a number of bits to one symbol of the constellation points. The most important characteristic of these schemes is the constellation capacity. However, each bit level has a distinct capacity as depicted in Fig. 2, i.e. different amount of protection. On the other hand, recursive systematic convolutional (RSC) codes or turbo codes, as a part of BICM-ID systems, provide information and parity bits. Therefore, the performance can be improved by assigning the information bits always to bits in constellation points with high bit-level capacity. By using

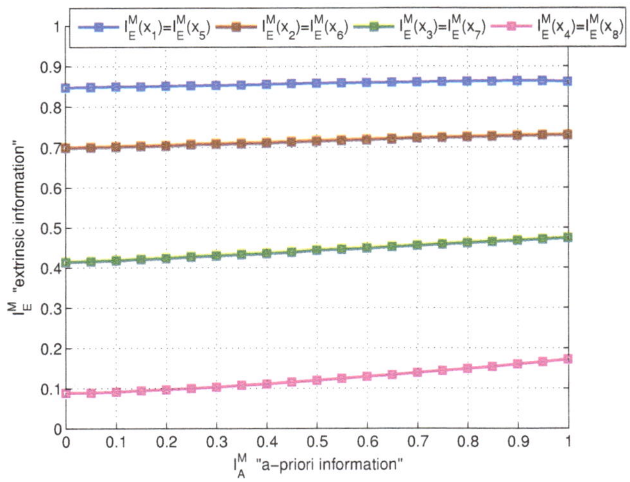

Fig. 2. EXIT curves of the individual bits in Gray 256-QAM mapping for AWGN channel at Eb/No=7.5 dB. The first and fifth bits (x_1 and x_5) have identical EXIT functions and similarly (x_2 and x_6), (x_3 and x_7), and (x_4 and x_8).

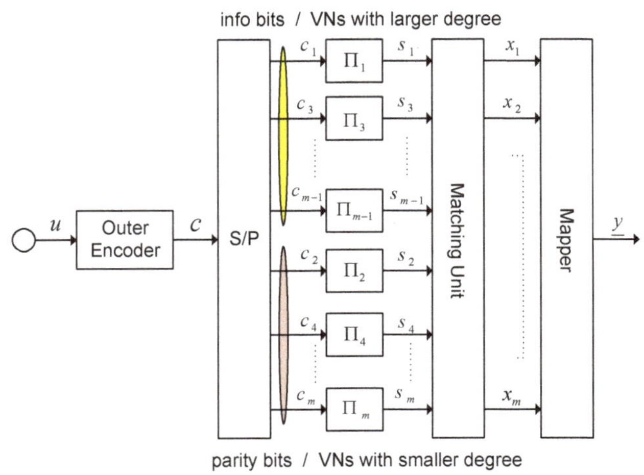

Fig. 3. Block diagram of BICM-ID transmitter with protection matching.

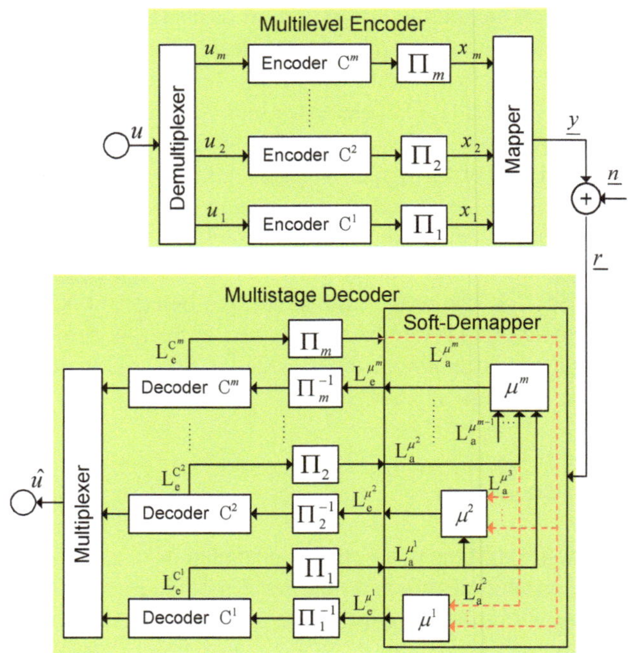

Fig. 4. Multilevel coding scheme with multistage decoder.

irregular LDPC codes as a part of BICM-ID system, other considerations should be taken into account for protection matching. The LDPC code ensembles are defined by degree distributions $(\lambda(x), \rho(x)) \equiv \left(\sum_{i=1}^{w_v} \lambda_i x^{i-1}, \sum_{i=1}^{w_c} \rho_i x^{i-1}\right)$ of the variable nodes (VN) and check nodes (CN) with λ_i (resp. ρ_i) being the fraction of edges connected to degree-i variable (resp. check) nodes. The LDPC code can be considered as a serial concatenation of outer repetition codes at the variable node, edge interleaver and inner parity check codes at the check node. Therefore, the degree of a variable node is equal to the number of repetitions. It is well known that more repetitions, i.e. larger variable node degree, lead to more reliable decoded bit. Based on this explanation, the performance of BICM-ID systems with LDPC codes can be improved by assigning the more reliable LDPC code bits always to the more protected bits in the constellation (Li and Ryan, 2005).

Figure 3 shows the block digram of BICM-ID transmitter with multiple interleavers and matching unit. This unit assigns the coded and interleaved bits to different bit positions in the M-QAM symbol.

3 Multilevel coding and multistage decoding

The structure of an MLC/MSD system is depicted in Fig. 4. In the MLC transmitter, a sequence of k information bits is partitioned into m sub-blocks of length k^i. Each partitioning level i has its own encoder C^i with code rate $R^i = k^i/n^i$ and is independently encoded and interleaved by a different interleaver Π_i where we assume the same block length n at all levels. The overall rate of the MLC system results from the sum of the rates of the component codes. Every channel coding concept is applicable. In this contribution binary LDPC codes are used. Then, one bit from each coded and interleaved level is grouped to the sequence $\mathbf{x} = \{x_1, x_2, ..., x_m\}$, $x_i \in \{0, 1\}$, which is mapped to one of the $M = 2^m$ signal points of the set $\mathbf{A}$. Set partitioning strategy as proposed by Ungerboeck is applied to maximize the minimum intra-subset Euclidean distance, see Fig. 5 for the case of 8-ASK (Ungerboeck, 1982).

The signal set $\mathbf{A}$ in the first step of set partitioning at level $i = 1$ is divided into the subsets $\mathbf{A}(x_1 = 0)$ and $\mathbf{A}(x_1 = 1)$. Then, all subsets $\mathbf{A}(x_1 \dots x_i)$ at partitioning level i are iteratively divided into two further subsets $\mathbf{A}(x_1 \dots x_i 0)$ and

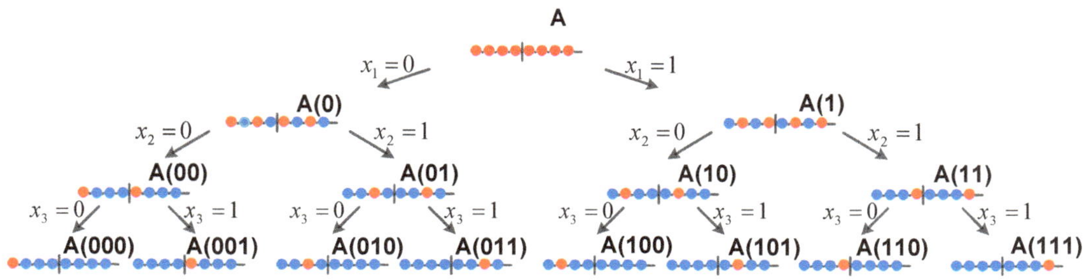

Fig. 5. Ungerboeck set partitioning for 8-ASK.

$\mathbf{A}(x_1 \ldots x_i 1)$ at partitioning level $i+1$. The iteration stops when each subset at level m contains only one signal point.

At MSD receiver, the soft-demapper firstly calculates the reliability information of the bits at level 1 with no a-priori information about the bits at the other levels $L_a^{\mu_i} = 0$, $i = 2, \ldots, m$. The codes C^i at the lowest level $i = 1$ can be decoded using soft information that was passed from the soft-demapper. At the next stage, the soft-demapper uses soft information that was passed from decoder C^1 as a-prior information for the bits at level 1 to calculate the reliability information of the bits at level 2 that will be passed to the decoder C^2. This procedure is carried out until the last code is decoded. The MSD structure is suboptimum because the decoders of lower levels are not provided with any information of the decoders of higher levels and error propagation may arise. Therefore, in some cases it is useful to iterate the decoding process with output information of higher levels, i.e. iterative MSD.

It was proved that MLC together with MSD can achieve the channel capacity at high spectral efficiency if the individual coding schemes with proper code rates are selected appropriately. The concept of designing the codes depends on the mutual information of the channel $I(\underline{Y}; \underline{R})$ between the transmitted signal points $\underline{y} \in \mathbf{A}$ and the received symbols $\underline{r}$ (Wachsmann et al., 1999). Using uppercase letters for the corresponding random variables, the mutual information of the channel input $\underline{y}$ is calculated with the bijective mapping according to $I((X^1, X^2, \ldots, X^m); \underline{R}) = I(\underline{Y}; \underline{R})$ which is equal to the discrete channel capacity C. By using the chain rule of mutual information

$$\begin{aligned} I(\underline{Y}; \underline{R}) &= I((X^1, X^2, \ldots, X^m); \underline{R}) \\ &= I(X^1; \underline{R}) + I(X^2; \underline{R}|X^1) + \ldots \\ &\quad + I(X^m; \underline{R}|X^1, X^2, \ldots, X^{m-1}), \end{aligned} \tag{2}$$

we can separate the transmission of the address vectors $x[k]$ over the physical channel into parallel transmissions of individual components $x_i[k]$ over equivalent channels, whereas the knowledge about the less significant components $x_1[k], x_1[k], \ldots, x_i[k]$ is a premise at the receiver. These equivalent channels are directly correspond to the individual levels in MLC as well as to the requirement of error-free decoding of the lower levels in the MSD. So, the components

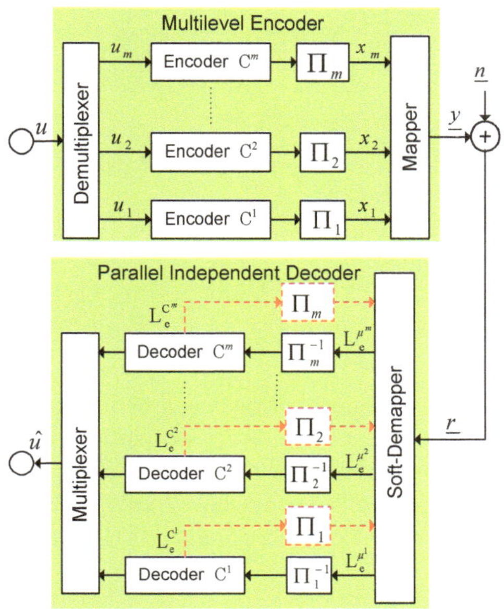

Fig. 6. Multilevel coding scheme with parallel decoder of the individual levels.

of the Eq. (2) are the capacities C^i of the equivalent channels, where the capacity of level i computes to

$$\begin{aligned} C^i &= I(X^i; \underline{R}|X^1, X^2, \ldots, X^{i-1}) \\ &= I(X^i, \ldots, X^m; \underline{R}|X^1, \ldots, X^{i-1}) \\ &\quad - I(X^{i+1}, \ldots, X^m; \underline{R}|X^1, X^2, \ldots, X^i). \end{aligned} \tag{3}$$

When MLC is decoded with MSD, the total channel capacity C can only be obtained if the code rate R^i at each level is equal to the capacity of the equivalent channel $R^i \equiv C^i$. This rate design method is called capacity rule.

4 Multilevel coding and parallel decoding on levels

An alternative decoding strategy for multilevel coding schemes is parallel decoding on the individual levels (PLD) (Schramm, 1997). In contrast to the MSD approach, all decoders are operating in parallel without using the decisions of other decoders as depicted in Fig. 6. The capacity C^i

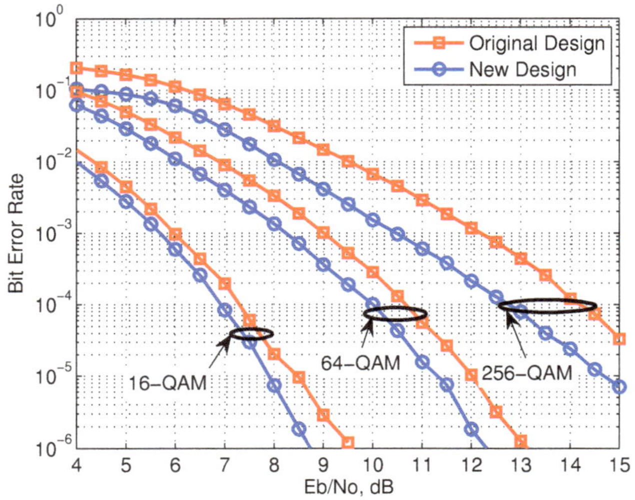

Fig. 7. BER of BICM-ID with a rate-1/2 convolutional code for AWGN channel after 5 iterations.

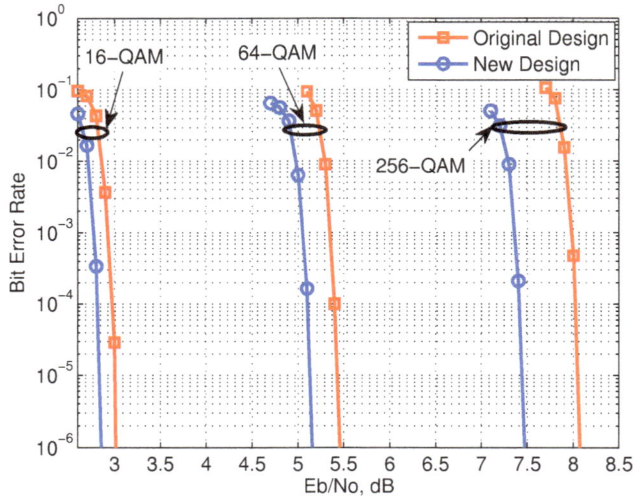

Fig. 8. BER of BICM-ID with a rate-1/2 turbo code for AWGN channel after 10 super iterations.

of the equivalent channel i with PDL is

$$C^i = I(X^i; \underline{R}). \tag{4}$$

The PDL capacity highly depends on the particular mapping. Gray mapping scheme can achieve a good performance for MLC/PDL systems. Therefore, we propose to use a strong FEC codes such as capacity-approaching LDPC codes. The individual code rates are designed using the EXIT charts for individual bit levels.

The soft-demapper EXIT curve can be estimated for a specific channel SNR by Monte-Carlo simulation where the a-priori LLR values can be assumed to be Gaussian distributed. This function can be expressed as

$$\begin{aligned} I_E^M(\mathbf{x}) &= T^M(I_A^M(\mathbf{x}), E_b/N_o) \\ &= \frac{1}{m}\sum_{i=1}^{m} T_i^M(I_A^M(\mathbf{x}), E_b/N_o), \end{aligned} \tag{5}$$

where T^M is the EXIT function of the soft-demapper and $\{I_E^M(x_i) = T_i^M(I_A^M(\mathbf{x}), E_b/N_o)\}$ is the EXIT function for the different bits, $i = 1, 2, \ldots, m$.

Based on the average EXIT function and bit-level EXIT functions, the individual code rates are designed as following. We construct the average EXIT function of mapping at a certain SNR, where the area under curve is

$$A = \frac{C}{m}. \tag{6}$$

Then, the area under each bit-level EXIT curve is $A^i = C^i$, $i = 1, 2, \ldots, m$. Therefore, the code rate at each level is $R^i \equiv C^i$.

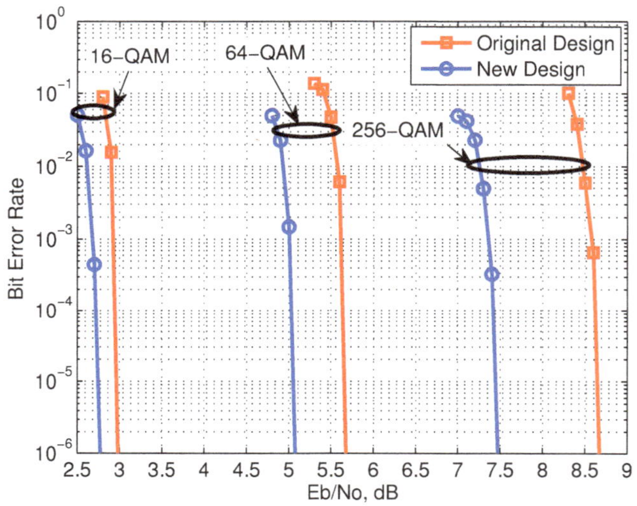

Fig. 9. BER of BICM-ID with a rate-1/2 LDPC code for AWGN channel after 10 super iterations.

5 Simulation results

In the following simulations we compare the performance of new BICM-ID design based on multilevel protection of high order modulation schemes versus basic design for different FEC coding schemes. Moreover, we introduce the performance of MLC/MSD applying Ungerboeck mapping and MLC/PDL applying Gray mapping.

Figure 7 shows the BER performance of the BICM-ID systems after 5 iterations applying Gray M-QAM mapping schemes and a rate-1/2 RSC code with generator polynomials in octal form $G = [23, 35]$. The block size of information bits is set to $k = 29\,996$ bits. As can be seen in Fig. 7, the new design with multiple interleavers considering protection matching can provide an improvement up to 1.5 dB

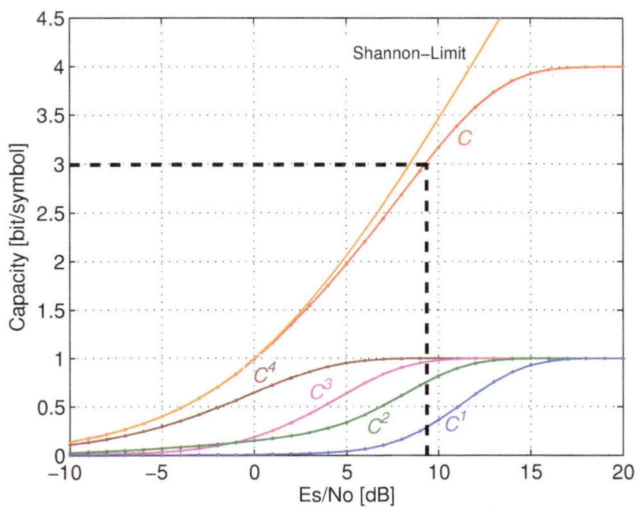

Fig. 10. Capacity $C(\mathbf{A})$ and capacities C^1, C^2, C^3 and C^4 of the equivalent channels for Ungerboeck 16-QAM Mapping.

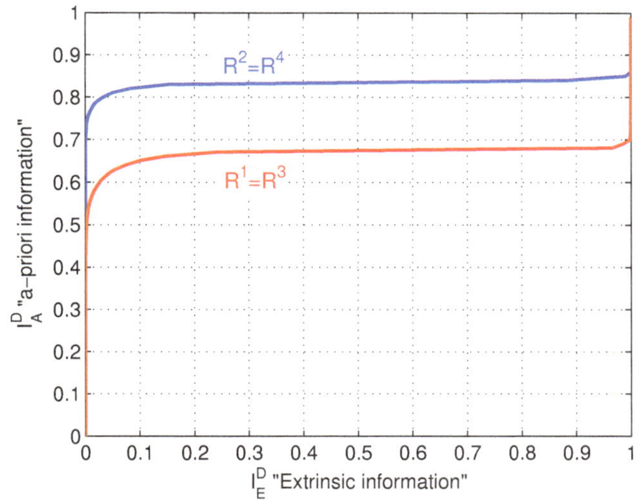

Fig. 12. EXIT chart of designed LDPC codes.

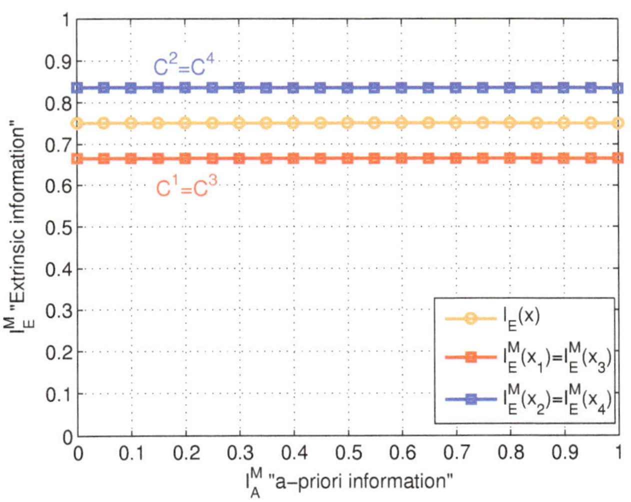

Fig. 11. Average EXIT function of Gray 16-QAM mapping and the bit-level EXIT curves for AWGN channel at $E_b/N_o = 9.3$ dB. The first and third bits have identical EXIT functions and similarly the second and fourth.

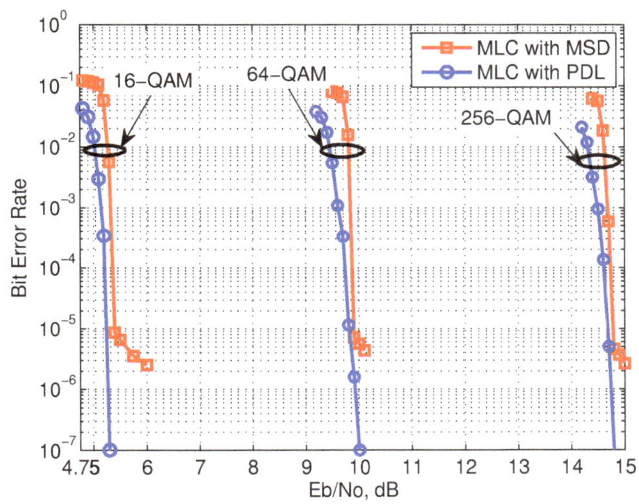

Fig. 13. BER of MLC/MSD and MLC/PDL with LDPC component codes. Different signal constellations including 16-QAM, 64-QAM and 256-QAM are used at spectral efficiencies of 3, 5 and 7 bit/sec/Hz, respectively.

for 256-QAM compared with the original system. We can also observe that the achievable improvement increases with growing constellation size.

Two BICM-ID systems are also considered. In the first one, the combination of Gray mapping scheme and a rate-1/2 turbo code, which is obtained by puncturing a rate-1/3 turbo code built from two parallel concatenated 16-state RSC codes with polynomials [23,37], is applied. The size of the pseudo-random interleaver that separates both RSC codes is 30 000. The other system uses a rate-1/2 (30 000,60 000) capacity-approaching LDPC code and Gray mapping. The number of internal iterations is set to 10 iterations and the number of super iterations between mapping and coding schemes is 10. In Figs. 8 and 9, simulation results show again that the new design of BICM-ID systems using protection matching for turbo code and LDPC code, respectively, can improve the performance for different constellation sizes.

On the other hand, we apply MLC systems that use capacity-approaching LDPC codes with proper code rates in combination with different constellation sizes $M = 16, 64, 256$ to get different spectral efficiencies $\eta = 3, 5, 7$ bit/sec/Hz. The codeword length for LDPC codes is set to 21000 bits. At the receiver, we use MSD approach for Ungerboeck mapping or PDL approach for Gray mapping.

The code rates for MLC/MSD systems can be designed according to capacity rules that was explained in Sect. 3. Considering the example of 16-QAM in Fig. 10 with a total rate $R = 3$, we obtain the rates design $R^1/R^2/R^3/R^4 = 0.29/0.75/0.96/1$. For MLC/PDL systems, the code rates are designed based on bit-level EXIT curves. We construct the average EXIT function of mapping with area under EXIT function C/m. Then, the areas under bit-level EXIT functions are identical to code rates. The rate design $0.67/0.83/0.67/0.83$ can be obtained for 16-QAM with a total rate $R = 3$ from Fig. 11. In this approach, the exit curves for individual codes have also to match the corresponding bit-level EXIT function at the demapper as shown in Figs. 11 and 12.

Figure 13 shows the simulation results for both MLC/MSD and MLC/PDL system at different spectral efficiencies. As we can see, the MLC/PLD systems provide a little improvement with eliminating error floor compared with MLC/MSD systems.

6 Conclusions

In this contribution we presented the method to improve the BICM-ID systems based on multilevel protection for different FEC coding schemes. An improvement up to 1.5 dB can be obtained using new BICM-ID design based on protection matching with multiple interleavers compared to the conventional BICM-ID system. On the other hand, we proposed to design the MLC/PDL systems according to bit-level EXIT charts. The performance of MLC/PDL with Gray mapping provides a slightly improvement and eliminate the error floor compared to MLC/MSD with Ungerboeck mapping. Moreover, the PDL approach can greatly decrease the time delay of decoding process at the receiver.

References

Hagenauer, J.: The Turbo Principle: Tutorial Introduction and State of the Arts, Symposium on Turbo Codes, Brest, France, September 1997, 1–11, 1997.

Imai, H. and Hirakawa, S.: A New Multilevel Coding Method Using Error-Correcting Codes, IEEE T. Inform. Theory, 23, 371–377, 1977.

Li, Y. and Ryan, W.E.: Bit-Reliability Mapping in LDPC-Coded Modulation Systems, IEEE Commun. Lett., 9, 1–3, Jan 2005.

Massey, J.: Coding and Modulation in Digital Communications, Proceedings of International Zurich Seminar on Digital Communication, Zurich, Switzerland, 12–15 March 1974.

Schramm, P.: Multilevel Coding with Independent Decoding on Levels for Efficient Communication on Static and Interleaved Fading Channels, IEEE International Symposium on Personal, Indoor and Mobile Radio Communications, Helsinki, Finland, 1–4 Sept 1997, 1361–1391, 1997.

ten Brink, S., Speidel, J., and Yan, R.-H.: Iterative Demapping and Decoding for Multilevel Modulation, IEEE International Conference on Global Communications (GLOBECOM), Sydney, Australia, 8–12 November 1998, 579–584, 1998.

ten Brink, S.: Convergence of Iterative Decoding, IEEE Electron. Lett., 35, 1117–1119, 1999.

Ungerboeck, G. and Csajka, I.: On Improving Data-Link Performance by Increasing Channel Alphabet and Introducing Sequence Coding, Proceedings of IEEE Int. Symp. Inform. Theory, Ronneby, Sweden, 21–24 June 1976.

Ungerboeck, G.: Channel Coding with Multilevel/Phase Signals, IEEE T. Inform. Theory, 28, 55–67, 1982.

Wachsmann, U., Fischer, R., and Huber, J.: Multilevel Codes: Theoretical Concepts and Practical Design Rules, IEEE T. Inform. Theory, 45, 1361–1391, 1999.

6

Design space exploration of high throughput finite field multipliers for channel coding on Xilinx FPGAs

C. de Schryver, S. Weithoffer, U. Wasenmüller, and N. Wehn

Microelectronic Systems Design Research Group, University of Kaiserslautern, Erwin-Schrödinger-Str., Germany

Correspondence to: C. de Schryver (schryver@eit.uni-kl.de)

Abstract. Channel coding is a standard technique in all wireless communication systems. In addition to the typically employed methods like convolutional coding, turbo coding or low density parity check (LDPC) coding, algebraic codes are used in many cases. For example, outer BCH coding is applied in the DVB-S2 standard for satellite TV broadcasting. A key operation for BCH and the related Reed-Solomon codes are multiplications in finite fields (Galois Fields), where extension fields of prime fields are used. A lot of architectures for multiplications in finite fields have been published over the last decades. This paper examines four different multiplier architectures in detail that offer the potential for very high throughputs. We investigate the implementation performance of these multipliers on FPGA technology in the context of channel coding. We study the efficiency of the multipliers with respect to area, frequency and throughput, as well as configurability and scalability. The implementation data of the fully verified circuits are provided for a Xilinx Virtex-4 device after place and route.

1 Introduction

Galois or finite field multiplications (FFMs) are key in a wide range of technical applications, for example in cryptography or channel coding. In particular BCH- and Reed-Solomon codes require a lot of Galois field (GF) multiplications to be performed in the decoding process. Nowadays, BCH codes are widely spread and employed in modern communication standards like DVB and in error correction units for flash memory devices (Liu et al., 2006), for instance. In Sect. 2 we summarize the theoretical background of FFM and BCH decoding.

However, the specific FFM requirements vary over the application range: high Galois field dimensions are used in cryptography, whereas for FFMs in channel coding the focus lies on high throughput and low latency. FFMs are key components in BCH decoder implementations and consume a significant amount of hardware resources in the overall design (Chen et al., 2009; Zhang et al., 2010).

Up to now, a lot of research has been made on area efficient and fast FFM architectures. However, Ahlquist et al. haven shown in 1999 (Ahlquist et al., 1999) that not every VLSI optimized multiplier architecture performs optimal on FPGAs.

Already in 1986, Scott et al. have proposed a scalable FFM that is easily adaptable to different GF sizes and primitive polynomials (Scott et al., 1986). Their bit-slice architecture makes efficient use of area providing flexibility for different field sizes. Building on the work of Scott et al., Kitsos et al. in 2003 proposed a flexible, reconfigurable FFM architecture for extension fields $GF(2^m)$ (Kitsos et al., 2003). Their bit-serial polynomial basis multiplier features reconfigurability for field sizes up to m and reduced power consumption due to gated clocking. The low hardware complexity promises good area performance.

In this work, we investigate the potential benefit of these published FFM architectures as well as two "standard" architectures for a high-throughput DVB-S2 standard BCH decoder. For this purpose we have implemented the selected FFMs as described in Sect. 3 and fully validated them with a hardware-software co-simulation setup. We have synthesized our implementations and compare area consumption and achievable throughput for a common Xilinx Virtex-4 FGPA target device as described in Sect. 4. Furthermore, we evaluate them with respect to their suitability for FPGA target devices, area consumption, latency and scalability. Finally, we comment on the impact of using sophisticated pipelined FFMs in a DVB-S2 BCH decoder and show which parts can profit most from those.

2 Theoretical background

For a better understanding of the finite field multiplier architectures and the context of BCH coding, we give an overview on the underlying mathematics in this section. For more details we refer to available standard literature (Bossert, 1998; Friedrichs, 1995).

2.1 Finite fields and multiplication

The basis for the finite field multiplication architectures implemented in this work is a reformulation of the finite field multiplication in polynomial representation. The derivation of the algorithm originally proposed by Scott et al. (Scott et al., 1986) is outlined in the following.

The nonzero elements of the extension field $GF(2^m)$ can be constructed by powers of the primitive element α, where α is a root of a primitive polynomial $P(x) = x^m + p_{m-1}x^{m-1} + \cdots + p_1 x + p_0$ over $GF(2)$. Since $P(\alpha) = 0$ (α being root of P) $\alpha^m = p_{m-1}\alpha^{m-1} + \cdots + p_1\alpha + p_0$ and therefore nonzero the elements can be written in the canonical base representation:

$$\{a_{m-1}\alpha^{m-1} + \cdots + a_1\alpha + a_0 | a_i \in GF(2) \text{ for } 0 \le i \le m-1\}$$

2.1.1 Multiplication algorithm

Based on the work of Scott et al., an algorithm for FFM can be derived similar to the design proposed by Kitsos et al. as follows: Let $A, B \in GF(2^m)$ in canonical base representation and P primitive polynomial over $GF(2)$. Then let

$$C = A(x) \cdot B(x) \bmod P \tag{1}$$

$$= \sum_{k=0}^{m-1} A(x) \cdot b_k x^k \bmod P \tag{2}$$

$$= (\underbrace{(0 \cdot x + b_{m-1}A(x))}_{K^0(x)} x^{m-1} + \cdots + b_0 A(x)) \bmod P(x) \tag{3}$$

$$= (\underbrace{\left(K^0(x)x + b_{m-2}A(x)\right)}_{K^1(x)} x^{m-2} + \cdots + b_0 A(x)) \bmod P(x) \tag{4}$$

$$\vdots$$

$$= K^{m-1}(x) \bmod P(x) \tag{5}$$

Thus, the product $C(x)$ can be obtained by performing iteratively $m-1$ times:

$$K^i(x) = \left(K^{i-1}(x)x + b_{m-1-i}A(x)\right) \bmod P(x) \tag{6}$$

where $K^-1(x) = 0$.

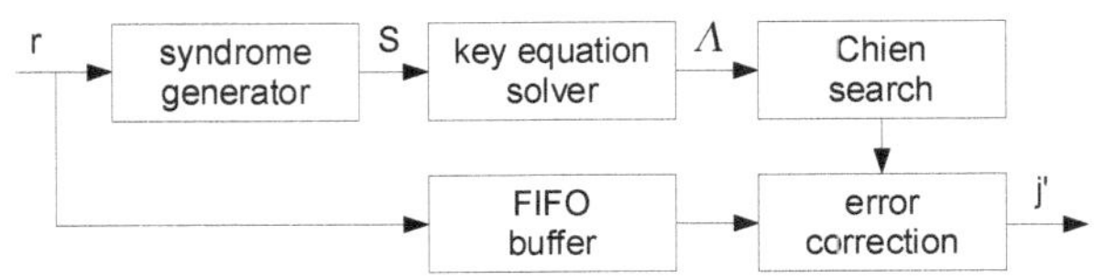

Fig. 1. Block Diagram of an BCH Decoder.

Equation (6) can be simplified by using the fact that $x^m \bmod P(x) = p_{m-1}x^{m-1} + \cdots + p_1 x + p_0$ and by writing $K^{i-1}(x)$ and $A(x)$ as $k_{m-1}^{i-1}x^{m-1} + \cdots + k_0^{i-1}$ and $a_{m-1}x^{m-1} + \cdots + a_0$ respectively and one obtains:

$$K^i(x) = \sum_{j=0}^{m-1} k_j^i x^j \tag{7}$$

with

$$k_j^i = k_{m-1}^{i-1} p_j + k_{j-1}^{i-1} + b_{m-1-i} a_j \quad \text{and} \quad k_{-1}^{i-1} = 0 \tag{8}$$

With Eqs. (6) and (8), the multiplication algorithm over $GF(2^m)$ is given, that is used for the architectures of the Scott- and Kitsos-multiplier.

2.2 BCH decoding

BCH codes have been invented in 1959 by Hocquenghem (Hocquenghem, 1959), and independently in 1960 by Bose and Ray-Chaudhuri (Bose and Ray-Chaudhuri, 1960). These codes provide high flexibility and predictable error correction ability, and hence are used in a wide range of technical applications. Details about their construction and decoding can be found in literature (Bossert, 1998; Friedrichs, 1995).

Two main categories of BCH codes exist: so-called *primitive* and *non-primitive* ones. Code words of primitive BCH codes are always of fixed length $n = 2^m - 1$, with n being the length of the code word and m the dimension of the Galois field 2^m. Non-primitive BCH codes provide flexibility in choosing the appropriate code word size. In contrast to code puncturing, where one or more positions in a code word are omitted, the minimum distance in a reduced code is not changed (Bossert, 1998). Non-primitive BCH codes are used in standards like DVB-T2, DVB-S2, DB-C2, DVB-H, ITU-T H.261 or ITU-T G.975.1, for example.

BCH codes are usually decoded by algebraic syndrome decoding. Figure 1 shows the generic structure of a syndrome based BCH decoder.

It is important to notice that the syndrome $S(x)$ only depends on a possible error vector e and not on the correct code word that has been sent. The decoder therefore has to find a possible error vector e with a minimum weight, that means an e with a minimum number of coefficients unequal to zero. In order to calculate e algebraically, the search for minimum weight is transformed into a search for a polynomial with minimum degree (Bossert, 1998).

The syndrome generator unit (SGU) computes the syndrome $S(x)$ that is fed into a so-called *key equation solver* (KES). The KES computes the *error locator polynomial* $\Lambda(x)$ out of the syndrome. The Chien search unit (CSU) afterwards checks for all elements in the specified Galois field $GF(2^m)$ if $\Lambda(x) = 0$. In this case, the error vector e has a 1 at the corresponding position; the error correction then flips the bits at all recognized error positions in the received code word r.

3 Hardware implementation

The hardware implementation of the FFMs should be flexible and generic with respect to the primitive polynomial and field size. Furthermore, they should exploit the underlying structures of the Virtex-4 FPGA-platform well to yield small area usage while providing sufficient throughput.

For the target DVB-S2 standard, BCH codes based on the Galois fields 2^{14} and 2^{16} are used. As this research was performed in the context of a more complex communication system, the common clock frequency of the whole system was predefined to 200MHz.

The hardware implementations considered in our work are based on (Scott et al., 1986) and (Kitsos et al., 2003). Furthermore, we give insight into the architecture of the plain combinatorial FFM implementations used.

3.1 Scott-Multiplier

The architecture of the Scott-multiplier resembles the multiplication algorithm derived in Sect. 2.1. Equation (8) is mapped to a basic cell (see Fig. 2). It consists of two 2-input AND-gates, two 2-input XOR gates and a flip-flop. Additionally, our implementation contains flip-flops for storing the coefficients of the A-operand during a calculation and control logic for that. Still, such cells fit comfortably into a Virtex-4 FPGA slice that offers two 4-input LUTs, two multiplexers, two 1-bit registers and arithmetic logic.

For a multiplier over $GF(2^m)$, m basic cells are connected into an array that calculates the product $C = A \cdot B \bmod P$ in m clock cycles. This array can be viewed as a linear feedback shift register and is illustrated in Fig. 3. The primitive polynomial of the Galois field can arbitrarily be set at design time.

3.2 Kitsos-multiplier

This architecture extends the Scott-multiplier by a reconfigurable feedback path, gated clocking and is also based on Eq. (8). The reconfigurable feedback path allows for a runtime reconfiguration of the multiplier. This reconfigurability aims to support not only arbitrary primitive polynomials but also field sizes $1 \leq i \leq m$ for an array length m. The

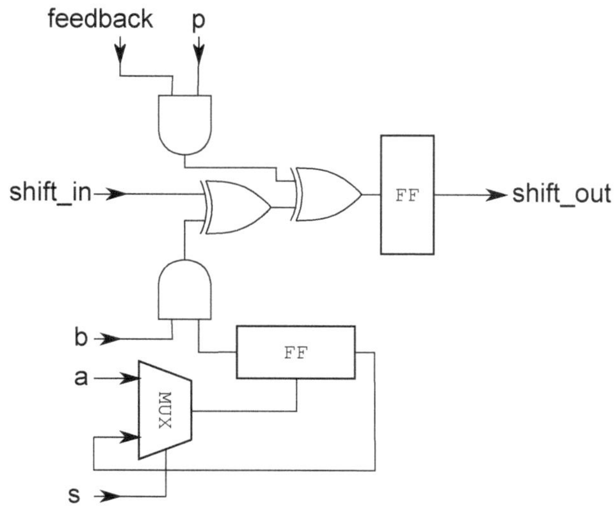

Fig. 2. Scott-Multiplier Basic Cell.

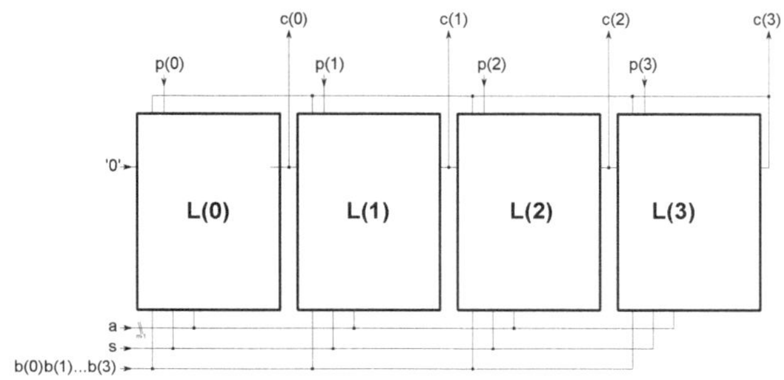

Fig. 3. Scott-Multiplier Array of Four Cells for $m = 4$.

gated clocking leads to an increased power efficiency when the multiplier is reconfigured to lower dimension fields.

In addition to the two 2-input AND-gates, two 2-input XOR gates and flip-flop required due to Eq. (8), the basic cells of the Kitsos-multiplier contain a demultiplexer and a two-input XOR and AND gate as shown in Fig. 4.

Similar to the Scott-multiplier, m basic cells form an array to allow for multiplication over $GF(2^m)$. The Kitsos-multiplier can be reconfigured at run time to support any field dimension j with $1 \leq j \leq m$ by selecting a shorter feedback path and disabling part of the flip-flops. Our implementation also features built-in default primitive polynomials and further control logic to be able to use an arbitrary primitive polynomial. An example for $m = 4$ is given in Fig. 5.

3.3 Plain multiplier architectures

In the original BCH decoder implementation, the multipliers used were based on a pure polynomial division algorithm. As a first step, plain unrolling of this polynomial division led to pure combinatorial FFMs. In the second step we have

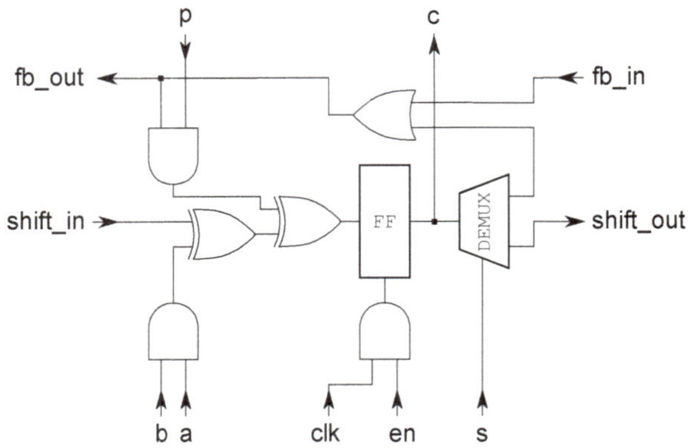

Fig. 4. Kitsos-Multiplier Basic Cell.

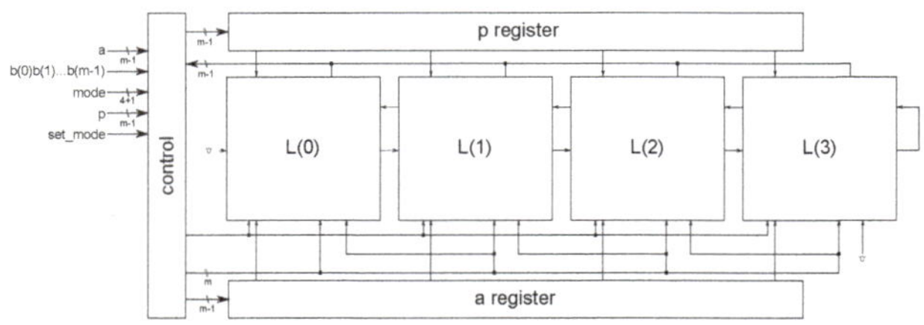

Fig. 5. Kitsos-Multiplier Array for Maximum $m = 4$.

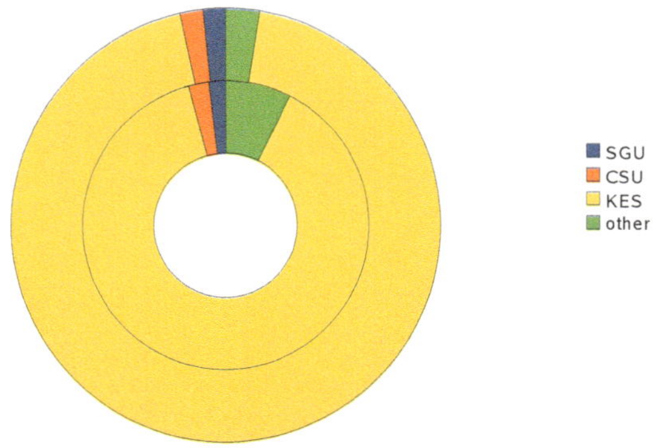

Fig. 6. Slice Distribution of BCH Decoder Components.

introduced pipeline registers to shorten the critical path of the circuit.

We use both plain FFM types as a comparison for our results. Detailed numbers for all multipliers are given in Fig. 2 and Table 2.

3.4 BCH decoder architecture

Our original decoder implementation is derived from the generic template shown in 1. The key equation solver is based on a fully parallel Euclidean implementation that uses non-pipelined combinatorial FFMs as described in Sect. 3.3. SGU and CSU use the same type of FFM. Since we intend to enhance our decoder to Reed-Solomon codes in the future, we have decided to go for a fully parallel Euclidean based key equation solver. The decoder can be configured at design time with the following parameters:

- n: length of the code word
- m: dimension of the extension field 2^m
- t: numbers of errors that can be corrected
- p: number of bits that are processed in parallel in the syndrome generator unit and the Chien search unit

In all three units in the upper part of Fig. 1, FFMs are the key components (Chen et al., 2009; Zhang et al., 2010). However, the complexity of some FFMs can be reduced at design time, since they have one input fixed to a constant value out of the specific Galois field used. This is the case for all $2 \times t$ multipliers in the syndrome computation unit and for all $t \times p$ multipliers in the Chien search unit.

The key equation solver only contains full FFMs with non-constant inputs. To support both GF dimensions (2^{14} and 2^{16}) in the DVB-S2 BCH decoder, every FFM instance internally consists of two FFMs, one for each GF dimension. They share the same wires and are selected via multiplexers at the inputs and outputs.

Figure 6 clearly shows for two explicit configuration cases that these $8 \times t + 2$ FFMs in total consume the vast majority of the overall area required by the complete BCH decoder.

InTable 1 the latency of the single decoder stages is given for combinatorial and Scott FFMs based on the decoder parameters. Due to the high parallelism in the KES, the majority of decoding time is spend in SGU and CSU. Nevertheless, these units can easily be speeded up by increasing p, reducing the latency in these units by $1/p$. However, this will also increase the consumed area by a factor of nearly p.

4 Results

4.1 Validation

With increasing complexity in current hardware designs, permanent evaluation and verification is crucial throughout the design process. To speed up the otherwise very time consuming simulation process, the VHDL models of the FFMs have been exhaustively co-simulated with C++ reference code for GF-multiplication. This co-simulation has been carried out using Mentor Graphics ModelSim and a SystemC test bench as illustrated in Fig. 7. As only field sizes up to 2^{16} are of interest for our BCH decoder, all meaningful stimuli have been applied and the models have therefore been fully verified.

Table 1. Latency Distribution of BCH Decoder Components.

Component	Latency Combinatorial FFMs	Latency Scott FFMs
Key Equation Solver	$4\times t+2$	$(4\times t+2)\times m$
Syndrome Generator Unit	$n/p-1$	$(n/p-1)\times m$
Chien Search Unit	$n/p+1$	$(n/p+1)\times m$

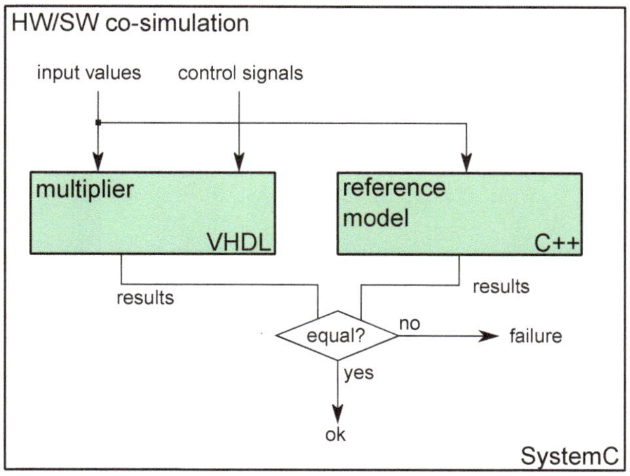

Fig. 7. HW/SW Co-Simulation Schematic.

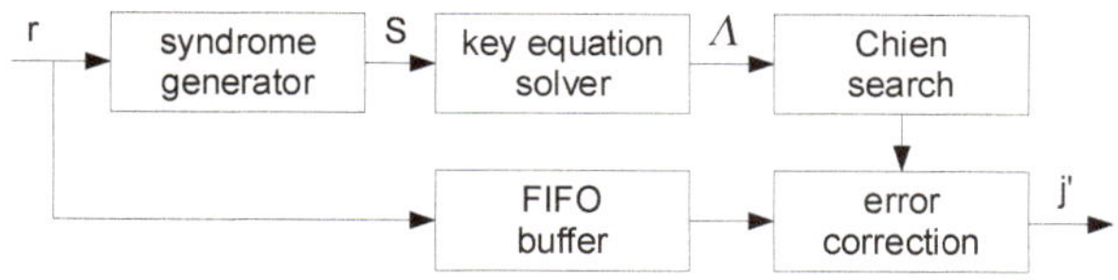

Fig. 8. Throughput per Slice @200 MHz.

4.2 Synthesis results

All multiplier architectures described in 3 have been implemented in VDHL. They have been synthesized for low area consumption on a Xilinx Virtex 4 FPGA (XC4VLX100, package FF1148). The applied tool chain was Xilinx ISE IDE Release 11.3 and the optimization effort for place and route was set to high. Figure 2 gives detailed synthesis results (post place and route).

Due to the given system requirements, the main focus lay on a resulting small footprint on the FPGA for given throughputs. Therefore the metric considered most important was throughput per slice at a fixed clock frequency. Figure 2 shows that the required target clock frequency of 200 MHz could be achieved by all implementations, except the Kitsos-multiplier for field dimensions $m \geq 15$.

For the Scott-multiplier, we can achieve possible clock frequencies of more than 1 GHz due to the short critical path (see Fig. 2). Furthermore, the Scott-multiplier outperforms the other multiplier architectures and even the plain combinatorial multiplier implementation with respect to the metric introduced above. This is shown in Fig. 2.

When comparing the Scott-multiplier and the plain combinatorial multiplier, the area saving outweighs the lesser throughput (factor $\frac{1}{m}$) by a factor from 14.75 (for $m = 12$) up to 18.1 (for $m = 16$).

The overhead of the Kitsos-multiplier implementation heavily impacts the area usage. While still less area is needed as for either one of the polynomial-division-based multipliers, the Kitsos-multiplier requires roughly ten times more area than the Scott-multiplier yet providing the same throughput. A BCH decoder for DVB-S2 requires FFMs for two different GF dimensions: 2^{14} and 2^{16}. As a result, two instances of the Scott-multiplier (one for each dimension) outperform the Kitsos-multiplier clearly.

4.3 Impact on BCH decoder

With respect to the consumed area,Table 2 shows that the the total number of slices occupied by the Scott-multiplier is 9 for GF 2^{14}. That are 5.4 % of the 168 slices required by the plain combinatorial FFM for this GF dimension. For GF 2^{16}, the Scott-FFM only consumes 5.5 % of area. For the FFM pairs in the DVB-S2 decoder as explained in Sect. 3.4, this results in a total area saving of around 94 % for the FFMs. With respect to Fig. 6, we thus expect a very high potential for area saving by employing Scott-FFMs in the key equation solver part. Nevertheless, the area occupied by the multiplexers and registers will not be reduced. Furthermore, the area reduction of FFMs with constant inputs as in the SGU and CSU will also be significantly lower.

By using pipelined multipliers, the latency in each decoder component is increased by a factor of m (see Table 1). For the configurations shown in Fig. 6, the latencies of SGU and CSU are already massively dominating over the KES latency. Using pipelined FFMs in SGU and CSU would there lead to a significant decrease of the overall decoder throughput, by only small area reduction in total.

However, Fig. 6 shows that the KES is the dominating part in the decoder with respect to area, but hardly contributes to the overall latency. This means that adding additional latency in the KES will only insignificantly increase the overall latency of the decoder. On the other hand, smaller FFMs in the KES will show a considerable impact on the overall

Table 2. Detailed Synthesis Result.

Type	m	slice-FFs	4-input LUTs	slices total	throughput $[\frac{results}{sec}]$@200MHz	throughput per slice $[\frac{results}{sec \cdot slice}]$@200MHz	max. freq. [MHz]
Scott	12	12	15	8	1.66×10^7	2.08×10^6	1034.1
	13	13	16	8	1.53×10^7	1.92×10^6	1177.9
	14	14	17	9	1.42×10^7	1.58×10^6	1189.1
	15	15	16	8	1.33×10^7	1.66×10^6	1116.1
	16	16	19	10	1.25×10^7	1.25×10^6	1083.4
Kitsos	12	59	136	83	1.66×10^7	0.2×10^6	226.50
	13	64	160	100	1.53×10^7	0.15×10^6	212.27
	14	69	186	110	1.42×10^7	0.12×10^6	209.03
	15	74	191	114	$*$	$*$	169.81
	16	79	169	104	$*$	$*$	198.77
Plain (pipelined)	12	392	177	200	20×10^7	10^6	833.33
	13	457	205	246	20×10^7	0.81×10^6	840.34
	14	527	236	270	20×10^7	0.74×10^6	889.68
	15	602	240	323	20×10^7	0.61×10^6	871.84
	16	682	302	348	20×10^7	0.57×10^6	701.26
Plain (combinatorial)	12	53	169	115	20×10^7	1.73×10^6	373.13
	13	61	182	130	20×10^7	1.53×10^6	394.01
	14	75	237	168	20×10^7	1.19×10^6	365.63
	15	66	220	148	20×10^7	1.35×10^6	401.61
	16	71	281	181	20×10^7	1.1×10^6	340.14

$*$: Maximum clock frequency was below 200 MHz.

consumed area, since those FFMs use up most of the hardware resources (see Sect. 3.4).

In order to achieve a well-balanced decoder design, we therefore suggest to employ two different FFM types in the decoder: combinatorial FFMs in the SGU and CSU, and small pipelined FFMs in the KES. This setup will exploit the benefits of both FFM types and reduce the overall area of the decoder, by only adding an insignificant amount of total latency.

5 Conclusions

Finite field multiplications are key in many technical applications, particularly in decoding BCH codes. In this work we have investigated available pipelined architectures with respect to scalability, throughput and area consumption on FPGAs. We have implemented and synthesized those designs for a common target device, a Xilinx Virtex-4 XC4VLX100 FPGA. Our implementations have been fully verified in a hardware-software co-simulation. We have evaluated and compared the selected architectures and show that for the choosen effiency metric the Scott-multiplier features in principle the best performance. For application of the considered multipliers in a DVB-S2 BCH decoder the Kitsos-multiplier is outperformed by the other multipliers, despite the inherent architectural flexibility of the Kitsos-multiplier. Furthermore, we conclude that the key equation solver part can profit most from pipelined finite field multipliers, and that simple combinatorial multipliers provide a higher benefit in the syndrome computation and the Chien search.

References

Ahlquist, G. C., Nelson, B. E., and Rice, M.: Optimal Finite Field Multipliers for FPGAs, in: FPL '99: Proceedings of the 9th International Workshop on Field-Programmable Logic and Applications, pp. 51–60, Springer-Verlag, London, UK, 1999.

Bose, R. C. and Ray-Chaudhuri, D. K.: On a class of error correcting binary group codes, Inform. Control, 3, 68–79, doi:10.1016/S0019-9958(60)90287-4, 1960.

Bossert, M.: Kanalcodierung, B. G. Teubner Stuttgart, 2nd edn., 1998.

Chen, Z., Zhang, Y., Ying, Y., Wu, C., and Zeng, X.: An Area-Efficient and Degree-Computationless BCH Decoder for DVB-S2, in: Proceedings of the IEEE 8th International Conference on ASIC, (ASICON) 2009. , 489–492, doi:10.1109/ASICON.2009.5351625, 2009.

Friedrichs, B.: Kanalcodierung, Springer Verlag Berlin Heidelberg, 1995, ISBN 3-540-59353-5.

Hocquenghem, A.: Codes correcteurs derreurs, Chiffres, 2, 147–56, 1959.

Kitsos, P., Theodoridis, G., and Koufopavlou, O.: An efficient reconfigurable multiplier architecture for Galois field GF (2m), Microelectr. J., 34, 975–980, 2003.

Liu, W., Rho, J., and Sung, W.: Low-Power High-Throughput BCH Error Correction VLSI Design for Multi-Level Cell NAND Flash Memories, in: Proceedings of the IEEE Workshop on Signal Processing Systems Design and Implementation, (SIPS) 2006. , 303–308, doi:10.1109/SIPS.2006.352599, 2006.

Scott, P., Tavares, S., and Peppard, L.: A Fast VLSI Multiplier for GF(2m), in: IEEE Journal on Selected Areas in Communications (J-SAC) 1986, 4, 62–66, 1986.

Zhang, B., Liu, D., Wang, S., Chen, X., and Liu, H.: Design and Implementation of Area-Efficient DVB-S2 BCH Decoder, in: Computer Engineering and Technology (ICCET), 2010 2nd International Conference on, 3, V3–179–V3–184, doi:10.1109/ICCET.2010.5485823, 2010.

7

On parallel random number generation for accelerating simulations of communication systems

C. Brugger, S. Weithoffer, C. de Schryver, U. Wasenmüller, and N. Wehn

Microelectronic Systems Design Research Group, University of Kaiserslautern, 67663 Kaiserslautern, Germany

Correspondence to: C. Brugger (brugger@eit.uni-kl.de)

Abstract. Powerful compute clusters and multi-core systems have become widely available in research and industry nowadays. This boost in utilizable computational power tempts people to run compute-intensive tasks on those clusters, either for speed or accuracy reasons. Especially Monte Carlo simulations with their inherent parallelism promise very high speedups. Nevertheless, the quality of Monte Carlo simulations strongly depends on the quality of the employed random numbers. In this work we present a comprehensive analysis of state-of-the-art pseudo random number generators like the MT19937 or the WELL generator used for parallel stream generation in different settings. These random number generators can be realized in hardware as well as in software and help to accelerate the analysis (or simulation) of communications systems. We show that it is possible to generate high-quality parallel random number streams with both generators, as long as some configuration constraints are met. We furthermore depict that distributed simulations with those generator types are viable even to very high degrees of parallelism.

1 Introduction

The ongoing research activities and technological progress over the last years have led to a tremendous increase in the complexity of current systems and scientific challenges. More and more realistic and highly sophisticated models require numerical computations, since analytical solutions are not available in many cases. For example in the analysis of communications systems, independent random numbers are required in a lot of places, e.g. for the generation of payload data, additive white gaussian channel noise, or the modeling of fading channels. Furthermore, we can mention modeling carrier offsets and timing offsets, and the multiple channels between transmit and receive antennas in multiple-input and multiple-output (MIMO) systems.

Monte Carlo (MC) simulations remain a widely employed universal method to solve problems when other approaches are not applicable (Korn et al., 2010). It is important to note that the result quality of a MC simulation strongly depends on the quality of the RNs used. When only one random number generator (RNG) is used for the complete simulation (for example for distributing one random number (RN) stream to only one or a smaller number of processing elements (PEs)), it is sufficient to instantiate a single generator with the desired quality properties.

Nevertheless, in order to obtain the results fast and with the desired accuracy, parallel simulations are mandatory. MC simulations are perfect candidates for parallel simulation, since they rely on a large number of independent experiments that can be carried out simultaneously. However, the generation of high-quality independent and identically distributed (i.i.d.) RNs is a big challenge. With huge computing clusters and the *Cloud* becoming accessible for more and more researchers and professionals, it is time to have a closer look on possible pitfalls and quality issues arising from highly parallel simulations.

Various approaches for creating i.i.d. parallel RN streams have been proposed up to now that we study in detail in this paper. We give a short summary about available generation methods for parallel RN streams and how they can be tested, together with an overview about related work. For the famous MT19937 Mersenne Twister (MT) and the well-equidistributed long-period linear (WELL) generator proposed by Pierre L'Ecuyer, we show that randomly seeding parallel RNGs is safe, also for very high degrees of parallelism. We have carried out a comprehensive analysis of possible inter-stream correlations and substreams in the popular MT19937 pseudo random number generator (PRNG) using the TestU01 test suite, a popular test framework for RN quality analysis. We show that for a well-defined seeding schemes no peculiarities are observed. In summary we demonstrate that the use of standard state-of-the-art PRNGs like the MT and the WELL is safe, also for parallel simulations.

2 Background

The generation of high-quality RNs in high volumes has long been a hot topic in research. In general, three types of RNs exist:

- *True RNs*,
- *quasi RNs*, and
- *pseudo RNs*.

At a first glance, *true RNs* seem to be most desirable, since they are really random and can not predicted in any way. However, true RNs come with some drawbacks: Firstly, their generation typically relies on a physical process and therefore needs a special device that is in general not available in generic computing systems like standard infrastructure as a service (IaaS) or platform as a service (PaaS) clusters or data centers. Secondly, it is quite likely that the untreated output of a true RNG is biased. Therefore the output stream needs to be monitored continuously, and a bias correction is needed that is adjusted continuously (Hill et al., 2013). Thirdly, true RNs are not repeatable, and forbid re-executing a simulation several times with generating exactly the same output – a knock-out criterion for science in general.

Quasi RNs are not random at all, but try to cover a (multi dimensional) space with certain characteristics. Prominent examples are Sobol or Halton sequences used for MC simulations (Korn et al., 2010). We do not consider true and quasi RNs further in this work, details can be found in literature (Knuth, 1997; Korn et al., 2010).

For the main reason of keeping simulations and calculations reproducible, PRNGs are the most favorable choice for simulation purposes. They try to generate a sequence looking like "good" RNs by a deterministic algorithm, this means i.i.d. from a statistical viewpoint (L'Ecuyer, 2007). PRNGs have a number of advantages: they can be run on different execution platforms, require no additional hardware support, and produce exactly the same output streams when seeded with equal values. In the following we will therefore focus on PRNGs only.

2.1 Generating parallel random numbers

A comprehensive summary about generation methods for parallel RN streams has been given recently by Hill et al. (2013).

This section will give a brief overview and comments on some of the methods for generating multiple independent RN streams. We focus on approaches that can be realized in hardware by either utilizing multiple RNGs or a fast, parallelized RNG and use WELL and MTs generators as examples. Parallel RN streams can be generated from:

- Multiple identical RNGs starting from different positions in the period,
- multiple RNGs of different type,
- a single parallelized RNG, or
- combinations of the above.

Implications of different approaches on the quality of statistics are discussed in detail in Sects. 4.1 to 4.3.

2.1.1 Multiple generators of the same type

Using identical RNGs and starting them from a different position in the generator period corresponds to partitioning a single RN stream into blocks (*Blocking*) and is suitable when the number of consumed RNs for each of the parallel streams is known and bounded. Multiple instances of one RNG type can then be initialized to a precomputed state. This so-called *jump-ahead* technique (Haramoto et al., 2008) allows for defined distances between the streams of the RNGs (w.r.t. positions in the RNG period). The initialization of the multiple generators can also be done by employing other RNGs. WELL and MT generators, for example, are commonly initialized (seeded) by other small RNGs (seed generators). This approach is called *Random Spacing*.

As these also have to be seeded, we realize a random seeding of the WELL and MT by seeding the seed generators differently.

2.1.2 Multiple generators of different type

The straight-forward way of getting RNGs of different type is to use a different algorithm for each of the RNGs. Because of the limited number of known algorithms with each having varying statistical properties and implementation complexities, such an approach poses strict limits on the number and quality of RNs generated in parallel for obvious reasons and is not considered here. The WELL and MT generators, however, allow the use of the same algorithm with different parameter sets that correspond to different underlying recurrence polynomials. Correlations between the RN streams generated by following this approach are considered highly unlikely. As the parameter sets for the WELL and MT generator types have to be found by extensive search, parallelism by using this so-called *Parametrization* is also limited (Tian and Benkrid, 2009).

2.1.3 Single, parallelized generator

Generating parallel streams of RNs by using a single parallelized RNG corresponds to parsing a single RN stream into substreams. If the RNs from a single stream are distributed according to a round-robin scheme (i.e. the first RN to the first substream, the second RN to the second substream, ...), the resulting partitioning is called *Leap Frog* partitioning (see Hill et al., 2013).

2.2 Quality issues

As stated before, PRNGs do not produce real RNs, but try to create a "good" output streams that looks random. Evaluating the quality of PRNGs has been an active topic for a long time. To a special degree, this research was driven by Pierre L'Ecuyer, Michael Mascagni, and others who investigated novel RNG approaches. For example, L'Ecuyer has formulated several key criteria for "good" RNGs: a long period, repeatable outputs, portability to different execution platforms, and the ability to split the output sequence into multiple independent blocks (what means they should implement an efficient jump-ahead strategy) and each block again into substreams with the leap frog approach (L'Ecuyer, 2007; L'Ecuyer and Panneton, 2005). D. E. Knuth has summarized the basic test strategies for single-stream RNGs in his famous book "The Art of Computer Programming" (Knuth, 1997), with the first edition already released in 1969. Nowadays, the TestU01 suite developed by L'Ecuyer in 2007 (L'Ecuyer and Simard, 2007) is considered the most comprehensive one and is preferred by many researchers (Salmon et al., 2011; Hill et al., 2013).

A first requirement for a good parallel PRNG is that it also has to be a good sequential PRNG (Srinivasan et al., 2003). Obviously, when testing different substreams of one PRNG, it is mandatory to ensure that exactly the same substream can be outputted again to allow debugging and reproducibility of the test results (Hill et al., 2013).

In 2003, Srinivasan et al. have stressed the importance of test concepts for parallel RNGs, in particular for high-performance computing (HPC) applications (Srinivasan et al., 2003). They have introduced the terms *intra-stream* and *inter-stream correlations* to highlight that both characteristics are crucial for a "good" parallel RNG: the quality of each substream itself, and the independence of all substreams from each other. They also presented first test strategies for parallel RNs.

In order to test inter-stream correlations, Salmon et al. (2011) have concatenated blocks of multiple substreams with a round-robin scheme to a single test stream that they have fed into the TestU01 suite (Salmon et al., 2011). Due to the correlation tests included in TestU01, correlations between the substreams coming from different generators are detected. We also follow this approach in our work, but show that there may be problems when not carefully selecting the right assembly configuration (see Sect. 4.3).

Table 1. Ressource consumption of implementations for FPGAs.

RNG	BRAMs	Slices	max. Freq. [MHz]	Throughput [Gbps]
MT19937[1]	1	57	100.0	3.20
TinyMT[1]	0	65	100.0	3.20
MT19937[2]	2	330	24.2	0.77
3-CP MT[3]	2	207	258.3	24.03
4-IP MT[3]	4	290	277.7	35.54
8-IP MT[3]	8	566	283.5	72.58

[1] On a MaxWorkstation (Virtex-6) from Maxeler Technologies Inc.
[2] (Chandrasekaran and Amira, 2008) on a Virtex-E.
[3] (Dalal and Stefan, 2008) on a Virtex-4.

3 Related work

A hardware implementation of the MT algorithm has, for example, been reported by (Chandrasekaran and Amira, 2008).

Parallel RN generation by multiple MT generator cores with different parameter sets was described by (Tian and Benkrid, 2009). They reported their design to outperform CPU and GPU implementations of the MT algorithm by a factor of 25× and 9× respectively w.r.t. throughput. An investigation of the statistical properties of the generated numbers, however, was not performed.

Dalal and Stefan presented *Interleaved Parallelization* (IP) and *Chunked Parallelization* (CP), two methodologies for parallelizing RNGs of MT and WELL type (Dalal and Stefan, 2008). There, the state vector of the RNG is partitioned in a way that multiple numbers can be generated in parallel. A corner case of this scheme, a fully parallel implementation of the MT, was reported by Sriram and Kearney (Sriram and Kearney, 2009). Both parallelization schemes result in a *Leap Frog* partitioning at the outputs of the parallelized RNGs. However, they are only practical for grades of parallelism that are a divider of the state vector size (that is 624 for the MT19337).

Table 1 lists the resource consumption of selected MT implementations on field programmable gate array (FPGA). Even if the different target FPGAs are taken into account it becomes clear that IP and CP are more hardware efficient than using multiple instances of the respective serial generator. We have been able to confirm these numbers in our own investigations. While BRAMs can become a critical resource in highly parallel RNG architectures we also considered the TinyMT with an internal state of 127 bits, far smaller than the MT with 19937 bits. The TinyMT only uses registers for the state storing, but is characterized by a smaller period.

L'Ecuyer and Panneton have compared throughput and jump-ahead computation times for LFSRs, MT, and the

Table 2. Upper bound probability of a collision for random number generators running at 5 GHz for 100 years.

RNG	2 generators	Chip with 1000 RNGs	System of 1000 of such chips	System of one million chips
TinyMT	$<10^{-18}$	$<10^{-13}$	$<10^{-7}$	$<10^{-1}$
MT19937	$<10^{-5982}$	$<10^{-5976}$	$<10^{-5970}$	$<10^{-5964}$
WELL44497b	$<10^{-13375}$	$<10^{-13369}$	$<10^{-13363}$	$<10^{-13357}$

WELL generator in 2005 (L'Ecuyer and Panneton, 2005). They have shown that the jump-ahead computation time exponentially increases with the number of internal states in the PRNG.

Hill et al. (2013) have carried out an analysis on various Mersenne Twister for Graphic Processors (MTGP) configurations with the TestU01 Big Crush battery in 2012 and have found only a few weaknesses (Hill et al., 2013). They concluded that the use of the MTGP with longer periods therefore is safe.

At SC11, Salom et al. (2011) have presented an evaluation of so-called *counter-based PRNGs*. They are highly scalable, can be implemented on central processing unit (CPU), graphics processor unit (GPU), and FPGA based architectures, and therefore seem to be very promising for distributed HPC simulations. The presented PRNGs pass the Big Crust test battery of the TestU01 suite without failures and even beat the MT19337. At the same time, the counter-based generators can achieve around the same throughput as the MT19337 in software. However, counter-based PRNGs have not been considered for hardware implementation up to now, this is ongoing research due to their higher implementation complexity.

4 New investigations

We have seen that a lot of approaches have been followed to generate and test parallel RNs. However, we had difficulties to evaluate the scalability and efficiency of these methods for highly parallel settings. For example, the leap frog approach does not scale above the dimensionality of the PRNG, i.e. 627 for the MT19937 (Hill et al., 2013). Splitting the RN stream into blocks in advance with the jump-ahead algorithm is critical if it cannot be ensured that the amount of consumed numbers never exceeds the defined block sizes. Furthermore, also the jump-ahead computations are quite compute intensive and do not scale very well (Hill et al., 2013). Instantiating multiple generators with different parameters sounds promising, but also in this case the available number of configuration is limited (Tian and Benkrid, 2009).

Therefore we have investigated partitioning schemes that aim at high degrees of parallel RN streams, in particular random seeding approaches with standard generators. The results are given in the following sections.

4.1 Mersenne twister substream analysis

Since the MT19937 is one of the most popular PRNGs and implemented in many available libraries and software tools, we have analyzed the quality of MT19937 substreams generated with the leap frog method. We have used the MT19937 C implementation as provided by Matsumoto and Nishimura, refined by Richard Wagner in 2009[1]. The seed was fixed to 0x00001571.

In total we have run 88 test on a selection of substreams from up to 64 total substreams. Due to the high runtimes even on the employed compute cluster especially for higher numbers of substreams, we were not able to check all configurations. In 58 of 88 cases, only tests number 80 and 81 (that are always failed by the MT19937) have failed. In 27 cases one additional test was suspicious, and only in 3 cases two additional tests showed suspicious outputs. In summary, the obtained results have shown no suspicious values at all. Therefore we conclude that it is safe to split standard MT19937 streams into up to 64 substreams, what should be sufficient for supplying up to 64 PEs with one generator instance.

4.2 Investigating seed collisions

One method of generating independent random number generators is to use multiple instances of the same generator with different seeds. Each obtained output stream is then a subsequence of one long period sequence, while the starting point depends on the seed. For random streams, the distance between two streams is unknown and they might even overlap. Overlapping streams are highly correlated and have to be avoided. Before considering this, let us investigate the probability it is that two substreams overlap.

It is helpful to visualize the period as a circle in which each instance of a random number generator draws consecutive numbers on this circle, see Fig. 1. When two seeds are too close together on this circle, the streams of random numbers overlap and we have a stream collision. In the following setup we consider n independent RNGs with a period length of L. For each RNG we draw l consecutive RNs. Let us consider two RNGs. The probability for the second one to not collide

[1] http://www.comp.nus.edu.sg/~noi/tasks/2010/PKMATCH/MersenneTwister.h, last access: 5 February 2014.

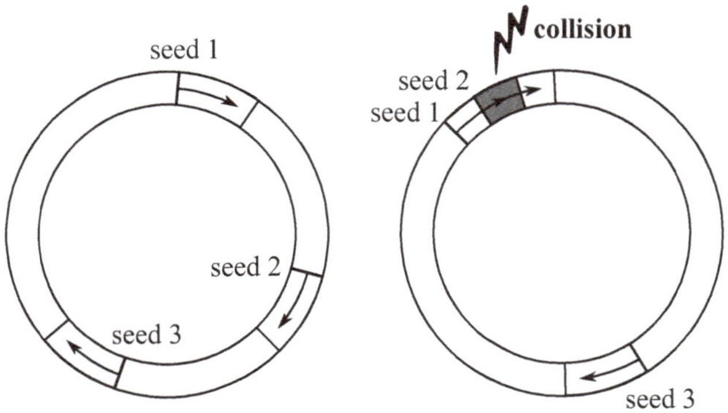

Figure 1. The circles represent the whole period of the random number generator. We consider three instances of the same random number generator. Each generator is seeded with a random number and a specific amount of random numbers are drawn from it. In the left case the streams do not overlap, while on the right side the first random number generator collides with the second.

with the first one is:

$$P_2 = \frac{L - 2l + 1}{L}. \qquad (1)$$

When we add one more, the probability of this third generator not colliding with the first two depends on their actual position. The probability is the least when both are far apart. In general it is:

$$P_3 \geq P_2 \frac{L - 2(2l - 1)}{L}. \qquad (2)$$

We can extend this series and get an upper bound estimate for the collision probability $\hat{P}_n$:

$$\hat{P}_n = 1 - P_n \leq 1 - \prod_{i=1}^{n} \frac{L - (i-1)(2l-1)}{L}. \qquad (3)$$

We assume that our random number generators run with 5 GHz, generating one RN per clock cycle, running for 100 years. This results in an l of about 10^{18}. For the MT the state vector has 19 937 bits resulting in a period of about 10^{6002}. We consider chips with 1000 of such generators. We are interested in the collision rate for such chips and huge system containing up to one millions of them. To calculate the fractions containing such large numbers we have used the *fractions* package in Python. To reduce the number of multiplications we estimate the collision rate with an even higher upper bound as:

$$\hat{P}_n \leq 1 - \left(\frac{L - (n-1)(2l-1)}{L} \right)^{n-1}. \qquad (4)$$

This power can be efficiently computed with only $\log_2(n-1)$ multiplications instead of $n-1$ required for Eq. (3). For $n = 10^6$ these are 20 instead of 999 999 multiplications.

Table 3. Test failing for our robust parallel seeding procedure.

number of RNGs	failing tests for block size of 1	4	8	12	30	1 k	10 k	100 k	1 M
1	2	2	2	2	2	2	2	2	2
2	2	2	–	–	–	–	–	2	2
4	2	–	–	–	–	–	–	2	2
8	2	–	–	–	–	–	–	2	2
16	–	–	–	–	–	–	–	2	2
64	–	–	–	–	–	–	–	2	2
256	–	–	–	–	–	–	–	2	2
1024	–	–	–	–	–	–	–	2	2

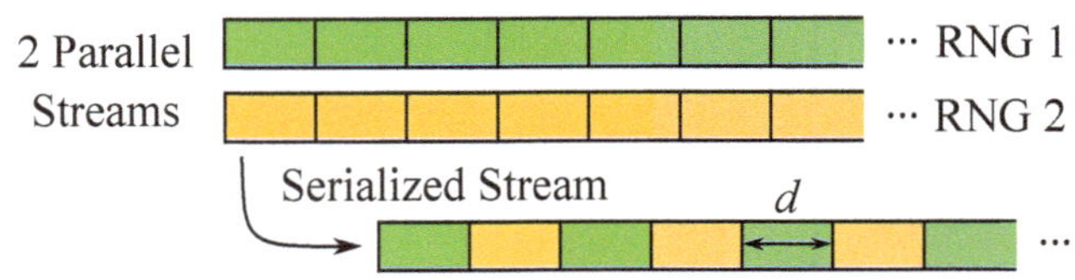

Figure 2. Showing how parallel streams of two independent RNGs can be serialized when run on a serial processor. In the shown case the parallel program operates on equal blocks d of random numbers.

The resulting probabilities for various RNGs are shown in Table 2. For RNGs with reasonably large state vectors the probability of a seed collision is below 10^{-5964} even for incredibly large systems containing one billion RNGs running at 5 GHz for 100 years. We can conclude that when the seed is chosen randomly a collision is negligibly unlikely for the MT19937 and WELL44497b. However, for the TinyMT more advanced parallelization methods should be utilized, like e.g. parameterization.

4.3 Inter-stream correlations

In the last section we have seen that stream collisions are highly unlikely for different seeds. Based on this insight it seems reasonable to realize parallel random number generation by using independent RNGs each with a random seeds. However then question remains what the quality of these streams is. The generator might for example show long-term correlations resulting in inter-stream correlations, as discussed in Sect. 2.2.

We now derive a methodology on how to analyze inter-stream correlations. For this it is helpful to keep the application in mind that uses the parallel RNG. In general it is possible to execute a parallel program on a serial processor for which only one instruction is executed at a specific time. That means a single stream of random numbers exists that is equal to how the parallel RNGs are accessed during execution. Thus we can analyze parallel RNGs also with the standard test suites for single streams by using this serialized stream.

While in general the serialized stream is an arbitrarily interleaved stream from the parallel streams, it is helpful to

Table 4. Test failing with Tausworthe seeding.

RNG count:	1	2	4	8	16	32	64	128
MT19937	2	2	2	41	64	79	100	106
WELL44497b	2	2	2	102	88*	89*	89*	

* Tested with smaller test-suite Crush instead of Big Crush.

consider some representative cases. We consider that the program operates on blocks of random numbers of the same size d. This is clearly the case for Monte Carlo simulations, used very widely in communication system simulations. Then we can serialize the parallel streams as shown in Fig. 2.

4.4 Random seeding

Random seeding requires the use of a RNG itself to generate numbers of all the seed values. In Sect. 4.2 the probabilities for seed collisions are only valid when all starting points are equally likely. This means the seeding procedure has to be carefully validated, such that it does not favor specific starting points.

For our robust seeding procedure, we use the seeding algorithm integrated in the MT19937 C implementation (version 2002) and seed the RNG n with seed_0+n. We have tested the serialized stream for eight independent seed_0 and calculated the mean of failing tests. All investigations have been made with the Big Crush battery from the TestU01 suite v1.2.3. The result is shown in Table 3.

For one generator already two test are failing, for which the MT is known for. However, for increasing numbers of RNGs and block sizes, the serialized stream gets even better. For block sizes of 100 k and higher the same TestU01 results are as for single MT stream. This is due to the limited analysis window of the TestU01 suite as sample sizes of the tests that have been conducted n have been in the order of $10^9 \geq n \geq 10^6$.

As a counterexample we have implemented a naive seeding procedure based on the Tausworthe 88 RNG. It seeds the complete state vector n of the MT with the first values from a Tausworthe RNG. This Tausworthe 88 is itself seeded with the vector [0x88cb47c9 + 2 (seed_0+n), 0x8a9cdf65 + 4 (seed_0+n), 0xcaf40ed9 + 8 (seed_0+n)]. In this case only a block size d of one has been considered for eight independent seed_0. The number of failing test are shown in Table 4. Starting from eight RNGs the number of failing tests dramatically increases, rendering this naive procedure useless. This stresses the importance of the seeding method.

For the robust seeding procedure we can conclude that no inter-stream correlations are observable. Further the use of independent MT generators even improves their quality.

5 Conclusion

The recent increase in available computation power on highly parallel computing clusters or the *Cloud* have enabled researchers and professionals to run distributed Monte Carlo simulations easier than ever before. However, those parallel simulations require high-quality parallel random number streams. In this work we have analysed to which extent state-of-the-art PRNG qualify for their employment in parallel settings. We show that the popular MT19937 Mersenne Twister and the superior WELL random number generator are able to create high-quality parallel random number streams, in particular with the random seeding method. Furthermore we reveal that seed collisions for those two generator types are very unlikely, even for high degrees of parallelism, and that no inter-stream correlations are observable. Nevertheless, the parallel seeding method has to be carefully selected. We present a working method and show how to test it. All in all, we conclude that applications involving parallel random number streams need special attention on this point, though available generators are well-applicable for these tasks.

Acknowledgements. We gratefully acknowledge the partial financial support from the Center of Mathematical and Computational Modelling $(CM)^2$ of the University of Kaiserslautern and from the German Federal Ministry of Education and Research under grant number 01LY1202D. The authors alone are responsible for the content of this paper.

Edited by: J. Götze

References

Chandrasekaran, S. and Amira, A.: High Performance FPGA Implementation of the Mersenne Twister, in: Electronic Design, Test and Applications, DELTA 2008, 4th IEEE International Symposium on, 482–485, doi:10.1109/DELTA.2008.113, 2008.

Dalal, I. L. and Stefan, D.: A Hardware Framework for the Fast Generation of Multiple Long-period Random Number Streams, in: Proceedings of the 16th international ACM/SIGDA symposium on Field programmable gate arrays, FPGA '08, 245–254, ACM, New York, NY, USA, doi:10.1145/1344671.1344707, 2008.

Haramoto, H., Matsumoto, M., Nishimura, T., Panneton, F., and L'Ecuyer, P.: Efficient Jump Ahead for F2-Linear Random Number Generators, INFORMS J. Comput., 20, 385–390, doi:10.1287/ijoc.1070.0251, 2008.

Hill, D. R. C., Mazel, C., Passerat-Palmbach, J., and Traore, M. K.: Distribution of random streams for simulation practitioners, Concurr. Comp.-Pract. E., 25, 1427–1442, doi:10.1002/cpe.2942, 2013.

Knuth, D. E.: Seminumerical Algorithms, vol. 2 of The Art of Computer Programming, Addison-Wesley, Reading, Massachusetts, 3 Edn., 1997.

Korn, R., Korn, E., and Kroisandt, G.: Monte Carlo Methods and Models in Finance and Insurance, Boca Raton, FL: CRC Press, 2010.

L'Ecuyer, P.: Pseudorandom Number Generators, Tech. rep., DIRO, Université de Montreal, C.P. 6128, Succ. Centre-Ville Montréal (Québec), Canada, H3C 3J7, available at: http://www.iro.umontreal.ca/~lecuyer/myftp/papers/eqf.pdf (last access: 26 August 2014), 2007.

L'Ecuyer, P. and Panneton, F.: Fast random number generators based on linear recurrences modulo 2: overview and comparison, in: Simulation Conference, 2005 Proceedings of the Winter, 110–119, doi:10.1109/WSC.2005.1574244, 2005.

L'Ecuyer, P. and Simard, R.: TestU01: A C library for empirical testing of random number generators, ACM Trans. Math. Software, 33, 22-1–22-40, doi:10.1145/1268776.1268777, 2007.

Salmon, J., Moraes, M., Dror, R., and Shaw, D.: Parallel random numbers: As easy as 1, 2, 3, in: High Performance Computing, Networking, Storage and Analysis (SC), 2011 International Conference for, 1–12, 2011.

Srinivasan, A., Mascagni, M., and Ceperley, D.: Testing Parallel Random Number Generators, Parallel Comput., 29, 69–94, doi:10.1016/S0167-8191(02)00163-1, 2003.

Sriram, V. and Kearney, D.: An FPGA Implementation of a Parallelized MT19937 Uniform Random Number Generator, EURASIP Journal on Embedded Systems, 2009, 507426, doi:10.1155/2009/507426, 2009.

Tian, X. and Benkrid, K.: Mersenne Twister Random Number Generation on FPGA, CPU and GPU, in: Adaptive Hardware and Systems, 2009. AHS 2009, NASA/ESA Conference on, 460–464, doi:10.1109/AHS.2009.11, 2009.

8

Analytical calculation of intrinsic shielding effectiveness for isotropic and anisotropic materials based on measured electrical parameters

M. Kühn[1], **W. John**[2], **and R. Weigel**[3]

[1]AUDI AG, 85045 Ingolstadt, Germany
[2]SIL System Integration Laboratory GmbH, Technologiepark 32, 33100 Paderborn, Germany
[3]Lehrstuhl für Technische Elektronik, Friedrich-Alexander-Universität Erlangen-Nürnberg, Cauerstrasse 9, 91058 Erlangen, Germany

Correspondence to: M. Kühn (michael.kuehn@audi.de)

Abstract. This contribution contains the mechanisms for calculation of magnetic shielding effectiveness from material samples, based on measured electrical parameters. For this, measurement systems for the electrical conductivity of high and low conductive material samples with respect to the direction of current flow are presented and discussed. Also a definition of isotropic and anisotropic materials with electrical circuit diagrams is given. For prediction of shielding effectiveness for isotropic and anisotropic materials, several analytical models are presented. Also adaptions to gain a near field solution are part of this contribution. All analytical models will also be validated with an adequate measurement system.

1 Introduction

In common, shielding effectiveness of material samples is measured with a capacitive coupled Transmission-Line-Holder (TLH) based on ASTM D4935-10 standard. This TLH is a widened coaxial cable, where the surface of the material sample is coupled perpendicular to the inner and outer conductor. In other words, the coaxial cable is separated by the material sample. This means, that the field vectors are oriented parallel to the surface of the material sample and the far field shielding effectiveness is measured. In theory this is described by the Impedance Concept of Schelkunoff (see Schelkunoff, 1938) with the physical processes of reflection, transmission and absorption of electromagnetic waves at material boundaries. In a vehicle with combustion engine or electrical drive train are in most of the cases electromagnetic source and victim placed in a closed distance (r) to each other, that the near field condition $r \ll \frac{\lambda}{2\pi}$ is fulfilled (Schwab and Kürner, 2011). Separation of the field components (electrical and magnetic field) for analysis of shielding effectiveness is necessary. The focus in this contribution is laid on magnetic shielding effectiveness in the near field area, reasoned by high switching currents generated by the power electronics in an electrified vehicle. Strong magnetic fields with fundamental frequencies in the lower frequency range (10 kHz to 20 kHz) with a wide spectrum up to the MHz range are the result, see Michel (2011). Based on measured electrical parameters (electrical conductivity), a method will be presented to calculate magnetic shielding effectiveness of material samples in the near field area. For isotropic materials (i.e. Aluminum) there are several analytical models in the literature available. Among others a modification of Schelkunoff's Impedance Concept (see Wolfsperger, 2008) for the near field area and the analytical model from Moser (see Moser, 1967). Also considered in this contribution, materials like Carbon Fibre Reinforced Polymers (CFRP). Anisotropy of CFRP will be discussed with electrical circuit diagrams. An approach for calculating magnetic shielding effectiveness based on measured electrical properties for the near field area will also be presented.

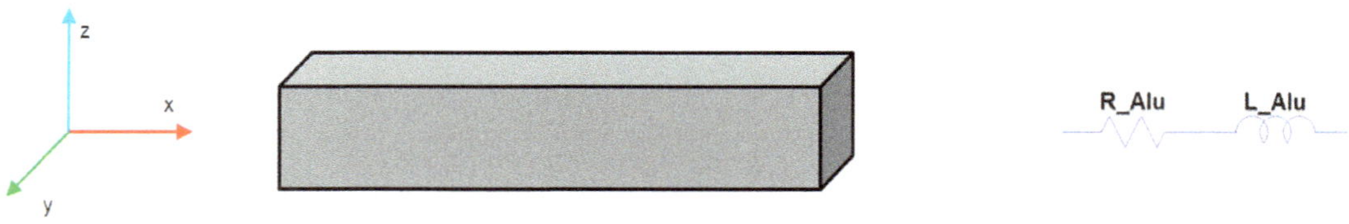

Figure 1. Isotropic material and equivalent circuit diagram.

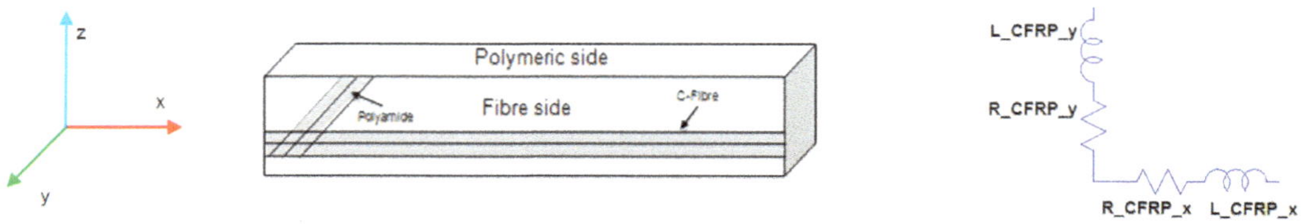

Figure 2. Anisotropic material and equivalent circuit diagram for x and y direction.

2 Near field shielding effectiveness

This section describes the physical processes accountable for shielding of electrical and magnetic fields in the frequency domain. First of all, a definition of near field is necessary. Fields are defined in the frequency range as near fields, when the wavelength λ is large compared to the distance (r) from field source to observing point. The following condition (1) must be fulfilled (Schwab and Kürner, 2011):

$$r \ll \frac{\lambda}{2\pi} \quad (1)$$

The near field shielding effectiveness of electrical fields on enclosure level is nearly infinity and the shielding process is described by the Faraday cage. The electrical field bears a force on charge carriers in a conductive material, Near field shielding effectiveness should be considered separately for the electrical and magnetic field. At first, the physical processes for electrical shielding effectiveness will be described.

2.1 Electrical fields

A finite piece of material (thickness d and area $A = a \cdot b$) is penetrated by an incident electrical field, displacement currents transported by the electrical field lines are closing in the conductive shield material. A part of electrical field lines encompasses the shield and closes behind it. So, electrical shielding effectiveness on material level depends strongly on the geometry of the shield (area A) and the distance from the source to the shield (r). For an enclosure or housing, shielding process is described by the Faraday cage structure (principle: charge carrier separation). In theory electrical shielding effectiveness of infinite extensive material samples is nearly infinity.

2.2 Magnetic fields

The magnetic field will be shielded respectively reduced by two physical processes (see Fig. 1), the generation of eddy currents in the shield and skin effect. The incident magnetic field leads to a voltage in the shield, which results in eddy currents in the shield. These eddy currents are generating an inverse oriented magnetic field, which weakens the incident magnetic field. The skin effect describes the reduction of the current density of a conductor or shield by raising frequency. In summary, the magnetic shielding effectiveness depends on the shield material (σ, μ_r) and the thickness of the shield.

3 Material catalog and measurement methods

The material catalog contains two types of materials:

- Isotropic materials
- Anisotropic materials

which will be described in the following subsections. Also two different methods for measuring conductivity will be presented and discussed.

3.1 Isotropic materials

An isotropic material is defined as a material with a constant electrical conductivity (σ) in all three spatial directions (x, y, z). Figure 1 shows an isotropic material (i.e. Aluminum) and the corresponding equivalent circuit diagram, which is valid for all three spatial directions.

The equivalent circuit diagram consists of a resistor and an inductance in series, which are geometry dependent.

3.2 Anisotropic materials

An anisotropic material distinguishes from an isotropic material in that way, that the electrical conductivity is not constant

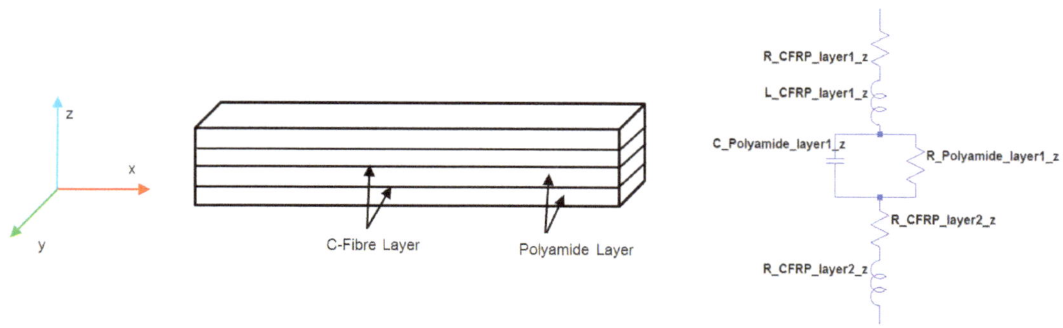

Figure 3. Anisotropic material and equivalent circuit diagram for z direction.

in all three spatial directions. For example CFRP (Carbon Fibre Reinforced Polymers) which consists of C-fiber layers (90° woven) and a Polyamide 6 matrix. Obviously, the conductivity should be similar in fiber direction and differs through the layered structure. Both is visualized with equivalent circuit diagrams, in Fig. 2 for x, y direction and Fig. 3 for z direction.

In fiber direction, the equivalent circuit diagram corresponds to the isotropic material and consists of a resistor and inductance in series.

Through the layered structure (assumption: alternating layers of C-Fiber and Polyamide 6) the equivalent circuit diagram also contains a capacitor for the non conductive Polyamide 6 and parallel a high ohmic resistor. The C-Fiber is also modeled by a resistor and inductance in series.

3.3 Four-Point-Probe-Measurement

Four-Point-Probe-Measurement method could be used for measurement of electrical conductivity at 0 Hz (DC). A known current (I_{in}, (1), (4)) was fed into the material sample (MUT) and the voltage drop (U_{meas}, (2), (3)) was measured. The measurement setup is shown in Fig. 4.

With the input current I_{in}, the voltage drop U_{meas} and the geometry of the material sample (area A, length l) it is possible to calculate the electrical conductivity σ with the following Eqs. (2) based on the specific resistance ρ (Bowler, 2006):

$$\rho = \frac{U_{\text{meas}}}{I_{\text{in}}} \cdot \frac{A}{l}$$
$$\sigma = \frac{1}{\rho} \tag{2}$$

Four-Point-Probe-Measurement method was performed for Aluminium in x, y and z direction CFRP in x and y direction. For CFRP was silver paste used for better connection of the electrodes and a homogenous current distribution in the MUT. Electrical conductivity of Aluminum is constant for all three spatial directions: $\sigma_{\text{Alu},x,y,z} \approx 37.9\,\text{MS/m}$ (comparison to literature value: 37.7 MS/m, see Vinaricky et al., 2002). Electrical conductivity in x and y direction of

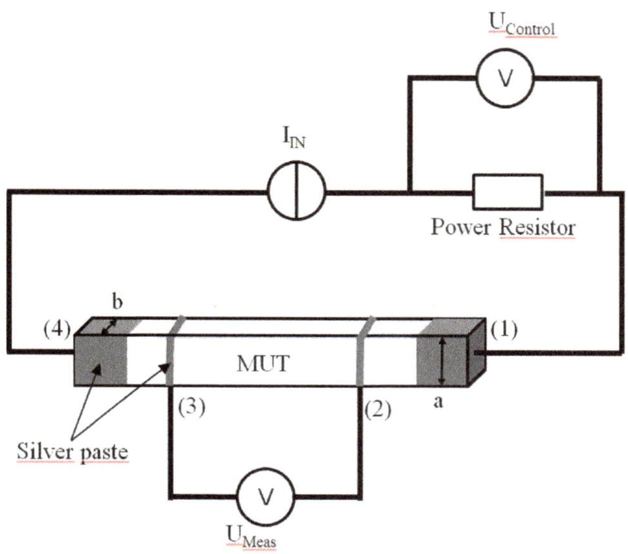

Figure 4. 4-Point-Probe-Measurement method.

$\sigma_{\text{CFRP},x,y} \approx 12.54\,\text{kS/m}$ was measured for CFRP. Additional is to say, that the conductivity of CFRP depends strongly of the quality and consistency of the MUT like fibre position and fibre concentration.

3.4 Capacitive measurement method

The missing z direction of an anisotropic material respectively of CFRP could be measured by a capacitive measurement method with an Impedance Analyser. This measurement method is usually used for measuring permittivity of dielectric materials respectively for materials with a very low conductivity (Impedance Spectroscopy). Also preparation of MUT with silver paste was necessary, for a better connection of the electrodes. With a Four-Point-Probe measurement were the capacitance (C') and the real part of the conductance (G') measured. Figure 5 shows the connection of the electrodes and the preparation of the MUT.

Based on geometry of the MUT, the measured capacitance C' and conductance (G') calculation of electrical

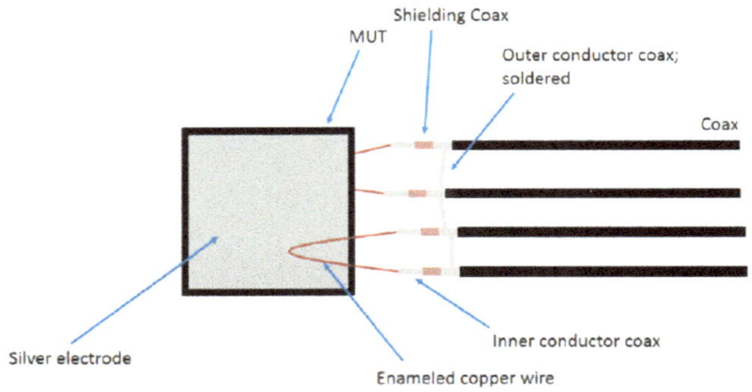

Figure 5. Capacitive measurement method and preparation of the MUT.

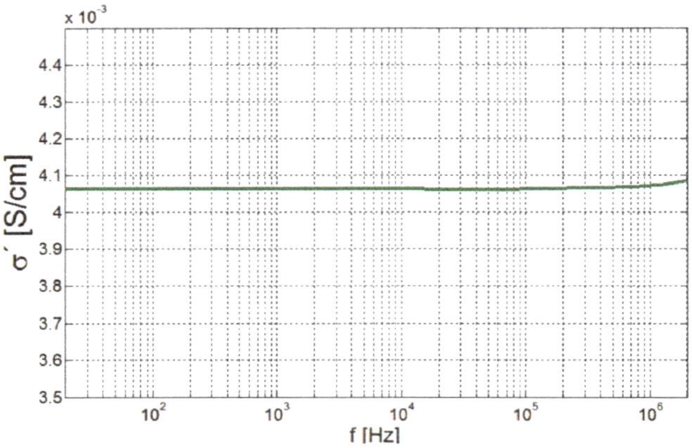

Figure 6. Measurement results of CFRP in z direction with capacitive method.

conductivity is possible, see Eqs. 3. This method is only suitable for materials with a very low conductivity, because higher conductive materials occur a short between the electrodes respectively the capacitor. The measurement was performed for CFRP in z direction in the frequency range $40\,\mathrm{Hz} < f < 10\,\mathrm{MHz}$.

$$C_0 = \frac{\epsilon_0 \cdot A}{d}$$
$$\epsilon' = \frac{C'}{C_0}$$
$$C'' = \frac{G'}{\omega}$$
$$\epsilon'' = \frac{C''}{C'' \cdot \epsilon'}$$
$$\sigma' = \epsilon \cdot \epsilon'' \cdot \omega \tag{3}$$

The measurement results in the considered frequency range is shown in Fig. 6.

The electrical conductivity measured with the capacitive method was for CFRP in z direction $\sigma_{\mathrm{CFRP},z} \approx 0.41$ S/m. The results have shown, that single Carbon fibre layers are electrically insulated. An assumption, that CFRP is a multilayered material with insulated layers could be made. All presented measurement results in this chapter will be used in the next chapter for the analytical calculation of the shielding effectiveness.

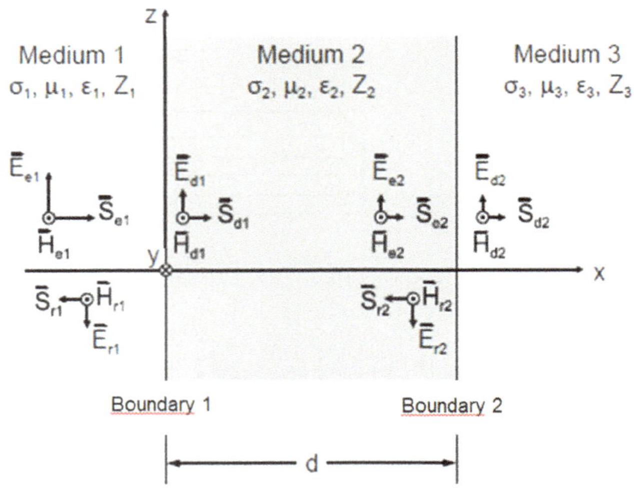

Figure 7. Impedance Concept by Schelkunoff, Wolfsperger (2008).

4 Analytical models for calculating shielding effectiveness

This section describes analytical models for calculating shielding effectiveness of isotropic and anisotropic materials.

4.1 Impedance Concept by Schelkunoff

The Impedance Concept from Schelkunoff (see Schelkunoff, 1938) describes the physical processes reflection, absorption and transmission of plane waves at material boundaries. This model is based on transmission line theory and the change of wave impedance at material boundaries. It is valid for isotropic, single layered materials. Figure 7 visualises the impedance concept for a plane wave (TEM-Wave). With the following equations calculation of free space impedance Z_0 (Eq. 4) and impedance of the material sample Z_M (Eq. 5) is possible. Z_M depends on the permeability μ, the permittivity ϵ, the frequency ω and the electrical conductivity σ. For dia- and paramagnetic materials, like Aluminum or CFRP, is $\mu_r \approx 1$. With this configuration, calculation of far field shielding effectiveness for material samples is possible.

$$Z_0 = \sqrt{\frac{\mu_0}{\epsilon_0}} \tag{4}$$

$$Z_M = \sqrt{\frac{i\omega\mu}{\sigma + i\omega\epsilon}} \tag{5}$$

As mentioned above, shielding effectiveness in the near

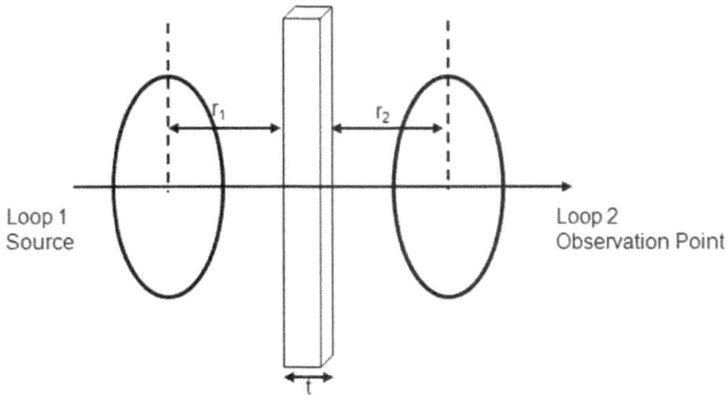

Figure 8. Analytical Model from Moser, Moser (1967).

field area is more interesting respectively the use case in a vehicle. In Wolfsperger (2008) is an approach for calculation shielding effectiveness with the Impedance Concept by Schelkunoff (Schelkunoff, 1938) in the near field area. This approach is based on substitution of the free space impedance Z_0 by the impedance of an loop antenna/dipole antenna in the near field area (Z_L Eq. 14; Z_C Eq. 7), Wolfsperger (2008).

$$Z_L = \frac{Z_0 \cdot \lambda}{2\pi \cdot r} \tag{6}$$

$$Z_C = \frac{Z_0 \cdot 2\pi \cdot r}{\lambda} \tag{7}$$

4.2 Model from Moser

The model from Moser (see Moser, 1967) is valid for the calculation of magnetic shielding effectiveness in the near field area. The schematic draw in Fig. 8 shows two coils (source and observation point) seperated by an infinite material sheet.

The model contains an exact solution of the vector wave Eq. (8).

$$\nabla^2 \vec{A} + k_0^2 \cdot \vec{A} = 0 \tag{8}$$

Validation with a measurement and comparison to the Impedance Concept by Schelkunoff was done by Moser (1967). For validation of the measurement method, analog to the setup in Fig. 8, comparison will be done again in Sect. 5.

4.3 Enhanced impedance concept for multi layered structures

In Imo et al. (1993) is a model for calculation shielding effectiveness of a parallel-plate-shield (see Fig. 9). This parallel-plate-shield consists of two infinite extensive plates with a defined thickness (d) and a distance to each other of $2 \cdot x_0$. This shield structure is penetrated by a plane wave (TEM-Wave) from outside and shielding effectiveness is calculated in the middle of the plates with a chain matrix method. Every single matrix could be understand as a quadripole, which

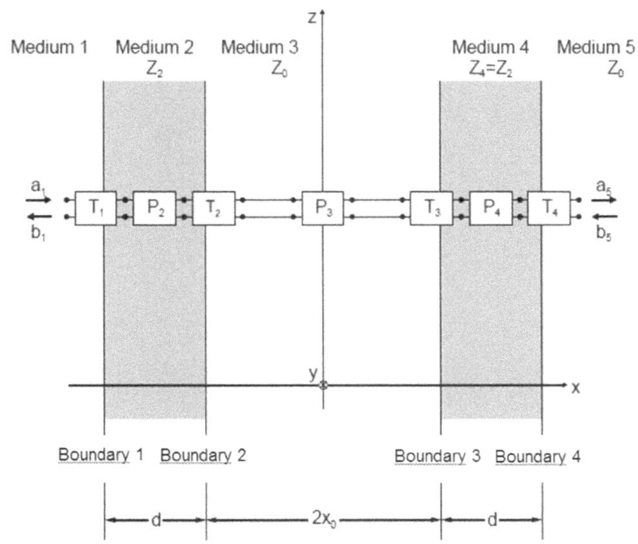

Figure 9. Enhanced Impedance Concept, Wolfsperger (2008).

consists of transmission- and reflection factors for every material and material boundary. The chain matrices could be enhanced for multilayered shields easily, also for CFRP materials. In the first approximation every Carbon fibre layer with 90° fibre direction could be seen as isotropic (see conductivity measurements in Sect. 3.3) and electrically insulated from the next Carbon fibre layer by a Polyamide layer. The validity of this approximation needs to ensured with measurements, see Sect. 5. The measurement results in Sect. 3.4 have shown a very low conductivity in z direction through the layered structure for CFRP ($\sigma_{\mathrm{CFRP},z} \approx 0.41$ S/m).

For calculation of shielding effectiveness in the near field area of multilayered materials (e.g. CFRP) adaptions of the model from Imo et al. (1993) were necessary:

- Shielding effectiveness behind the shield respectively multilayered material
- Solution for near field sources

Due to the simple structure of the quadripoles and the matrices, calculation of shielding effectiveness behind the screen was easy. In the following, the method will be called: Chain matrix method for multilayered materials (CMM). For the following example ($n = 2$ layers of Carbon fibre and $m = 1$ layer of Polyamide, see Fig. 10) the chain matrix looks like (Eqs. 9 and 10):

$$\begin{bmatrix} a_1 \\ b_1 \end{bmatrix} = [T_1][P_2][T_2][P_3][T_3][P_4][T_4]\begin{bmatrix} a_5 \\ 0 \end{bmatrix} \tag{9}$$

$$\begin{bmatrix} a_1 \\ b_1 \end{bmatrix} = \begin{bmatrix} s & t \\ u & v \end{bmatrix}\begin{bmatrix} a_5 \\ 0 \end{bmatrix} \tag{10}$$

The parameter $b_5 = 0$ because there are no reflections behind the material, there is just free space. Shielding effectiveness

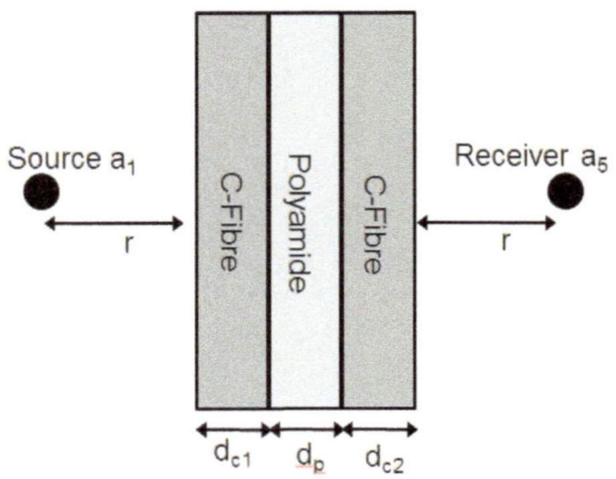

Figure 10. Example for the chain matrix method for calculation shielding effectiveness of CFRP.

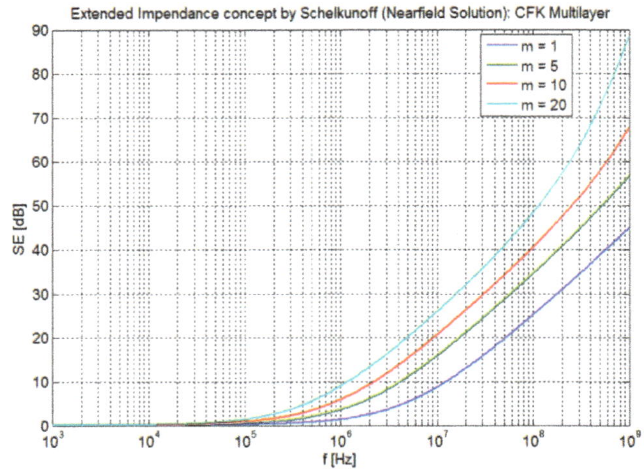

Figure 11. Shielding effectiveness calculated with chain matrix method for $n = m + 1$ layers Carbon fibre.

could be calculated by the following Eq. (11):

$$s^{-1} = \frac{a_5}{a_1} \tag{11}$$

For calculation of the near field shielding effectiveness, a detailed view on the transmission matrix (T_1, see Eq. 12) respectively there elements ($d_{II,2}$ transmission factor equation and $r_{II,2}$ reflection factor) was necessary. As described in Sect. 4.1 a substitution of the free space impedance (Z_0) with the impedance of the inductive field (Z_L) or capacitive field (Z_C) it is also possible to get the near field solution.

$$[T_1] = \frac{1}{d_{II,1}} \begin{bmatrix} 1 & r_{II,1} \\ r_{II,1} & 1 \end{bmatrix} \tag{12}$$

$$d_{II,1} = \frac{2 \cdot Z_{L,C}}{Z_M + Z_{L,C}} \qquad r_{II,1} = -\frac{Z_M - Z_{L,C}}{Z_M + Z_{L,C}} \tag{13}$$

$$Z_L = \frac{Z_0 \cdot \lambda}{2\pi \cdot r} \quad Z_C = \frac{Z_0 \cdot 2\pi \cdot r}{\lambda} \quad Z_M = \sqrt{\frac{i\omega\mu}{\sigma + i\omega\epsilon}} \tag{14}$$

The following Fig. 11 shows the results for the magnetic near field shielding effectiveness for $m = 1, 5, 10, 20$ layers of Polyamide respectively $n = m + 1$ layers of Carbon fibre.

For raising m or n magnetic shielding effectiveness in the near field is also raising. In the lower frequency range ($< 10\,\text{kHz}$) is no shielding functionality available, this is reasoned by the low conductivity of the Carbon fibre layer ($\sigma_{\text{CFRP},x,y} \approx 12.5\,\text{kS/m}$).

5 Validation with measurements

For validation of the previous presented models a suitable measurement method was designed, The previous presented models need to be validated with measurements. For this, a suitable measurement system was designed. The measurement system with two coils (source and observation point) wounded on a PVC tube (radius $d = 0.0375\,\text{m}$, $N = 5$ Windings, distance between the coils $D = 0.085\,\text{m}$) was inspired by Fig. 8. The material sample separates the coils in the middle. Shielding effectiveness (SE) could be calculated after measuring the induced voltage in the observation coil with two different configurations (U_{indmax} without material sample, $U_{\text{indsample}}$ without material sample), see Eq. (15):

$$SE = 20 \cdot \log \frac{U_{\text{indmax}}}{U_{\text{indsample}}} \tag{15}$$

The MUT (Aluminum, thickness $d = 300\,\mu\text{m}$) was chosen with adequate area ($A = 0.3\text{m} \cdot 0.2\text{m}$) to minimize the scattered field (magnetic field outside the MUT), which would distort the measurement result. The frequency range were the measurement was performed: $500\,\text{Hz} < f < 300\,\text{kHz}$. The comparison with the Impedance Concept of Schelkunoff (near field modification), the model from Moser and the measurement for Aluminum is shown in Fig. 12.

The measurement (blue crosses) and the analytical model from Moser (green curve) show, especially in the lower frequency range ($f < 1\,\text{kHz}$), a very good match. The Impedance Concept from Schelkunoff (red curve) matches by $f > 2.5\,\text{kHz}$ the measurement or the model from Moser. The frequency range $f < 2.5\,\text{kHz}$ is not described correctly by the Impedance Concept, negative values are not plausible. For $f > 225\,\text{kHz}$ a deviation from measurement to the models is recognisable, this is reasoned by the limits of the measurement system (dynamic problem). The same measurement was also performed with a CFRP Material. The measurement was compared with the results of the Chain Matrix Method. Parametrisation of the model was appropriate to the corresponding CFRP material. Figure 13 shows the measurement results compared to the Chain Matrix Method results.

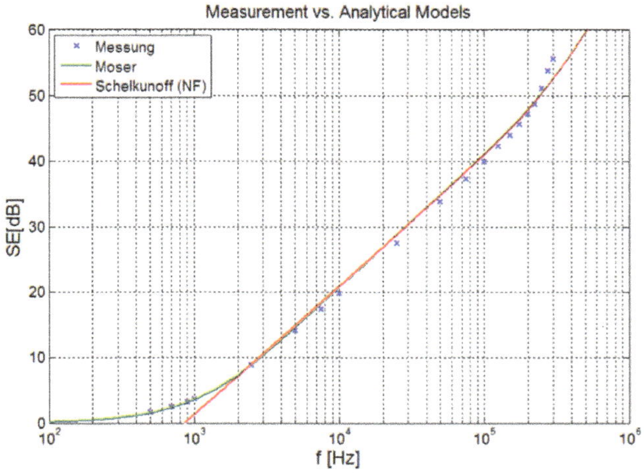

Figure 12. Comparison of Schelkunoff's Impedance Concept and the model from Moser with a measurement.

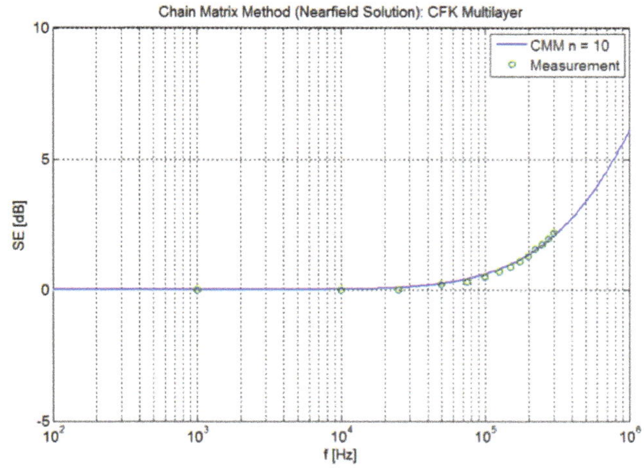

Figure 13. Comparison of chain matrix method with measurement.

6 Conclusions

In this contribution, the electrical behaviour of isotropic (Aluminum) and anisotropic (CFRP) based on electrical circuit diagrams, was analysed. Also measurement techniques for low and high conductive materials, with respect to the current flow direction, were presented. Those measurements have shown, that CFRP could be handled as multilayered material (Carbon fibre layers insulated by Polyamide). For prediction of magnetic shielding effectiveness were different analytical models presented and discussed, the Impedance Concept by Schelkunoff and the analytical model from Moser. The Impedance Concept was adapted for the near field solution and the model from Moser contains an exact solution for the magnetic near field area. To calculate shielding effectiveness of anisotropic materials, the enhanced Impedance Concept for calculation of shielding effectiveness of parallel-plate-shields, was modified and adopted. Validation of all analytical models was done with an adequate measurement setup, which was inspired by the model from Moser. Measurement results showed a good match with the analytical model from Moser in the considered frequency range. The Impedance Concept (near field solution) correlated for $f > 2.5\,\mathrm{kHz}$ with the measurement. Also the chain matrix method showed a good match with the magnetic shielding effectiveness measurement of CFRP and consideration of CFRP as a multilayered material with electrically insulated layers was valid. Only disadvantage was the very low shielding effectiveness of CFRP against magnetic fields, this could be enhanced by modifying the material with a high conductive material i.e. copper. Modifications like this, could also be calculated with the chain matrix method.

Acknowledgements. This contribution was developed within the scope of the project EM4EM (Electromagnetic Reliability of Electronic Systems for Electro Mobility – Subproject: Entwurfs- und EMZ-Messmethodik für EMZ-Analysen auf EV-Gesamtsystemebene) which is funded by the BMBF (Bundesministerium für Bildung und Forschung) under the grant number 16M3092A. The responsibility for this publication is held by the authors only.

Edited by: W. Mathis

References

Bowler, N.: Theory of four-point direct-current potential drop measurements on a metal plate, Res. Nondestruct. Eval., 17, 29–48, 2006.

Imo, F., Fuchs, C., and Schwab, A. J.: Erweitertes Impedanzkonzept zur Berechnung von geschlossenen Schirmen, Arch. Elektrotech., 76, 437–441, 1993.

Michel, M.: Leistungselektronik: Einführung in Schaltungen und deren Verhalten, Springer-Verlag, 5th edn., 2011.

Moser, J.: Low-Frequency Shielding of a Circular Loop Electromagnetic Field Source, IEEE T. Electromagn. C., 9, 6–18, 1967.

Schelkunoff, S. A.: The Impedance Concept and its Application to Problems of Reflection, Refraction, Shielding and Power Absorption, Bell Syst. Tech. J., 17, 17–48, doi:10.1002/j.1538-7305.1938.tb00774.x, 1938.

Schwab, A. J. and Kürner, W.: Elektromagnetische Verträglichkeit, VDI-Buch, Springer, 2011.

Vinaricky, E., Keil, A., and Merl, W.: Elektrische Kontakte, Werkstoffe und Anwendungen: Grundlagen, Technologien, Prüfverfahren, Engineering online library, Springer Berlin Heidelberg, 2002.

Wolfsperger, H.: Elektromagnetische Schirmung: Theorie Und Praxisbeispiele, Vdi-buch, Springer, 2008.

9

A relation between algebraic and transform-based reconstruction technique in computed tomography

S. Kiefhaber, M. Rosenbaum, W. Sauer-Greff, and R. Urbansky

Technical University of Kaiserslautern, Chair of Communications Engineering, Germany

Correspondence to: S. Kiefhaber (s.kiefhaber@t-online.de)

Abstract. In this contribution a coherent relation between the algebraic and the transform-based reconstruction technique for computed tomography is introduced using the mathematical means of two-dimensional signal processing. There are two advantages arising from that approach. First, the algebraic reconstruction technique can now be used efficiently regarding memory usage without considerations concerning the handling of large sparse matrices. Second, the relation grants a more intuitive understanding as to the convergence characteristics of the iterative method. Besides the gain in theoretical insight these advantages offer new possibilities for application-specific fine tuning of reconstruction techniques.

1 Introduction

Computed tomography is a well-established method in medicine, material science and quality control. It is an imaging technique used to create a cross-sectional image of the interior of a body to be examined. In contrast to classical imaging where the image corresponds directly to the measurements, in computed tomography the image is generated in an indirect manner. Therefore a series of transmission measurements, e.g. utilising x-rays, is taken. The projection data can be used to reconstruct an image of the inner structure of an object.

Concerning algorithms, the key part of every tomographic imaging system is the reconstruction technique. There are two major reconstruction techniques that are considered in this paper. The first, transform-based approach, was introduced in 1917 by the Austrian mathematician Johann Radon (Radon, 1917). The Radon transform and its inverse are analytical expressions of the projection process during the transmission measurements and the reconstruction technique, respectively. They are well understood from a signal processing point of view. The second, algebraic reconstruction technique is attributed to Bender et al. and, in principal, is an accomplishment of the Polish mathematician Stefan Kaczmarz who proposed an iterative method to approximate solutions of systems of linear equations (Kaczmarz, 1937). In terms of linear algebra this technique is well known.

The main contribution of this paper is the reinterpretation of the algebraic reconstruction technique by means of two-dimensional signal processing. Starting with the analysis and reinterpretation of the description of the measurement process using a system of linear equations in Subsect. 3.1 this leads to a mathematical relation between the algebraic and the transform-based reconstruction technique.

Knowledge about this relation has two advantages. First, the algebraic reconstruction technique can now be implemented efficiently as to the usage of memory avoiding the necessity to handle large sparse matrices. Second, the closed form of the iteration exactly describes the convergence characteristics of the iterative method. In addition to the academical benefit these advantages could offer new possibilities for application-specific fine tuning of reconstruction techniques.

The remainder of this publication is organized as follows. A short introduction to the process of data acquisition in computed tomography along with the mathematical description of the measured data, namely the Radon transform, is given in Subsect. 2.1. The essential facts about the inverse Radon transform are revisited in Subsect. 2.2 in a intuitive manner based on Morneburg (1995) avoiding the indeed exact but also cumbersome mathematical derivation. The principal ideas of Kaczmarz' method are summarised in Subsect. 2.3. The reinterpretation of the algebraic reconstruction technique is presented in Sect. 3. The proof of the relation between the two methods is given in Subsect. 3.2. The implications of the introduced relation between algebraic and transform-based reconstruction technique are discussed in

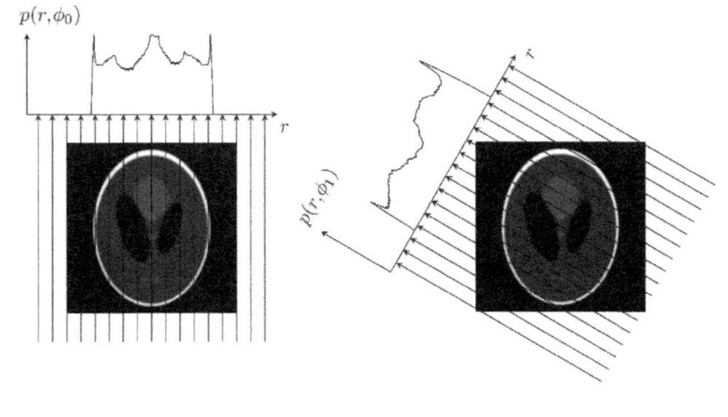

Fig. 1. Measurement of the attenuation of x-rays on their paths along parallels through the phantom repeatedly for different directions, here exemplary for two different angles.

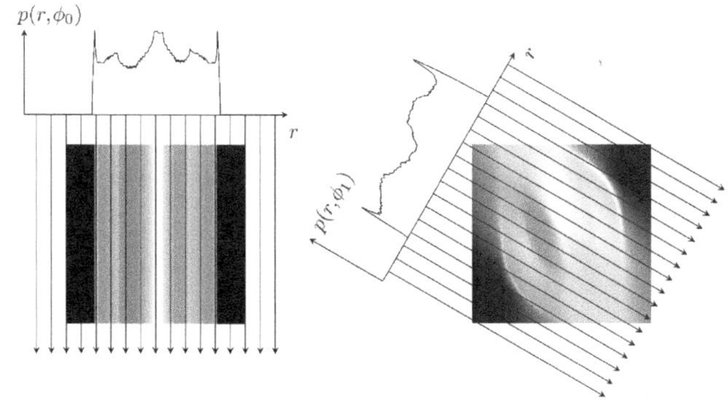

Fig. 2. Illustration of the backprojection operation.

Subsect. 3.4. A concluding summary of the basic ideas and insights of this contribution is given in Sect. 4.

2 Preliminary Considerations

In this section a short recapitulation of the underlying ideas and the necessary preliminary considerations concerning computed tomography, the transform-based and the algebraic reconstruction technique are given.

2.1 Computed tomography

Computed tomography was first successfully conducted by Sir Godfrey Newbold Hounsfield in 1971 who in 1979 received the *Nobel Price in Physiology or Medicine* for his efforts together with Allan McLeod Cormack. With his device he measured the accumulated attenuation of x-rays on their paths along parallels through the phantom repeatedly for different directions of equally distributed angles between zero and 180 degrees. This is illustrated in Fig. 1.

Mathematically the measured attenuation can be expressed as the Radon transform of the function describing the spacial distribution of the attenuation coefficient. The local attenuation coefficient in a plane $(x, y) \in \mathbb{R}^2$ is expressed as a continuous function $g(x, y)$. The paths are expressed as straight lines $G(r, \phi) : \{(x, y)|0 = r - x\cos(\phi) - y\sin(\phi)\}$ where r is the perpendicular distance of G to the origin and ϕ is the angle between G and the y-axis. The ensemble of integrals over $g(x, y)$ along the paths defined by $G(r, \phi)$ is the Radon transform

$$p(r, \phi) = \mathcal{R}\{g(x, y)\} = \int_{G(r,\phi)} g(x, y)\mathrm{d}x\mathrm{d}y \tag{1}$$

and corresponds to the measured data. The main goal in computed tomography is to reconstruct the function describing the original spacial distribution of the attenuation coefficient $g(x, y)$ from the integral values measured along parallel paths from different directions as represented by the Radon transform $p(r, \phi)$. The practical means to solve this problem is given by so-called reconstruction techniques.

For further details on the Radon transform as well as its inverse which is subject to Subsect. 2.2 please cf. Cho et al. (1993); Morneburg (1995); Poularikas (1996).

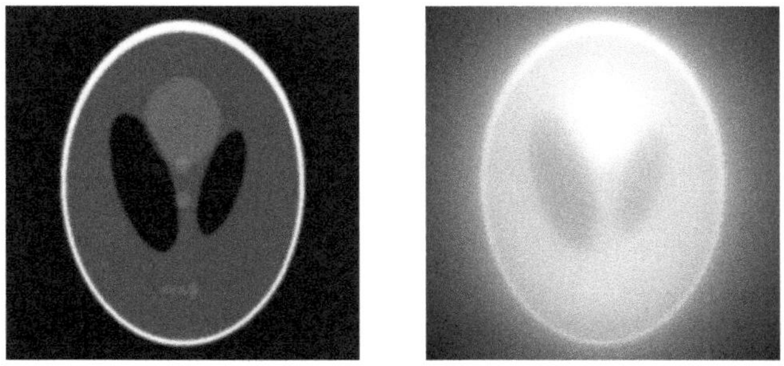

Fig. 3. Comparison of original and result of plain backprojection.

2.2 Transform-based reconstruction technique

Having the Radon transform in mind, the first idea to solve the reconstruction problem is quite manifest, namely inversion of the measurement process by backprojecting the integral values onto their respective paths and accumulating the portions from each direction. Mathematically, this backprojection can be written as the integral of $p(r, \phi)$ over ϕ.

$$\mathcal{B}\{p(r, \phi)\} = \int_0^{\pi} p(r, \phi)\mathrm{d}\phi \tag{2}$$

The backprojection operation is illustrated in Fig. 2. Figure 3 shows the original image as well as the result of the consecutive Radon transform and backprojection. Obviously, the reconstruction resulting from the plain backprojection differs significantly from the original image by a certain blurring effect.

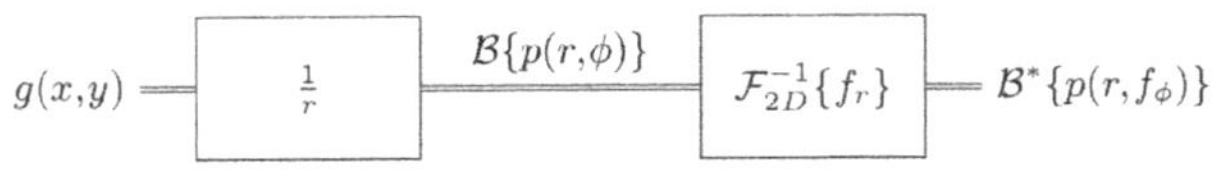

Fig. 4. 2-D transmission system consisting of Radon transform, backprojection and compensating filter.

The next step towards the solution of the reconstruction problem is to treat the concatenation of Radon transform and backprojection as a two-dimensional linear transmission system. Thereby it is possible to identify the transmission behaviour of the system and compensate for it by adding a filter with the inverse behaviour to the signal processing chain.

According to Deans (1983), the impulse response as well as the transfer function of the linear transmission system consisting of the concatenation of Radon transform and backprojection decay according to the reciprocal radial distance from the origin, i.e. $h_{\mathcal{RB}} = \frac{1}{r}$ and $H_{\mathcal{RB}} = \frac{1}{f_r}$ in the original and the Fourier domain, respectively. The functional identity of the impulse response $h_{\mathcal{RB}}$ and the transfer function $H_{\mathcal{RB}}$ is called a fix point of the Hankel transform of order zero which for rotationally symmetric functions is identical to the Fourier transform in two dimensions. The transfer function of the composite system $H_{\mathcal{RB}} = \frac{1}{f_r}$ will play a key role in the derivation of the relation between the transform-based and the algebraic reconstruction technique.

The impulse response of the compensating filter follows from inverting the transfer function $H_{\mathcal{RB}}$ and then taking its two-dimensional inverse Fourier transform. Mathematically, the filtering corresponds to a convolution which, incorporated into Eq. (2), leads to the so-called filtered backprojection.

$$\mathcal{B}^*\{p(r,\phi)\} = \int_0^\pi p(r,\phi) * \mathcal{F}_{2D}^{-1}\{f_r\} \mathrm{d}\phi \tag{3}$$

The filtered backprojection is one representation of the inverse Radon transform and an analytical solution to the reconstruction problem. Discretization and implementation leads to the transform-based reconstruction technique. Figure 4 illustrates the described construction of the filtered backprojection. The inverse Radon transform can also be derived in a strictly mathematical manner as, e.g., shown in Poularikas (1996). The strictly mathematical derivation allows to identify the transfer function of the compensating filter f_r as the determinant of the Jacobian matrix altered due to the transform from Cartesian to polar coordinates which is inherent to the geometry of the measurement process.

2.3 Algebraic reconstruction technique

In comparison to transform-based methods, the derivation of the algebraic reconstruction technique follows a somewhat different approach. The reconstruction problem is stated directly in a discretised form rather than continuously as for the transform-based reconstruction technique. Figure 5 shows a four by four pixel area to be reconstructed which is overlayed with six corridors corresponding to the parallel paths for one exemplary direction. The commonly used approach (see, e.g., Kak et al., 2001) is to set up one linear equation for each path considering all directions

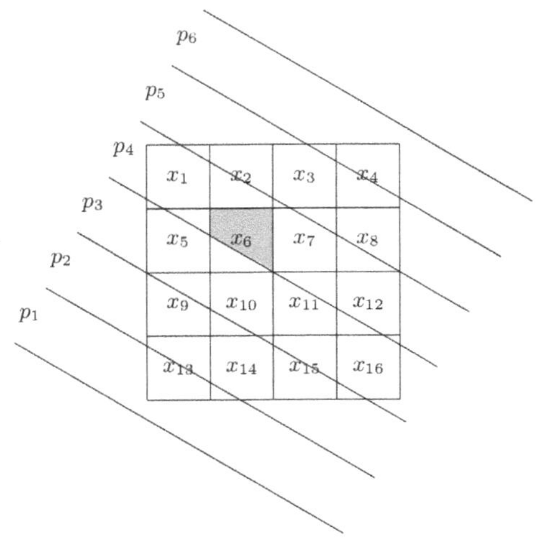

Fig. 5. Setting up the equation system by defining one linear equation per ray.

$$x_1 a_{1,j} + x_2 a_{2,j} + \ldots + x_N a_{N,j} = p_j \tag{4}$$

and to combine them to a system of linear equations written here using a vector matrix formalism.

$$\mathbf{Ax} = \mathbf{p} \tag{5}$$

The individual coefficients $a_{i,j}$ may for example be chosen to account for the intersectional area of the pixel i with corridor j. Also common practice is to use a hit-or-miss approach in order to reduce the computational effort. If a pixel intersects with the corridor at hand the respective coefficient is set to one, otherwise to zero.

In the linear equation system (5) vector $\mathbf{p}$ corresponds to the measured data, matrix $\mathbf{A}$ accounts for the geometry of the tomographic system and vector $\mathbf{x}$ corresponds to the wanted spacial distribution of the attenuation coefficient. On first sight the solution of the reconstruction problem is obtained by solving the system of linear equations for $\mathbf{x}$. However, there are certain difficulties that in general prohibit the straight forward solution of Eq. (5).

First, matrix $\mathbf{A}$ does not necessarily need to be square or, second, invertible at all for that matter. Even if $\mathbf{A}$ is a square matrix possible measurement noise might disrupt an exact solution. Third, the mere size of the equation system for any real tomographic system makes a solution, for example using

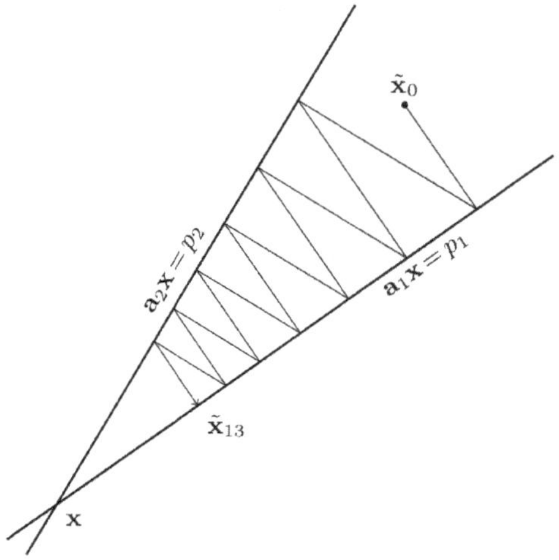

Fig. 6. Move arbitrary point $\tilde{\mathbf{x}}_0$ towards the intersectional point of hyperplanes through successive orthogonal projections to find a solution of system of linear equations. Exemplary for first 13 iterations.

the Moore-Penrose pseudoinverse to approximate a solution, computationally extremely expensive.

Therefore it is advisable to utilize an iterative method to directly determine (in general) an approximation $\tilde{\mathbf{x}}$ to a solution $\mathbf{x}$ of the system of linear equations. In computed tomography there are several methods used such as the *algebraic reconstruction technique* as suggested by Bender et al., the *simultaneous iterative reconstruction technique* or the *simultaneous algebraic reconstruction technique* (cf. e.g. Kak et al., 2001). All of them are in principle based on a method proposed by (Kaczmarz, 1937).

His idea was to interpret $\mathbf{x}$ as a point in an n-dimensional space. Therewith every line of the linear equation system can be used to define a hyperplane in that space. The intersection of all hyperplanes, provided it exists, corresponds to the solution of the system of linear equations. So, to solve the system of linear equations means to find the intersectional point of all hyperplanes. Kaczmarz suggested to do so by choosing an arbitrary point $\tilde{\mathbf{x}}_0$ in the n-dimensional space and moving the initially guessed point towards the real solution through an iterative process of successive orthogonal projections on the hyperplanes . This process is illustrated for a two-dimensional solution-space in Fig. 6.

Mathematically, Kaczmarz' method can be written as follows. For a detailed derivation cf., e.g., Kak et al. (2001).

$$\tilde{\mathbf{x}}_j = \tilde{\mathbf{x}}_{j-1} + \frac{\mathbf{a}_j^\mathsf{T}}{|\mathbf{a}_j|}(p_j - \tilde{p}_j) \qquad \text{where} \quad \tilde{p}_j = \mathbf{a}_j \cdot \tilde{\mathbf{x}}_{j-1} \quad (6)$$

Here $\tilde{\mathbf{x}}_j$ is the approximation to the solution of the system of linear equations after j orthogonal projections, $\mathbf{a}_j$ is the j-th line of the coefficient matrix $\mathbf{A}$, and p_j is the j-th measured value and $\tilde{p}_j$ is the result of the linear mapping defined in (5) assuming $\tilde{\mathbf{x}}_j$ is the wanted spacial distribution of the attenuation coefficient. According to Kaczmarz (1937); Tanabe (1971), if for every value of n one full projection cycle is conducted it follows that in the limit n to infinity $\tilde{\mathbf{x}}_n$ approximates the solution with arbitrary accuracy.

$$\lim_{n\to\infty} \tilde{\mathbf{x}}_n = \mathbf{x} \qquad (7)$$

Note that if there is no unique solution, the estimate oscillates within the neighbourhood of the intersections.

3 Relation of methods

The first step to derive a relation between the inverse Radon transform and Kaczmarz' method is to reexamine the initial model on which the setup of the system of linear equations is based. This leads to a reinterpretation of Kaczmarz' method as iterative backprojection. Thereafter the transmission behaviour of the iterative backprojection will be closely examined by presenting a closed form representation which also exactly describes the convergence characteristics of the iterative method.

3.1 Reinterpretation of linear mapping as discretised Radon transform

The representations of the distribution function and the paths in terms of pixels and corridors, respectively, correspond to the usage of a Haar basis; in terms of sampling and interpolation theory it corresponds to interpolation with B-splines of order zero or, colloquial, to a *nearest neighbour* interpolation (cf., e.g., Unser, 2000). By disregarding the particular method of interpolation the setup of the linear equation system can be done assuming the elements of vectors $\mathbf{x}$ and $\mathbf{p}$ being the weights of a Dirac basis representations, i.e. the samples of the continuous function describing the spacial distribution of the attenuation coefficients and their directed sums as discretisation of the path integrals, respectively. This is illustrated in Fig. 7. The particular method of interpolation is accounted for by the values of the coefficient matrix $\mathbf{A}$.

From that point of view it becomes comprehensible that depending on the kind of interpolation the system of linear Eq. (5) is essentially a discretisation of the Radon transform. By applying this idea on Eq. (6) Kaczmarz' method can be understood as iterative backprojection of the difference between virtual projection data $\tilde{p}_n$ computed using the estimated distribution function $\tilde{\mathbf{x}}_{n-1}$ and the actually measured data p.

Based on the above considerations Eq. (6) can be rewritten using the Radon transform and backprojection operators on the continuously defined distribution function.

$$\begin{aligned}\tilde{g}_{n+1} &= \tilde{g}_n + \mathcal{B}(p - \mathcal{R}\tilde{g}_n) \qquad \text{where} \quad p = \mathcal{R}g \\ &= \tilde{g}_n + \mathcal{B}(\mathcal{R}g - \mathcal{R}\tilde{g}_n) \\ &= \tilde{g}_n + \mathcal{B}\mathcal{R}g - \mathcal{B}\mathcal{R}\tilde{g}_n \qquad (8)\end{aligned}$$

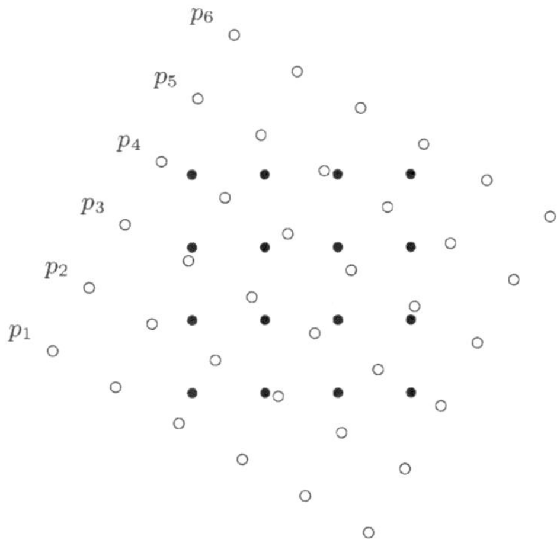

Fig. 7. Dirac representation of model on which the set up of the system of linear equations is based.

At this point it is obvious that both reconstruction techniques are not completely independent of each other as, e.g., asserted in Kak et al. (2001). Using the means of two-dimensional signal processing, in particular sampling and interpolation theory Kaczmarz' method has indeed been shown to be a kind of iterative backprojection using the Radon transform to determine the estimation error. In the following Subsect. 3.2 the iterative backprojecting will be shown, for all practical means, to actually have the same transmission behaviour as the inverse Radon transform. This will also yield an explicit description of the convergence characteristics of the iterative method.

3.2 Transmission behaviour of iterative backprojection

In order to understand the transmission behaviour of the iterative backprojection it is self-evident to take a closer look at the Fourier space representation of Eq. (8). From Subsect. 2.2 it is known that a linear two-dimensional transmission system consisting of consecutive Radon transform and (unfiltered) backprojection can be described by an impulse response as well as a transfer function following $\frac{1}{x}$. Bearing that in mind the Fourier transform of Eq. (8) can be written as follows.

$$\tilde{G}_{n+1} = \tilde{G}_n + \frac{1}{f_r}G - \frac{1}{f_r}\tilde{G}_n \tag{9}$$

Here f_r is the spacial frequency in radial direction and G and $\tilde{G}$ are the spectra of the actual and estimated distribution functions, respectively.

The idea for deriving a closed form representation of the iterative method arises from taking a look at the first four iteration steps in Fourier space. The initial guess has been chosen according to $\tilde{G}_0 = 0$.

$$\begin{aligned}
\tilde{G}_0 &= 0\\
\tilde{G}_1 &= \frac{1}{f_r}G\\
\tilde{G}_2 &= \left(\frac{2}{f_r} - \frac{1}{f_r^2}\right)G\\
\tilde{G}_3 &= \left(\frac{3}{f_r} - \frac{3}{f_r^2} + \frac{1}{f_r^3}\right)G\\
\tilde{G}_4 &= \left(\frac{4}{f_r} - \frac{6}{f_r^2} + \frac{4}{f_r^3} - \frac{1}{f_r^4}\right)G
\end{aligned}$$

The term in brackets can be identified as part of an alternating binomial series for which the following relationship is known.

$$\sum_{k=0}^{n}\binom{n}{k}a^{n-k}b^k = (a+b)^n \tag{10}$$

This leads to the assertion that for $f_r \geq 1$ and $\tilde{g}_0 = 0$ the following relation holds, where $a = 1$ and $b = -\frac{1}{f_r}$ have been used in Eq. (10).

$$\tilde{G}_n = \left(1 - \left(1 - \frac{1}{f_r}\right)^n\right)G$$

This can be shown by mathematical induction.

3.2.1 Base case ($n = 0$)

$$\tilde{G}_0 = \left(1 - \left(1 - \frac{1}{f_r}\right)^0\right)G = 0 \tag{11}$$

3.3 Inductive step

($n \rightarrow n+1$) using (11) in (9)

$$\begin{aligned}
\tilde{G}_{n+1} &= \left(1 - \left(1 - \frac{1}{f_r}\right)^n\right)G + \frac{1}{f_r}G - \frac{1}{f_r}\left(1 - \left(1 - \frac{1}{f_r}\right)^n\right)G\\
&= \left(1 - \left(1 - \frac{1}{f_r}\right)^n + \frac{1}{f_r}\left(1 - \frac{1}{f_r}\right)^n\right)G\\
&= \left(1 - \left(1 - \frac{1}{f_r}\right)\left(1 - \frac{1}{f_r}\right)^n\right)G\\
&= \left(1 - \left(1 - \frac{1}{f_r}\right)^{n+1}\right)G \quad \text{Q.E.D.}
\end{aligned} \tag{12}$$

(Note that the above identity can also be derived making use of the fact that Eq. (9) is a linear inhomgenous recursion formula of first order for which there is a general closed form correspondence.)

Taking the limit of Eq. (12) for n to infinity clarifies the transmission behaviour of the iterative backprojection. By rewriting the term in brackets it becomes obvious that if f_r is greater than one the bracketed term vanishes in the limit and the spectrum of the estimated solution $\tilde{G}_n$ approximates the spectrum G of the actual solution with arbitrary accuracy.

$$\lim_{n\to\infty} \tilde{G}_n = \lim_{n\to\infty} \left(1 - \left(\frac{f_r - 1}{f_r}\right)^n\right) G = G \tag{13}$$

In the original space it then follows for the estimated and actual solutions that

$$\lim_{n\to\infty} \tilde{g}_n = g. \tag{14}$$

Hence, for frequencies $f_r > 1$ the iterative backprojection converges to the solution as the term in brackets vanishes in Eq. (13).

3.4 Implications

The first implication regards the stability of the numerical implementation of the algebraic reconstruction technique. For the numerical computation the constrain $f_r > 1$ means that in terms of linear algebra the biggest eigenvalue of the concatenated projection and backprojection has to be lower or equal to one to guarantee convergence. In terms of the Radon and backprojection operators the operator-norm of the concatenation has to be lower or equal to one. Practically, the square sum of the pixel values , i.e. the energy of the reconstruction may not increase from one iteration step to the next. This can, e.g., be ensured by weighting the whole reconstruction with a constant factor. The closer the implementation is to the stability bound the faster the convergence.

The second implication regards the runtime of the reconstruction techniques. As for every cycle of the iterative backprojection a Radon transform and a backprojection has to be performed it is obvious that the transform-based reconstruction technique delivers its result in a fraction of the time that is needed for the algebraic reconstruction technique. Albeit, the exact time needed for the algebraic reconstruction technique depends on the required accuracy of the reconstruction which can be utilized as a criterion to abort the iteration process.

Last, the introduced relation allows to implement the algebraic reconstruction technique independently of the vector-matrix formalism avoiding the necessity to deal with handling large sparse matrices.

4 Summary

In this contribution the derivation of a relation between the algebraic and the transform-based reconstruction technique in computed tomography was discussed in a very general way. Therefore, based on preliminary considerations, the model for setting up the system of linear equations was reinterpreted using the means of sampling and interpolation theory which lead to understand Kaczmarz' method as iterative backprojection. The transmission behaviour of the iterative backprojection was examined and characterised offering a closed form representation of the iteration. Finally the implications and advantages of the new relation besides the academical benefit where discussed. The insights offered in this contribution give the means for further application-specific refinement and tuning of reconstruction techniques.

References

Bender, R., Gordon, R., and Herman, G.: Algebraic Reconstruction Techniques (ART) for Three-Dimensional Electron Microscopy and X-Ray Photography, in J. Theoret. Biol., 29, 471–481, 1970.

Cho, Z.-H., Jones, J. P., and Singh, M.: Foundations of medical imaging, New York: Wiley, 1993, ISBN 0-471-54573-2, 1993.

Deans, S.: The Radon transform and some of its Applications, New York, Wiley, ISBN 0-471-89804-X, 1983.

Kaczmarz, S.: Angenäherte Auflösung von Systemen linearer Gleichungen, Bull. Acad. Pol. Sci. Lett., A35, 355–357, 1937.

Kak, A., and Slaney, M.: Principles of Computerized Tomographic Imaging, New York, Society of Industrial and Applied Mathematics, ISBN 0-87942-198-3, 2001.

Morneburg, H. (Hrsg.): Bildgebende Systeme für die medizinische Diagnostik: Röntgendiagnostik und Angiographie, Computertomographie, Nuklearmedizin, Magnetresonanztomographie, Sonographie, integrierte Informationssysteme. 3., wesentlich überarb. und erw. Aufl. München: Publicis-MCD-Verl., ISBN 3-89578-002-2, 1995.

Poularikas, A., D. (Hrsg.): The transforms and applications handbook, Boca Raton, Fla.: CRC Press [u.a.], ISBN 0-8493-8342-0, 1996.

Radon, J.: Über die Bestimmung von Funktionen durch ihre Integralwerte längs gewisser Mannigfaltigkeiten, in: Berichte Sächsische Akademie der Wissenschaften, Math.-Phys. Kl., Leipzig, 69, 262–267, 1917.

Tanabe, K.: Projection Method for Solving a Singular System of Linear Equations and its Applications, in: Numer. Math., 17, 1971, 203–214, Springer-Verlag, 1971.

Unser, M.: Sampling–50 Years after Shannon, in: Proceedings of the IEEE, 88, 4, 569–587, Publisher Item Identifier S-0018-9219(00)01299-7, 2000.

10

The road surveying system of the federal highway research institute – a performance evaluation of road segmentation algorithms

R. Streiter and G. Wanielik

Chemnitz University of Technology, Professorship of Communications Engineering , Reichenhainer Str. 70, 09126 Chemnitz, Germany

Correspondence to: R. Streiter (robin.streiter@etit.tu-chemnitz.de)

Abstract. The construction of highways and federal roadways is subject to many restrictions and designing rules. The focus is on safety, comfort and smooth driving. Unfortunately, the planning information for roadways and their real constitution, course and their number of lanes and lane widths is often unsure or not available. Due to digital map databases of roads raised much interest during the last years and became one major cornerstone of innovative Advanced Driving Assistance Systems (ADASs), the demand for accurate and detailed road information increases considerably. Within this project a measurement system for collecting high accurate road data was developed. This paper gives an overview about the sensor configuration within the measurement vehicle, introduces the implemented algorithms and shows some applications implemented in the post processing platform. The aim is to recover the origin parametric description of the roadway and the performance of the measurement system is being evaluated against several original road construction information.

Fig. 1. The measurement vehicle is based on a Volkswagen Multivan witch was prepared to incorporate all sensors (5 cameras in front, lane mark detection system, high accurate positioning module with inertial measurement unit and a laserscanner).

1 Introduction

The structure of federal roads and highways is subject to many construction rules that ensure safe and comfortable driving even at high velocities. To verify and monitor the state and the course of roads the highway research institute commissioned a multisensory equipped measurement vehicle including Cameras, lane mark detection system, laser scanner and high precision positioning components with inertial measurement unit in order to observe the state of roads. The aim of that work is to give an overview and a performance evaluation of the implemented algorithms that recover the parameters of the roads from the raw data collected by the measurement system. Several segmentation approaches were evaluated – Unscented Kalman Filter, Hough Transformation and (RANdom SAmple consensus) RANSAC. For several roads original planning information was available as ground truth that leads to an absolute performance evaluation of all approaches. The concept and all technical details about the measurement system are introduced in Streiter et al. (2012). This paper is focused on the software algorithms that were implemented in order to recover the original parameters that can describe the characteristics of the road parametrically.

2 Measurement system

The whole system is divided into two separate units. There is the actual measurement vehicle equipped with a multisensor platform where all data is recorded and there is a postprocessing unit where all data is visualized and mapped to existing road meta-information. That platform offers many opportunities to evaluate different data fusion algorithms. Within that work the road segmentation algorithms are introduced. The course of a road will be estimated by a parametrical description of its road elements. To ensure a continuous curvature profile of a roadway, prescribed road elements are used for the planning and dimensioning of a road. Here, the elements straight line, clothoid and circular arc are used for designing federal roads and highways. These elements are joined in that order with the function of the clothoid to fit the curvature of the circular arc to the straight line with respect to the curvature continuity.

3 Segmentation algorithms

There are several profiles of the road which parameters can be determined along the runlength of a given segment. The descripting parameters for each profile of a road segment are shown in Table 1.

Most challenging task is to determine the transition points between the segments. In order to do this, several approaches were implemented. On the one hand there are consecutive algorithms, where the position measurements are given consecutively. An Unscented Kalman Filter was used here to estimate the parameters of a road segment, including all measurements arrived up to the last time step. When the generated segment violates a given error threshold, all arrived measurements are discarded and the last estimated segment is accepted. All following measurements are assigned to the new segment. These approaches have the disadvantage that available knowledge about the future is ignored and the transmission points are recognized too late. That makes the algorithm find more small segments in a given sequence than it is actually containing. The currently used algorithm will process all position measurements from the whole sequence. Number and describing parameters of the containing segments are estimated by a RANSAC Xu und Lu (2012) approach that – thanks to the given future knowledge is able to optimize the transmission points in both directions. Due to the algorithm is implemented in a post processing application there is no need for an online solution.

3.1 Unscented Kalman filtering

The main condition for estimating the curvature profile of a road segment is the continuous curvature along the whole path. That requirement is satisfied by matching the path trough a set of clothoids. A fundamental introduction into clothoid mathematics is given in Wilde (2009) and Shimizu et al. (2006). The special cases straight line and arc are also special cases of clothoids with $c_0 = 0$ and $\mathrm{dc} = 0$ for straight lines and $c_0 = \mathrm{const}$ and $\mathrm{dc} = 0$ for circle arcs. In road designing rules these two elements are connected by clothoids with a given $c_0 = \mathrm{const}$ and $\mathrm{dc} = \mathrm{const}$. That means, that the whole path can be described through a set of consecutive clothoids. The task is, to consecutively put the position measurements to the filter and let it decide weather to set up a new clothoid or to assign it to the current one.

3.1.1 Measurement model

The measurement model input are position measurements x_n in Universal Transverse Mercator (UTM) coordinate system[1] $x_{\mathrm{meas}}, y_{\mathrm{meas}}$. In order to transform the position measurements of the positioning module into parameters, the following measurement model is used. The state vector $\boldsymbol{X}$ consists of the clothoid parameters g_0, c_0 and dc. The task of the measurement model is to transform the measurements into the model, which parameters should be determined. As shown in Fig. 3 the corresponding point s on the clothoid to the point x_n must be calculated. The given clothoid in the image is generated from the UKF which proposes several values for $\boldsymbol{X}$. The distance between x_n and s is a measure for the fittnes of the proposed clothoid. That approach leads to promising results, but the curvature transition between the estimated clothoids is not continuous and the real transition points are not being recognized correctly.

To meet the continuous curvature condition the curvature at the end of the last clothoid must be equal to the curvature of the beginning of the new clothoid. One approach is to extend the measurement model to a set of two clothoids with a state vector containing the parameters of both clothoids. There are many approaches like in Schwartz (2003) that stated, data fusion algorithms are not able to estimate a set of parameters fitting the clothoid. It was shown that, due to the strong nonlinearity of the filtering problem the UKF is not able to converge, too. Therefore, the condition is implemented as shown in Gackstatter et al. (2010). The idea is to estimate two consecutive clothoids and then to fit the transition curvatures as proposed. That approach has the advantage to estimate clothoid elements, that are fitting well the given path. The disadvantage is, that it cannot detect the transition points between the original elements, correctly. That means a given path of several known consecutive elements is estimated to have more short elements than less long elements because they are fitting the given trajectory much better, due to more degrees of freedom. In order to find the correct number of elements requires an algorithm that gets all position measurements at once. The most common approach is to optimize the required parameters for a set of given measurements with the help of an optimizer (i.e. nelder-mead). Unfortunately, due to the unknown number of elements these approaches do also not deliver sufficient results. The approach introduced in the next section however, can handle all the requirements.

[1] Global coordinate system with plane cartesian projection.

Fig. 2. A cornerstone of the measurement system is the postprocessing application, where the road profiles are calculated. The image shows a screenshot of the camera images, where the laserscans are transformed to. A well-known RANdom SAmple consensus (RANSAC) implementation delivers the slope of the road from every laserscan. All slope measurements of a given road segment together are the input for the estimation algorithm which delivers a parametric description of the entire course of the segment.

Table 1. Profiles of the road, which parameters are estimated along the runlength of the given set of measurements (segment).

profile	description
curvature	The curvature profile contains the elements straight line, circle arcs and clothoids, the curvature values are calculated from the positioning module and are plotted over the length of the segment
height	Contains the elements straight line and circle arcs, the height values are derived from the height information of the positioning module and are plotted over the length of the segment
width	The width profile contains straight lines, the width values are calculated from the lane mark detection system and are plotted over the length of the segment
slope	The slope profile contains straight lines, the slope values are calculated from the laserscans and are plotted over the length of the segment

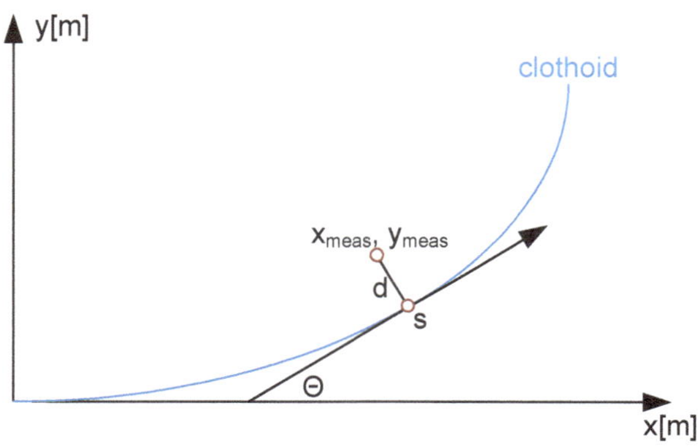

Fig. 3. The image depicts the measurement model of the Unscented Kalman Filter. The point $x_{\text{meas}}, y_{\text{meas}}$ is the position of the measurement in local clothoid coordinates, s is the corresponding base point of the measurement on the clothoid and θ is the gradient of the clothoid in s.

3.2 Hough transform

The Hough transform is a feature extraction technique used in image analysis, computer vision, and digital image processing. The purpose of the technique is to find imperfect instances of objects within a certain class of shapes by a voting procedure. This voting procedure is carried out in a parameter space, from which object candidates are obtained as local maxima in a so-called accumulator space that is explicitly constructed by the algorithm for computing the Hough transform. An illustrating application of the Hough transform is shown in Fardi und Wanielik (2004) where it is used for road border detection in infrared images. The idea behind the Hough transform was, to not directly estimate the parameters of the clothoids but to calculate the curvature in every point of the measurement set and then estimate the transition points from the resulting curvature graph. For that approach the Hough transform is implemented as edge-detection in images. Figures 4 and 5 show the results of the Hough implementation. The upper images show the transformed curvature measurements of a segment from the A72 and the L288. These images are the basis for the Hough Transformation. The lower images show the Hough image. Every segment in the Hough image can now be parametrically described as part of a linear function (How the clothoids are obtained from the linear segments is described in Sect. 3.3.

Due to the resulting elements in the Hough image are ambiguous that approach is also not suitable for a fully automated evaluation of several road sections.

3.3 RANSAC implementation

In RANSAC, the goal is to determine a set of inliers from an input data set of points. Inliers are data points consistent with the best model. Inliers and their corresponding model can be estimated optimally by repeatedly drawing samples from the input data set. A fundamental introduction was given in Fischler und Bolles (1981) and some common examples are proposed in Xu und Lu (2012). In order to find the dedicated segments in the given data sets of different road profiles, different models were implemented, that fit the data points. In

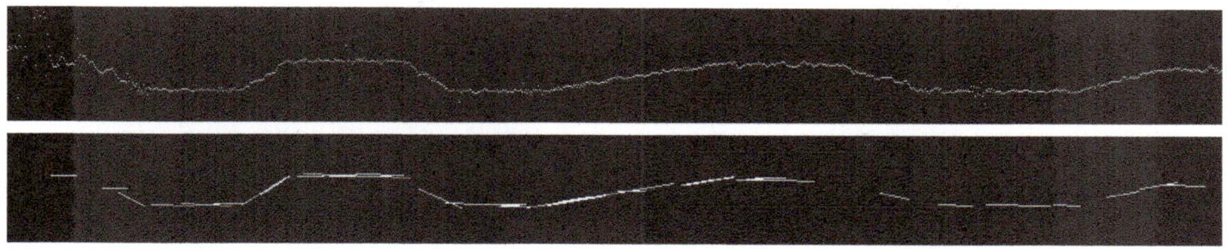

Fig. 4. The upper image shows the transformed curvature measurements of a segment from the A72. That image is the basis for the Hough Transformation. The lower image shows the image after the Hough Transformation. Every part of the image can now be parametrically described as part of a linear function.

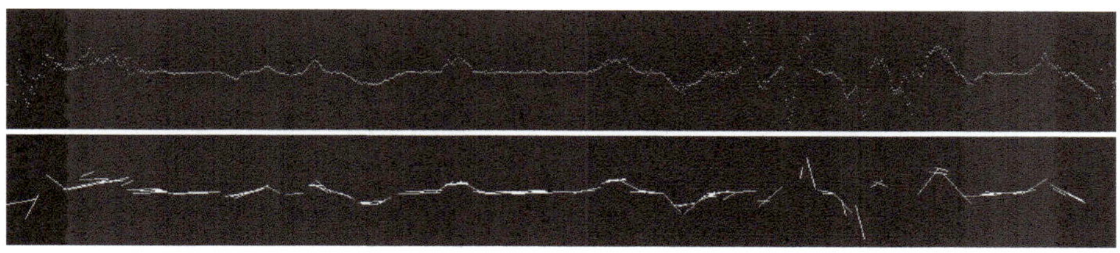

Fig. 5. The upper image shows the transformed curvature measurements of a segment from the L288. That image is the basis for the Hough Transformation. The lower image shows the image after the Hough Transformation. Every part of the image can now be parametrically described as part of a linear function.

the case of the curvature profile it is a quadratic fit model. In the case of the height profile it is a circle fit model and in the case of the slope and width profile a simple linear fit model is implemented. The following lines show the application of the RANSAC implementation on the curvature profile. Due to the strong nonlinearity of the operation $c = \frac{1}{r}$ in every point the calculation of the curvature of a given set of position measurements will always cause a very noisy plot of the curvature along the run length of the segment. This transformation is shown in the upper plot of Fig. 4. Obviously, that result makes it hard for any segmentation algorithm to determine the containing elements. Due to the noisy graph of the calculated curvature profile, the alternate approach is to obtain the curvature profile from the heading information of the positioning module. An example of the heading measurements is shown in Fig. 6. It is composed of different polynomials of 2nd order where the the curvature information can be derived from. The implemented RANSAC algorithm is able to find these polynomials in the following manner. Randomly heading measurements are drawn from the whole set of measurements until the best fitting polynomial (it is always the largest element in the group) is found. All measurements belonging to the element are deleted from the measurement list and the algorithm starts from the beginning until all measurements are erased. Figure 7 shows an overlay of Fig. 6 with the result of the RANSAC algorithm, where the 2nd order polynomials $h_n(x) = A_n x^2 + B_n x + C_n$ are plotted. The derivates of these polynomials are linear functions $c_n(x)$ where each is representing a clothoid with the following parameters. c_0 is the function value of $c_n(s)$ where s is the length on the segment in the beginning of the current element and dc is the gradient of $c_n(x)$. Figure 8 shows $c_0(x)$ to $c_{28}(x)$ of the given example.

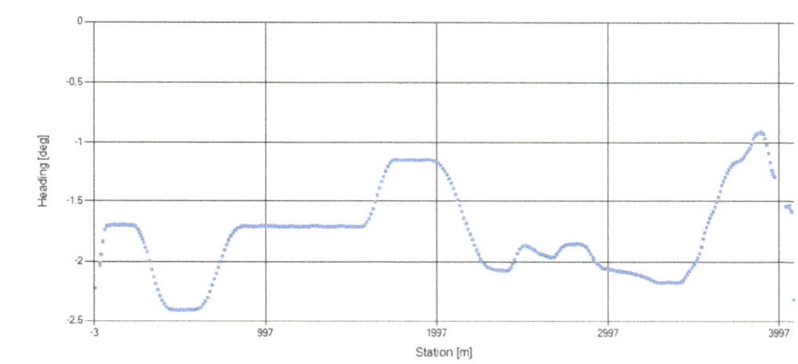

Fig. 6. The diagram shows the heading measurements of the positioning module. In contrast to the calculated curvature profiles, the heading information is a direct output of the positioning module and has a high quality of accuracy.

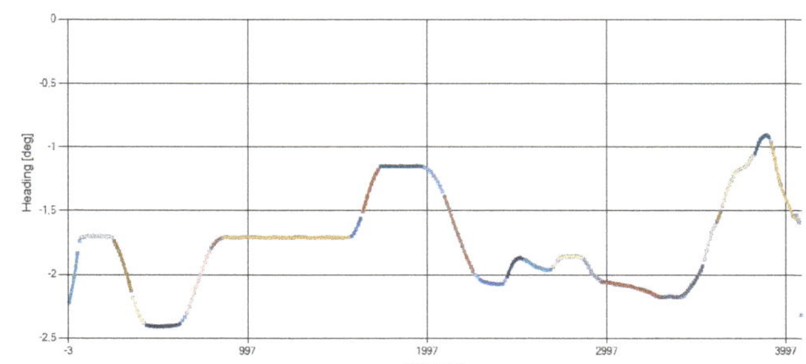

Fig. 7. The overlay to Fig. 6 shows the 2nd order polynomials as a result of the RANSAC algorithm.

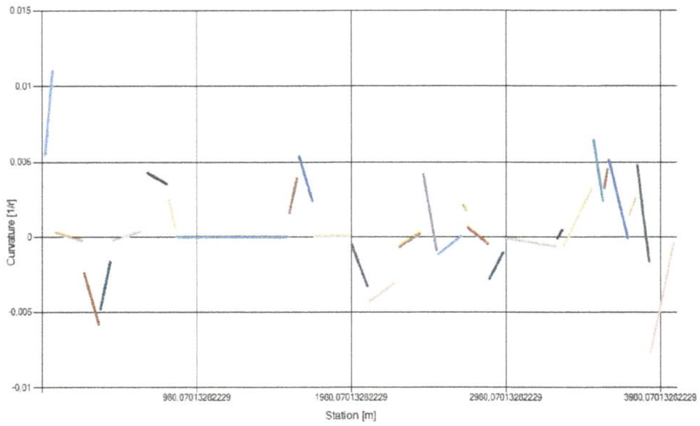

Fig. 8. The derivatives of the polynomials shown in Fig. 7 are shown in that diagram. The resulting graph is unambiguous and does – after some simple connection heuristic is applied – directly represent the curvature profile of the given road segment.

4 Conclusions

After an heuristic postprocessing of the generated plots of the RANSAC implementation (Application of some designing rules for roads), the results can directly (without any human revision) be included into the road database. The road segmentation algorithms and the road representation models will be applied in future research for robust relative positioning. Proposed road segments and the knowledge from the complex road models will be used to estimate the course of roads for several seconds and integrate that knowledge to new lane mark detection algorithms and for position smoothing of low cost GNSS receivers.

Acknowledgements. This work was done in the order of the Federal Highway Research Institute (BASt) as a sub-department of the Federal Ministry of Traffic, Building and Urban Development. The authors would like to thank the BASt and all partners of the project, namely TRANSVER GmbH, Chemnitz University of Technology, Ruetz Technologies and Intenta GmbH for the good cooperation and the individual contributions. For the evaluation and generation of the ground truth, precise realtime corrections provided by ascos (http://www.ascos.de) were used. Data recording and playback framework is based on BASELABS Connect and implementation of probabilistic filter algorithms are based on BASELABS Create (www.baselabs.de).

References

Fardi, B. and Wanielik, G.: Hough transformation based approach for road border detection in infrared images, Intelligent Vehicles Symposium, 2004 IEEE, 549–554, June 2004.

Fischler, M. A. and Bolles, R. C.: Random sample consensus: a paradigm for model fitting with applications to image analysis and automated cartography, Commun. ACM 24, 381–395, doi:10.1145/358669.358692, 1981.

Gackstatter, C., Heinemann, P., Thomas, S., Rosenhahn, B., and Klinker, G.: Fusion of clothoid segments for a more accurate and updated prediction of the road geometry, Intelligent Transportation Systems (ITSC), 2010 13th International IEEE Conference on, 1691–1696, ISSN 2153-0009, September 2010.

Schwartz, D. A.: Clothoid road geometry unsuitable for sensor fusion clothoid parameter sloshing, Intelligent Vehicles Symposium, Proceedings, IEEE, June 2003, 484–488, 2003.

Shimizu, M., Kobayashi, K., and Watanabe, K.: Clothoidal Curve-based Path Generation for an Autonomous Mobile Robot, SICE-ICASE, 2006, International Joint Conference, 478–481, October 2006.

Streiter, R., Mattern, N., and Wanielik, G.: Development of a multisensor road surveying measurement system, Systems, Signals and Devices (SSD), 2012 9th International Multi-Conference on, 1–5, March 2012.

Wilde, D. K.: Computing clothoid segments for trajectory generation, Intelligent Robots and Systems, 2009, IROS 2009, IEEE/RSJ International Conference on, 2440–2445, October 2009.

Xu, M. and Lu, J.: Distributed RANSAC for the robust estimation of three-dimensional reconstruction, Computer Vision, IET, 6, July, Nr. 4, 324–333, ISSN 1751-9632, 2012.

11

A wireless MIMO CPM system with blind signal separation for incoherent demodulation

O. Weikert and U. Zölzer

Helmut Schmidt University – University of the Federal Armed Forces Hamburg, Department of Signal Processing and Communications, Holstenhofweg 85, 22043 Hamburg, Germany

Abstract. A multiple-input multiple-output (MIMO) wireless transmission system with continuous phase modulation (CPM) is considered. A novel MIMO CPM receiver with incoherent modulation is presented. The incoherent demodulation of CPM allows an uncomplicated handling of a frequency offset compared to coherent approaches requiring exact knowledge of the carrier frequency. Blind signal separation (BSS) is applied in the proposed MIMO CPM receiver to separate the signals without any knowledge of the MIMO channel. The BSS permits the demodulation of each separated signal by an incoherent CPM demodulator. For bandwidth efficient transmission partial response CPM and non-binary modulation is applied. The applicability of the proposed system is verified by simulation results.

1 Introduction

Continuous phase modulation (CPM) offers a remarkable amount of benefits. The constant envelope is well suitable for high frequency components. Incoherent demodulation of CPM allows an uncomplicated handling of frequency offsets compared to coherent approaches requiring exact knowledge of the carrier frequency. Multiple-input multiple-output (MIMO) systems provide higher data rates that data-demanding applications require. Applying continuous phase modulation in wireless MIMO systems provides the opportunity to combine the benefits of higher data rates with the capability of incoherent demodulation. As CPM belongs to the class of non-linear modulation schemes the separation of the mixed receive signals is a well-known challenge.

Training based channel estimation is not directly applicable for MIMO systems with continuous phase modulation. Regarding the training symbols for linear modulation schemes the principle of superposition holds. This is not the case for nonlinear modulation schemes like continuous phase modulation. At the MIMO CPM receiver each receive antenna receives a linear superposition of all transmit signals. However these receive signals are not a linear superposition of the training symbols. The training symbols have to be inserted into the transmit signal prior to the nonlinear modulation. The circumstance is explained in detail later on.

Binary CPM signals can be represented as a sum of a finite number of amplitude modulated pulses (Laurent, 1986). Using the decomposition in Laurent (1986) a layered MIMO system with binary CPM signals is proposed in Zhao and Giannakis (2005). The system in Zhao and Giannakis (2005) is a coherent demodulator which requires exact knowledge of the carrier frequency. The demodulation has a very high complexity due to the representation of CPM signals by amplitude modulated pulses. In the MIMO CPM receiver in Zhao and Giannakis (2005) the perfect knowledge of the MIMO channel is assumed. The practical realization of any MIMO CPM system requiring knowledge of the MIMO channel in the receiver is questionable as training based channel estimation is not applicable for MIMO CPM systems.

Space-Time Codes for MIMO CPM are proposed with its design criteria in Zhang and Fitz (2003), a soft decision iterative receiver is described (Zhang and Fitz, 2002). The system in Zhang and Fitz (2002) is a coherent demodulator which requires exact knowledge of the carrier frequency. The MIMO CPM receivers with Space-Time Codes have a high complexity.

In this paper a MIMO CPM receiver with an incoherent demodulation of the CPM signals is presented. Blind signal separation (BSS) is applied in the MIMO CPM receiver to separate the transmitted signals. A carrier frequency mismatch in the receiver can be handled by the BSS. The BSS permits the demodulation of each separated signal by an incoherent CPM demodulator. On the separated signals classical algorithms for timing and carrier frequency synchronization can be applied. In the proposed MIMO CPM receiver neither any knowledge of the MIMO channel nor the exact carrier frequency is required. The system can cope with any modulation filter and is not restricted to binary modulation. For bandwidth efficient transmission partial response

Correspondence to: O. Weikert
(oomke.weikert@hsu-hamburg.de)

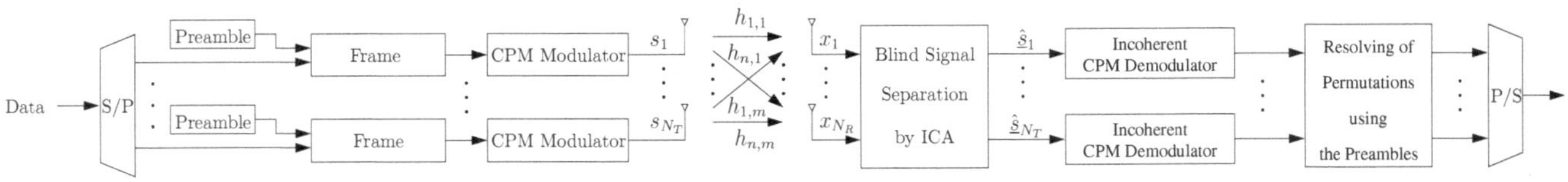

Fig. 1. Proposed wireless MIMO CPM system with incoherent demodulation.

CPM and non-binary modulation is applied. As the order of the transmitted signals cannot be determined by blind signal separation the permutations are resolved by transmission of preamble symbols.

The paper is organized as follows. The MIMO system and the MIMO channel model with the frequency offset introduced is described in Sect. 2. The MIMO receiver for continuous phase modulation with blind signal separation and incoherent demodulation is presented in Sect. 3. Simulation results are given in Sect. 4. A summary and conclusion marks can be found in Sect. 5.

2 System description

2.1 MIMO CPM transmitter

We consider a MIMO wireless system with N_T transmit and N_R receive antennas (see Fig. 1). The serial bit stream is split into N_T parallel substreams, one for every transmit antenna. The substreams are organized in frames. Each frame of length N_F consists of N_I payload bits and N_P preamble bits. The preamble sequences are orthogonal to each other. The CPM modulators generating the transmit signals are described in the next section. The signal substreams are subsequently transmitted over the N_T antennas at the same time. The symbol transmitted by antenna m at time instant k is denoted by $s_m(k)=e^{j\varphi_m(k)}$. The transmitted symbols are arranged in vector

$$\begin{aligned} s(k) &= [s_1(k),\ \ldots,\ s_{N_T}(k)]^T \\ &= [e^{j\varphi_1(k)},\ \ldots,\ e^{j\varphi_{N_T}(k)}]^T \end{aligned} \tag{1}$$

of length N_T, where $(\cdot)^T$ denotes the transpose operation.

2.2 CPM modulator

With continuous phase modulation (CPM) a constant signal envelope with a continuous phase are obtained. The CPM signal is described by Anderson et al. (1986)

$$s(t) = e^{j(\varphi(t)+\varphi_0)} \tag{2}$$

with an arbitrary phase offset φ_0 and the instantaneous phase

$$\varphi(t) = 2\pi h \int_0^t \sum_{i=0}^{l} d(i) g(\tau - iT)\, d\tau \tag{3}$$

with modulation index h, the transmit filter $g(t)$, the symbols $d(l)$ and the symbol interval T. The discrete CPM modulator is given in Fig. 2. The symbols $d(l)$ are taken from the M-ary alphabet $d(l)\in\{\pm 1, \pm 3, ..., \pm(M-1)\}$ with M even.

Before pulse shaping the symbols are upsampled by the factor SPS specifying the number of samples per symbol. Between two successive symbols $SPS-1$ zeros are inserted. The transmit filter

$$g(k) = \begin{cases} \frac{1}{2L\cdot SPS}\left(1-\cos\left(\frac{2\pi k}{L\cdot SPS}\right)\right) & 0 \le k \le L\cdot SPS \\ 0 & \text{elsewise} \end{cases} \tag{4}$$

is a raised cosine impulse (L-RC) (Anderson et al., 1986) with the length of the impulse L in symbol intervals. The transmit filter fulfills the normalization condition

$$\sum_{k=0}^{L\cdot SPS} g(k) = \frac{1}{2}. \tag{5}$$

A full response CPM has a transmit filter of length $L{=}1$. With overlapping transmit pulses that is a transmit filter with $L{>}1$ the scheme is denoted as partial response. By introducing symbol interferences a higher bandwidth efficiency is obtained (Anderson et al., 1986). After pulse shaping the signal is multiplied by $2\pi h$ and integrated.

2.3 MIMO channel

The non-frequency selective MIMO channel can be described by a complex channel matrix

$$\mathbf{H} = \begin{bmatrix} h_{1,1} & \cdots & h_{1,N_T} \\ \vdots & \ddots & \vdots \\ h_{N_R,1} & \cdots & h_{N_R,N_T} \end{bmatrix} \tag{6}$$

of the dimension $N_R \times N_T$. We suppose that the channels remain constant over the transmission of a frame and vary independently from frame to frame (*block fading channel*). The elements $h_{n,m}$ are complex random variables with a Gaussian distributed real and imaginary part, zero mean and variance σ^2. The element $h_{n,m}$ includes also the direct component with the amplitude $p_{n,m}$ and the power $p_{n,m}^2{=}c_R{\cdot}\sigma^2$, where c_R is the Rice factor.

The channel energy is normalized by the condition $E\left\{|h_{n,m}|^2\right\}=p_{n,m}^2+\sigma^2:=1$. We assume additive white Gaussian noise (AWGN) with zero mean and variance σ_n^2 per receive antenna.

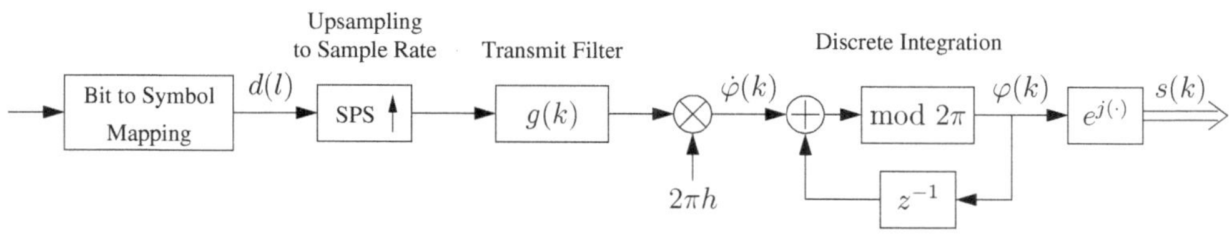

Fig. 2. CPM modulator.

2.4 Frequency offset

The discrepancy of the carrier frequency at the receiver from the one used at the transmitter is specified by the frequency offset Δf. The impact is described in the continuous case by the multiplication of the signal by $e^{j2\pi\ \Delta f\ t}$. By sampling of the signal the frequency offset Δf is related to the sampling frequency $f_A{=}1/T_A$ and the normalized frequency offset $\Delta F{=}\Delta f/f_A{\in}[-0.5, 0.5]$ is introduced. Due to the sampling theorem the determinable range of the frequency offset is limited to $[-\frac{f_A}{2}, +\frac{f_A}{2}]$.

2.5 Receive signals

The symbol received by antenna n at time instant k is denoted by $x_n(k)$. Calculation of $x_n(k)$ can be expressed as a discrete sum

$$x_n(k) = e^{j2\pi\Delta Fk} \sum_{m=1}^{N_T} h_{n,m}\, e^{j(\varphi_m(k)+\varphi_0)} + n(k), \quad (7)$$

with $\varphi_m(k)$ as a function of the symbols $d(l)$ where $d(l)$ denote the symbols in front of the CPM modulator. $n(k)$ is the additive white Gaussian noise at time instant k. As the receive signals $x_n(k)$ depend in a nonlinear manner on the symbols $d(l)$, linear methods for channel estimation are not applicable.

The symbols received by the N_R antennas are arranged in a vector $\boldsymbol{x}(k){=}[x_1(k),\ \ldots,\ x_{N_R}(k)]^T$ of length N_R, which can be expressed with Eqs. (1), (6) and $\boldsymbol{n}(k)$ as noise vector of length N_R as

$$\boldsymbol{x}(k) = e^{j2\pi\Delta Fk}\, \mathbf{H}\, \mathbf{s}(k) + \boldsymbol{n}(k). \quad (8)$$

3 MIMO CPM receiver with incoherent demodulation

The MIMO receiver for continuous phase modulation in Fig. 1 uses blind signal separation to estimate and separate the independent components, i.e. the transmitted signals. The blind signal separation is performed at the sampling rate f_A which is by the factor SPS higher than the symbol rate. A frequency offset is supposed. Each separated signal corresponds to one of the N_T transmitted signals. The BSS permits the demodulation of each separated signal by an incoherent CPM demodulator. The incoherent CPM demodulator does not require any knowledge about the frequency offset. As the order of the transmitted signals cannot be determined by blind signal separation preamble sequences are transmitted to resolve the permutations. Subsequently the transmitted signals are extracted from the frames of the N_T parallel substreams.

3.1 Blind signal separation

3.1.1 Whitening

The whitening transformation

$$\mathbf{z} = \mathbf{V}\mathbf{x} = \mathbf{V}\mathbf{H}\ \underline{\mathbf{s}} \quad (9)$$

with

$$\underline{\mathbf{s}}(k) = e^{j2\pi\Delta Fk}\ \mathbf{s}(k) \quad (10)$$

$$\underline{s}_m(k) = e^{j2\pi\Delta Fk}\ s_m(k) \quad (11)$$

and the linear transformation matrix $\mathbf{V}$ (Hyvärinen et al., 2001) of dimension $N_T{\times}N_R$ decorrelates the observed mixtures $\mathbf{x}$. The number of observed mixtures N_R is reduced to the number of transmitted signals N_T.

3.1.2 Independent Component Analysis

Independent Component Analysis (ICA) separates the observed mixtures $\mathbf{z}$ of the mixing model

$$\mathbf{z} = \mathbf{V}\mathbf{H}\ \underline{\mathbf{s}} \quad (12)$$

into the symbols $\underline{\mathbf{s}}$, called independent components (IC), while the mixing matrix $\mathbf{VH}$ is unknown.

Using ICA the following assumptions hold (Hyvärinen et al., 2001):

- ICs must have *nongaussian* distributions (e.g. leptokurtic or platykurtic)
- ICs have to be statistically *independent*
- ICs are assumed to have zero mean
- The number of observed mixtures has to be at least equal to the number of ICs

The below-mentioned ambiguities regarding the estimated ICs remain (Hyvärinen et al., 2001):

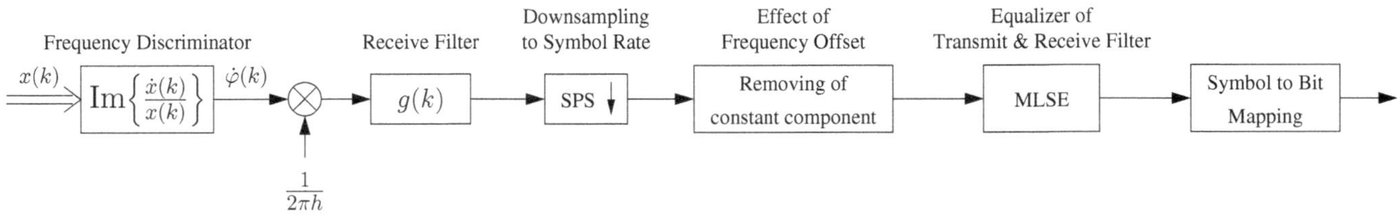

Fig. 3. Incoherent CPM demodulator.

- The estimated ICs can be arbitrarily permuted
- The phases of the estimated ICs are unknown
- The variances (energy) of the estimated ICs can not be determined

Ordinary complex ICA algorithms like JADE (Cardoso and Souloumiac, 1993; Hyvärinen et al., 2001) are applied to obtain estimates of the separated signals $\underline{\hat{s}}_m(k)$. The ICA algorithms compute the unitary matrix $\mathbf{W}^H$ to calculate an estimate

$$\underline{\hat{\mathbf{s}}} = \mathbf{W}^H \mathbf{z} = \mathbf{W}^H \mathbf{V} \mathbf{H}\ \underline{\mathbf{s}} \tag{13}$$

of the independent components. As the channel is assumed to be a block fading channel the received data is processed in frames. Recalling the restrictions of ICA, the estimated ICs are arbitrary permuted.

3.2 Incoherent CPM demodulator

The incoherent CPM demodulator is given in Fig. 3. The frequency discriminator demodulates the receive signal. The receive filter corresponds to the transmit filter $g(k)$ in Eq. (4). The receive filter is followed by the downsampling to symbol rate. The impact of the frequency offset can be eliminated by removing the signals constant component. The constant component is estimated and subtracted from the signal. Based on the receive signal $x(t){=}s(t){\cdot}e^{j2\pi\ \Delta f\ t}$ the signal after the frequency discriminator can be written as

$$\begin{aligned}
&\mathrm{Im}\left\{\frac{\dot{x}(t)}{x(t)}\right\} \\
&= \mathrm{Im}\left\{\frac{j\dot{\varphi}(t)s(t)e^{j2\pi\Delta f t}+s(t)j2\pi\Delta f e^{j2\pi\Delta f t}}{s(t)e^{j2\pi\Delta f t}}\right\} \\
&= \dot{\varphi}(t) + 2\pi\ \Delta f
\end{aligned} \tag{14}$$

with $\dot{\varphi}(t){=}2\pi h \sum_{i=0}^{l} d(i) g(t{-}iT)$. The frequency offset causes a constant component in the demodulated signal. In the discrete case the signal after the frequency discriminator follows to

$$\mathrm{Im}\left\{\frac{\dot{x}(k)}{x(k)}\right\} = \dot{\varphi}(k) + 2\pi\ \Delta F \tag{15}$$

with the normalized frequency offset ΔF.

Using a partial response CPM with overlapping transmit pulses the received signal has to be equalized. For the equalization of the overlapping transmit pulses a maximum likelihood sequence estimation (MLSE) implemented as a Viterbi equalizer is used. For a full response CPM no equalizer is needed since no overlapping transmit pulses are existing.

3.3 Resolving permutations

The ambiguities left by blind signal separation are solved by orthogonal preamble sequences (Weikert and Zölzer, 2006). The preambles are the orthogonal rows of a Hadamard matrix $\mathbf{S}_P$. Hadamard matrices are binary orthogonal matrices. To ensure the orthogonality the preamble length N_P per transmit antenna should be $N_P{\geq}N_T$. The received preambles are extracted and inserted line by line in the matrix $\hat{\mathbf{S}}_P$. A hard decision is applied, subsequently the preamble matrix is labeled as $\tilde{\mathbf{S}}_P$. The unitary cross correlation matrix

$$\mathbf{R}_{\tilde{\mathbf{S}}\mathbf{S}} = \frac{1}{N_P}\tilde{\mathbf{S}}_P \mathbf{S}_P^H, \tag{16}$$

where $(\cdot)^H$ denotes the complex-conjugate (Hermitian) transpose, will be an identity matrix if no ambiguities exist. Elsewise the ambiguities are resolved by

$$\mathbf{s}(k) = \mathbf{R}_{\tilde{\mathbf{S}}\mathbf{S}}^{-1} \cdot \hat{\mathbf{s}}(k) = \mathbf{R}_{\tilde{\mathbf{S}}\mathbf{S}}^{H} \cdot \hat{\mathbf{s}}(k). \tag{17}$$

4 Simulation results

The simulations are performed using a non-frequency selective MIMO system with a line of sight between the transmit and receive antennas. Thereby the Rice coefficient is chosen to be $c_R{=}1$. For the MIMO CPM system $M{=}4$-ary symbols, a modulation index $h{=}0.5$ and $SPS{=}8$ samples per symbol are chosen. The blind signal separation is performed with the JADE algorithm, $N_F{=}100$ symbols per frame i.e. 800 samples per frame. For 2 and 4 transmit antennas $N_P{=}2$ and 4 preamble symbols per antenna are used respectively. The discrete differentiation is performed by a FIR filter of order 3.

In Figs. 4 and 5 one can see the Bit Error Rate (BER) versus the Signal-to-Noise-Ratio (SNR) for the MIMO CPM system with incoherent demodulation. The results are shown for a normalized frequency offset of $\Delta F{=}0.05$ and for zero

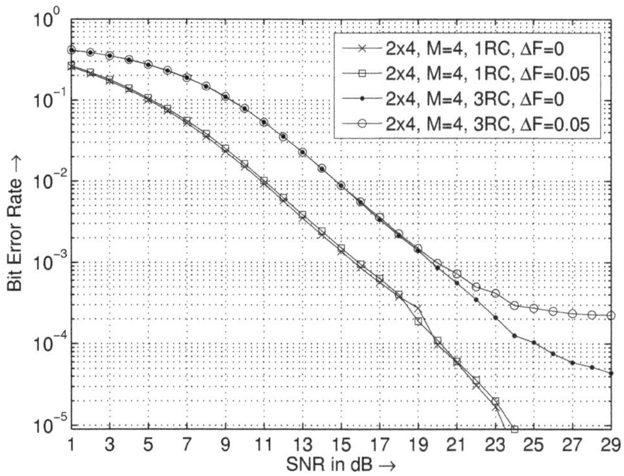

Fig. 4. Bit Error Rate (BER) versus the Signal-to-Noise-Ratio (SNR) for the MIMO CPM system with N_T=2 transmit and N_R=4 receive antennas.

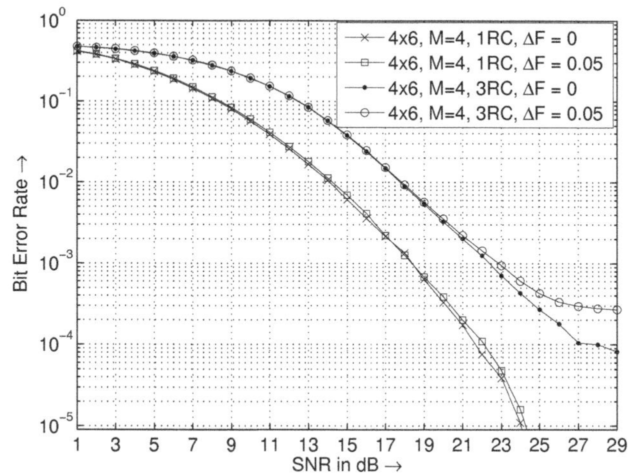

Fig. 5. Bit Error Rate (BER) versus the Signal-to-Noise-Ratio (SNR) for the MIMO CPM system with N_T=4 transmit and N_R=6 receive antennas.

frequency offset. For bandwidth efficient transmission a partial response MIMO CPM (3RC) with a raised cosine transmit filter of length 3 (in symbol intervals) is used. Furthermore the results for a full response MIMO CPM (1RC) are shown. In Fig. 4 the results with N_T=2 transmit and N_R=4 receive antennas (2×4) are presented. Figure 5 shows the results with N_T=4 transmit and N_R=6 receive antennas (4×6). In comparison to a GSM system with binary symbols and a single transmit antenna, the data rate is increased by a factor of eight for the MIMO CPM system with 4 transmit antennas and M=4-ary symbols.

The presented MIMO CPM system achieves a Bit Error Rate less than $1\cdot10^{-5}$ for a full response MIMO CPM (1RC) with a SNR higher than 25 dB even in the presence of a frequency offset. An increased frequency offset results in a slightly increased Bit Error Rate. A bandwidth efficient partial response MIMO CPM (3RC) requires a 5 dB increased SNR to achieve a Bit Error Rate of $5\cdot10^{-3}$ compared to the full response MIMO CPM.

5 Conclusions

We proposed a semi-blind MIMO wireless system with continuous phase modulation. The presented MIMO CPM receiver separates the received mixtures into the transmitted CPM signals without knowledge of the MIMO channel. The separated signals are demodulated with incoherent CPM demodulators. The presented MIMO CPM system achieves a Bit Error Rate less than $1\cdot10^{-5}$ for a full response MIMO CPM (1RC) with a SNR higher than 25 dB. In case of a carrier frequency mismatch the approach shows almost the same performance.

References

Anderson, J. B., Aulin, T., and Sundberg, C.-E.: Digital Phase Modulation, Plenum Press, London, 1986.

Cardoso, J.-F. and Souloumiac, A.: Blind Beamforming for Non Gaussian Signals, in: IEE-Proceedings-F, 140(6), 362–370, 1993.

Hyvärinen, A., Karhunen, J., and Oja, E.: Independent Component Analysis, John Wiley & Sons, New York, 2001.

Laurent, P. A.: Exact and Approximate Construction of Digital Phase Modulations by Superposition of Amplitude Modulated Pulses (AMP), IEEE Transactions on Communications, 34, 150–160, 1986.

Weikert, O. and Zölzer, U.: New Approach for Resolving Ambiguities for Semi-blind Equalization of MIMO Frequency Selective Channels, in: Proc. of 17th Annual IEEE International Symposium on Personal, Indoor and Mobile Radio Communications (PIMRC'06), Helsinki, Finland, 2006.

Zhang, X. and Fitz, M.: Soft-Output Demodulator in Space-Time-Coded Continuous Phase Modulation, IEEE Transactions on Signal Processing, 50, 2589–2598, 2002.

Zhang, X. and Fitz, M.: Space-Time Code Design with Continuous Phase Modulation, IEEE Journal on IEEE Journal on Selected Areas in Communications, 21, 783–792, 2003.

Zhao, W. and Giannakis, G.: Reduced Complexity Receivers for Layered Space-Time CPM, IEEE Transactions on Wireless Communications, 4, 574–582, 2005.

12

Validation of a measurement method for magnetic shielding effectiveness of a wire mesh enclosure with comparison to an analytical model

M. Kühn[1], W. John[2], and R. Weigel[3]

[1]AUDI AG, 85045 Ingolstadt Germany
[2]SIL System Integration Laboratory GMBH, Technologiepark 32, 33100 Paderborn Germany
[3]Lehrstuhl für Technische Elektronik, Friedrich-Alexander-Universität Erlangen-Nürnberg, Cauerstrasse 9, 91058 Erlangen Germany

Correspondence to: M. Kühn (michael.kuehn@audi.de)

Abstract. This paper deals with the validation of a measurement method for determining the magnetic shielding effectiveness of a wire mesh enclosure in the frequency range from 10 kHz to 150 kHz. The comparison with an analytical model (parallel-plate-shield, see Kaden, 1959) for magnetic shielding effectiveness of wire mesh is also part of this contribution. To measure the shielding effectiveness of an enclosure in general, two steps are necessary:

1. Reference measurement of the incident electro-magnetic field.
2. Measurement of the incident electro-magnetic field in the enclosure by same conditions.

This method presented in the contribution uses a Helmholtz coil as magnetic field source, controlled by a signal generator and an amplifier in voltage mode. At first, a field meter was used to measure the frequency dependent magnetic field in the center of the Helmholtz coil. The magnetic field strength was also analytically calculated and compared to the measurement results. The next step was to measure the induction voltage caused by the magnetic field with a magnetic field probe, which was later used for the measurements of the shielding effectiveness either. A signal amplifier was needed to raise the signal-to-noise-ratio (SNR). The gain of the signal amplifier was also determined and measured. Two positions of the enclosure in the Helmholtz coil were considered, horizontal and vertical. The vertical position of the enclosure approaches the analytical model of Kaden (1959) closely and the results show a good match with the analytical model.

1 Introduction

Vehicles with electric drive have enormous potential to reduce the growing dependence on fossil fuels, for sustainable economic growth and to support environmentally-friendly energy usage. The German government has set a goal for Germany to be the market leader in the development of these technologies in the next years. However, huge technological challenges, which can only be addressed through an extraordinary degree of interdisciplinary research, are standing in the way of Germany achieving the above said goal. Apart from some individual solutions that represent a certain level of vehicle functionality, the requirements for a mass market product are immense. Customer demands in reference to driving safety, comfort and quality, over and above, the perseverance of electromagnetic compatibility (EMC) are obligatory requirements (such as CISPR 25) which need to be met. The key challenge here is the electro-magnetic decoupling of the electric powertrain, which is composed largely of coaxial cables, connectors, a battery, the inverter and the electric motor. Within the machine, a rotating field is generated through the pulse-width modulated (PWM) voltages of the inverter, which typically have fundamental frequencies between the range of 8 kHz to 24 kHz, Michel (2011). Strong magnetic fields are the result with a wide spectrum up to the MHz range. The entire high-voltage wiring is performed in a completely closed shielding to comply with the requirements for electrical safety and the reduction interference emissions Hillmer et al. (2009). Conventional shielding against magnetic fields are made of solid metal plates with thicknesses from 1 mm upward, (Schwab and Kürner, 2010).

Evidently, the resulting weight gain is counterproductive to an efficient hybrid or electric vehicle. Therefore, lightweight fabrics, composite materials or wire mesh screens come to the fore. For predicting the shielding effectiveness of those materials it is necessary to validate measurement methods and their results with analytical models.

2 Magnetic shielding effectiveness

The magnetic shielding effectiveness is a part of the shielding effectiveness in the frequency domain. The frequency range is divided in two parts of shielding effectiveness. These are magnetic shielding effectiveness and electromagnetic shielding effectiveness. The magnetic shielding effectiveness represents the behaviour of the shield in the frequency range where the wavelength is large compared to the dimensions of the shield or enclosure. The magnetic field will be reduced by two physical processes (see Fig. 1), the generation of eddy currents in the shield and skin effect. The incident magnetic field leads to a voltage in the shield, which results in eddy currents in the shield. These eddy currents are generating an inverse oriented magnetic field, which weakens the incident magnetic field. The skin effect describes the reduction of the current density of a conductor or shield by raising frequency. In summary, the magnetic shielding effectiveness depends on the shield material (σ, μ_r) and the thickness of the shield. In the frequency range to be considered here, the incident electrical field could be neglected due to Faraday cage effect. When the wavelength reaches the dimensions of the shield or enclosure (approximately $\lambda/10$, see also Schwab and Kürner, 2010) the electromagnetical shielding effectiveness gets dominant. The shielding effectiveness is influenced by apertures and the shield dimensions itself. Therefore, resonances occur at discrete frequencies, which are represented by a drop in shielding effectiveness. The complete shielding behaviour in the frequency range is shown in Fig. 1, (Wolfsperger, 2008).

Two analytical models, based on a parallel-plate-shield from Kaden (1959), for calculating the magnetic shielding effectiveness in the frequency range of a homogeneous material and a wire mesh will be used for comparison.

3 State of the art

This chapter describes two analytical models selected from the literature. The main focus was laid on analytical models for calculating magnetic shielding effectiveness of enclosures. As above described, the wavelength is large compared to the dimensions of the enclosure or shield. Two major models from Kaden (Kaden, 1959) are described for a homogeneous material (for example: copper, aluminum) and a wire mesh screen. There are other models available for calculating the coupling through the apertures of the mesh, for example Casey (1988) or Wolfsperger (2008).

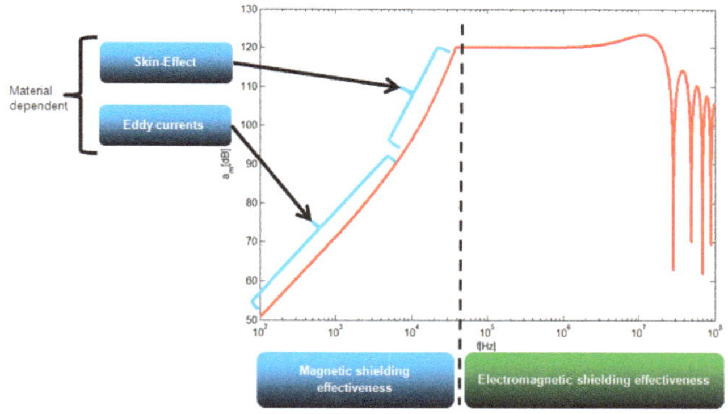

Fig. 1. Principle behaviour of shielding effectiveness a_m [dB] in the frequency domain.

3.1 Analytical model for homogeneous materials

At first we analyse the magnetic shielding effectiveness of a parallel-plate-shield based on homogeneous material defined by a model from Kaden (1959). The parallel-plate-shield represents a closed shield structure (like a conductor loop) with the condition, that the distance between the plates is even smaller than the length of the plates ($|y| >> x_0$, see Fig. 2). The model is based on the ratio of the magnetic field component inside and outside of the parallel-plate-shield. This model is only valid for the magnetic shielding effectiveness, so the wavelength is even larger than the screen dimensions. The skin effect is also considered by this model. The magnetic shielding effectiveness is calculated by Eq. (1) and (2) for a parallel oriented vector of the magnetic field to the parallel-plate-shield, so the incident field penetrates the conductor loop (Kaden (1959), page 74, equation 11, 12).

$$K = \frac{k_w \cdot x_0}{\mu_r} \tag{1}$$

$$a_s = 20 \cdot \log|\cosh(k_w d) + K \sinh(k_w d)| \tag{2}$$

The parameter k_w describes the eddy current constant, x_0 is the half distance between the plates and d is the shield thickness. The main parameters for influencing the shielding effectiveness are the conductivity (σ), the permeability (μ_r), both considered in the eddy current constant k_w, and the shield geometry. Figure 3 shows the magnetic shielding effectiveness of a parallel-plate-shield (copper and aluminum, $x_0 = 0.15$ m, $d = 1$ mm) with respect to frequency.

Figure 3 describes in the frequency domain ($100\,\text{Hz} \leq f \leq 1\,\text{MHz}$) the behaviour of the shielding effectiveness of a parallel-plate-shield structure penetrated by a magnetic field. The shielding effectiveness was limited to 120 dB, because higher values are not realistic and the electromagnetic shielding effectiveness was also not considered. Theoretically the

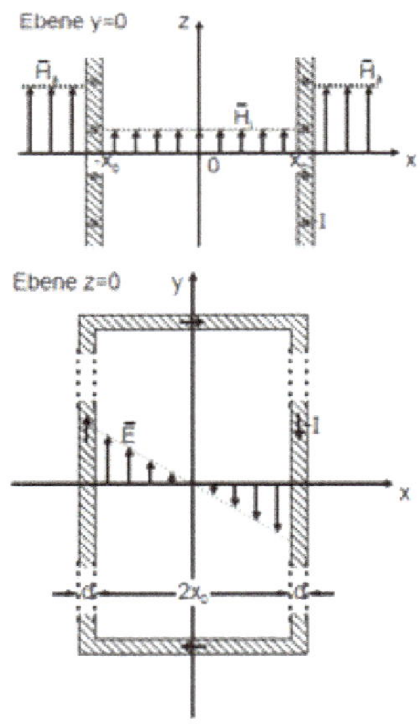

Fig. 2. Structure of a parallel-plate-shield for homogeneous materials (Kaden, 1959).

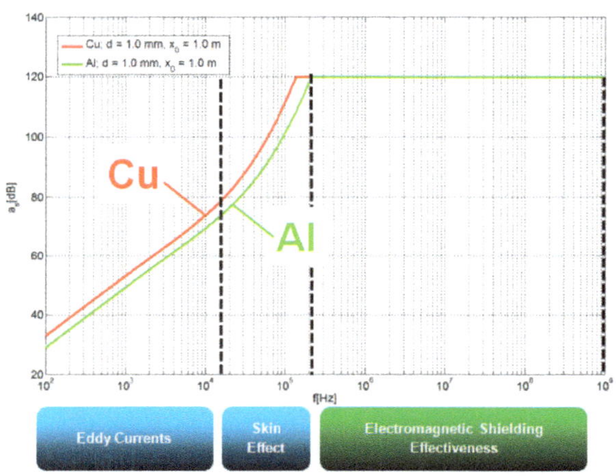

Fig. 3. Calculated shielding effectiveness of a parallel-plate-screen using the Kaden model.

magnetic shielding effectiveness would converge to infinity by raising frequency. Figure 3 shows also, that the magnetic shielding effectiveness of the copper parallel-plate-shield is approximately 3 dB higher than the shielding effectiveness of the same structure build by aluminum material. Another model of the parallel-plate-shield for homogeneous materials from Kistenmacher et al. (1995) is available and similar to the above described model from Kaden. The main difference is the consideration of the area and the circumference (ratio) instead of the distance between the plates.

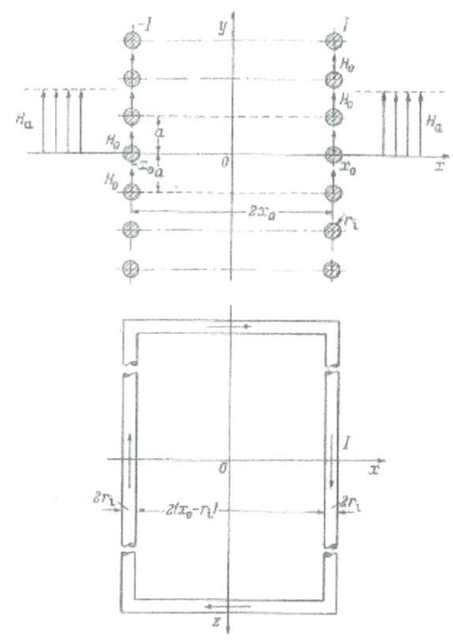

Fig. 4. Structure of a parallel-plate-shield based on wire mesh (Kaden, 1959).

3.2 Analytical model for wire mesh screens

In order to describe the magnetic shielding effectiveness of a parallel-plate-shield based on a wire mesh, a model from Kaden (1959) will also be used for comparison of measurement results. In this case the homogeneous material is substituted by a wire mesh. The wire mesh to be investigated consists of equidistant wires, which are defined about the distance to each other (mesh size) and the wire radius. Figure 4 shows the model structure of the parallel-plate-shield for a wire mesh used for the analysis. The physical shielding effects are similar to the parallel-plate-shield for homogeneous materials. The difference is, that opposite wires are forming a short connection, where eddy currents are induced and generating an inverse directed magnetic field, which weakens the incident magnetic field. The variation of the wire resistance and inductance in dependency of the frequency is also, due to the skin effect, part of the Kaden model (Kaden (1959), page 276, equation 16).

Figure 5 shows the calculated magnetic shielding effectiveness as a function of frequency. As in Fig. 3, two materials (copper and aluminum) are compared. The mesh size (a_s) is 1 mm and the wire radius (r_w) is 10 μm. As in Fig. 3 shown, copper has a higher magnetic shielding effectiveness in comparison to aluminum by the same shield conditions (approximately 4 dB). This is reasoned by the higher conductivity of copper. The field penetration through the mesh is not considered, the model is only valid for the magnetic shielding effectiveness determination. (The characteristic parameters of the parallel-plate-shield model for wire mesh are: the conductivity (σ), the permeability (μ_r), the distance between the plates (x_0), the mesh size (a_s) and the wire radius (r_w)).

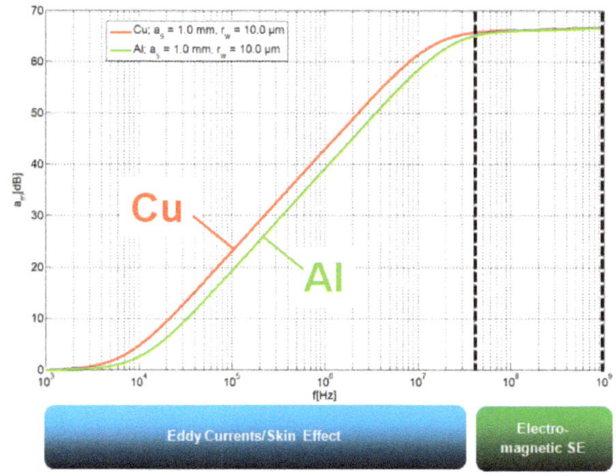

Fig. 5. Calculated shielding effectiveness of a wire mesh parallel-plate-shield using the Kaden model.

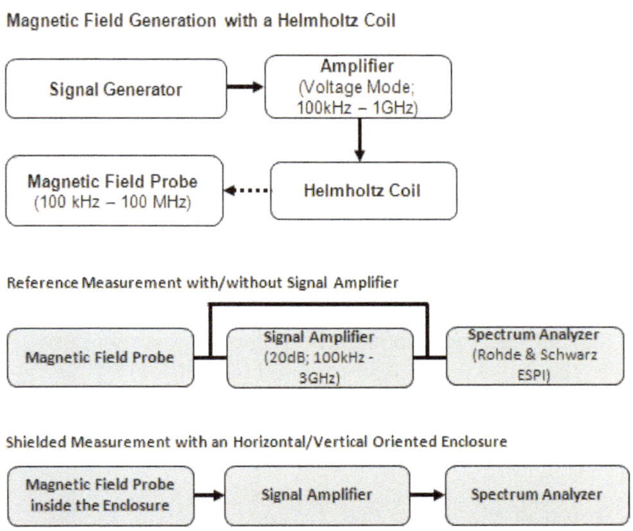

Fig. 6. Setup for shielding effectiveness measurements.

4 Measurement setup

This chapter describes the setup and the achieved results of the magnetic shielding effectiveness measurement. A Helmholtz coil was used to generate the magnetic field. The field generation was controlled by a signal generator and an amplifier in voltage mode. A constant amplitude (4 V, sine wave, peak-to-peak) in the frequency range from 10 kHz to 150 kHz was adjusted in the signal generator and boosted by the amplifier (40 dB voltage gain). Figure 6 shows a block diagram of the measurement setup for magnetic field generation by the Helmholtz coil, reference measurement and shielded measurement.

Consequently, there was no constant magnetic field amplitude. It was dependent on the impedance of the Helmholtz coil. The magnetic field in the center of the Helmholtz coil

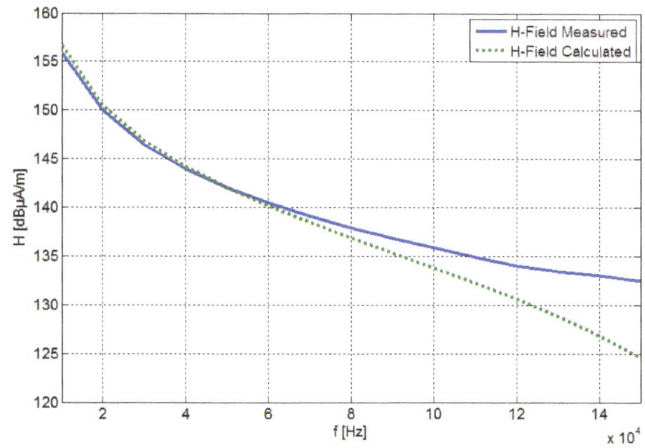

Fig. 7. Measured and calculated magnetic field in the center of the Helmholtz coil.

was measured with a field meter (solid line in Fig. 7) and analytically calculated (dotted line in Fig. 7). The difference of measured and calculated values occurs due to the complex resistance of the coil pair and the influence of the capacitive coupling between the layers of the coil. The complex resistance was calculated for this application with a simplified circuit model of the Helmholtz coil.

For the measurement of the induced voltage, a magnetic field probe from Rohde & Schwarz (6 cm diameter, specified frequency range: 100 kHz to 100 MHz) and a signal amplifier from Langer EMV (20 dB gain, specified frequency range: 100 kHz to 1 GHz) to raise the SNR were used. Additionally a spectrum analyzer also from Rohde & Schwarz was applied. The signal amplifier was validated with measurements of the induced voltage with and without the signal amplifier. In Fig. 8 the induced voltage measured with and without the signal amplifier as a function of frequency is visualized. The graph shows, that the induction voltage for both measurements is nearly constant, except for the lower frequencies, where the signal amplifier does not work properly. By subtracting both curves, the gain of the signal amplifier could be calculated and for approximately 19 dB in the frequency range from 30 kHz to 150 kHz.

With the magnetic field measurement results shown in Fig. 7 and the induced voltage shown in Fig. 8, the transfer function of the field probe can be calculated. For calculating the shielding effectiveness of the wire mesh enclosure, two measurements are necessary:

1. Reference measurement with magnetic field probe in the center of the Helmholtz coil.
2. Shielded measurement with magnetic field probe in the geometric center of the enclosure in the Helmholtz coil. center

The measurement procedure applied here is based on the standard DIN EN 61000-5-7.

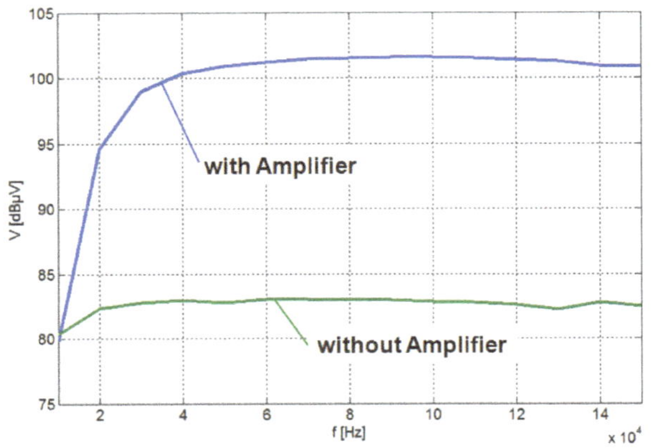

Fig. 8. Measured voltage of the field probe located in the center of the Helmholtz coil.

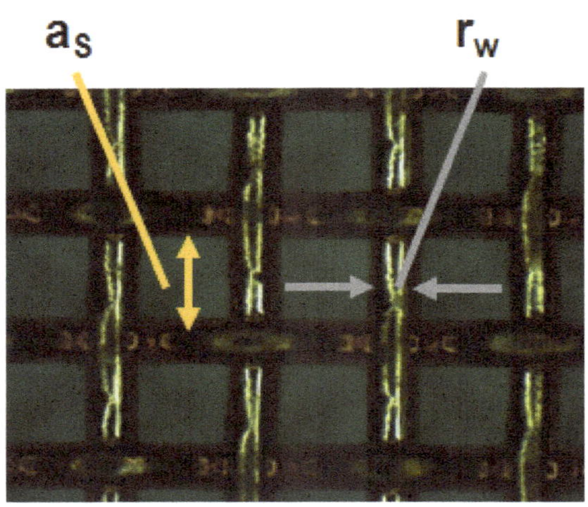

Fig. 9. Microscopic view of the wire mesh.

4.1 Wire mesh enclosure

For the magnetic shielding effectiveness measurements a wire mesh enclosure was chosen, too. The enclosure based on a wooden frame with overlapping copper wire mesh on each site for an electrical connection of the sides, so that a closed shield structure is guaranteed. The size of the enclosure is $30 \cdot 30 \cdot 10$ cm, the mesh material is copper with a mesh size of ~364 μm and wire radius of ~100 μm, see microscopic view in Fig. 9. Two positions in the Helmholtz coil are possible for the enclosure: horizontal (leads to small areas for the incident field) and vertical (large areas for the incident field). The enclosure in the vertical position approaches the above introduced parallel-plate-shield model for wire mesh nearly. The following Fig. 10 shows the wire mesh enclosure with opened cover, the magnetic field probe was located in the geometrical center and was rotated to the field direction accordingly.

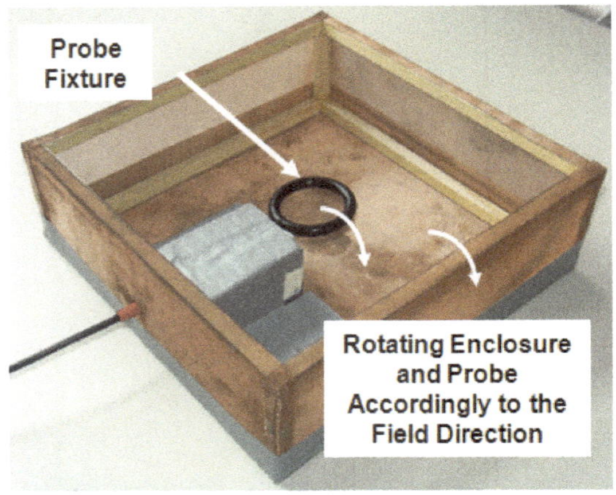

Fig. 10. Wire mesh enclosure and position of the magnetic field probe.

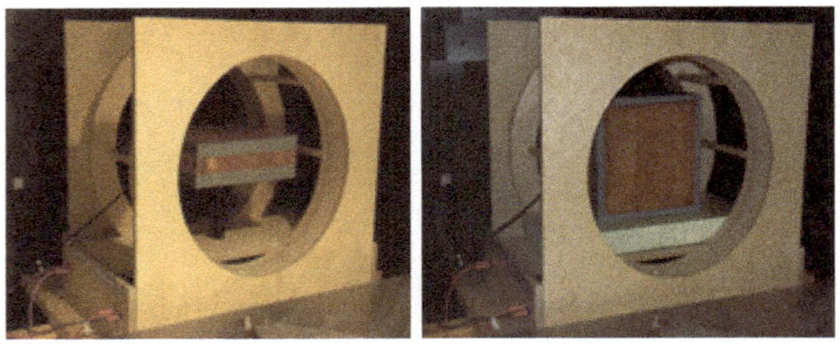

Fig. 11. Horizontal (left) and vertical (right) position of the enclosure inside the Helmholtz coil

For the shielded measurement horizontal and vertical positions of the enclosure were considered (see Fig. 11).

5 Measurement results

A discussion of the results and a comparison with the analytical model (Kaden (1959), page 276, equation 16) will be presented. The shielding effectiveness was calculated by subtracting the induced probe voltage values of the shielded measurement from the reference measurement. Equation 3 shows this relation and must be calculated for each frequency step. All units are in dB and depend on the frequency.

$$SE = V_{\text{ref}} - V_{\text{shield}} \tag{3}$$

5.1 Reference measurement of magnetic shielding effectiveness

The reference measurement was done by placing the magnetic field probe in the center of the Helmholtz coil and measuring the induced voltage. The frequency depended induced voltage is shown in Fig. 8. As above described, the induced voltage falls about 20 dBμV by 10 kHz, because the signal amplifier is not specified for the frequency range below 100 kHz, this does not affect the shielding effectiveness

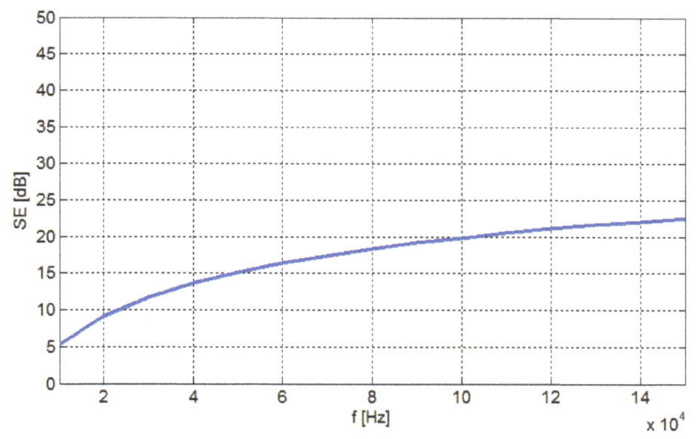

Fig. 12. Measurement result of the shielding effectiveness (SE) of a horizontal oriented wire mesh enclosure (see also Fig. 11 – left)

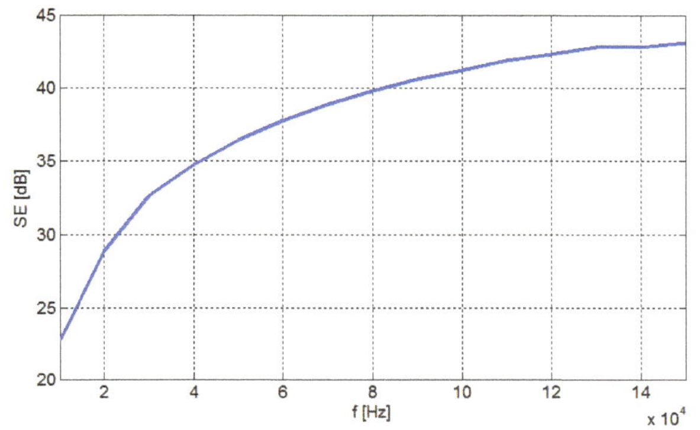

Fig. 13. Measurement result of the shielding effectiveness (SE) of a vertical oriented wire mesh enclosure (see also Fig. 11 – right)

measurement. Due to the fact that the conditions are the same for all measurements (with and without enclosure).

5.2 Shielded measurement

The results for the measurement in the horizontal position are shown in Fig. 12. The magnetic shielding effectiveness is very low (from 5 dB to 23 dB in the frequency range to be considered), this is reasoned by a small area for the incident field (10·30 cm), so the propagation of eddy currents was limited. Therefore the compensation of the incident field caused by the inverse directed magnetic field of the eddy currents is very low.

The results for the measurement in the vertical position are shown in Fig. 12. The magnetic shielding effectiveness of the vertical position is in comparison with the results of the horizontal position larger (~20 dB). This is due to larger area (30 cm·30 cm) for the incident field and the occurred eddy currents in field direction. Therefore, there was a better compensation of the incident field by inverse directed magnetic field resulting from the eddy currents.

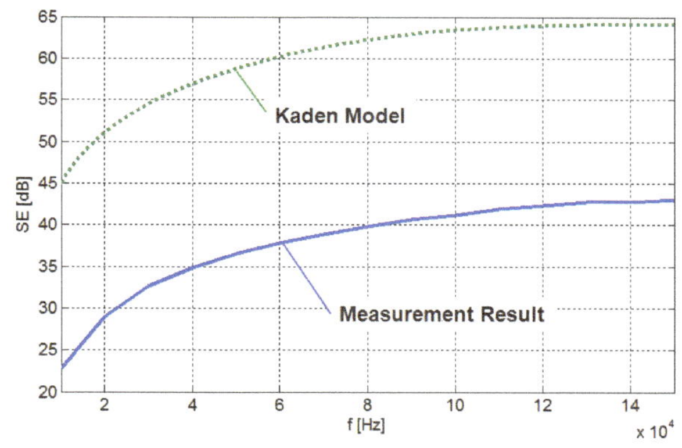

Fig. 14. Measured and calculated (Kaden, 1959) shielding effectiveness (SE) of a vertical oriented wire mesh enclosure.

5.3 Comparison with analytical model

For the comparison with the analytical model of Kaden, the results of the vertical shielding effectiveness measurement were used. There is an even better approach to the parallel-plate-shield. The comparison with the analytical model and the measurement results of the vertical oriented enclosure is shown in Fig. 14. In the literature (for example Wolfsperger (2008); page 207) it was also reported, that the theoretical calculated shielding effectiveness will be according to experience maximal 10–20 dB higher than the practical reached values. This could also be observed for the measurement results presented in this paper. The measured results and the model of Kaden are showing an almost constant deviation for the frequency range to be covered (approximately 22 dB). The model uses the exact material parameters of the wire mesh enclosure (mesh size: $\sim a_s = 364\,\mu$m, wire radius: $\sim r_w = 100\,\mu$m). This deviation is mainly due to:

1. Geometric properties of the model and the enclosure
2. Electrical connections between the plates and the sides

The geometric properties differ. The model is based on a conductor loop and the wire mesh enclosure is a fully shielded box (6 sides for shielding). Subsequently the aspect ratio of the model is assumed to be $|y_0| >> 2 \cdot x_0$ and for the wire mesh enclosure $y_0 = 2 \cdot x_0$. The effective area of the conductor loop from Kaden is even larger than the effective area of the wire mesh enclosure and the induced voltage depends on the size of the area, which interacts the magnetic field. The geometric dimensions of the shield, like the effective shielding area are not considered in the model from Kaden. Just the distance between the plates has been considered. The increasing resistance by raising the distance between the plates is also not considered in the model. Although the electrical connections between the plates of the wire mesh enclosure are not perfect.

6 Conclusions

A simple and reproducible measurement method for the magnetic shielding effectiveness in frequency range from 10 kHz to 150 kHz of small enclosures was presented in this paper. The results have shown, that the magnetic shielding effectiveness is also a function of the area size, where the incident field is interacting (eddy currents). This effect could be used to reduce the electromagnetic interference of cables near sensitive circuits in enclosures. The comparison with an analytical model (Kaden, 1959) has shown, that the physical behaviour is described correctly. The deviation to the measured results is approximately 22 dB. The main reason for the difference are the geometric assumptions, such as no consideration of the exact shield geometry in the model from Kaden. Another reason could be the quality of the electrical connection between the sides of the wire mesh enclosure. With a better electrical connection (low-ohmic) improved shielding effectiveness could be achieved.

7 Outlook

As shown in conclusions, the general physical behaviour of the analytical model in comparison to the measurements is described suitable. Nevertheless there is a deviation by ~22 dB for the vertical position of the enclosure in the frequency range. Therefore, the next step is the comparison of the measurement results (vertical and horizontal position) with results gained by numerical field calculations. Based on a numerical field calculations it will be possible to add a correction term to the used analytical model based on an empirical approach. This correction term will describe the influence of the geometrical properties in a better way. As shown in this paper, the magnetic shielding effectiveness of an enclosure depends on the size of the area, where the incident magnetic field vector is oriented perpendicular. Also, sharp edges and corners (see corner effect, Kaden, 1959) have an influence on the the shielding effectiveness and need to be analyzed within the next steps. The connection of the enclosure parts have also an impact to increase or decrease the shielding effectiveness.

Acknowledgements. This contribution was developed within the scope of the project EM4EM (Electromagnetic Reliability and Electronic Systems for Electro Mobility – Subproject: Entwurfs- und EMZ-Messmethodik für EMZ-Analysen auf EV-Gesamtsystemebene) which is funded by the BMBF (Bundesministerium für Bildung und Forschung) under the grant number 16M3092A. The responsibility for this publication is held by the authors only.

References

DIN EN 61000-5-7; Elektromagnetische Verträglichkeit (EMV) – Teil 5–7: Installationsrichtlinien und Abhilfemaßnahmen; Schutzarten durch Gehäuse gegen elektromagnetische Störgrößen (EM-Code), Beuth Verlag, 2001.

CISPR 25; Vehicles, boats and internal combustion engines-Radio disturbance characteristics-Limits and methods of measurement for the protection of on-board receivers, Edition 3, 2007.

Casey, K.: Electromagnetic shielding behavior of wire-mesh screens, IEEE Transactions on Electromagnetic Compatibility, 30, 298–306, doi:10.1109/15.3309, 1988.

Hillmer, C., Mayer, M.-O., and Reinhardt, U.: EMV von Elektro-Hybridfahrzeugen, in: Vortrag der 5. GMM-Fachtagung München, 2009.

Kaden, H.: Wirbelströme und Schirmung in der Nachrichtentechnik, Springer-Verlag, 2. Auflage, 1959.

Kistenmacher, P., Kimmel, M., and Schwab, A.: Schirmdämpfung inhomogener Metallgehäuse beliebiger Kontur im quasistationären Magnetfeld, Electrical Engineering 78 Springer-Verlag-Verlag, 339–344, 1995.

Michel, M.: Leistungselektronik: Einführung in Schaltungen und deren Verhalten, Springer-Verlag, 5. Auflage, 2011.

Schwab, A. and Kürner, W.: Elektromagnetische Verträglichkeit, VDI-Buch, Springer-Verlag, 6. Auflage, 2010.

Wolfsperger, H.: Elektromagnetische Schirmung: Theorie Und Praxisbeispiele, VDI-Buch, Springer-Verlag, 2008.

13

Reproducing system-level bulk current injection test in direct power injection setup for multiple-port DUTs

S. Miropolsky and S. Frei

TU Dortmund, Dortmund, Germany

Correspondence to: S. Miropolsky (sergey.miropolsky@tu-dortmund.de)

Abstract. Many investigations have been published on the transferability of RF immunity test results between system and IC-levels. The RF signal level at DUT (Device under Test) inputs, i.e. either RF voltage amplitude or RF input current, is used as a reference value for the load on the DUT. Existing approaches analyze the DUT response as a function of the RF signal level at a single input pin, e.g. supply voltage. Sufficient accuracy of such an approach could be shown in several cases, but results are not sufficient as a general solution for complex DUT. This paper proposes both theoretical analysis and practical implementation of a DPI setup, where a disturbance, equivalent to system-level BCI setup, can be delivered to multiple DUT input ports.

1 Introduction

Among the main trends of today's electronic industry, the safety and reliability, and in particular the EMC is of higher importance. The devices are tested for both EM interference (EMI) and susceptibility (EMS). The Direct Power Injection (DPI, IEC, 2006) is a common method for IC testing, while at system-level various methods, e.g. Bulk Current Injection (BCI, ISO, 2005) or Absorber Line Shielded Enclosure, (ALSE), are used.

A system-level EMC test requires large and complex test setups with cable harnesses and special RF equipment, and thus can only be performed in specialized EMC laboratories, while the DPI test can be more easily performed in a common test laboratory. Multiple investigations have been published to find the correlation between system- and IC-level tests. Simulation models for BCI setups are developed since at least 1986 (Sultan, 1986), and the investigations continue up to now (Grassi et al., 2007; Lafon et al., 2007; Alaeldine et al., 2007; Miropolsky et al., 2011). The theory of common into differential mode RF signal conversion at the test PCBs is actively discussed e.g. in Crovetti and Fiori (2011). The RF signal level at the DUT pins is proposed to be a reference value for comparative analysis of test results (Miropolsky and Frei, 2011; Oguri and Ichikawa, 2012). Several proposals have been made to calculate the equivalent DPI forward power necessary to produce the same RF signal at DUT as in the original BCI test (Miropolsky and Frei, 2011; Durier et al., 2011). The test results are either compared using the equivalent DPI power level, or the calculated equivalent forward power is injected into the DUT in the DPI setup and the DUT response is compared to that of the BCI test. The summarized goal of these investigations is to find a method to test the ICs so that the success of the consequent system-level RF immunity test is assured.

Existing approaches analyze the DUT response as a function of the RF signal at a single input pin, e.g. at supply pin. High accuracy could be observed in several cases (Miropolsky and Frei, 2011; Durier at al., 2011; Oguri and Ichikawa, 2012), but this cannot be applied as a general solution. In case of BCI test of two-pin modules reproducing the differential RF disturbance at a single pin pair (supply vs. ground) is sufficient. However, in case of three pin sensors (Durier at al., 2011; Oguri and Ichikawa, 2012), where signal and supply lines are separate, the RF signal coupled to a supply line might influence the immunity threshold at a signal pin and vice versa. In a BCI test on a CAN or FlexRay transceiver network both common and differential RF signals at the two IC pins can be responsible for an IC failure. Thus in the discussion of equivalent conversions the RF injection into multiple pins has to be considered.

Moreover, the IC-EMC-tests claim to test the ICs, while the measurement is technically performed on the PCB where the IC is only one of the components. The transfer function of the PCBs might not be neglected, especially in discussions regarding test method equivalence.

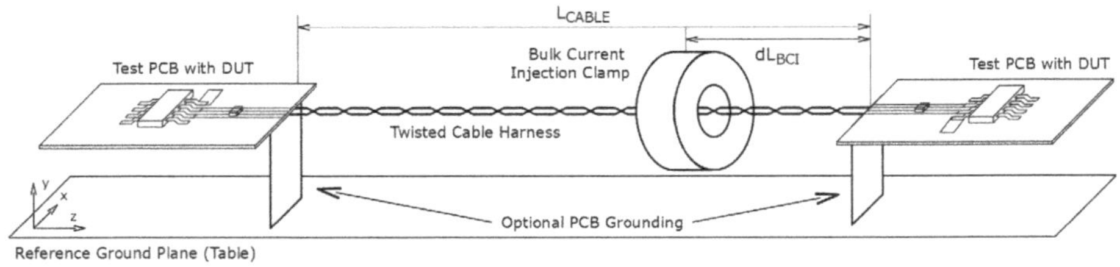

Fig. 1. Sample BCI test setup structure and overview.

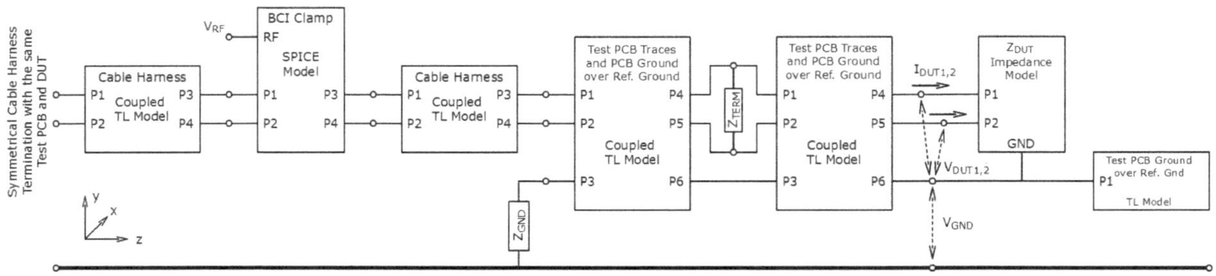

Fig. 2. Sample BCI test setup simulation circuit.

This paper proposes a DPI testing method, where the RF disturbances, equivalent to an original BCI setup, can be delivered to multiple DUT pins.

2 Sample study case: a system-level Bulk current injection setup

2.1 Test setup structure and overview

The system level test setups usually contain of one or more electronic control units (ECU, here called EUT), connected to each other or to the support circuitry with a cable harness. A sample system discussed in current paper consists of two CAN transceiver units. Each EUT is a small PCB containing the transceiver IC (DUT) with necessary periphery, e.g. bus terminations of 120 Ω, traces, ports, supply voltage protection, etc. These EUT PCBs are located in 5 cm over the reference ground plane. Depending on the application the PCBs may optionally be grounded to the main reference ground with a low impedance connection (e.g. metal tape). The EUTs are connected to each other with a twisted pair cable of 1 m length in 5 cm above the table. The transceivers communicate over this cable. The RF disturbance of specified amplitude is injected into the cable harness using one of the mentioned methods, and safe operation of the system under RF disturbance should be assured.

Theoretically, any RF immunity test method can be used for such network. In the current work the BCI method is considered, and the setup shown in Fig. 1 is discussed. The BCI injection clamp is placed at a specified position along the cable harness.

According to the method (BCI, ISO, 2005), the BCI clamp has to be calibrated in advance, and the forward power necessary to inject a certain common mode current amplitude into the cable harness must be measured. To simplify the analysis, this step is avoided and constant forward power level is injected. For real application, the calibration curve of the BCI clamp has to be considered. This however can be introduced by just another scaling coefficient in the calculation procedure.

According to a general assumption supported by many authors (Miropolsky and Frei, 2011; Durier at al., 2011; Oguri and Ichikawa, 2012), the DUT response is dependent on the RF signal delivered through the test setup to the IC pins. The test setup is no more than a small-signal transfer function from the system RF port to the DUTs. To obtain the transfer function a small-signal model of the setup is necessary.

2.2 Small-signal simulation model

The simulation model shown in Fig. 2 is assembled according to the original setup.

The FCC F140 BCI clamp is modelled with a circuit similar to one shown in Miropolsky and Frei (2011). The twisted cable harness is modelled as lossy coupled transmission line (TL). The cable parameters were obtained by analyzing the cable cross-section with HSPICE 2-D field solver. The twistedness effect is neglected in the model for simplicity purposes.

The test PCB is also modelled as a coupled transmission line over reference ground plane, similar to an approach shown in Crovetti and Fiori (2011). Top level PCB traces are first and second conductors, the PCB ground plane is the third conductor. The exact properties of the cables and PCB

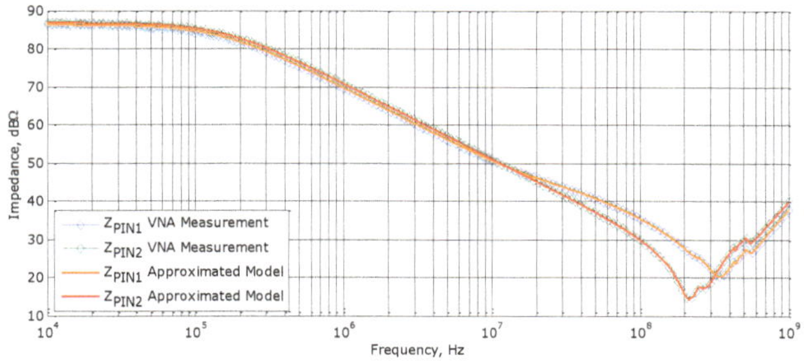

Fig. 3. Sample DUT input impedances, measurement and macromodel.

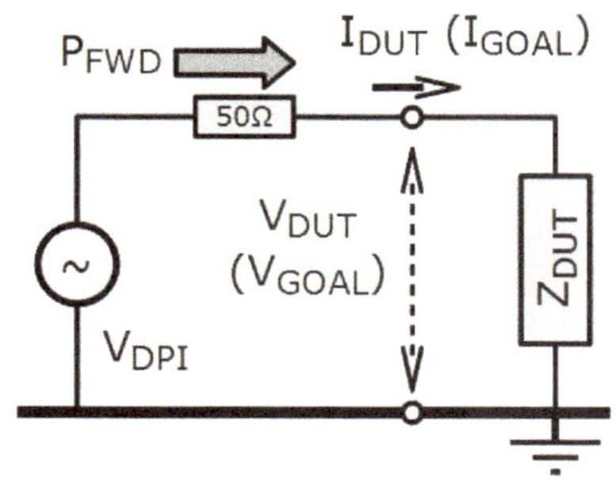

Fig. 4. One-port equivalent RF source.

traces are study-case-dependent, and thus are irrelevant for the general discussion purpose.

The 120Ω CAN bus termination is modeled as simple resistor (Z_{TERM}) in the middle between the PCB input ports and DUT (see Fig. 2). Since the DUT is located in the center of the test PCB, the remaining PCB ground plane provides additional coupling to the reference ground, this effect is considered with one more TL device.

The model is used to simulate RF signal levels at DUT pins in the BCI test. These levels are used as a reference to generate and verify the equivalent DPI setup.

2.3 DUT input impedance

The DUT impedance can often be modeled as a RC circuit, with additional package effects visible at higher frequencies. Commonly the impedances may be considered to be close to RC up to approx. 100 MHz. More complex models have to be considered above this frequency.

In the analysis of multiple-port DUTs, under the assumption of possible internal coupling between the pins, representing DUT as a set of pin-to-ground impedances is insufficient. A full port-to-port impedance matrix must be considered.

In current work a sample CAN transceiver IC is analyzed as example DUT. The RF injection into differential interface consisting of two pins, CANH and CANL, is considered. A linear impedance model is created as a numerical approximation of measurement data. The two-port impedance dataset is captured with VNA and deembedded with the Matlab RF Toolbox. The data is approximated with vector fitting (Gustavsen and Semlyen, 1999), and an equivalent SPICE model for the approximation is generated. The impedances are shown in Fig. 3.

3 Equivalent DPI source calculation procedure

3.1 Motivation

The development of an equivalent DPI test follows a single goal: the RF disturbances that appear in the complex system-level setups, e.g. BCI, should be reproduced in a simplified low-cost DPI setup. Among the initial conditions we must note the fixed 50Ω impedance of RF sources, and significant limitations on reproducing arbitrary impedances in physical setups. On the other side significant freedom degree is available in controlling RF amplitudes and phase shifts of several 50Ω sources.

3.2 Equivalent DPI source calculation

According to (Durier at al., 2011), the equivalent DPI power level can be calculated with a commercial calculator from the complex RF voltage and current values, which must be extracted from system level setup simulation or measurement as a table vs. frequency. This implies the known value of the DUT input impedance. In this case the internal RF amplitude of the source can be simply calculated at each frequency with Eq. (1).

$$V_{\mathrm{DPI}} = V_{\mathrm{GOAL}} + R_{50\Omega} \cdot I_{\mathrm{GOAL}} \tag{1}$$

$$P_{\mathrm{DPI}} = V_{\mathrm{DPI}}^2 / 4R_{50\Omega} \tag{2}$$

The procedure is illustrated in Fig. 4. The index "GOAL" refers to the signal to be reproduced at RF source output. The equivalent DPI source level is valid only for current DUT impedance. Attaching another DUT to the setup results in different RF signal values at DUT ports, and another calculation of an equivalent DPI source is necessary. This is a significant disadvantage comparing to e.g. an equivalent Norton source, since the latter generates the signal at the load in strict correlation to original setup for any load impedance.

The equivalent DPI source is easily extendable for a multiport DUT. A separate independent 50Ω RF source must be attached to each of the DUT ports, i.e. the number of sources has to be the same as the amount of DUT nodes to be driven. The source voltage or forward power value for each port can be calculated with the Eq. (1) too. For multiple-port networks the relative signal phases are important and must be considered.

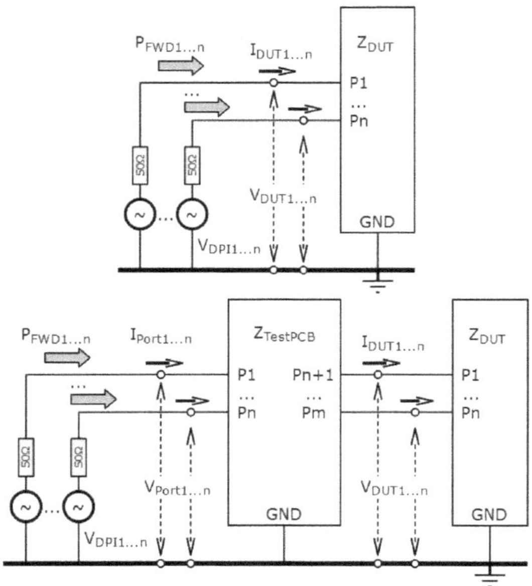

Fig. 5. Multiple-port equivalent DPI RF source **(a)** direct injection into DUT (virtual case) **(b)** accounting for the PCB transfer function (realistic case).

An example of an equivalent setup for a single two-port DUT is shown in Fig. 5a.

3.3 DPI PCB properties

In physical DPI test the RF source is never connected to the DUT directly, but to the RF port of the test PCB. The transfer function from the PCB RF port to the DUT pins must be considered. Since the DPI test is performed on a grounded PCB, only the differential signal transfer must be considered, the PCB ground plane is used as the signal reference.

The easiest approach to introduce the transfer function is to use a multiport **ABCD** matrix of the PCB structure (3):

$$\begin{bmatrix} \boldsymbol{V}_{\mathrm{PORT}} \\ \boldsymbol{I}_{\mathrm{PORT}} \end{bmatrix} = \begin{bmatrix} \mathbf{A}_{\mathrm{PCB}} & \mathbf{B}_{\mathrm{PCB}} \\ \mathbf{C}_{\mathrm{PCB}} & \mathbf{D}_{\mathrm{PCB}} \end{bmatrix} \cdot \begin{bmatrix} \boldsymbol{V}_{\mathrm{DUT}} \\ \boldsymbol{I}_{\mathrm{DUT}} \end{bmatrix} \tag{3}$$

where $\boldsymbol{V}$ and $\boldsymbol{I}$ are vectors of differential voltages and currents, i.e. with PCB ground plane used as a reference. The indices correspond to signals at DUT pins ($\boldsymbol{V}_{\mathrm{DUT}}, \boldsymbol{I}_{\mathrm{DUT}}$) and signals at PCB ports at cable harness side ($\boldsymbol{V}_{\mathrm{PORT}}, \boldsymbol{I}_{\mathrm{PORT}}$), as shown in Fig. 5b. Each A, B, C and D element is a sub-matrix with a size corresponding to the number of ports at each PCB side. The sub-matrices can be found by solving the linear equation system of the admittance representation (Eq. 4) for vectors $\boldsymbol{V}_{\mathrm{PORT}}$ and $\boldsymbol{I}_{\mathrm{PORT}}$.

$$\begin{bmatrix} \boldsymbol{I}_{\mathrm{PORT}} \\ \boldsymbol{I}_{\mathrm{DUT}} \end{bmatrix} = \begin{bmatrix} \mathbf{Y}_{\mathrm{PCB}} \end{bmatrix} \cdot \begin{bmatrix} \boldsymbol{V}_{\mathrm{PORT}} \\ \boldsymbol{V}_{\mathrm{DUT}} \end{bmatrix} \tag{4}$$

The values of the equivalent DPI sources ($\boldsymbol{V}_{\mathrm{DPI}}$) can then be calculated with Eqs. (1) and (2) using RF signals at PCB ports ($\boldsymbol{V}_{\mathrm{PORT}}$ and $\boldsymbol{I}_{\mathrm{PORT}}$) as "goal" values. When the 50Ω RF sources with calculated amplitude and phase are attached to the PCB ports, the RF signals $\boldsymbol{V}_{\mathrm{DUT}}$ and $\boldsymbol{I}_{\mathrm{DUT}}$ will be reproduced at DUT pins.

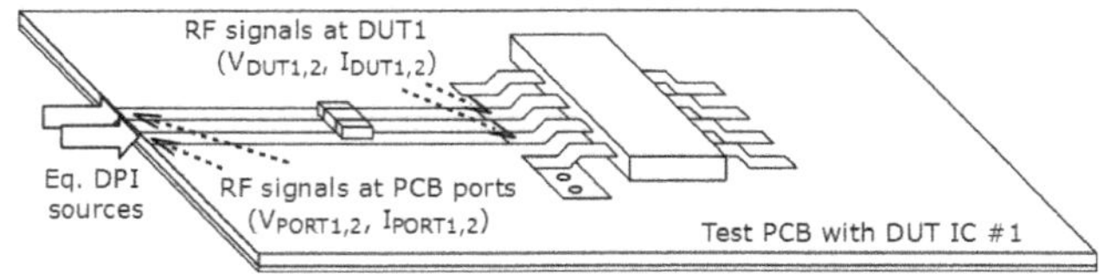

Fig. 6. Equivalent DPI test setup for sample two-port DUT, overview.

The impedance matrix of the test PCB can be obtained by either passive modeling approach with discrete circuit elements or with e.g. 3-D EM field simulations or with VNA measurements.

4 Equivalent DPI source application

The general procedure for equivalent DPI source development is described in Sect. 3. In the current application, the "original" RF signal levels at DUT ($V_{\mathrm{DUT1,2}}, I_{\mathrm{DUT1,2}}$) are obtained by AC simulation of a BCI setup model discussed in Sect. 2.2 with DUT impedances discussed in Sect. 2.3.

4.1 Single-DUT testing with equivalent DPI source

The same test PCB as in BCI test can be used. The RF ports are connected between the PCB traces and ground as shown in Fig. 6. The PCB model is developed in a similar way as for the main BCI test setup. The ABCD-matrix of the PCB is calculated. The test PCB port signals (V_{PORT} and I_{PORT}) are calculated with Eq. (3). The RF source amplitude is calculated with Eq. (1) using port signals as goal values. Attaching the RF sources to PCB ports will reproduce the original RF signal levels at DUT ($V_{\mathrm{DUT1,2}}, I_{\mathrm{DUT1,2}}$), both in simulation and, more important, in measurement. The resulting circuit is the same as shown in Fig. 5b.

The single-DUT configuration can be efficiently used for other DUT types, e.g. sensor ICs. In the case discussed here at least two ICs are necessary to establish the communication. A two-DUT configuration must be considered.

4.2 Two-DUT equivalent DPI configuration

The configuration shown in Fig. 8 was investigated. Two test PCBs with DUTs are symmetrically connected to each other, and equivalent DPI sources are attached to central nodes. The decoupling network (not shown in the figure) can be included on necessity. Due to the symmetry of such DPI circuit the RF signals will be generated symmetrically at both DUTs. The signal levels at input ports of the test PCBs can be calculated in the same way with Eq. (2) as for the single DUT case.

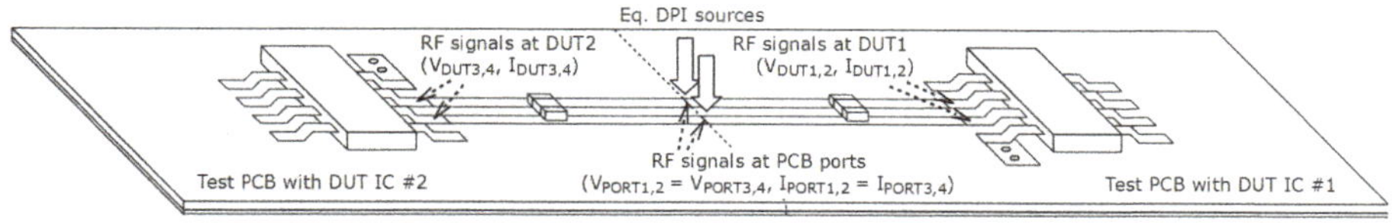

Fig. 7. Equivalent DPI test setup for two sample two-port DUTs, overview.

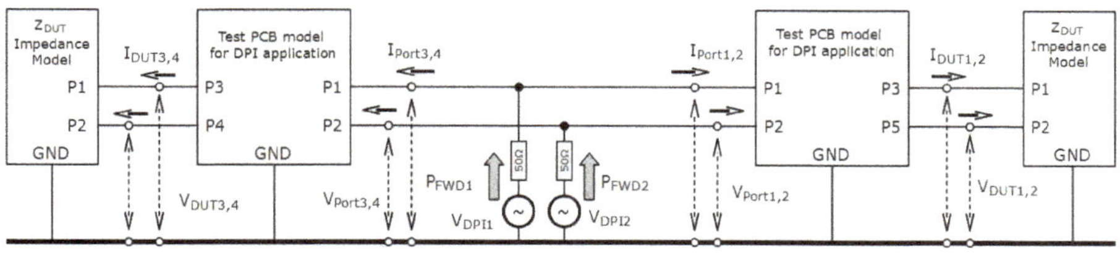

Fig. 8. Equivalent DPI test setup for two sample two-port DUTs, simulation.

From the RF source point of view, the two test PCBs are just the loads connected in parallel, thus the RF sources amplitudes can be calculated with Eq. (1) using $V_{PORT1,2}$ voltages and *double* $I_{PORT1,2}$ currents as "goal" values.

Obviously, in the real BCI test such symmetry cannot be expected. Different RF signals have to be reproduced at both DUTs. Theoretically, it can be performed with a 4-port equivalent RF source of the same or similar structure. However, in real application such configuration is rather unrealistic. Each physical RF source consists of a signal generator followed by an amplifier. Equivalent DPI implies the control of both amplitude and phase. Hence the complexity of such configuration with *four* physical sources is very high, and running a BCI test would be more effective.

To avoid unnecessary complexity, the RF levels of each DUT in the BCI setup can be reproduced sequentially. The first set of equivalent RF sources is calculated to reproduce the original RF levels $V_{DUT1,2}$ and $I_{DUT1,2}$ at both DUTs. The same procedure is performed for the second set of RF signals $V_{DUT3,4}$ and $V_{DUT3,4}$. This way after two test iterations each of the DUT ICs is subdued once to a RF disturbance equivalent to that from the BCI test.

4.3 Simulation results

The calculated forward power levels (P_{DPI}) is shown in Fig. 9. Generally, the equivalent source forward power is much lower than the original forward power injected into the BCI setup, i.e. only a small part of the produced RF power is delivered to the DUTs. However, at resonance locations the values reach and even exceed the original BCI forward power. Since the two sources are injecting RF signals simultaneously, the total forward power is even higher.

The RF signal levels at both DUTs in the original BCI setup and in equivalent DPI setup are shown in Figs. 10 and 11. In small-signal simulation the RF signals at the first DUT pins are reproduced with high accuracy. The RF source forward power is calculated to reproduce the RF signals at the first DUT. Same RF signal levels are generated at the second DUT too, what obviously doesn not correspond to the original BCI setup. During the consecutive second run, the eq. forward power is calculated for the second DUT, and both DUTs are tested with this second set of RF signals. Altogether, each DUT is at least tested once with a RF disturbance level equivalent to that in original BCI setup. Also none of the DUTs is being overtested, i.e. in no case the RF signals exceed the maximum that might appear at the nodes in original BCI configuration.

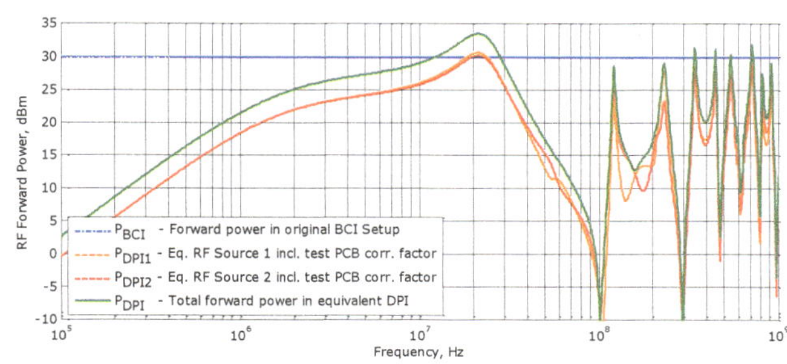

Fig. 9. RF forward power levels in original BCI and equivalent DPI.

4.4 Advanced testing strategies

The main issue of BCI RF immunity testing in the high frequency range is the dependency of the test results on test setup transfer function, which is itself strongly dependent on smaller factors. This leads to instability of corresponding BCI RF signal levels and uncertain reproducibility of the test results above 200–300 MHz.

This issue can be avoided by extraction of maximum RF levels for a sample DUT. The BCI test setup has to be simulated in small-signal mode with variation of its parameters in allowed ranges. Critical here are cable cross-section, length and clamp position. Less critical but still important are the

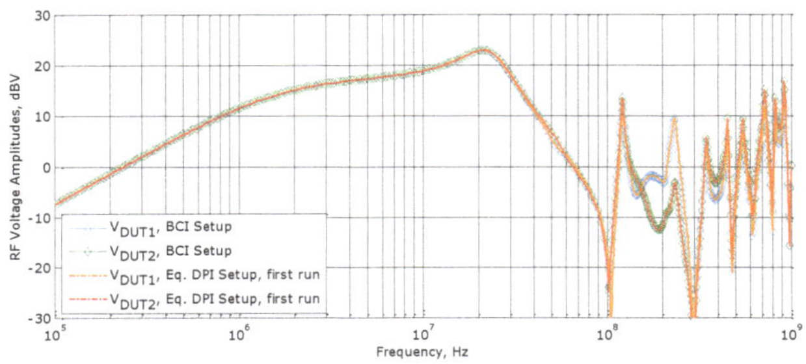

Fig. 10. RF signal levels at DUT 1 under equiv. DPI, first run.

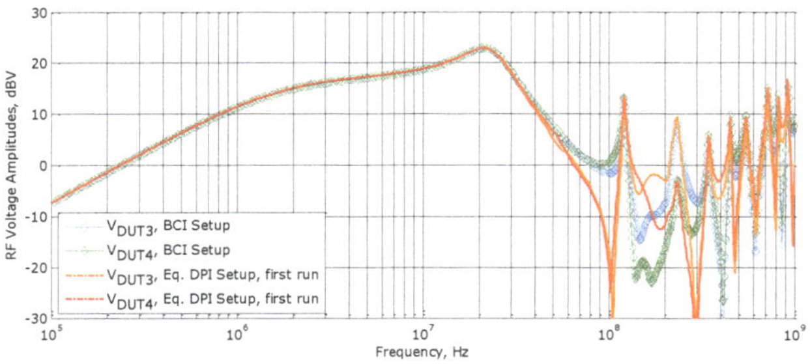

Fig. 11. RF signal levels at DUT 2 under equiv. DPI, first run.

PCB layout, termination impedances, and the DUT location. The DUT impedance may be varied in the expected range. By running e.g. Monte-Carlo AC analysis or manually sweeping the parameters the maximum RF levels and corresponding equivalent forward powers can be obtained. By injecting this forward power in an equivalent DPI setup, the maximum possible RF disturbance in BCI can be covered. The success of almost any consequent real BCI test can be assured.

5 Conclusions

A method to reproduce the RF disturbances from a system-level BCI test in a less complex DPI test is proposed. The DPI sources may be attached to the same test PCB as in BCI test. By injecting the equivalent RF forward power (with both amplitude and relative phase being controlled) into the test PCB the same RF disturbance at the DUT can be achieved, and the results of a BCI test can be predicted. The transfer function of the test PCB must be considered in the calculation of the equivalent DPI forward power.

The method is applicable for the systems of multiple DUTs, e.g. for testing communication systems. A test has to be repeated several times, the RF disturbances are sequentially reproduced at each DUT in the system, with neither devices being over- nor under-tested after the entire test sequence is performed.

Acknowledgements. The reported R+D work was partially carried out within the CATRENE project CA310 EM4EM (Electromagnetic Reliability and Electronic Systems for Electro Mobility). This particular research is supported by the BMBF (Bundesministerium fuer Bildung und Forschung) of the Federal Republic of Germany under grant 16 M3092 I. The responsibility for this publication is held by the authors only.

References

Alaeldine, A., Cordi, J., Perdriau, R., Ramdani, M., and Levant, J. L.: Predicting the Immunity of Integrated Circuits through Measurement Methods and Simulation Models, 18th Int. Zurich Symposium on EMC, Munich 2007.

Crovetti, P. S. and Fiori, F.: Distributed Conversion of Common-Mode Into Differential-Mode Interference, IEEE Tran. on Microwave Theory, Vol. 59, No. 8, August 2011.

Durier, A., Pues, H., Vande Ginste, D., Chernobryvko, M., Gazda, C., and Rogier, H.: Novel Modeling Strategy for a BCI setup applied in an Automotive Application, EMC Compo 2011, Dubrovnik, Croatia, 2011.

Grassi, F., Marliani, F., and Pignari, S. A.: Circuit Modeling of Injection Probes for Bulk Current Injection, IEEE Tran., EMC 49, No. 3, 563–576, 2007.

Gustavsen, B. and Semlyen, A.: Rational Approximation of Frequency Domain Responses by Vector Fitting, IEEE Tran. on Power Delivery, 14, 1052–1061, 1999.

IEC 62132-4: Integrated Circuits – Measurement of Electromagnetic Immunity 150 kHz to 1 GHz – Part 4: Direct RF Power Injection Method, 2006.

ISO 11452-4: Road Vehicles – Component Test Methods for Electrical Disturbances from Narrowband Radiated Electromagnetic Energy – Part 4: Bulk Current Injection (BCI), ISO, 2005.

Lafon, F., Belakhouy, Y., and De Daran, F.: Injection Probe Modeling for Bulk Current Injection Test on Multiconductor Transmission Lines, IEEE Symp. on Embedded EMC Proceedings, Rouen, France, 2007.

Miropolsky, S. and Frei, S.: Comparability of RF Immunity Test Methods for IC Design Purposes, EMC Compo 2011, Dubrovnik, Croatia, 2011.

Miropolsky, S., Frei, S., and Frensch, J.: Modeling of Bulk Current Injection Setups for Virtual Automotive IC Tests, EMC Europe 2011, Wroclaw, Poland, 2011.

Oguri, Y. and Ichikawa, K.: Simulation Method for Automotive Electronic Equipment Immunity Testing, EMC Europe 2012, Rome, Italy, 2012.

Sultan, M. F.: Modeling of a Bulk Current Injection Setup for Susceptibility Threshold Measurements, IEEE Int. Symp. on EMC Proc., 188–195, 1986.

14

Analytic investigations on the susceptibility of nonlinear analog circuits to substrate noise

C. Widemann, S. Stegemann, W. John, and W. Mathis

Institut für theoretische Elektrotechnik, Leibniz Universität Hannover, Appelstraße 9A, 30167 Hannover, Germany

Correspondence to: C. Widemann (widemann@tet.uni-hannover.de)

Abstract. This work deals with the conducted susceptibility of nonlinear analog circuits with respect to substrate noise. The substrate coupling mechanism is modeled by a passive three-terminal network that is obtained by means of the finite element method with a subsequently performed model order reduction. Applying this substrate model to the bulk terminal of MOS transistors in integrated analog circuits, it is possible to examine the influence of substrate noise on the circuit's functionality. By means of a block-oriented approach, analytic expressions for the output behavior of the circuits are found. The utilized multi-input Wiener model separates the linear dynamic from the nonlinear static circuit properties. Due to this separation the frequency response of both signals, i.e. input signal and substrate noise, respectively, can be identified, and hence, the frequency range in which the circuit is most susceptible to substrate noise. Since the nonlinear static behavior of each MOS transistor depends on two signals, truncated multivariate Taylor series expansions of the nonlinear elements are performed on the basis of the EKV model description (Enz et al., 1995). The proposed modeling is illustrated by a simple example.

1 Introduction

The electromagnetic compatibility (EMC) of electronic systems and components became the focus of attention in research and development during the last two decades. The omnipresence of wireless communication systems with high frequencies caused an increasing electromagnetic pollution that endanger the proper functionality of sensitive electronic systems. In this consideration, the role of integrated circuits (IC) which are the core elements of electronic systems can be source and victim of electromagnetic interference (EMI). As source, the ICs radiate or conduct electromagnetic signals that are mainly caused by simultaneous switching noise to other ICs. In addition to this external EMI, internal chip coupling via the supply network or the shared substrate exists in mixed-signal ICs (cf. Fig. 1). There are several guidelines for IC designers in order to reduce the external and internal coupling, e.g. with decoupling capacitors (Dhia et al., 2006). The impact of substrate noise coupling can be reduced by so-called p^+ guard rings that isolate the active area of NMOS transistors (PMOS are isolated by the n-well) by draining the noise to the off-chip ground (Bronckers, 2009). However, this technique does not attenuate the noise perfectly. Thus, time-varying bulk potentials cause a shift of the DC operating point. Hence, the analog part of mixed-signal ICs is more susceptible to substrate noise than the digital part due to its continuous dynamic range. As a result of interfering signals, intermodulation products in the output spectrum of the nonlinear analog circuits emerge. In order to identify circuit parameters that have direct impact on the emerging intermodulation products, methods for distortion analysis have to be applied. At first, the substrate coupling mechanism is examined in the following section. The resulting model is applied to an exemplary circuit that is evaluated by means of the block-model presented in Sect. 3.

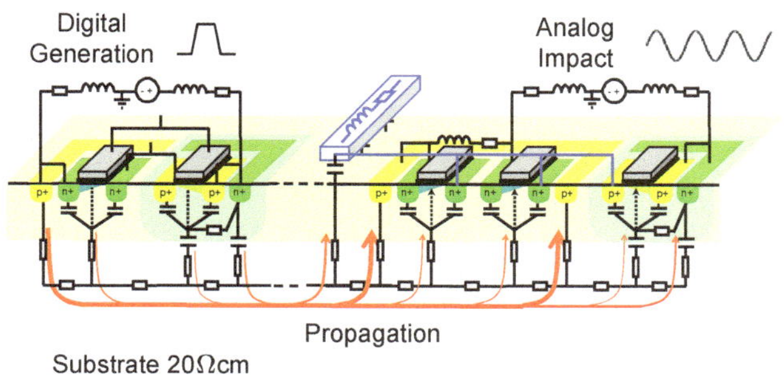

Fig. 1. Substrate coupling in mixed-signal ICs (Bronckers, 2009).

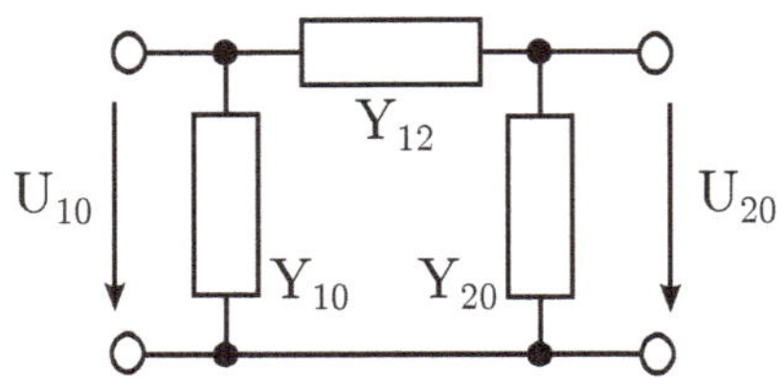

Fig. 2. Three-terminal substrate model (Lin et al., 2009).

2 Modeling the substrate coupling

In mixed-signal ICs, the simultaneous switching noise which is the main cause for electromagnetic interference influences the more sensitive analog circuitry. The noise generated by the switching of the digital gates propagates via the shared substrate and impacts the analog active devices. Bronckers (2009) examines the noise coupling mechanism, describes model approaches and introduces possible solutions for the reduction of unwanted effects due to the substrate noise. As depicted in Fig. 1, the substrate possesses resistive and capacitive properties that are modeled by a lumped network. An approximation to the electromagnetic field in the substrate can be found using this lumped network by means of FEM. However, a more practical substrate model is required for further analysis that yield analytic expressions. Thus, exploiting a model order reduction technique leads to a reduced *three-terminal substrate model* that consists of three delta connected impedances (cf. Fig. 2). Depending on the regarded frequency range, the three impedances are either purely resistive or parallel-connected resistive and capacitive elements (Lin et al., 2009). In the following, this three-terminal model is used in combination with a block model approach for distortion analysis of analog circuits that is reviewed in the next section.

3 Block oriented EMI analysis

Due to electromagnetic interference (EMI) the functionality of electronic circuits can be distorted significantly depending on the disturbing signal's frequency and the magnitude. In order to identify circuit parameter that influence the susceptibility to EMI analytical expressions for the unwanted intermodulation products that lie in the signal band are necessary. In general, these analytic expressions can be found by extending the known distortion analysis methods (e.g. Wambacq and Sansen (1998)) with the distortion as additional input signal. The most commonly-used method is a truncated functional series expansion. If nonlinear dynamic systems are regarded, the only possible functional series is the well-known *Volterra series*. However, even in a multi-tone single input case the so-called *Volterra kernels* (Weiner and Spina, 1980) are hard to calculate. Hence, for a general multi-input distortion analysis a block-oriented approach is used to model each stage of a multi-stage circuit.

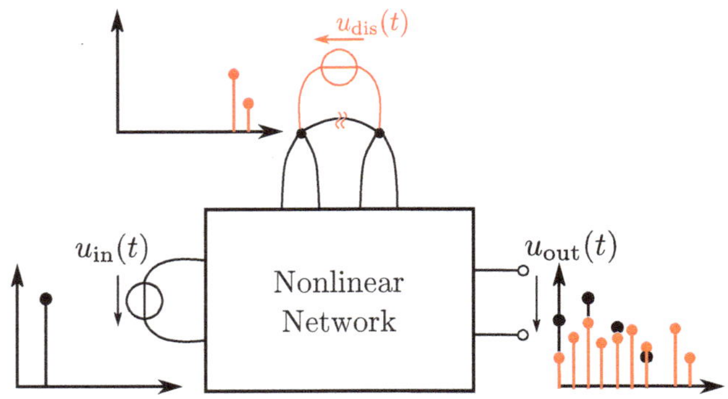

Fig. 3. Modeling the disturbance as independent sources.

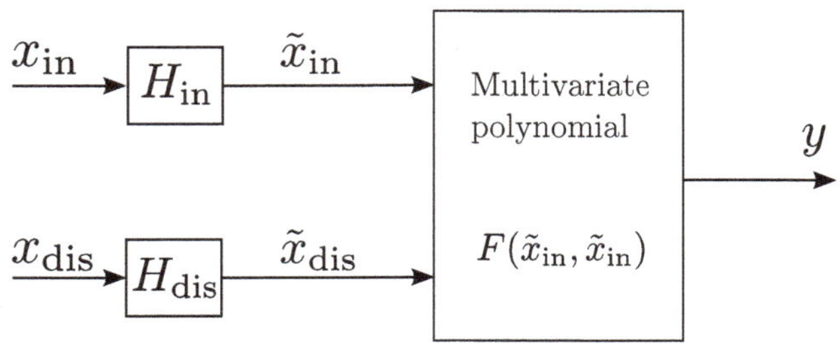

Fig. 4. Block model for distortion analysis (Stegemann et al., 2012).

This approach is based on the classical power series approach for weakly nonlinear single input systems presented e.g. in Weiner and Spina (1980). Moreover, the EMI analysis of nonlinear analog circuits using the known methods for distortion analysis requires that the electromagnetic disturbance can be reduced to a network problem. Thus, the disturbing signals have to be represented by circuit elements, e.g. independent voltage or current sources applied to an arbitrary circuit branch (cf. Fig. 3). As a result, the nominal and the disturbing signal possess their own transfer characteristic with respect to the circuit's output. Hence, a Wiener model approach with multiple input paths is chosen that separates the system into purely linear dynamic blocks and a static nonlinear multi-input block (Stegemann et al., 2012). This block approach is shown in Fig. 4, wherein x_{in} and x_{dis} are the generalized input and disturbing signal and H_{in} and H_{dis} the corresponding purely linear dynamic transfer characteristic, respectively. The multi-input block $F(\tilde{x}_{in}, \tilde{x}_{dis})$ represents the static nonlinear behavior of the analog circuit by which intermodulation products between the linearly distorted signals $\tilde{x}_{in}$ and $\tilde{x}_{dis}$ emerge in the spectrum of the circuit's output signal $y(t)$. The following sections review the methods to obtain analytic expressions for the different blocks presented in Stegemann et al. (2012) and reveal problems in adapting the method for determining the static nonlinearity $F(\tilde{x}_{in}, \tilde{x}_{dis})$ for the circumstance of a disturbed substrate.

3.1 Linear dynamic block

The linear dynamic blocks H_{in} and H_{dis} are completely described by their frequency domain transfer functions

$$H_{\text{in}}(s) = \frac{\tilde{X}_{\text{in}}(s)}{X_{\text{in}}(s)} \quad \text{and} \quad H_{\text{dis}}(s) = \frac{\tilde{X}_{\text{dis}}(s)}{X_{\text{dis}}(s)}. \tag{1}$$

For the calculation of analytical expressions of these transfer functions the modified nodal analysis (MNA) (Vlach and Singhal, 1983) is applied that results in the frequency domain to the network description

$$(s\mathbf{C}+\mathbf{G})\mathbf{Z}(s) = \mathbf{B}\boldsymbol{X}(s). \tag{2}$$

The matrices **C** and **G** includes the matrix stamps of the dynamic (capacitances and inductances) and the resistive elements, respectively. The vector $\mathbf{Z}(s)$ contains the Laplace transforms of the unknown node potentials and currents of the inductances and voltage sources. The right hand side is the product of the matrix **B** and the vector $\boldsymbol{X}(s) = (X_{\text{in}}(s), X_{\text{dis}}(s))^T$ that denotes the frequency description of the signals $x_{\text{in}}(t)$ and $x_{\text{dis}}(t)$. The former maps the independent sources to their corresponding node voltage or branch current (depending on the source type of $x_{\text{in}}(t)$ and $x_{\text{dis}}(t)$). The output signals of the linear transfer blocks are obtained by the product of the vector **Z** and the mapping matrix **L**:

$$\begin{bmatrix} \tilde{X}_{\text{in}}(s) \\ \tilde{X}_{\text{dis}}(s) \end{bmatrix} = \mathbf{L}\mathbf{Z}(s). \tag{3}$$

Regarding the rows of Eq. (3) including the solution of the linear system of equations (2) separately leads to the both transfer functions

$$H_{\text{in}}(s) = \boldsymbol{l}_1\,[s\mathbf{C}+\mathbf{G}]^{-1}\,\boldsymbol{b}_1 \tag{4}$$

$$H_{\text{dis}}(s) = \boldsymbol{l}_2\,[s\mathbf{C}+\mathbf{G}]^{-1}\,\boldsymbol{b}_2, \tag{5}$$

wherein $\mathbf{l}_1$, $\mathbf{l}_2$ and $\mathbf{b}_1$, $\mathbf{b}_2$ denote the first and second row vector of **L** and column vector of **B**, respectively. Applying these transfer functions to the harmonic Q-tone input signals x_{in}, x_{dis} result in (Weiner and Spina, 1980)

$$\tilde{x}(t) = \sum_{q=1}^{Q} \hat{x}_q |H(f_q)| \cos\left(2\pi f_q t + \Theta_q + \psi_H(f_q)\right). \tag{6}$$

Therein, $\hat{x}_q$ and Θ_q denote amplitude and phase of the qth tone of the initial signal $x(t)$ and $\psi_H(f_q)$ the phase of the transfer function at the frequency f_q.

3.2 Static nonlinear block

The multivariate polynomial block $F(\tilde{x}_{\text{in}}, \tilde{x}_{\text{dis}})$ represents the static nonlinear behavior of the circuit, i.e. without any memory effects of dynamic elements such as capacitances or inductances.

In Stegemann et al. (2012), it was presented how the method of nonlinear current sources (cf. e.g. Schetzen (1985)) for voltage controlled nonlinear circuit elements $i = f_g(u)$ can be used for the Nth order approximation of $F(\tilde{x}_{\text{in}}, \tilde{x}_{\text{dis}})$. Therein, the dependency f_g of each nonlinear element was assumed to be analytic. Hence, the functions were approximated by a truncated power series

$$i(t) = f_g(u) \approx \sum_{n=1}^{N} a_n u^n(t), \tag{7}$$

wherein the factors a_n denote the nonlinearity coefficients of the nonlinear element of the order n and u the controlling incremental voltage. Applying the iterative method of nonlinear current sources, the MNA leads to systems of equations, each one that corresponds to the order n. Thereby, the excitation vector of the nth order system only depends on the solutions of the systems of orders $m < n$. Hence, an analytic approximation of the order N for the nonlinear output signal $y(t)$ is calculated successively by

$$y_N(t) = \sum_{i=0}^{N} \sum_{j=0}^{N-i} d_{ij} \tilde{x}_{\text{in}}^i(t) \tilde{x}_{\text{dis}}^j(t). \tag{8}$$

Rearranging this expression leads to a description with terms that depend either only on $\tilde{x}_{\text{in}}$ or $\tilde{x}_{\text{dis}}$ and mixed terms of both signals

$$\begin{aligned} y_N(t) =& d_{00} + \sum_{i=1}^{N} d_{i0} \tilde{x}_{\text{in}}^i(t) + \sum_{j=1}^{N} d_{0j} \tilde{x}_{\text{dis}}^j(t) + \\ &+ \sum_{l=2}^{N} \sum_{k=1}^{l-1} d_{(l-k)k} \tilde{x}_{\text{in}}^{l-k}(t) \tilde{x}_{\text{dis}}^k(t). \end{aligned} \tag{9}$$

However, in the present problem, the nonlinear elements depend on more than one controlling variable, i.e. $i = f_g(u, v, ...)$ due to the time-varying bulk potential. Hence, the approach using Eq. (7) is not applicable. Instead, another description is necessary that considers the multi-dimensional dependency such as the multivariate Taylor series (Wambacq and Sansen, 1998) at an operating point $(u_0, v_0, ...)$ that is given for the two-dimensional case by

$$\begin{aligned} i(u,v) =& \sum_{n=0}^{\infty} \frac{1}{n!} \left(\frac{\partial}{\partial u}\Delta u + \frac{\partial}{\partial v}\Delta v \right)^n f(u_0, v_0) \\ =& f(u_0, v_0) + \Delta u \left.\frac{\partial f}{\partial u}\right|_{u_0, v_0} + \Delta v \left.\frac{\partial f}{\partial v}\right|_{u_0, v_0} + \\ &+ \frac{1}{2}\left(f_{uu}\Delta u^2 + 2 f_{uv} \Delta u \Delta v + f_{vv} \Delta v^2 \right) + .. \end{aligned} \tag{10}$$

Therein, $\Delta u = u - u_0$, $\Delta v = v - v_0$ denote the deviation from the operating point $(u_0, v_0, ...)$ and $f_{uu} = \left.\frac{\partial^2 f}{\partial u^2}\right|_{u_0, v_0}$ etc. the partial derivatives. In the following, the influence of substrate noise on a single transistor stage is examined by the use of the proposed model (Fig. 4).

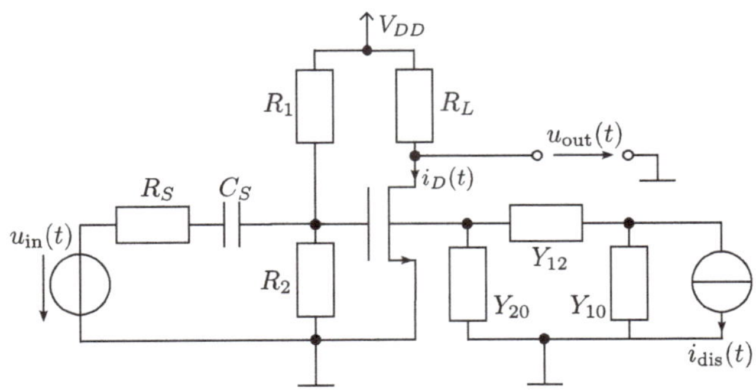

Fig. 5. Common source amplifier stage with gate voltage divider.

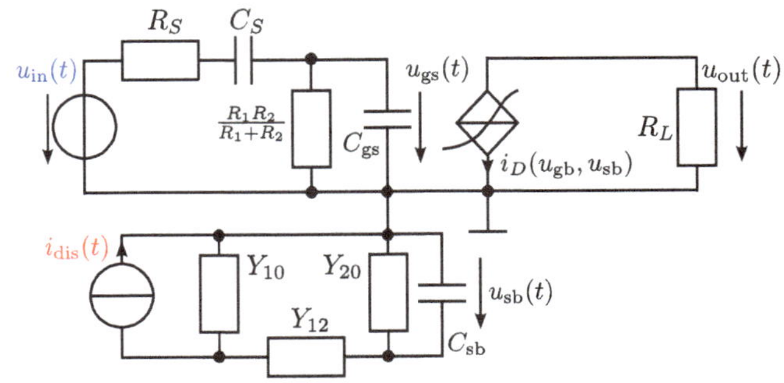

Fig. 6. Small signal model of the disturbed common source stage.

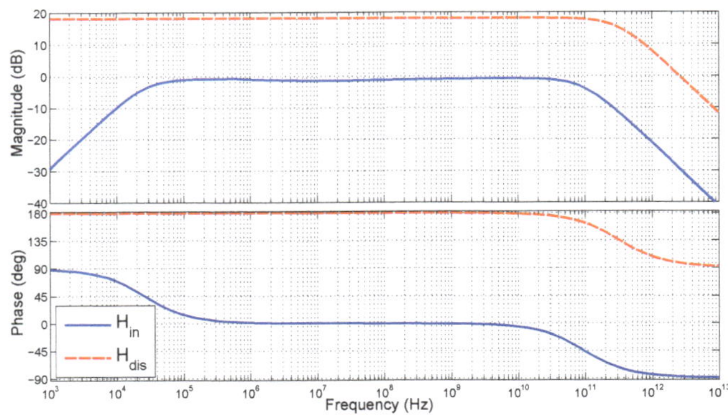

Fig. 7. Frequency response of H_{in} and H_{dis}.

4 Example

In the following, the proposed method is examined by a simple and hence vivid example. The considered circuit is the common source stage in Fig. 5. The MOS transistor is disturbed by the substrate noise that is modeled as independent current source. The noise couples into the bulk node via the examined three-terminal substrate model (cf. Fig. 2). The gate-source voltage of the MOSFET is set by a gate voltage divider.

4.1 Characterizing the linear dynamic block

The characterization of the linear transfer functions requires an equivalent linear small signal model of the circuit (cf. Fig. 6). The transistor is modeled as voltage controlled current source depending on the gate-bulk and source-bulk voltage and capacitances C_{gb} and C_{sb}. Further elements such as the drain-gate capacitance are neglected. Hence, the individual blocks can be analyzed separately (Weiner and Spina, 1980). The controlling voltages u_{gb} and u_{sb} are influenced by both signals, $u_{\text{in}}(t)$ and $i_{\text{dis}}(t)$. The approach shown in Sect. 3.1 leads to an analytic description of

$$H_{in}(s) = \left.\frac{\tilde{U}_{\text{gs}}(s)}{U_{\text{in}}(s)}\right|_{I_{\text{dis}}(s)=0}, \tag{11}$$

$$H_{dis}(s) = \left.\frac{\tilde{U}_{\text{sb}}(s)}{I_{\text{dis}}(s)}\right|_{U_{\text{in}}(s)=0} \tag{12}$$

in dependency of the network elements. Fig. 7 shows the frequency responses of $H_{\text{in}}(s)$ and $H_{\text{dis}}(s)$ of the analytic expressions. The input transfer function exhibits a bandpass characteristic and the distortion transfer function low-pass characteristic. Due to its impedance character, $H_{\text{dis}}(s)$ has a gain of approximately 20 dB. Thus, the frequency range of largest susceptibility can be identified which is in the present example up to 100 kHz wherein the nominal input signal is attenuated and the distortion signal is not. The input signals of the nonlinear block $\tilde{u}_{\text{in}}$ and $\tilde{u}_{\text{dis}}$, respectively, are determined using (6) assuming only periodic signals that can be expanded into a Fourier series.

4.2 Characterizing the nonlinear static block

Since the nonlinear block only represents the static behavior of the circuit, the capacitances are neglected and the regarded transconductance $i_D = f(\tilde{u}_{\text{in}}, \tilde{u}_{\text{dis}})$ is controlled by the previously determined signals (cf. Fig. 8). As an analytic approach to the static dependency, the drain current is assumed to be the forward current of the EKV model (Enz et al., 1995)

$$I_D(U_{\text{gb}}, U_{\text{sb}}) \approx \left[\ln\left(1 + e^{\frac{V_P - U_{\text{sb}}}{2U_t}}\right)\right]. \tag{13}$$

Here, $U_t = \frac{kt}{q}$ denotes the thermal voltage and V_P the pinch-off voltage given by

$$V_P = U_{\text{gb}} - V_{\text{th0}} - \gamma \left[\sqrt{U_{\text{gb}} - V_{\text{th0}} + \left(\sqrt{\Psi_0} + \frac{\gamma}{2}\right)^2} - \left(\sqrt{\Psi_0} + \frac{\gamma}{2}\right)\right]. \tag{14}$$

with the extrapolated threshold voltage V_{th0}, the body effect factor γ and the surface potential for strong inversion at equilibrium Ψ_0. Transferring the static expression for the drain current (13) to the multivariate Taylor series (10), an analytic expression for the time-varying drain current in dependency of the intermediate signals with $u_{\text{gb}} = \tilde{u}_{\text{in}} + \tilde{u}_{\text{dis}}$ and

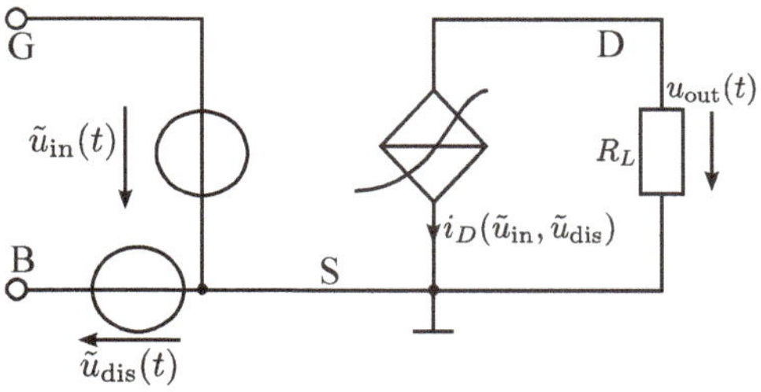

Fig. 8. Two-dimensional voltage controlled current source.

$u_{\rm sb} = \tilde{u}_{\rm dis}$ can be determined

$$i_{D_N}(t) = f(\tilde{u}_{\rm in}, \tilde{u}_{\rm dis}) \quad (15)$$
$$= \sum_{m=1}^{N}\sum_{n=0}^{m}\left(\frac{\tilde{u}_{\rm in}^{n}(t)}{n!}\frac{\tilde{u}_{\rm dis}^{m-n}}{(m-n)!}\frac{\partial^m i_D(\tilde{u}_{\rm in}, \tilde{u}_{\rm dis})}{\partial \tilde{u}_{\rm in}^{n}\partial \tilde{u}_{\rm dis}^{m-n}}\bigg|_{\rm OP}\right).$$

As a result, a dependency of the output current on the input signals and the circuit parameters that influence the Taylor coefficients and the linear transfer functions is found. Comparative simulations are accomplished using these expressions whereby for the use of simplicity the signals $u_{\rm in}(t)$ and $i_{\rm dis}(t)$ are single tones, respectively. The frequencies are set to $f_{\rm in} = 10\,{\rm kHz}$ and $f_{\rm dis} = 3\,{\rm kHz}$ since in this frequency range the input signal $u_{\rm in}(t)$ is attenuated noticeably. For validation, a comparison between a harmonic balance simulation in Spectre and the analytic expression found by the block approach is performed, both for amplitudes of $\hat{u}_{\rm in} = 100\,{\rm mV}$ and $\hat{i}_{\rm dis} = 25\,{\rm mA}$. The resulting output spectra for terms up to the third order are shown in Fig. 9 and reveal a very good match for the block model approach.

5 Conclusions

In this contribution, an approach for the susceptibility analysis of nonlinear analog circuits due to substrate noise has been presented. The coupling mechanism of digitally generated noise has been modeled as a three-terminal network excited by an independent current source. This model has been used in combination with a multi-input Wiener model proposed in Stegemann et al. (2012) in order to get an analytic approximation of the input-output characteristic of the disturbed system. Hence, circuit designers are equipped with equations for the estimation of most crucially influencing parameter. In addition, the frequency range in which the circuits are the most susceptible to distortion signals can be found. The block-oriented approach is based on classic methods for the distortion analysis, e.g. Wambacq and Sansen (1998). It has been discussed how the blocks can be determined in order to examine the influence on a single transistor. Nevertheless, it has been noted that the MNA based method presented in Stegemann et al. (2012) cannot be adopted for complex circuits due to the multivariate power series approach for each nonlinear element. Thus, future work will deal with this

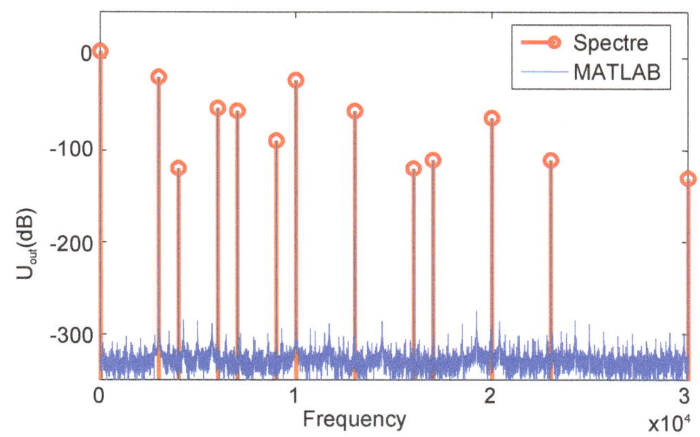

Fig. 9. Spectrum of the output voltage $u_{\rm out}$.

drawback. One possible approach could be the use of one-dimensional power series expansions with the consideration of time-variant operating points due to the substrate noise.

Acknowledgements. This contribution was developed within the scope of the project *MAS* which is funded by the BMBF (Bundesministerium für Bildung und Forschung) under the project number *16SV4059*. The responsibility for this publication is held by the authors only.

References

Bronckers, S.: Substrate Noise Coupling in Analog/RF Systems, Ph.D. thesis, Vrije Universiteit Brussels, 2009.

Dhia, S. B., Ramdani, M., and Sicard, E. (Eds): Electromagnetic compatibility of integrated circuits techniques for low emission and susceptibility, Springer, New York, 2006.

Enz, C. C., Krummenacher, F., and Vittoz, E. A.: An analytical MOS transistor model valid in all regions of operation and dedicated to low-voltage and low-current applications, Analog Integr. Circuits Signal Process., 8, 83–114, http://dx.doi.org/10.1007/BF01239381, 1995.

Lin, L., Xiong, J., and Mathis, W.: Modellierung und Simulation des Substrat-Rauschens in integrierten RF CMOS-Schaltungen, Adv. Radio Sci., 7, 163–168, doi:10.5194/ars-7-163-2009,2009.

Schetzen, M.: Multilinear theory of nonlinear networks, Journal of the Franklin Institute, 320, 221–247, 1985.

Stegemann, S., John, W., C.Widemann, and Mathis, W.: A Block-Oriented Approach to EMI-induced Distortion Analysis for Nonlinear Analog Circuits, in: EMC EUROPE 2012, Roma, 2012.

Vlach, J. and Singhal, K.: Computer Methods for Circuit Analysis and Design, Van Nostrand Reinhold, 1983.

Wambacq, P. and Sansen, W.: Distortion Analysis of Analog Integrated Circuits, Kluwer Academic Publishers, 1998.

Weiner, D. D. and Spina, J. F.: Sinusoidal Analysis and Modeling of weakly nonlinear Circuits, Van Nostrand Reinhold, 1980.

15

Substituting EMC emission measurement by field and cable scan method using measured transfer function

D. Rinas, J. Jia, A. Zeichner, and S. Frei

Dortmund University of Technology, Dortmund, Germany

Correspondence to: D. Rinas (denis.rinas@tu-dortmund.de)

Abstract. Today EMC emissions of automotive components are often measured in anechoic chambers by an antenna at fixed position according to CISPR 25 (ALSE-method). The antenna voltage often cannot sufficiently describe the behaviour of the measured electronic components and systems. Furthermore space requirements and costs are very high for the ALSE-method. Field- and cable-scan methods combined with near-field to far-field transformation techniques might be a good alternative.

Residual reflections from the walls, the metallic floor, the measuring table, interaction of the antenna with the environment, and other factors affect the measurements. Thus, models which only regard the current distribution for near- and far field calculation cannot produce results equal to a chamber measurement.

In this paper methods for computing transfer functions for the substitution of EMC antenna measurements with field- and cable scans in a specified calibration area are introduced. To consider influences of the environment, the environment is characterized in a first step and included with transfer functions in the calculation process for the equivalent ALSE-field.

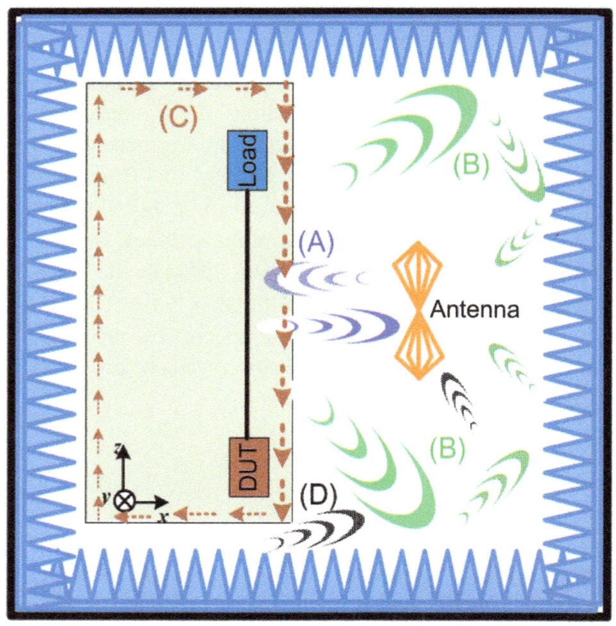

Fig. 1. Influencing factors of CISPR 25 ALSE method.

1 Introduction

Standardized component field measurement methods, like the ALSE antenna method defined e.g. in CISPR 25 (2007) for evaluation of electro-magnetic emissions from automotive systems, suffer from the need of large and expensive anechoic chambers. Also a single field strength value is often not sufficient to characterize the EMI behavior of a complex system. Residual reflections from the walls, the metallic floor, the measuring table, interaction of the antenna with the environment, and other influences affect the measurements and reduce reproducibility. Furthermore it is not possible to use the measurement data for behavioral simulation model creation. Alternative methods are necessary.

Basically the electromagnetic emission can be distinguished in the emission of printed circuit boards with housing and the emission of the connecting cable bundles. The behavior of the cable can be obtained by cable scan methods where the dominant common-mode current is used to create an emission model (Jia et al., 2012). Time domain measurements give the possibility to get amplitude and phase information. Alternatively phase retrieval algorithm based on current distribution measurements could identify in some cases the phase based on amplitude-only frequency domain measurements. To compute an emission model of PCBs

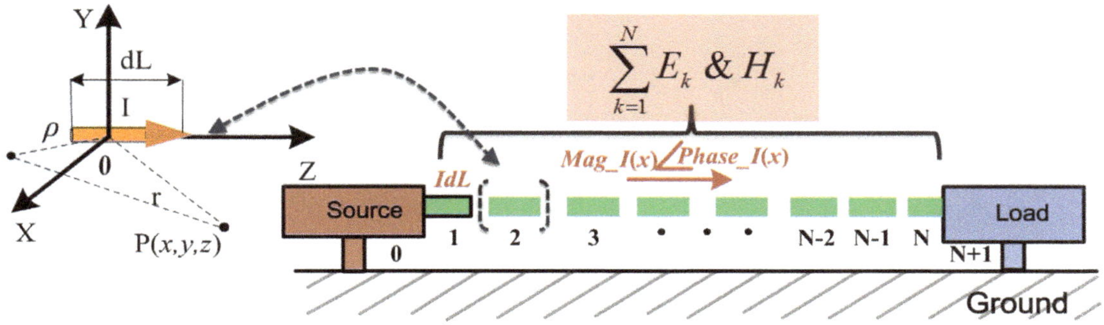

Fig. 2. Multiple-dipole-model for a cable bundle.

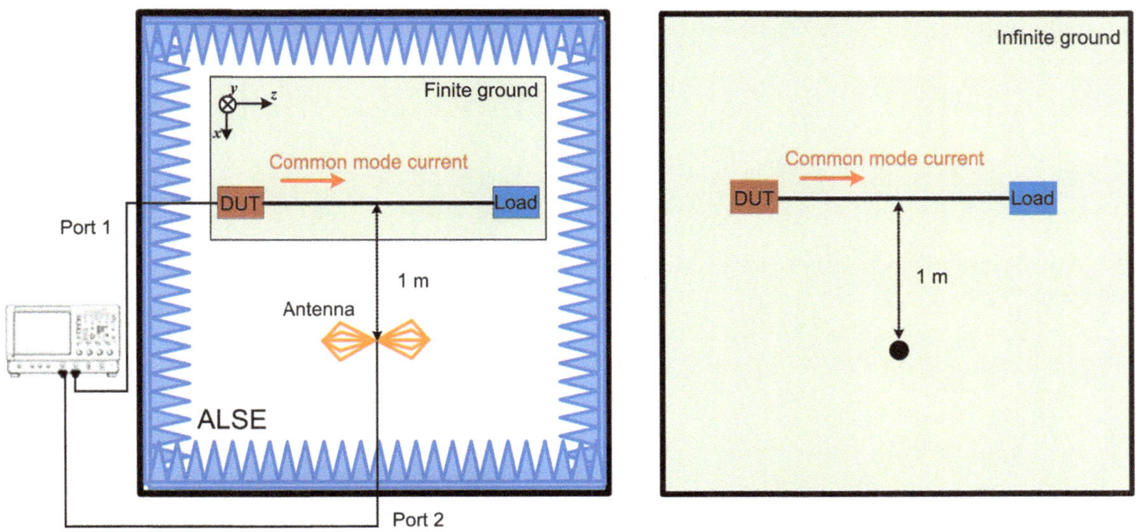

Fig. 3. CISPR 25 setup and simulation model. **(a)** Antenna measurement from cable bundle. **(b)** Field simulation from cable bundle by MoM.

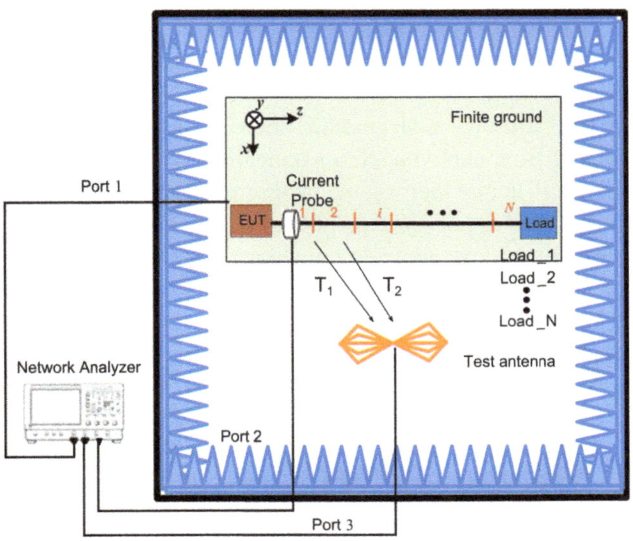

Fig. 4. Transfer factor measurement setup.

respectively planar structure field scan methods are applicable (Rinas and Frei, 2012; Chen et al., 2009; Thomas et al., 2012). Here the current distribution on the PCB can be found through measuring the spatial magnetic field above the PCB. Including geometry properties and knowledge of possible current paths the radiation model can be optimized.

As the cable and PCB models from near-field scan methods allow emission calculation in free space but the standardized ALSE antenna measurements are done in chamber measurement environment with influencing factors listed above, the alternative methods must consider in the first step the chamber effects in order to produce comparable results (Smith and Frazier, 1998). Influence factors are interaction with antenna (A), reflections from the walls of anechoic chamber (B), currents in table edges (C) and emission from measurement equipment cables (D), as shown in Fig. 1. Methods to determine correction and transfer functions based on measurement data are presented in this paper.

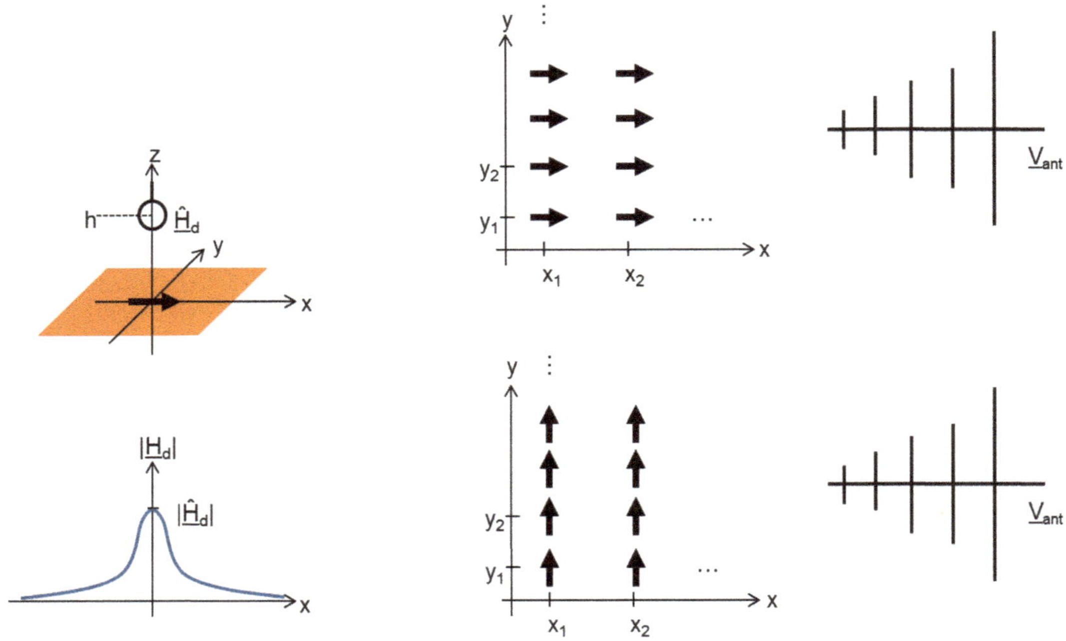

Fig. 5. Hertzian dipole configurations for calibration process. **(a)** Hertzian dipole and dipole near field. **(b)** Calibration in x- and y-direction for transfer function computation.

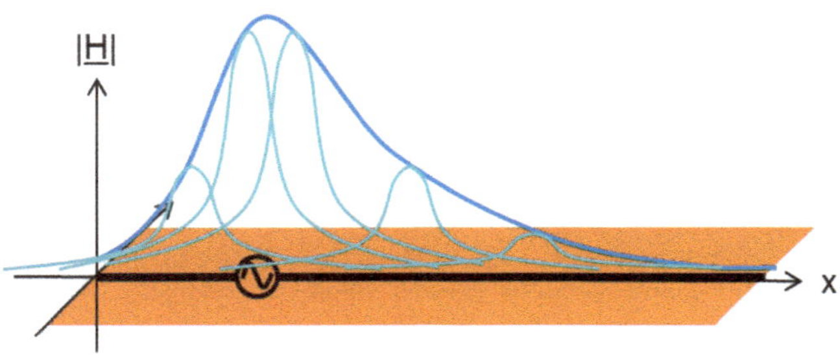

Fig. 6. Magnetic near field in each row and column of PCB can be approximated by varying position of a set of dipoles.

2 Methods for substituting of antenna emission measurement

In the following sections the basic methods for antenna emission measurement substitution are introduced. The differences between cable bundle and PCB measurements are discussed.

2.1 Cable bundles

In a radiation model a cable bundle can be divided into a set of short dipoles (Jia et al., 2012) as shown in Fig. 2. In this model, the influence of ground is incorporated into the radiation model using mirror theory which is actually different from real test configuration with more complex environment, i.e. metallic table of limited size, large test antenna, non-perfect absorbing materials, and so on. Aiming at integrating real influence factors in radiation models, we developed two methods to correct calculated field result.

The proposed method calculates the absolute electric field at a reference point, the antenna center point. However, the test antenna converts the electric field around the antenna to a receiving voltage. In order to get the antenna field, the measured voltage should be multiplied with the antenna factor (AF) provided by the antenna manufacturer. In real test configurations, the antenna is located in the near field from the cable bundle in a wide frequency range, while AF from manufacture is calibrated in far field in an open area test site. Therefore this antenna factor will not be accurate when used in near field test. We first measure antenna receiving voltage radiated from cable bundle in standard configuration as shown in Fig. 3a and then use AF from manufacturer to obtain electric field at reference point (1 m distant to cable bundle). Subsequently we simulate the radiation from a verified MoM cable model to acquire electric field at the same point, as shown in Fig. 3b. Finally we can calculate the correction factor according to Eq. (1):

$$\begin{aligned} E_{\text{measure}} &= V_{\text{measure}} + AF(\text{dB}) \\ K &= E_{\text{measure}} - E_{\text{simulation}}(\text{dB}) \end{aligned} \tag{1}$$

Another method is based on a set of measured transfer functions from known currents and measured antenna voltages, as shown in Fig. 4. We scan the currents at N different nodes along the cable bundle when the cable is terminated with N different loads. $N \times N$ currents and N antenna voltages can be acquired, and then we can solve an equation system (4) to get a set of transfer functions $[T_1, T_2, \ldots T_i, \ldots, T_N]$. In

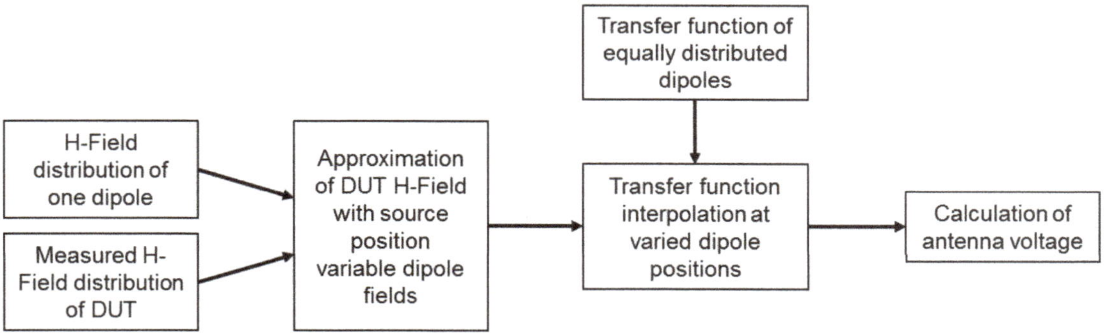

Fig. 7. Process diagram of antenna emission measurement substitution.

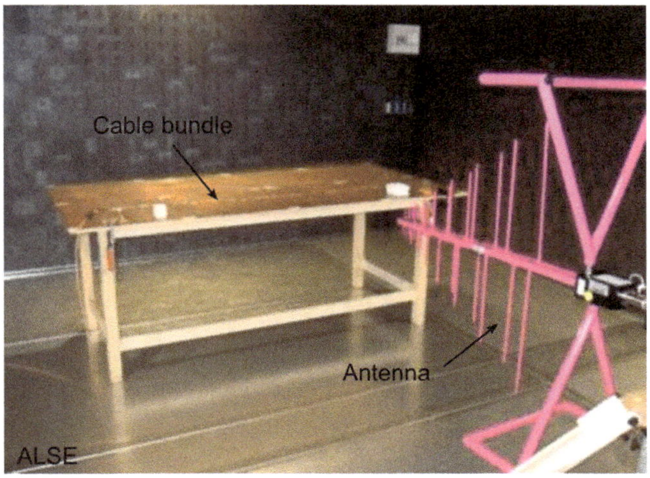

Fig. 8. Setup for measurement of transfer functions and verification of the method.

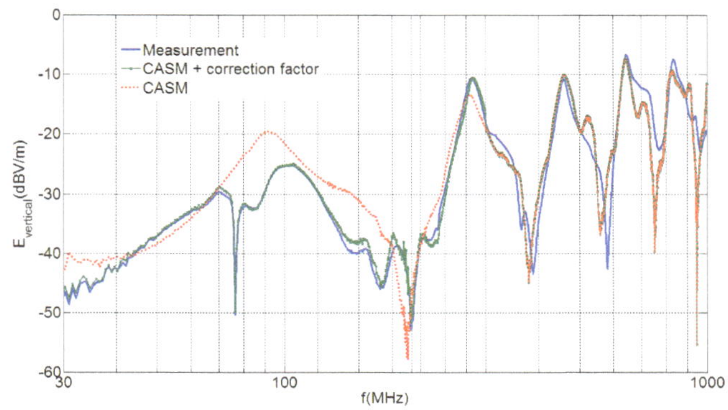

Fig. 9. Electric field in vertical direction of measurement and CASM.

real test, we can directly evaluate the antenna voltage based on scanned currents of cable bundle (amplitude and phase) and transfer functions T_i.

$$S_{21} = 2 \times \frac{V_2}{V_S} = 2 \times \frac{I_i \times Zt}{V_S} \tag{2}$$

$$S_{31} = 2 \times \frac{V_3}{V_S} = 2 \times \frac{U_i}{V_S}(i = 1 \cdots N) \tag{3}$$

$$\begin{bmatrix} I^1_{\text{Load\_1}} & I^2_{\text{Load\_1}} & \cdots & I^N_{\text{Load\_1}} \\ I^1_{\text{Load\_2}} & I^2_{\text{Load\_2}} & \cdots & I^N_{\text{Load\_2}} \\ \vdots & \vdots & \vdots & \vdots \\ I^1_{\text{Load\_N}} & I^2_{\text{Load\_N}} & \cdots & I^3_{\text{Load\_N}} \end{bmatrix} \begin{bmatrix} T_1 \\ T_2 \\ \vdots \\ T_N \end{bmatrix} = \begin{bmatrix} U_1 \\ U_2 \\ \vdots \\ U_N \end{bmatrix} \tag{4}$$

2.2 PCB

In Sect. 2.1 it is shown that the calibration between cable current and test antenna voltage can be computed by dividing a cable in N segments and determining a transfer function for each segment. In the same way a PCB can be discretised in $N \times M$ squares and transfer functions can be computed for each square in x- and y-current direction, as shown in Fig. 5b.

As reference for calibration process and theoretical considerations a dipole antenna is applied. The magnetic near field in a plane above the dipole can be calculated or measured (Rinas et al., 2011), as shown in Fig. 5a. The transfer function for each square is calculated by varying the position of the reference dipole along the squares, as shown in Fig. 5b.

$$T_{nm,x} = \frac{V_{\text{ant}}}{H_{nm,x}} \tag{5}$$

$$T_{nm,y} = \frac{V_{\text{ant}}}{H_{nm,y}} \tag{6}$$

where $T_{nm,x}$ and $T_{nm,y}$ are the transfer functions for the squares in x- and y-direction, V_{ant} is the corresponding antenna voltage and $H_{nm,x}$and $H_{nm,y}$the measured magnetic fields.

To get the magnetic near field of the PCB it is scanned with near field scanner in x- and y-direction. The resolution of near field scanning is $K \times L$ with $K \gg N$ and $L \gg M$. With the use of an optimization algorithm a distribution of

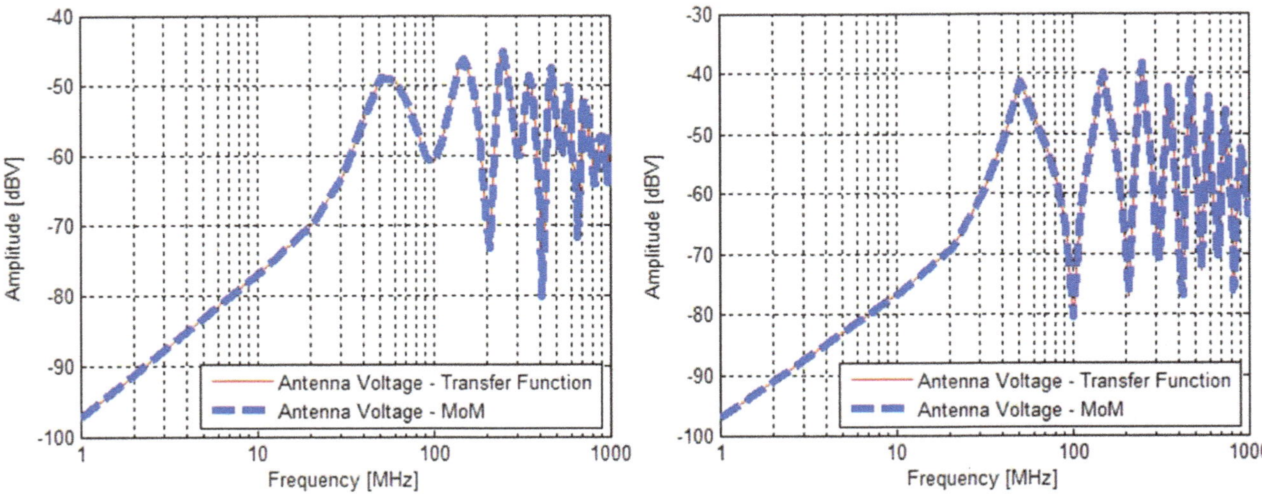

Fig. 10. Antenna voltage from MoM and transfer function method. **(a)** Load: 1600 Ω. **(b)** Load: 400 Ω − 500 μH − 200 pF.

a dipoles can be found which best fits the magnetic field of each scanned row and column, as shown in Fig. 6. After that the transfer function for each dipole at new positions is interpolated. It should be noted that approximating the magnetic field in amplitude and phase close to the source structure is equal to approximating the current distribution. Knowing current distribution voltage distribution is known to and magnetic and electric field can be calculated. Superposition of transfer function calibrated squares leads to calculation of resulting antenna voltage.

$$V_{\text{ant}} = \sum T_{nm} H_{nm} \tag{7}$$

The process diagram for PCBs is shown in Fig. 7.

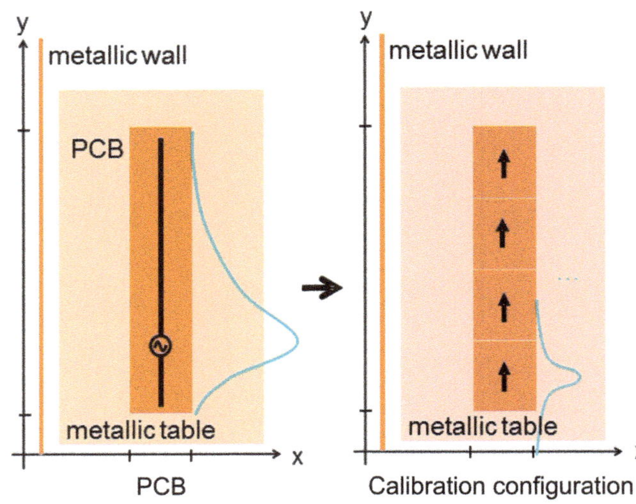

Fig. 11. Transfer function calibration of PCB with set of dipoles.

3 Results

To show the applicability of the presented methods parts of results are based on Method of Moments (MoM) simulation data.

3.1 Cable bundles

In order to verify the proposed methods of correction factor and transfer function, we use an EMC antenna to test radiation from a simple cable, as shown in Fig. 8. The CASM method (Jia et al., 2012) was used to acquire amplitude and phase of currents along the cable and then electric field was calculated at the reference point using dipole model. We compared the results from measurement (measured antenna voltage plus antenna factor), CASM with mirror theory and CASM based on correction factor, as shown in Fig. 9. It can be seen that the resonances around 70 MHz and 350 MHz originating from test environment can be considered through correction factor. The deviation in high frequency is mainly from measured current amplitude error and CASM algorithm itself.

For transfer function method, we use here simulation results by MoM method. The cable has a length of 1.5 m, a height to metallic table of 0.05 m, and is terminated by 17 different loads: short, open, 50 Ω, $100 \times M$ Ω ($M = [1, 2, \ldots, 14]$). We simulate the currents at different nodes of cable terminated by 17 different loads; also a MoM model of a biconical antenna was created to simulate antenna voltages. According to Eq. (4) the transfer function T_N can be determined. In order to verify the transfer functions, we simulated two cases: the cable is terminated by 1600 Ω or a series impedance of 400 Ω − 500 nH − 200 pF. Figure 10 presents antenna voltage measured directly and antenna voltage computed using the transfer function. It can be seen, the results from transfer function method agree well with direct antenna

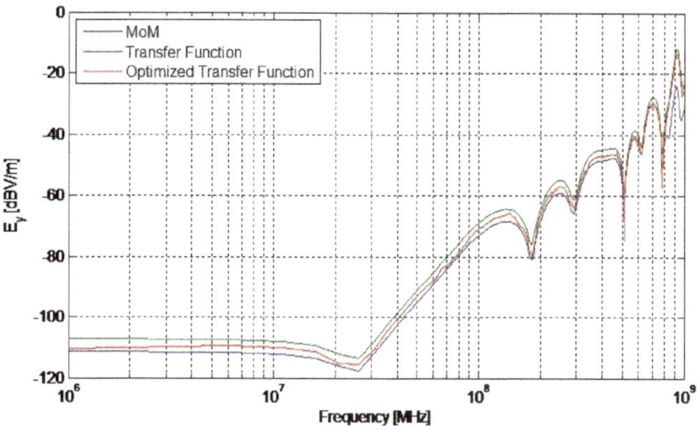

Fig. 12. Electric field calculated with transfer function and optimized transfer function in comparison to electric field from MoM.

voltage measurements. But it is noted that the scanned current and voltage antenna here are based on simulation results, which are free from noise, which can not be avoided in real measurements. The current and antenna voltage are both complex numbers in Eq. (4) which need amplitude and phase simultaneously.

3.2 PCB

For transfer function substitution, we present simulation results by MoM method. The test PCB consists of a single cable with a length of 0.16 m, a height over metallic plane of 5 mm, and two open ends. It is fed at length of 0.04 m with a supply voltage of 1 V. Transfer function calibration is done with four dipoles of finite length in a height of 5 mm above metallic plane, as shown in Fig. 11. According to Eqs. (5)–(7) transfer functions T_{nm} are computed and antenna voltage is calculated. To get better results approximation to near field distribution and transfer function interpolation is done.

Figure 12 presents the results of the substitution process. It can be seen that the electric field computed with transfer function shows good results up to a frequency of 800 MHz. Using near field approximation and transfer function interpolation the results can be improved in the entire frequency range.

4 Conclusions

In this paper methods for substituting antenna emission measurements with field and cable scan methods are introduced. Influencing factors like interaction with antenna, wall reflections, table currents and emission from measuring cable can be considered by use of measurement data from "calibration structures" and transfer functions. Cable bundles and PCBs are divided in segments and the transfer function can be computed for each of these segments. Superposition of the emissions gives the desired antenna voltage.

First results, partly based on MoM simulation data, are shown which proves the applicability of the proposed method. The investigations have to be extended in future.

Acknowledgements. The reported R + D work was carried out within the CATRENE project CA310 EM4EM (Electromagnetic Reliability and Electronic Systems for Electro Mobility). This particular research is supported by the BMBF (Bundesministerium fuer Bildung und Forschung) of the Federal Republic of Germany under grant 16 M3092 I. The responsibility for this publication is held by the authors only.

References

Chen, Q., Chakarothai, J., and Sawaya, K.: Estimation of Current Distribution by Near-Field Measurement, CEEM, China, 2009.

CISPR 25 Ed. 3: Vehicles, boats and internal combustion engines – Radio disturbance characteristics – Limits and methods of measurement for the protection of on-board receivers, IEC, 2007.

Jia, J., Rinas, D., and Frei, S.: Prediction of Radiated Fields from Cable Bundles based on Current Distribution Measurements, EMC Europe, Rome, 2012.

Smith, W. T. and Frazier, R. K.: Prediction of Anechoic Chamber Radiated Emissions Measurements Through Use of Empirically-Derived Transfer Functions and Laboratory Common-Mode Current Measurements, EMC, Denver, 1998.

Thomas, D. W. P., Obiekezie, C., Greedy, S., Nothofer, A., and Sewell, P.: Characterisation of Noisy Electromagnetic Fields from Circuits using the Correlation of Equivalent Sources, EMC Europe, Rome, 2012.

Rinas, D. and Frei, S.: Methoden zur Optimierung von Störaussendungsmodellen für Platinenstrukturen auf Basis von Nahfeldmessdaten, EMV Düsseldorf, Germany, 2012.

Rinas, D., Niedzwiedz, S., Jia, J., and Frei, S.: Optimization Methods for Equivalent Source Identification and Electromagnetic Model Creation based on Near-Field Measurements, EMC Europe, York, 2011.

16

DMT transmission in the context of industrial telecontrol applications

S. Edinger, C. Bauer, and N. J. Fliege

University of Mannheim, Chair of Electrical Engineering, Germany

Abstract. In this paper, we discuss the use of Discrete Multi Tone (DMT) modulation in the context of industrial telecontrol applications. We highlight the specific requirements and characteristics of the telecontrol settings and present methods to cope with the environmental challenges posed. It turns out that DMT is ideally suited for the tasks at hand. Further enhances enable our proposed system to provide superior connection stability even under the most adverse conditions.

1 Introduction

DMT is the wireline application of the well-known Orthogonal Frequency Division Multiplexing (OFDM) transmission scheme. In recent years, commercial applications of DMT such as Asymmetric Digital Subscriber Line (ADSL) have gained immense popularity. Those applications are characterized by high data rates and relatively short transmission loops. The use of DMT in industrial process data communications and metering applications has not found wide-spread acceptance yet, despite its obvious beneficial aspects. This is mainly due to the fact that conventional DMT systems suffer from relatively high transmission latency. Industrial settings are characterized by relatively low data rates but long transmission loops, require a high degree of reliability, and pose real-time constraints.

At our chair, a DMT transceiver for deployment in industrial settings was developed (Bauer, 2004) tackling the shortcomings of conventional DMT systems. Sophisticated signal processing methods lead to a latency-optimized design with online adaptation facilities to suddenly occurring deteriorations of the transmission channel. Thus, both low latency and high adaptivity are combined to demonstrate the viability of DMT transmission for telecontrol applications with superior connection reliability.

Correspondence to: S. Edinger
(stefan.edinger@ti.uni-mannheim.de)

2 System model

A basic model of the transmission system is depicted in Fig. 1. The binary input data stream is parallelized and partitioned into so-called sub-carriers. Each sub-carrier i can transmit a certain number of bits b_i. The block denoted QAM performs a quadrature amplitude modulation (QAM) of the bits b_i, yielding one complex value $u_i[k]$ for each sub-carrier i and DMT symbol k. The transmit signal in time domain is generated by computing the inverse discrete Fourier transform (IDFT) over the complex frequency domain signal. Transmission takes place in the baseband. Thus, only real-valued information can be transmitted. To generate a real IDFT result, a complex conjugation of the frequency-domain transmit signal has to be executed and appended to the original vector (Bingham, 1990).

The IFDT and DFT operations are usually executed with a power of 2 as input and output lengths which allows an efficient computation in the form of the fast Fourier transform (FFT). Specifically, we use an FFT length of 128. Thus, at most 64 sub-carriers can be modulated independently.

The resulting transmit vector of 128 real values is prepended by a so-called cyclic prefix of 32 values which is used to prevent inter-symbol and inter-carrier interference (ISI and ICI)due to the channel impulse response. The vector is converted to serial and transmitted over the channel with impulse response h_c. In this paper it is assumed that the length of the channel impulse response does not exceed the length of the cyclic prefix to preclude the occurrence of ICI and ISI.

Transmission is impaired by noise stemming from both additive white Gaussian noise (AWGN) and narrow bandwidth interference (NBI). At the receiver the operations per-

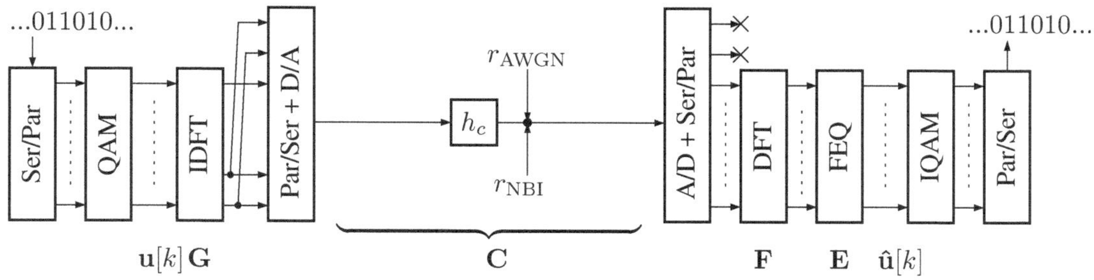

Fig. 1. DMT transmission scheme.

Fig. 2. Topology of example transmission channel.

formed at the transmitter are reversed. Additionally, a one-tap frequency domain equalizer removes distortions due to the transmission channel. QAM demodulation should yield the same binary sequence that was sent by the transmitter.

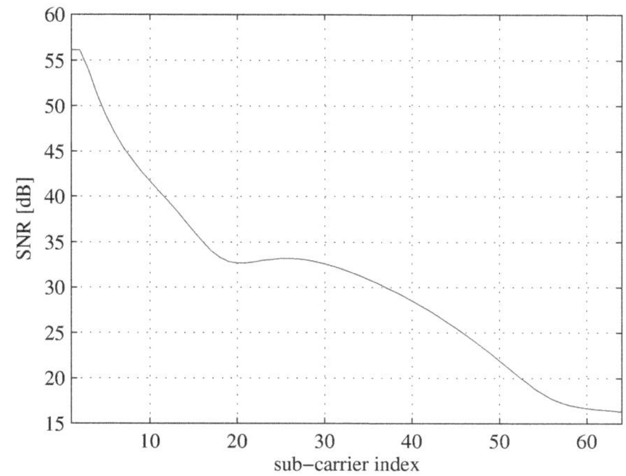

Fig. 3. SNR for example transmission channel.

3 Transmission channel

In telecontrol communications over wire lines one is faced with frequency selective channels. The impulse response of the transmission channel is determined by material constants of the medium and by the topology of the transmission network. Sophisticated models exist that allow derivation of the transfer function for arbitrary topologies. Usually, telecontrol applications in industrial settings use rather long transmission lines. Possibly, there are unterminated or imperfectly terminated bridge taps resulting in reflections of the transmitted data. The superposition of several reflected waves leads to a pronounced frequency selectivity of the encountered transfer functions. However, the transfer function is slowly time-variant for wire lines. This is a great advantage compared to wireless communication channels which fluctuate very quickly.

The slow time variance of the transfer function allows efficient use of the available resources by means of adaptation to the respective channel characteristics. Before actual data transmission starts, the channel is estimated during the initialization phase. A special training sequence with content known to both stations is transmitted and comparison with the received sequence allows estimation of the channel attenuation factors g_i and noise powers n_i per sub-carrier. Together with the reference power p_{ref} per sub-carrier used to transmit the training sequence, the signal to noise ratios (SNR) per sub-carrier i can be computed as

$$\text{SNR}_i = \frac{p_{\text{ref}} \cdot g_i}{n_i} \tag{1}$$

4 Initial adaptation

For the rest of the paper, we will consider two stations connected by the topology shown in Fig. 2. Both stations are terminated with a resistance of 120 Ω. One unterminated bridge tap is present. Lengths of the line segments are given in feet (1 ft = 30.48 cm). The transmission wire is 26AWG. This transmission channel will be used to illustrate the techniques explained in this paper.

We assume white Gaussian background noise with power $n_i = 120\,\text{dBm/Hz}$. The total transmit power of the system is limited to $p_{\text{tot}} = 25\,\text{dBm}$. This power is equally distributed over all sub-carriers during the channel estimation phase, thus $p_{\text{ref}} = p_{\text{tot}}/128$. The sampling frequency of the DMT system is $f_S = 1024\,\text{kHz}$. The rather high sampling frequency and the short FFT length effectively reduce the latency of the system due to block oriented signal processing. The DMT symbol rate equals $r_S = 6400\,\text{Hz}$.

With all these parameters, the SNR can be computed as described in the last section and are depicted in Fig. 3.

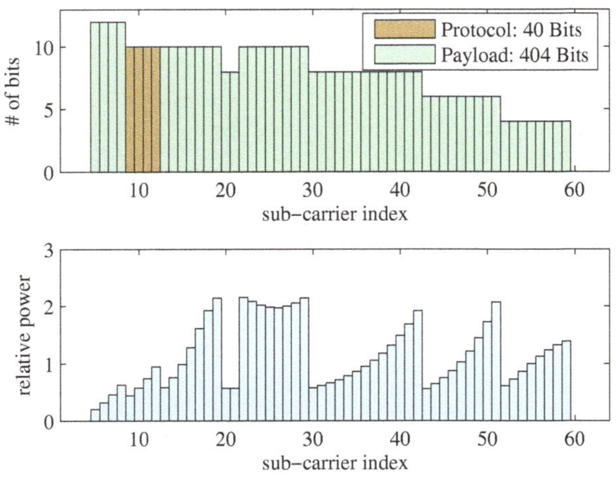

Fig. 4. SNR for example transmission channel.

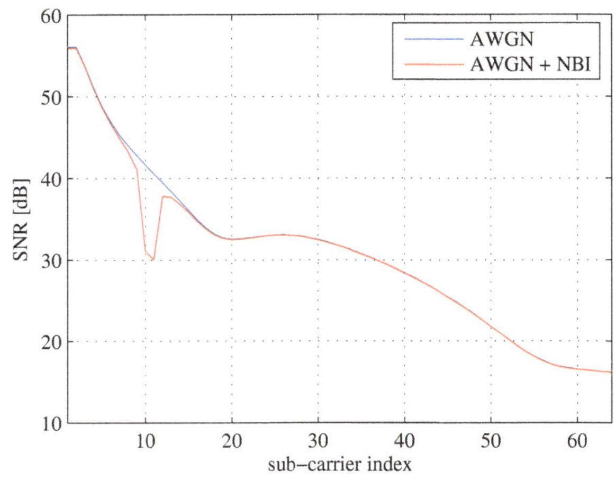

Fig. 5. SNR with and without additional NBI.

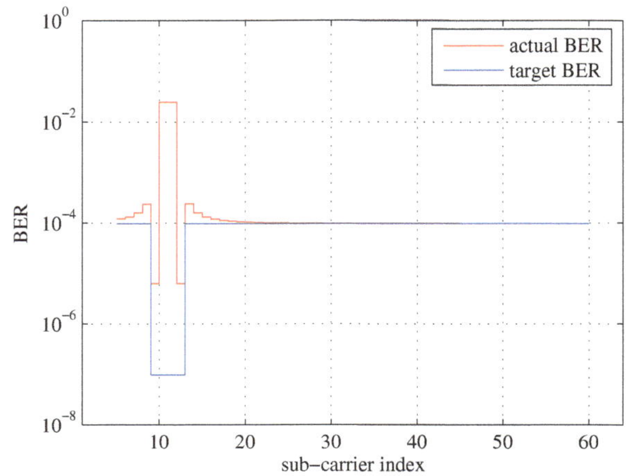

Fig. 6. Change of BER due to NBI.

According to the SNR values computed during the initialization phase, a so-called bit-loading algorithm determines how many bits should be transmitted and what transmit powers should be used per sub-carrier. The result of the loading algorithm are two tables, one indicating the bit-load (bit allocation table, BAT) and the other indicating the powers relative to the reference power (power allocation table, PAT).

Our DMT system applies a modified version of the well-known Hughes Hartogs bit-loading algorithm (Hughes-Hartogs, 1989) to compute these two tables. The modification allows to specify two data classes with different protection (Hoo et al., 1999b). In transmission systems, protection is usually measured and given in terms of bit error rate (BER), the number of erroneous bits divided by the total number of transmitted bits. Our system makes a distinction between protocol information used to establish and maintain the point-to-point connection of the two stations and between payload data that is transparently transmitted. Note that a temporary increase in the BER of payload data does not harm the connection itself. However, erroneous protocol information will lead to the use of defective transmission parameters and after a short time to the breakdown of the connection.

In telecontrol applications, high stability of the connection is of great concern. Hence, we provide increased protection for the protocol data. For illustration purposes, we fix the amount of protocol data to 40 bits per DMT symbol, and we call these data the embedded service channel (ESC). The target BER for the ESC is fixed to 10^{-7}. For payload data, we set the target BER to 10^{-4}. Again, these parameters are application dependent values. We chose values that allow for easy simulation of results. One might think of the payload information as voice data or video surveillance information, where rather high BER is acceptable.

Figure 4 shows the BAT and PAT for the above parameters that results in the maximum number of payload bits per DMT symbol $b_{\max}$, in this case $b_{\max} = 404$.

Note that due to the use of imperfect anti-imaging and anti-aliasing filters, some sub-carriers at low frequencies and high frequencies can not be used. The actual number of independently modulated sub-carriers is thus reduced to 55.

5 Appearance of disturbances

Once optimal BAT and PAT are computed and exchanged, the actual data transmission starts using these transmission parameters. The occurrence of a disturbance, e.g. an increase of noise power on some sub-carriers, can significantly deteriorate the transmission quality for the affected sub-carriers. An exemplary disturbance, NBI centered between sub-carriers 10 and 11, is shown in Fig. 5.

Due to the deterioration of the SNR of the affected sub-carriers, the BER for these carriers rises too. The increase of BER is shown in Fig. 6, together with the target BER.

It can be seen that despite the increased protection for the ESC, the disturbance heavily impairs the transmission system.

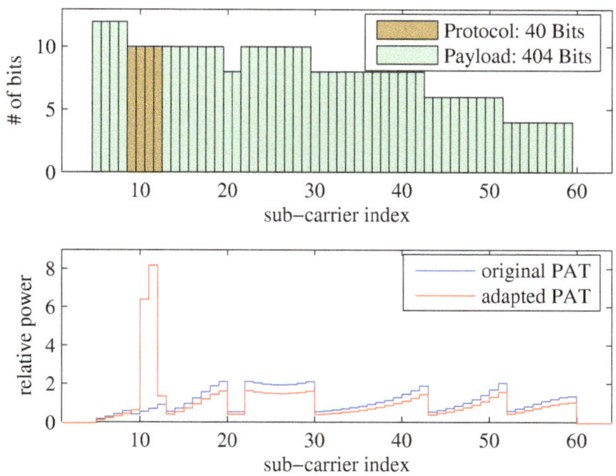

Fig. 7. Increase of transmit power.

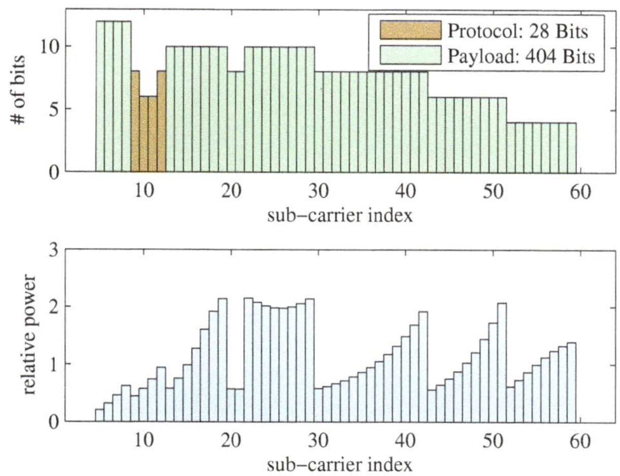

Fig. 8. Reduction of constellation size.

This situation calls for dynamic adaptation, that is, adaptation of the system to changed channel conditions during runtime. In ADSL, for example, a sequence of bit-swap commands can be exchanged to redistribute bits away from heavily disturbed sub-carriers to less disturbed sub-carriers (Hoo et al., 1999a). However, such methods require an exchange of protocol information over the ESC. If the ESC itself is affected by a disturbance, as is the case in our example, the protocol information might be faulty. In ADSL, the protocol data is protected by a five-out-of-nine repetition code (ADSL). We could show in Edinger et al. (2005b) that even such protection is not sufficient. The use of faulty transmission parameters will lead to a loss of connection.

The next section examines methods to restore acceptable BER for the ESC, and thus restoring reliable information exchange, by means of signal processing.

6 Securing the ESC

A number of methods can be applied, to restore ESC quality and reliability if it is impaired by a disturbance. Some of these methods can be applied blindly, others imply the exchange of information. Some methods work on individual sub-carriers, others work on several or all sub-carriers. The amount of information exchange should be as small as possible.

6.1 Increase of transmit power

If the noise power on a sub-carrier increases by factor δ_n, the transmit power on this sub-carrier can be increased by the same factor to restore the target BER. Fig. 7 shows the transmit powers necessary to restore a BER of 10^{-7} for the ESC in the presence of the example disturbance.

Note the large power peaks at affected ESC carriers. If power spectral density (PSD) constraints restrict the maximum transmit power level, the above constellation might violate those constraints. Additionally, since total transmit power is limited, the additional power for the ESC must be collected from payload sub-carriers, which further increases their BER. Another drawback of this method is that good estimates of the noise power must be available for determining the additional transmit power required. These estimates are not immediately available but must be gathered from observations of the noise process which adds to the time required for adaptation.

6.2 Reduction of constellation size

When the number of bits on a sub-carrier is reduced by 2, this corresponds to an increase of transmit power by $\approx 6\,\text{dB}$. Thus, the constellation size of disturbed ESC sub-carriers can be reduced until the amount of bits equals zero or the required transmission quality is met again. However, dynamically reducing the amount of ESC information per DMT symbol significantly increases the complexity of the underlying protocol. Information might need to be spread over several symbols and recombined after reception.

Figure 8 shows the changed BAT that is required to reach BER of the ESC of equal or less than target BER. A reduction of the ESC by 12 bits is necessary.

Figure 9 compares the two above methods in terms of the actual BER they obtain. It can be seen that power adaptation exactly reaches the target BER. Due to limited transmit power the BER of payload sub-carriers is increased. Note that the BAT does not change. Only the updated PAT values for the ESC must be exchanged, the remaining PAT entries can be computed from those values.

On the other hand, payload BER is not increased for the method of reducing ESC constellation sizes. The PAT is not changed. The only information that must be transmitted are the updated BAT values for the ESC which can be

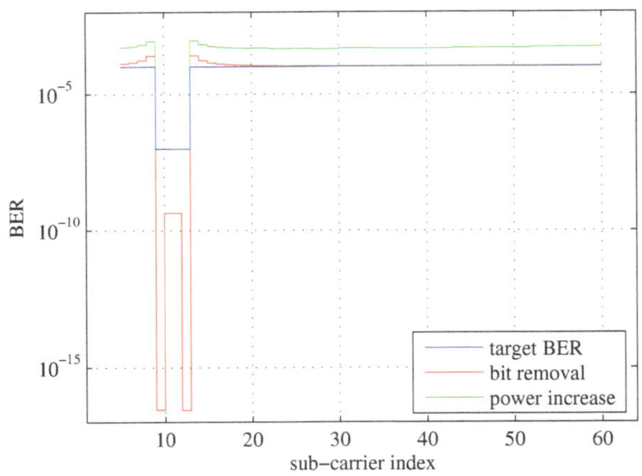

Fig. 9. Comparison between ESC protection methods.

represented more compactly than PAT entries. However, the resulting BER are much lower than actually required. This indicates an inefficient use of the available resources.

6.3 Change of ESC location

So far, we have only considered methods that leave the spectral location of the ESC untouched. Clearly, this is unfavorable if this spectral location is affected by a disturbance. Therefore, a mere re-allocation of the ESC at sub-carriers further away from the disturbed frequency range can significantly improve the BER for the ESC. One possible alternative ESC location is depicted in Fig. 10. Note that complete updated BAT and PAT are used. The spectral range previously occupied by ESC information is now filled with payload data.

Figure 11 shows the resulting BER. Requirements are almost perfectly met for the ESC. Now, payload data is heavily affected by the disturbance. However, this does not lead to a loss of connection. Additionally, due to the use of a new BAT, the amount of payload bits is reduced by 2. This payload rate reduction must either be handled by higher protocol layers, or one can insert dummy bits at the receiver which will have BER 0.5 in the mean.

6.4 Precomputation of alternative ATs

The main drawback of the last alternative was the necessity to exchange complete BAT and PAT. The amount of information is considerable and impractical if the connection is disturbed. One solution for this problem is to precompute a number of alternative BAT and PAT with the ESC located on disjunct and coherent sets of sub-carriers. These allocation tables (AT) are then exchanged and stored for later use in both stations. If the ESC is disturbed in its present location, an alternative AT can be used instead. The information required is then just an index pointing out the respective AT that should be chosen.

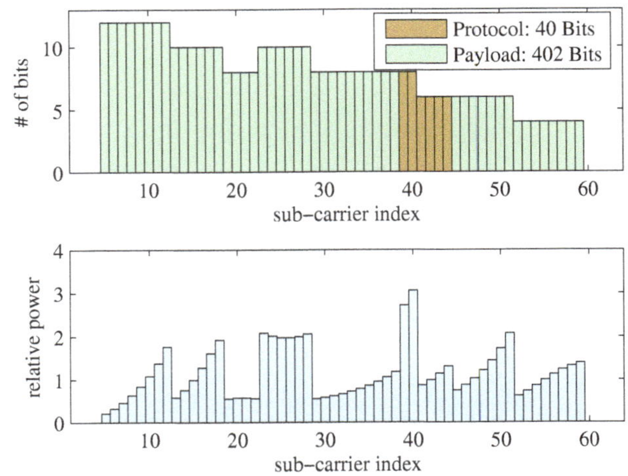

Fig. 10. New spectral location for ESC.

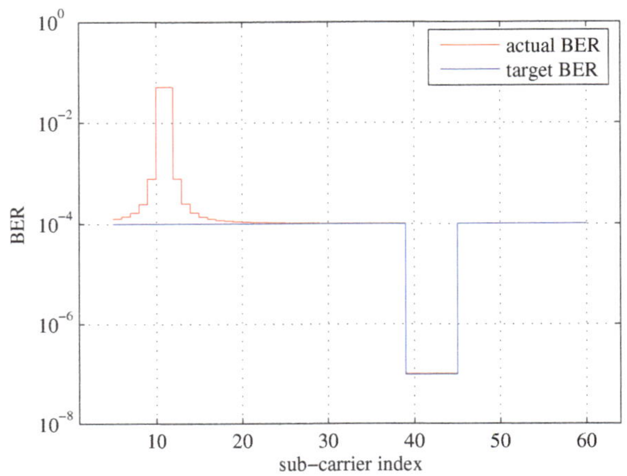

Fig. 11. BER with new spectral location for ESC.

Even if no reliable communication is possible for some time, both stations can cycle through all possible transmit and receive combinations to finally guess the optimal choice of AT blindly.

If some a-priori information is available about the types and likely locations of disturbances to be expected for the transmission channel or environment that the stations are presented with, this information can be used to design the alternative AT especially for those situations.

A more detailed description of the protocol design for our system can be found in Edinger et al. (2005a). It uses a special automatic repeat request (ARQ) protocol that protects only the ESC information. Faulty ESC bits can be detected and the respective information can be retransmitted. If errors occur, the protocol tries to increase the probability of correct reception by applying the methods described above. The protocol works incrementally, that means, it starts with simple methods changing as few parameters as possible and if these changes do not bring about the desired results, the

Table 1. Duration of adaptation tasks.

Task	# of DMT symbol clock cycles
Detection of disturbance	10
Securing the ESC	30
Channel estimation	400
BAT/PAT computation	200
BAT/PAT transmission	130
Total time	770

protocol continues with other methods, until the connection is finally secured.

7 Completing the adaptation process

After BER of the ESC has been restored to allow reliable information exchange, the adaptation process continues with estimating the transmission channel again. Data about the changed noise conditions can be gathered using decision feedback methods. However, this might lead to misestimation especially for heavily disturbed sub-carriers. A better but more costly alternative is to insert known training patterns into the DMT symbol. The effective payload data rate will be reduced by this method.

After information about the changed SNR conditions have been gathered, a new BAT and PAT with respect to these new SNR values are computed and exchanged over the ESC together with a time stamp indicating when the new transmission parameters will take effect. Most probably, the new amount of payload data that can be transmitted per DMT symbol will be reduced. Thus, higher protocol layers must provide some capability to accommodate changes in data rate. The BER for both payload and ESC data will perfectly meet the target values, provided the channel estimation is sufficiently accurate.

8 Simulation results

Our simulations have shown that the above protection methods in connection with the incremental ARQ are extremely robust and powerful. The complete adaptation process runs also very fast. Table 1 lists the individual tasks required for the adaptation together with the number of DMT symbol clock cycles typically required for these tasks.

At the rate of 6400 DMT symbols per second used by our system, the complete adaptation requires only 0.12 seconds. In comparison, re-establishing a lost connection requires at least one second. Additionally, during that time, no payload data at all can be transmitted while our system still allows transmission of considerable amounts of payload data even during adaptation.

For the example disturbance and power increase for the ESC, the mean BER for the payload data equals $5.0 \cdot 10^{-4}$. For the method of changing the ESC location, this mean BER equals $3.17 \cdot 10^{-3}$. However, the former method requires accurate data about the disturbance which will not be available immediately. Additionally, information exchange must take place over the disturbed ESC. Both factors require time which is not needed for the latter method. Thus, both will be approximately equal in the amount of erroneous bits received but the latter will complete sooner.

Whichever method is applied, it becomes clear that even in the presence of disturbances, the mean BER of payload might allow relatively unhampered transmission during the adaptation process.

9 Conclusions

Our proposed system makes effective use of the multiple benefits of DMT modulation and alleviates its few disadvantages. The system has superior connection stability even in the presence of strong disturbances. It can recover extremely quickly from such disturbances. Finally, it is very flexible and can easily be fine tuned to meet specific requirements.

Acknowledgements. This work was partly founded by the German Research Foundation (Deutsche Forschungsgemeinschaft, DFG).

References

ADSL: AMERICAN NATIONAL STANDARD FOR TELECOMMUNICATIONS: Asymmetric Digital Subscriber Line (ADSL) Metallic Interface Specification, Tech. rep., Alliance for Telecommunications Industry Solutions, 1998.

Bauer, C.: Mehrträger-Übertragungssysteme mit dynamischer Adaption an zeitvariante Kanaleigenschaften, Dissertation, Universität Mannheim, 2004.

Bingham, J. A. C.: Multicarrier Modulation for Data Transmission: An Idea Whose Time Has Come, IEEE Communications Magazine, Vol. 28, No. 5, 5–14, 1990.

Edinger, S., Bauer, C., Gaida, M., and Schwarz, J.: Quality of Service and Dynamic Adaptive Discrete Multitone Modulation, in: Proc. IEEE International Workshop on Spectral Methods and Multirate Signal Processing, SMMSP, Riga, Latvia, 2005a.

Edinger, S., Schwarz, J., and Fliege, N. J.: Protected Protocol Securing Transition Phases in Dynamically Adaptive OFDM System, in: Proc. of the International OFDM Workshop, InOWo, Hamburg, Germany, 2005b.

Hoo, L., Salvekar, A., Aldana, C., Cioffi, J. M., Chow, P., and Carlo, J.: Express and Confirmation AOC Swapping Commands for DMT DSLs, ANSI Contribution T1E1.4/99-118, 1999a.

Hoo, L. M. C., Tellado, J., and Cioffi, J. M.: Discrete Dual QoS Loading Algorithms for Multicarrier Systems, Proc. IEEE International Conference on Communications, ICC, Vol. 2, 796–800, 1999b.

Hughes-Hartogs, D.: Ensemble Modem Structure for Imperfect Transmission Media, U.S. Patents No. 4679227 (July 1987), 4731816 (March 1988), 4833706 (May 1989), 1989.

17

Adaptive decoding of convolutional codes

K. Hueske, J. Geldmacher, and J. Götze

Information Processing Lab - AG DT, University of Dortmund, Germany

Abstract. Convolutional codes, which are frequently used as error correction codes in digital transmission systems, are generally decoded using the Viterbi Decoder. On the one hand the Viterbi Decoder is an optimum maximum likelihood decoder, i.e. the most probable transmitted code sequence is obtained. On the other hand the mathematical complexity of the algorithm only depends on the used code, not on the number of transmission errors. To reduce the complexity of the decoding process for good transmission conditions, an alternative syndrome based decoder is presented. The reduction of complexity is realized by two different approaches, the syndrome zero sequence deactivation and the path metric equalization. The two approaches enable an easy adaptation of the decoding complexity for different transmission conditions, which results in a trade-off between decoding complexity and error correction performance.

1 Introduction

The Viterbi Decoder (VD) (Viterbi, 1967) is the standard approach for decoding convolutional codes. The decoder is based on the application of the Viterbi Algorithm (VA) to the trellis representation of the convolutional encoder. Forney Jr. (1973) furthermore showed that the Viterbi Decoder is an optimum Maximum Likelihood (ML) decoder, i.e. the valid code sequence with minimum distance to the received sequence is obtained.

The mathematical complexity only depends on the used code, i.e. it is not depending on the channel behaviour, which will be described as Signal to Noise Ratio (SNR) in this paper. This means, a constant high amount of decoding operations is required, even if few or no errors occurred. This is a disadvantage, especially for applications that require energy efficient implementations (e.g. mobile terminals).

This paper presents an alternative syndrome based convolutional decoder, whose complexity can be adaptively reduced in case of good transmission conditions. The decoder also uses the VA, but the algorithm is applied to the trellis representation of the syndrome former. While the VD determines the most likely transmitted code sequence directly, the syndrome based decoder determines the most probable error sequence first. It will be shown that both methods allow optimum ML decoding, but using the syndrome former trellis is advantegous in terms of adaptivity and complexity.

Syndrome based convolutional decoders were also described by Schalkwijk and Vinck (1975); Ariel and Snyders (1999); Reed and Truong (1985), but the presented decoder allows further adaptivity in terms of a trade-off between computational complexity and error correction performance. The reduction in complexity is realized using two approaches:
The syndrome zero sequence deactivation considers the dependencies between the syndrome and the error sequence, which allows the deactivation of the decoder for error free sequences. This leads to a reduction of decoding complexity for high SNR with no or marginal loss in decoding performance.

The path metric equalization reduces the required number of Add-Compare-Select (ACS) operations for the VA by using identical path metrics for different error patterns. The error correction performance is decreased but the decoding complexity can be reduced by 25%. This is especially useful for applications that require a BER threshold and do not benefit from further BER degradation. While the performance of the presented decoder without adaptation is equivalent to the standard VD, we can reduce the complexity of the decoding process if the BER is below the required threshold.

Section 2 presents the basic syndrome decoding algorithm, which is based on Schalkwijk and Vinck (1975). Furthermore the equivalence to the VD will be derived, which shows

Correspondence to: K. Hueske
(klaus.hueske@uni-dortmund.de)

that both are optimum ML decoders. In Sect. 3 two methods for complexity reduction are introduced, syndrome zero sequence deactivation and path metric equalization. Furthermore an estimation technique for the Bit Error Rate (BER) is introduced, which is used for an adaptive complexity reduction. For performance and complexity analysis simulation results are presented in Sect. 4. Conclusions are given in Sect. 5.

2 Syndrome decoding

In this section the syndrome decoding algorithm proposed by Schalkwijk and Vinck (1975) is presented.

2.1 Syndrome decoding

The syndrome decoding algorithm is presented for code rates $R{=}k/n$. Sequences and transfer functions are represented in frequency domain as power series in D with coefficients from $GF(2)$. For the sake of clarity, the delay operator D is omitted in the following.

An information sequence is encoded by multiplication with a generator matrix

$$\mathbf{v} = \mathbf{u}\mathbf{G}, \tag{1}$$

where $\mathbf{G}$ is a polynomial generator matrix with k rows and n columns, $\mathbf{v}$ is a row vector of length n containing the coded bits and $\mathbf{u}$ is a row vector of length k containing the information bits.

The sequence $\mathbf{v}$ is transmitted over a memoryless, noisy channel, causing the corrupted received sequence

$$\mathbf{r} = \mathbf{v} + \mathbf{e}, \tag{2}$$

where $\mathbf{e}$ is the error sequence resulting from channel noise.

A syndrome sequence $\mathbf{b}$ is computed from the received sequence as

$$\mathbf{b} = \mathbf{r}\mathbf{H}^T, \tag{3}$$

where $\mathbf{H}^T$ is a polynomial matrix with n rows and $n-k$ columns and is called the syndrome former matrix of $\mathbf{G}$. The syndrome former matrix is defined to be orthogonal to $\mathbf{G}$, thus the syndrome sequence $\mathbf{b}$ is the zero sequence if and only if $\mathbf{r}$ is a valid code sequence. It follows that the syndrome sequence only depends on the error sequence and is independent of the transmitted information:

$$\mathbf{b} = \mathbf{r}\mathbf{H}^T = (\mathbf{v} + \mathbf{e})\mathbf{H}^T = \mathbf{e}\mathbf{H}^T \tag{4}$$

The syndrome decoder has to map the syndrome sequence back to the error sequence. However the mapping from $\mathbf{b}$ to $\mathbf{e}$ is not unique, but there is a set of error sequences corresponding to one syndrome sequence. For ML decoding the decoder has to determine the code sequence with minimum distance to the received sequence, which corresponds to the error sequence with minimum weight. Thus the decoding problem can be formulated in terms of an optimization problem with constraint:

$$\min_{\hat{\mathbf{e}}} ||\hat{\mathbf{e}}|| \quad \text{with} \quad \mathbf{b} = \hat{\mathbf{e}}\mathbf{H}^T \tag{5}$$

The norm $||...||$ is defined as hamming distance for hard decision decoding and as 2-norm for soft decision decoding. The minimization can be done by searching the syndrome former trellis for the error sequence with minimum weight using the VA. The constraint is incorporated by only allowing the transitions belonging to the current syndrome value at every stage in the syndrome former trellis.

When the ML error sequence $\hat{\mathbf{e}}$ has been found, the transmission errors are corrected by subtracting the estimated error sequence from the received sequence

$$\hat{\mathbf{v}} = \mathbf{r} - \hat{\mathbf{e}}, \tag{6}$$

where $\hat{\mathbf{v}}$ is the estimated code sequence. Finally the information sequence can be obtained as the product of the estimated code sequence and the right inverse generator matrix

$$\hat{\mathbf{u}} = \hat{\mathbf{v}}\mathbf{G}^{-1}. \tag{7}$$

2.2 Syndrome former matrix and trellis representation

The syndrome former $\mathbf{H}^T$ can be calculated from the invariant factor decomposition (IFD) of the generator matrix $\mathbf{G}$. The IFD of a matrix $\mathbf{G}$ is defined as

$$\mathbf{G} = \mathbf{A}\Gamma\mathbf{B} \tag{8}$$

where $\mathbf{A}$ is a polynomial $(k \times k)$ matrix and $\mathbf{B}$ is a polynomial $(n \times n)$ matrix with $\det(\mathbf{A}){=}\det(\mathbf{B}){=}1$. Γ is a $(k \times n)$ matrix and is called Smith-Form of $\mathbf{G}$. An algorithm for the computation and the properties of the IFD are given in Johannesson and Zigangirov (1999).

The first k rows of $\mathbf{B}$ form an equivalent generator matrix of $\mathbf{G}$ and the last $n{-}k$ rows are the transpose of the inverse of $\mathbf{H}$:

$$\mathbf{B} = \begin{pmatrix} \mathbf{G}_b \\ (\mathbf{H}^{-1})^T \end{pmatrix} \tag{9}$$

As $\mathbf{B}$ is non-singular, it can be inverted. The right inverse equivalent generator matrix $\mathbf{G}_b^{-1}$ and the syndrome former $\mathbf{H}^T$ can be identified as the first k columns and the last $n-k$ columns of $\mathbf{B}^{-1}$, respectively:

$$\mathbf{B}^{-1} = \begin{pmatrix} \mathbf{G}_b^{-1} & \mathbf{H}^T \end{pmatrix} \tag{10}$$

For $R{=}k/n$ the syndrome former takes a sequence of n polynomials as input and produces a syndrome sequence represented by $n{-}k$ polynomials. As the number of memory elements required to realize the syndrome former is equal to the number of the memory elements of the encoder (Forney Jr., 1970), the syndrome former trellis has the same number of states as the encoder trellis. The syndrome trellis has

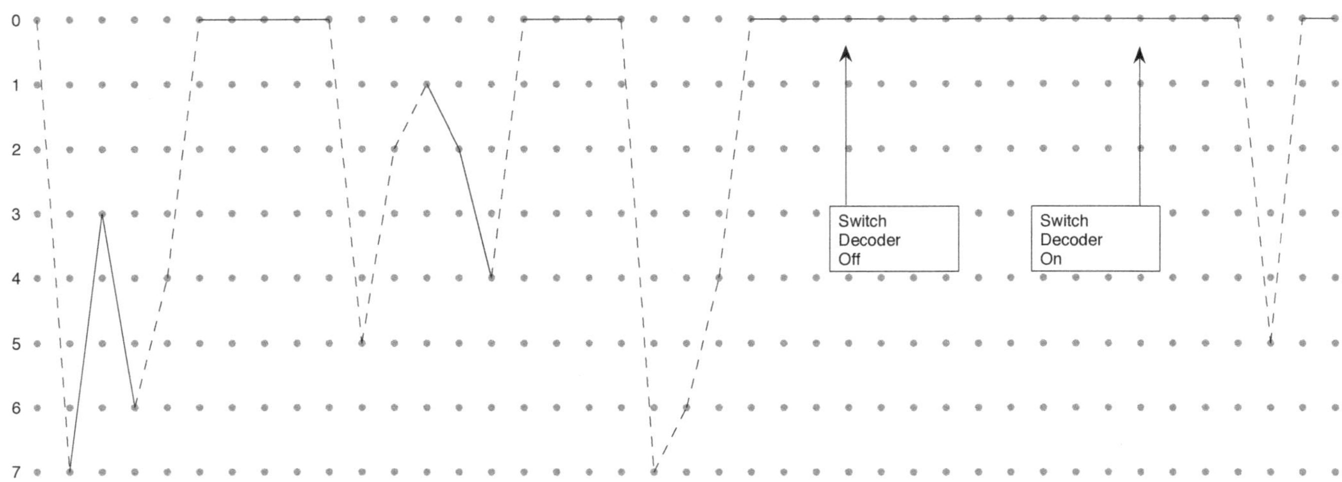

Fig. 1. Example of an error path, "0"-syndrome sequences are plotted as lines, "1"-syndrome sequences are plotted dashed.

2^v states, where v is the number of memory elements of the encoder, and 2^n edges leaving each state.

For $R=1/2$ the syndrome trellis has 4 edges leaving each state. The trellis can be split into two parts, with each part containing only the transitions of one syndrome symbol. This results in two trellises with each 2^v states and 2 edges leaving each state.

To find the minimum weight error sequence, a shortest path algorithm like the VA is used. As branch metric for the transition of a state to a successor state, the weight of the corresponding error pattern is taken. To incorporate the constraint from Eq. (5), at each decoding stage only the trellis part corresponding to the current syndrome symbol is considered.

2.3 Equivalence to the Viterbi Decoder

The equivalence of the syndrome based decoder to the VD can be shown by replacing the estimated error $\hat{\mathbf{e}}$ in Eq. (5) by $\mathbf{r}-\hat{\mathbf{v}}$. Expression (5) then becomes

$$\min_{\hat{\mathbf{v}}} ||\mathbf{r} - \hat{\mathbf{v}}|| \text{ with } \mathbf{b} = (\hat{\mathbf{v}} + \hat{\mathbf{e}})\mathbf{H}^T, \tag{11}$$

and using $\mathbf{b} = \hat{\mathbf{e}}\mathbf{H}^T$ leads to

$$\min_{\hat{\mathbf{v}}} ||\mathbf{r} - \hat{\mathbf{v}}|| \text{ with } \mathbf{0} = \hat{\mathbf{v}}\mathbf{H}^T, \tag{12}$$

which represents the decoding problem in "Viterbi sense". As both decoders are equivalent, the syndrome based decoder is also an optimum ML decoder.

We note that the syndrome computation in Eq. (3) and the mapping from estimated code sequence to information sequence in Eq. (7) can be realized as simple XOR operations. Thus the estimation of the error sequence has a similar complexity as the VD.

3 Adaptive decoding

3.1 Syndrome zero sequence deactivation

Good transmission conditions will lead to error free sequences in $\mathbf{r}$, i.e. to sequences where $\mathbf{e}=\mathbf{0}$. As the syndrome sequence only depends on the channel errors, as shown in Eq. (4), error free periods will lead to zero sequences in the syndrome $\mathbf{b}$, too. For zero sequences, no decoding is necessary, the zero path in the syndrome former trellis is taken.

The first proposed optimization of the syndrome decoder now consists of detecting zero sequences in the syndrome, and switching the decoder off for these sequences. A reasonable length of the period has to be assumed, to be sure, that the ML path in the trellis has returned to zero state. Decoding is done by always tracing back from zero state.

If the minimum zero period length is chosen long enough, the decoding performance can be as good as the performance of the unmodified decoder. There is a trade-off, decoding performance can be exchanged with decoding complexity by reducing or increasing the required length of the zero sequences.

There are two critical parameters for the adaptive decoder: The first parameter l_{off} is the number of syndrome zeros, after which the decoder can be safely switched off. The second parameter l_{on} is the number of stages before the first syndrome one, when the decoder has to be switched on again. The second parameter depends on the number of path registers v. Because an error event is propagated with maximum delay of v stages to the syndrome, the decoder has to be switched on v stages before the first syndrome one. The first parameter also depends on v but is also influenced by the received data in case of soft decision, which makes a general choice infeasible. Therefore, simulations are used to determine a reasonable value.

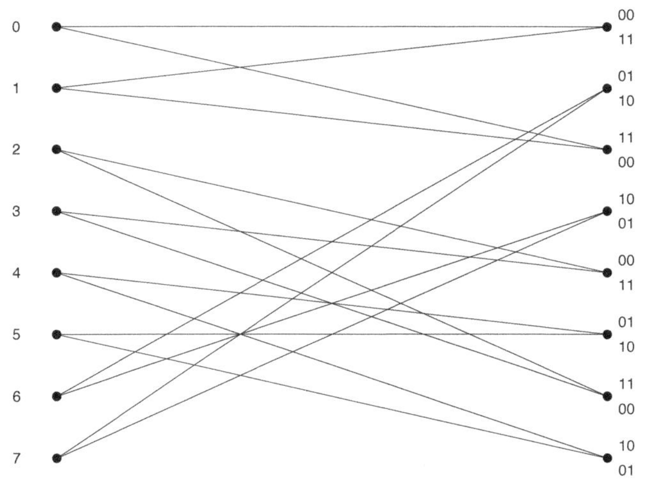

Fig. 2. Syndrome former trellis for the example code, Syndrome 0 part.

As an example, Fig. 1 shows an error path for a υ=3 Code. In this example the decoder is switched off, when six successive syndrome zeros are detected. The decoder is turned off after l_{off}=3 syndrome zeros and turned on again l_{on}=3 stages before the next syndrome one.

3.2 Path metric equalization

A second reduction of the computational complexity is based on a closer examination of the syndrome former trellis structure. As Schalkwijk et al pointed out in Schalkwijk and Vinck (1976), the syndrome former trellis has a special structure. We can identify pairs of states which have the same predecessor states and at the same time have identical weights of the corresponding transition in case of hard decision.

For example, Fig. 2 shows a syndrome former trellis for a code with memory length υ=3 and generator polynomials $\mathbf{G}=[D^3+D^2+1,\ D^3+D^2+D+1]$. For simplicity only the zero part is considered. The trellis is annotated on the left side with the state numbers and on the right side with the error input corresponding to the transition. One can see that the states 1 and 3 have the same predecessor states (6 and 7) with the transitions having the same weight 0+1=1 and 1+0=1. The same holds for the states 5 and 7 which have the predecessors 4 and 5.

Let $M^{(i)}(t)$ be the metric of state i at time t and $\mu^{(i,j)}(t)$ the branch metric for a transition from state i to state j. At every decoding stage and for every state the VA has to compute the metrics for all transitions from each predecessor state, compare the metrics and choose the survivor path as the path with minimum metric (ACS). For example, for the states 1 and 3 the metric is computed as

$$\begin{aligned} &M^{(1)}(t+1) = \\ &\min\{M^{(6)}(t) \quad +\mu^{(6,1)}(t), M^{(7)}(t)+\mu^{(7,1)}(t)\} \\ &M^{(3)}(t+1) = \\ &\min\{M^{(6)}(t) \quad +\mu^{(6,3)}(t), M^{(7)}(t)+\mu^{(7,3)}(t)\} \end{aligned} \tag{13}$$

If harddecision is applied, the branch metrics are computed as

$$\begin{aligned} \mu^{(6,1)}(t) &= \mu^{(7,3)}(t) = 0+1 = 1 \\ \mu^{(7,1)}(t) &= \mu^{(6,3)}(t) = 1+0 = 1 \end{aligned} \tag{14}$$

and Eq. (13) becomes

$$\begin{aligned} M^{(1)}(t+1) &= \min\{M^{(6)}(t)+1, M^{(7)}(t)+1\} \\ M^{(3)}(t+1) &= \min\{M^{(6)}(t)+1, M^{(7)}(t)+1\} \end{aligned} \tag{15}$$

The metric computation is identical for both state 1 and state 3 and thus for state 3 no computation is required. The same holds for states 6 and 7.

We can use this for reducing the complexity of the decoder and save one ACS operation for each of the $2^{\upsilon-2}$ pairs. If the decoder uses harddecision, 25% of the ACS operations can be saved without loss of decoding performance (Schalkwijk and Vinck, 1976).

However as hard decision is rarely used, we now discuss the metric computation for the soft decision case. Let $\mathbf{r}(t)=[r_1(t)\ r_2(t)]$ be the 3-Bit quantized vector received at time t, zero symmetric BPSK modulation and AWGN channel assumed. The absolute value of $\mathbf{r}(t)$ can be seen as a measure for the confidence of the received word. The lower the absolute value of $r_i(t)$, the more likely a transmission error occurred at this position. Thus the branch metric calculation for softdecision can formulated as

$$< \mathbf{e}(t), \mathbf{r}(t) >= [e_1(t)\ e_2(t)][|r_1(t)|\ |r_2(t)|]^T \tag{16}$$

to incorporate the confidence informations. The metric calculation for the example becomes

$$\begin{aligned} M^{(1)}(t+1) &= \min\{M^{(6)}(t)+|r_2(t)|, M^{(7)}(t)+|r_1(t)|\} \\ M^{(3)}(t+1) &= \min\{M^{(6)}(t)+|r_1(t)|, M^{(7)}(t)+|r_2(t)|\} \end{aligned} \tag{17}$$

So the metric calculation is not identical for soft decision and a simplification is no longer possible without decoding performance loss.

However, the metric calculation can be modified to allow a simplification analog to the hard decision case. This is realized by assigning the same branch metrics to the 01 and 10 error patterns, which can be done by choosing the minimum of $|r_1(t)|$ and $|r_2(t)|$ as branch metric for the 01 and 10 transitions:

$$\mu^{(i,j)}(t) = \begin{array}{ll} 0 & \text{for } \mathbf{e}(t) = [00] \\ \min\{|r_1(t)|, |r_2(t)|\} & \text{for } \mathbf{e}(t) = [10] \\ \min\{|r_1(t)|, |r_2(t)|\} & \text{for } \mathbf{e}(t) = [01] \\ |r_1(t)| + |r_2(t)| & \text{for } \mathbf{e}(t) = [11] \end{array}$$

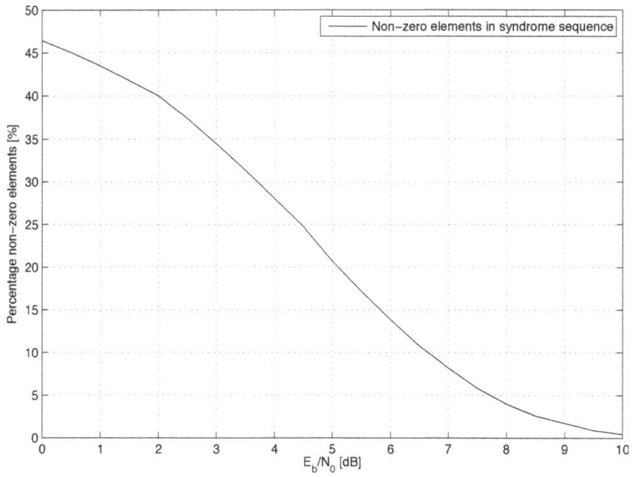

Fig. 3. Percentage of non-zero elements in syndrome vector.

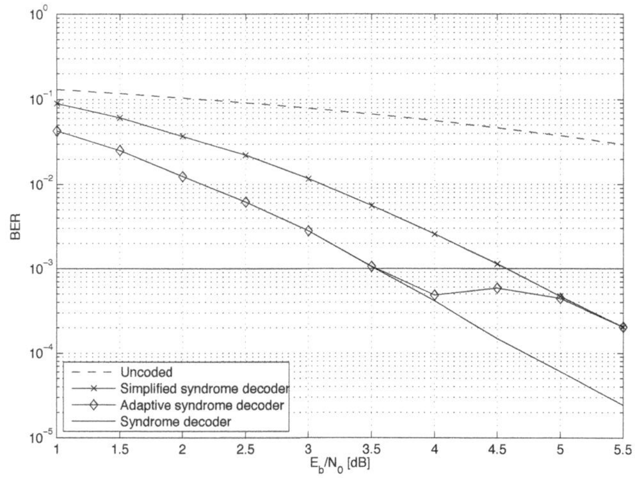

Fig. 4. Comparison of the Bit-Error-Rates of the reduced complexity decoder and the full complexity decoder.

The metric computation for the states 3 and 6 then becomes

$$\begin{aligned} M^{(1)}(t+1) &= \min\{M^{(6)}(t) + \min\{|r_1(t)|, |r_2(t)|\}, \\ &\qquad M^{(7)}(t) + \min\{|r_1(t)|, |r_2(t)|\}\} \\ M^{(3)}(t+1) &= \min\{M^{(6)}(t) + \min\{|r_1(t)|, |r_2(t)|\}, \\ &\qquad M^{(7)}(t) + \min\{|r_1(t)|, |r_2(t)|\}\} \end{aligned} \tag{18}$$

If the metric calculation is modified in this way, the reduction of complexity can be applied as in the hard decision case. However, there will be a loss of decoding performance compared to the full complexity soft decision decoder.

3.3 BER estimation

The path metric equalization is especially suitable for transmission systems that require a specific BER and do not benefit from better channel conditions. If the desired BER is reached, the complexity of the decoder can be reduced adaptively. The total error correction performance will be decreased, but the required BER threshold will always be achieved.

The problem that comes up with this approach is that the BER is generally unknown in the receiver. But using the syndrome vector we can roughly estimate the BER by calculating the syndrome vector weight. This means for good transmission conditions only few errors occur, which results in a syndrome vector with only few non-zero elements. On the other hand if the SNR is very low, we expect many transmission errors, which produces many non-zero elements. The dependencies between SNR and number of non-zero elements in the syndrome vector is depicted in Fig. 3.

Using this relation we can define a syndrome vector weight threshold in the receiver, which allows an adaptive switching between the normal and the reduced complexity syndrome based decoder.

3.4 Summary

There are two options for reduction of decoding complexity. The first option is an adaptive decoding approach, which is based on analyzing the syndrome sequence and turning off the decoder in error-free periods. A reasonable minimum length assumed, decoding costs can be reduced without loosing decoding performance. The second option is based on the syndrome former trellis structure and allows a reduction of decoding complexity by about 25% but resulting in a loss of decoding performance.

4 Simulation results

This section gives simulation results for both options. All simulations have been done for a code with generator matrix

$$\mathbf{G} = [D^3 + D^2 + 1,\ D^3 + D^2 + D + 1] \tag{19}$$

with v=3 path registers and corresponding syndrome former

$$\mathbf{H}^T = [D^3 + D^2 + D + 1,\ D^3 + D^2 + 1]^T \tag{20}$$

The codewords are transmitted over a memoryless AWGN channel using BPSK modulation. For the BER simulations soft decision was applied, considering 2000 simulated errors.

Figure 4 shows the results for the reduced complexity syndrome decoder using path metric equalization. The performance of the simplified syndrome decoder is about 1dB worse compared to the regular syndrome decoder, but it requires 25% less operations. The same figure shows the adaptive decoding approach with a desired BER of 10^{-3}. For a $E_b/N_0 < 3.5$ dB only the unmodified syndrome decoder is used, i.e. the decoding complexity is equivalent to the VD. For higher SNR the reduced complexity decoder is used, depending on the syndrome vector weight of the actual received

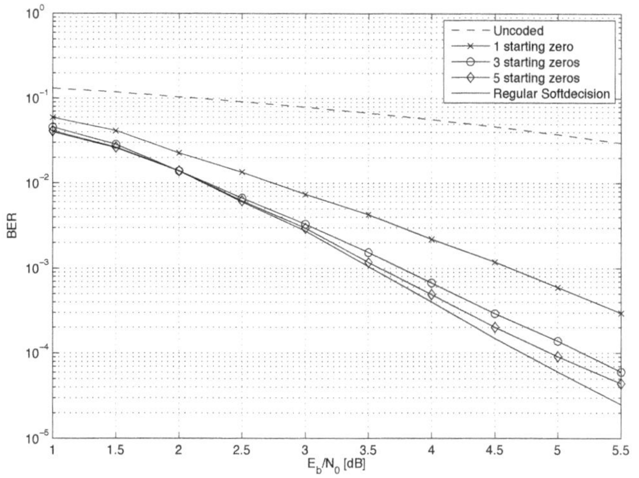

Fig. 5. Comparison of the Bit-Error-Rates of the adaptive decoders using 3 different minimum zero sequence lengths and the regular decoder for Softdecision.

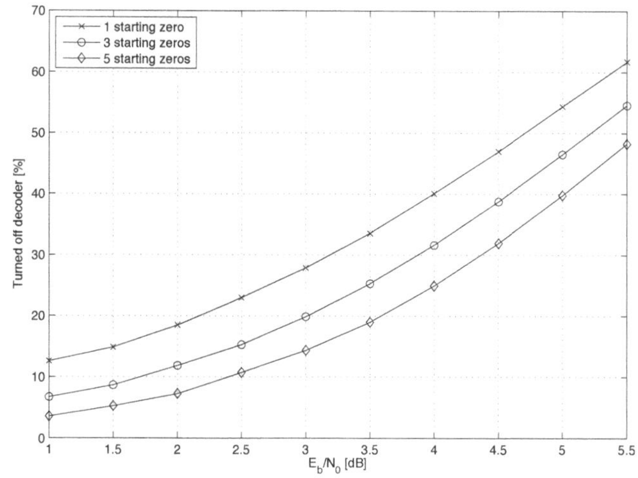

Fig. 6. The percentage of turned off decoders for th 3 different minimum zero sequence lengths.

data block. For $E_b/N_0>5.0$ dB all decoding operations are done by the reduced complexity decoder. This shows, that the decoding complexity can be easily adapted for different channel conditions.

Figure 5 shows the performance of the decoder using syndrome zero sequence deactivation. Simulations are shown for $l_{on}=3$ and $l_{off} \in \{1, 3, 5\}$. In Fig. 6 the percentage of the decoding time the decoder is turned off is plotted. For example at $E_b/N_0=4.5$ dB the decoder with $l_{off}=5$ is turned off about 33% of the time with about 0.15 dB loss in performance compared to the regular decoder. This allows a reduction of complexity with only marginal performance loss. If desired, the complexity can be further decreased by reducing l_{off}. At $E_b/N_0=4.5$ dB the decoder with $l_{off}=1$ is turned off about 48% of the time while taking a loss of about 1 dB.

5 Conclusions

While performance and computational complexity of the Syndrome Decoder are equivalent to the Viterbi Decoder for worst case transmission conditions, the syndrome based decoder allows an adaptive reduction of complexity for good channel conditions. With the proposed methods the adaptive reduction can be performed in two steps. First, the syndrome zero sequence deactivation will reduce the decoding complexity without significant loss in decoding performance. If desired, the required number of operations can be further decreased by either using a reduced length l_{off} or using the path metric equalization. The last step is especially applicable for transmission systems that require a BER threshold and do not benefit from further SNR improvement.

References

Ariel, M. and Snyders, J.: Error-trellises for convolutional codes .II. Decoding methods, Communications, IEEE Transactions on, 47, 1015–1024, doi:10.1109/26.774852, 1999.

Forney Jr., G.: Convolutional codes I: Algebraic structure, Information Theory, IEEE Transactions on, 16, 720–738, 1970.

Forney Jr., G.: The Viterbi Algorithm, IEEE Proceedings, 61, 268–278, 1973.

Johannesson, R. and Zigangirov, K. S.: Fundamentals of Convolutional Coding, IEEE press, 1999.

Reed, I. and Truong, T.: Error-trellis syndrome decoding techniques for convolutional codes, IEE Proceedings Pt. F, 132, 77–83, 1985.

Schalkwijk, J. and Vinck, A.: Syndrome Decoding of Convolutional Codes, Communications, IEEE Transactions on, 23, 789–792, 1975.

Schalkwijk, J. and Vinck, A.: Syndrome Decoding of Binary Rate-1/2 Convolutional Codes, Communications, IEEE Transactions on, 24, 977–985, 1976.

Viterbi, A.: Error bounds for convolutional codes and an asymptotically optimum decoding algorithm, Information Theory, IEEE Transactions on, 13, 260–269, 1967.

Optimized power allocation for iterative multiuser detection for a SC-CDMA uplink

P. Weitkemper[1], K. Zielinski[2], K.-D. Kammeyer[1], and R. Laur[2]

[1]Department of Communications Engineering, University of Bremen, Otto-Hahn-Allee, 28359 Bremen, Germany
[2]Institute for Electromagnetic Theory and Microelectronics, University of Bremen, Otto-Hahn-Allee, 28359 Bremen, Germany

Abstract. This paper deals with optimization of the received power profile for iterative parallel and successive interference cancellation (PIC/SIC) in coded CDMA systems. For practical implementation additional constraints should be applied. This paper focuses on the maximum tolerable bit error rate. It will be shown that optimized power profile can considerably gain the overall system performance. Due to unequal required receive powers an allocation to certain users can be done with respect to their individual power constraints. This is important especially in near-far scenarios. Beside these constraints also the maximum number of iterations is implemented due to limiting the computational complexity in the receiver.

1 Introduction

Nonlinear iterative multi-user detection exploiting channel decoding is a well known approach to increase the possible system load while still achieving single-user performance. Analysis of the PIC scheme has been done by different approaches (Schlegel and Shi, 2004; Boutros and Caire, 2002; Kuehn, 2004) and the possibility of power profile optimization has been shown in e.g. Caire et al. (2004). So far optimization of receive power distribution was done with a simple optimization goal and only few constraints. The focus was on minimizing the sum over all received powers which may be feasable for downlink but not for uplink scenarios. The constraint was ensuring all users to have a maximum tolerable bit error rate which is the most interesting point in practicle systems, but single user performance is not required implicitly. If the SINR is still higher than the SNR but sufficiently small for the desired BER, there is no need for increasing the power further only to reach this certain point. In the uplink of a cellular network, individual power constraints have to be considered. One possibility is to assume additional constraints for individual powers of the users which was presented in Weitkemper et al. (2006). A more general approach used in this paper is not to minimize the sum over all powers but the maximum power which results in a more homogeneous distribution. Furthermore some additional constraints are considered which take practical implemetation issues into account e.g. maximum number of iterations. For successive interference cancellation the analysis is more difficult because the statistics of the users differ from each other. The generalization of the analysis to SIC and the basics of power profile optimization were presented in Weitkemper et al. (2005).

The paper is organized as follows: Sect. 2 introduces the system model of the considered CDMA system. In Sect. 3 the analysis based on multi-user efficiency (MUE) (Boutros and Caire, 2002) is described and applied to the parallel interference cancellation. The generalized MUE analysis of SIC is given in Sect. 4. In Sect. 5 power profile optimization is introduced for both detection schemes. Results for different optimization objects and additional constraints are presented and discussed in Sect. 6. A conclusion is given in Sect. 7.

2 System model

In order to simplify derivations and notation we assume a synchronous single carrier- (SC-) CDMA system with a complex AWGN channel and pseudo-noise spreading sequences (Verdú, 1998). The number of active users is denoted by U. The information bit vector of the u-th user is denoted by $\mathbf{d}_u$, which is encoded by a convolutional code of rate $R_c=1/2$ that is identical for all users. The coded bit sequence is BPSK-modulated and interleaved by a user-specific random interleaver Π_u of length L. Finally the signals are spread with random spreading codes $s_u(k)\in\{-1/\sqrt{N}, +1/\sqrt{N}\}$. k

Correspondence to: P. Weitkemper
(weitkemper@ant.uni-bremen.de)

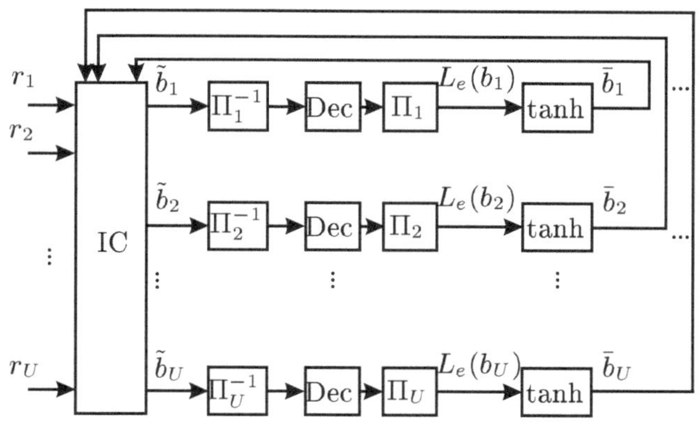

Fig. 1. Receiver structure of PIC.

denotes the chip and l the symbol index. The length N of the sequences $s_u(k)$ is called spreading factor and $\beta{=}U/N$ denotes the system load. **b** is assumed to be the vector comprising BPSK symbols of all users at a particular time instance and **C** is the $N \times U$ spreading matrix. It contains the vectors of spreading sequences as columns each multiplied with an individual phase term of the channel. The received vector containing the superposition of the spread signals of all users and the noise can be described in vector-matrix notation, yielding

$$\mathbf{y} = \mathbf{C}\mathbf{b} + \mathbf{n}\ . \tag{1}$$

For notational simplicity time indices have been dropped. The vector **n** represents the complex additive white Gaussian noise with covariance matrix $\sigma_n^2\,\mathbf{I}$. At the receiver a bank of matched filters (MF) is applied for despreading and the real-valued matched filter output can be written as

$$\mathbf{r} = \mathrm{Re}\left\{\mathbf{C}^H\mathbf{y}\right\} = \underbrace{\mathrm{Re}\left\{\mathbf{C}^H\mathbf{C}\right\}}_{\mathbf{R}}\mathbf{b} + \underbrace{\mathrm{Re}\left\{\mathbf{C}^H\mathbf{n}\right\}}_{\tilde{\mathbf{n}}}\ . \tag{2}$$

The elements of **R** contain the real part of the correlation coefficients $\varrho_{ij}{=}\mathrm{Re}\left\{\rho_{ij}\right\}$ between the i-th and the j-th user's signature sequence with $\mathrm{E}\{|\rho_{ij}|\}_{i\neq j}{=}1/N$. The multi-user interference characterized by these correlation coefficients degrades the performance significantly even for moderate system loads if individual decoding and hard decision is applied to these matched filter outputs. Interference cancellation techniques are able to improve performance significantly by estimating the multi-user interference and cancelling it before detection. Additional gain is achieved by iterative structures.

3 Multi-user efficiency

Figure 1 shows the structure of a parallel interference canceller. In order to get at least approximated extrinsic log-likelihood-ratios $L_e(b_u)$ at low computational cost, the Max-Log-MAP channel decoder is applied. The soft estimates

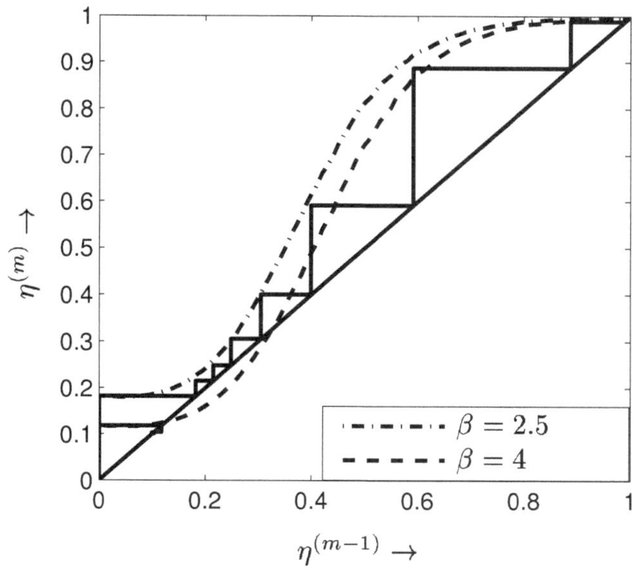

Fig. 2. Predicted transfer functions and trajectories for PIC, $N{=}4$, $U{=}10/16$, $E_b/N_0{=}6$ dB.

of the coded symbols $\bar{b}$ are calculated as $\bar{b}{=}\tanh(L_e/2)$. The signal-to-interference-plus-noise-ratio (SINR) of each branch is a parameter indicating the quality of the interference cancellation and is defined as $\mathrm{SINR}{=}2\sigma_d^2/(\sigma_n^2+\sigma_{\mathrm{MUI}}^2)$. It is equal to the $\mathrm{SNR}{=}2E_s/N_0$ in the case of perfect interference cancellation which is equivalent to the single-user bound (SUB). σ_d^2 and σ_{MUI}^2 describe the variance of the desired signal and of the remaining multi-user interference after cancellation respectively. The latter can be calculated as $\sigma_{\mathrm{MUI}}^2{=}\sigma_d^2 \cdot \mu(U{-}1)/N$. The remaining mean squared error of the estimated symbols after decoding is denoted as $\mu{=}\mathrm{E}\{|\bar{b}{-}b|^2\}$ which is approximately the same for each user in the case of PIC. The ratio of SINR and SNR is called multi-user efficiency (MUE) and is denoted by η (Boutros and Caire, 2002). Perfect interference cancellation is indicated by $\eta{=}1$ and therefore describes the case with no loss compared to the SUB. For the large system limit (N, $U{\rightarrow}\infty$) $(U{-}1)/N{\approx}U/N{=}\beta$, η can be written as

$$\eta = \frac{\mathrm{SINR}}{\mathrm{SNR}} = \frac{2\sigma_d^2/(\sigma_n^2+\sigma_{\mathrm{MUI}}^2)}{2\sigma_d^2/\sigma_n^2} = \frac{1}{1+\beta\mu E_s/N_0}\ . \tag{3}$$

The parameter η can be used to visualize and predict the behavior of an iterative detection scheme. In the initial iteration there is no a-priori information available for interference cancellation. The soft bits are initialized with zero, the variance μ is therefore equal to 1 and the MUE becomes $\eta^{(1)}{=}1/(1+\beta E_s/N_0)$. After simultaneously decoding all users, soft estimates $\bar{b}$ of the transmitted symbols are obtained which are used in the next iteration for interference cancellation. Since channel decoding is generally a nonlinear process μ cannot be calculated analytically, but has to be predetermined. The output error $\mu^{(m)}$ of the decoder in the

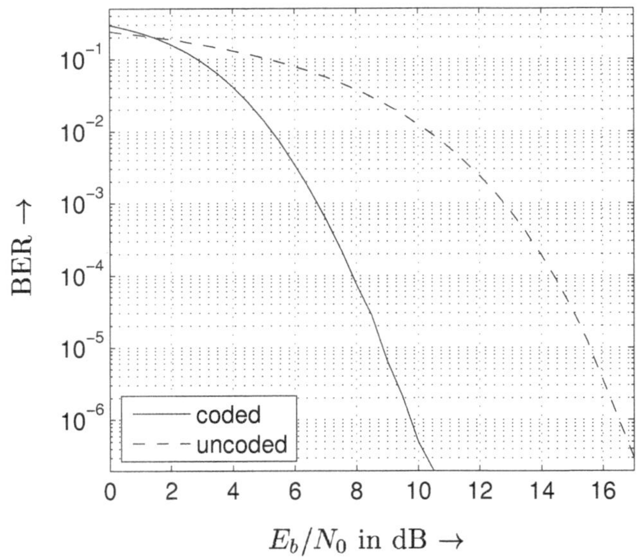

Fig. 3. Coded and uncoded error rate for a AWGN channel and [7 5] convolutional code.

m-th iteration depends on the SINR at the input

$$\mu^{(m)} = g\left(\text{SINR}\right) = g\left(\eta^{(m-1)}\text{SNR}\right) \tag{4}$$

and therefore on the MUE of the previous iteration $\eta^{(m-1)}$. Because $\eta^{(m)}$ itself depends on $\mu^{(m)}$ the behavior of the PIC at iteration m can be described by $\eta^{(m)}=f\left(\eta^{(m-1)}\right)$. This function is depicted in a two-dimensional plot in Fig. 2. The transfer function describes the theoretical behavior and the trajectory gives the measured values during simulation. The detection starts in the lower left corner and tends to the upper right corner. This point corresponding to $\eta = 1$ describes perfect interference cancellation. It can be seen that the behavior is predicted very precisely. This plot corresponds to a system with a spreading factor N=4, U=10/16 users, an E_b/N_0 of 6 dB and a convolutional code with generator polynomials $[7, 5]_8$. The system with 10 users will converge to the SUB within 7 iterations. In the case of 16 users there is an intersection between the transfer function and the bisecting line so the detection gets stuck at $\eta \approx 0.15$.

Since multi-user efficiency is a relative quantity, the performance assuming the uncoded BER has to be calculated by

$$P_b = \frac{1}{2} \cdot \text{erfc}\left(\sqrt{\frac{\text{SINR}}{2}}\right) = \frac{1}{2} \cdot \text{erfc}\left(\sqrt{\frac{P}{2\sigma_n^2}}\right) \tag{5}$$

η only includes information about the error probability before channel decoding. As the coded bit error rate is the important parameter for QoS requirements, a connection between these error probabilities have to be considered. The coded and uncoded BER for the [7 5] code is shown in Fig. 3.

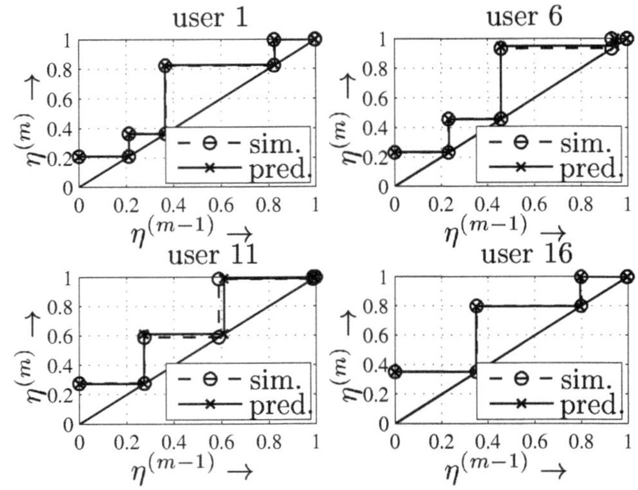

Fig. 4. Predicted and simulated trajectories for SIC, N=8, U=16 and E_b/N_0=6 dB.

4 Analysis of successive interference cancellation

The difference concerning successive interference cancellation is the processing of the users one after another. The parameters change not only after 1 iteration but during the iterations when decoding each individual user. Therefore the prediction in the same manner as for PIC is not possible. While for PIC the error variance μ is the same for all users in the large system limit, this is not the case for SIC. The U users have different variances $\mu_u^{(m)}$ at each iteration m. The remaining errors of the users are assumed to be independent. So a simple addition of their variances weighted with the corresponding correlation coefficient can be applied for calculating the resulting interference on the desired user signal. For that reason the MUE can be calculated by

$$\eta_u^{(m)} = \frac{1}{1 + \frac{1}{N}\left(\sum_{i=1}^{u-1} \mu_i^{(m)} + \sum_{i=u+1}^{U} \mu_i^{(m-1)}\right)\frac{E_s}{N_0}}. \tag{6}$$

To show how good the prediction works, Fig. 4 depicts the predicted and the simulated trajectories for the same system parameters as in Fig. 2 in one diagram per user. It can be seen that the prediction works well also in the successive interference cancellation case. A simple transfer function as shown in Fig. 2 cannot be drawn in these plots since the transfer function differs for each user and each iteration due to being conditioned on the current state of all the other users. In order to avoid a very complex diagram only the trajectories are depicted in Fig. 4.

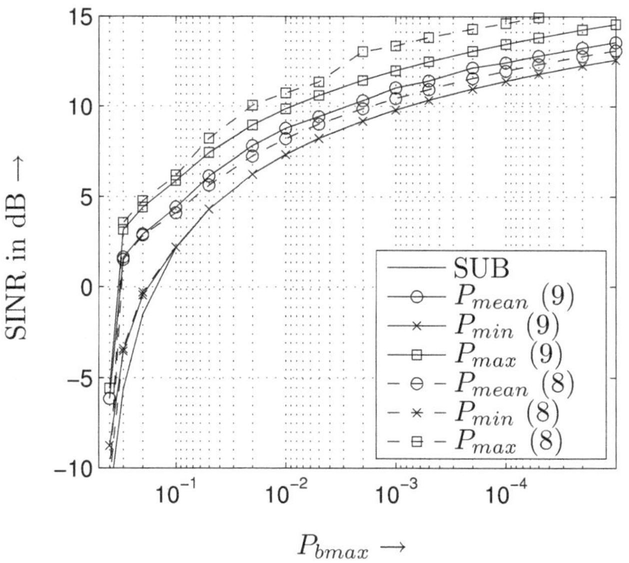

Fig. 5. Required receive power of all users dependent on the required BER for different expressions to be minimized (Eqs. 8 and 9).

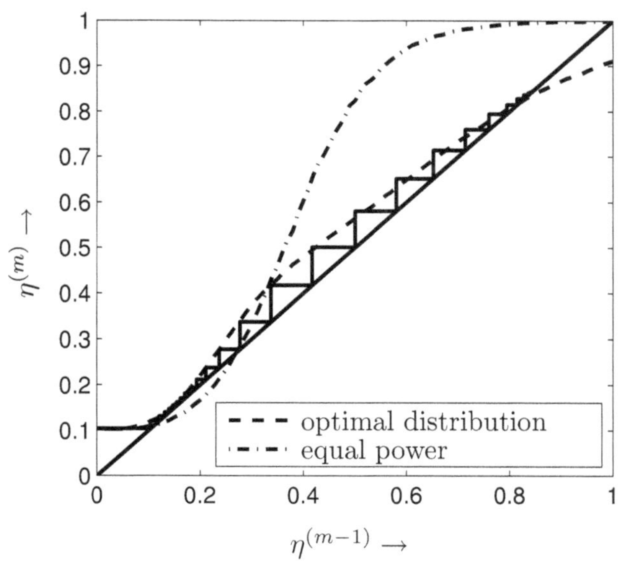

Fig. 6. Transfer function with optimized power profile for $P_{b\ \max}$=0.2 and β=4.

5 Power profile optimization

5.1 Formulation of the optimization problem

Up to now the analysis was based on uniformly distributed powers of the users. To describe an unequal power distribution by multi-user efficiency, a way to calculate a kind of average efficiency is necessary (Schlegel and Shi, 2004). Thus the resulting multi-user efficiency can be calculated more generally as

$$\eta_u = \frac{1}{1+\frac{\bar{E}_s}{N_0}\frac{1}{N}\sum\limits_{v\neq u}\mu_v\cdot P_v} \quad , \quad \sum_v P_v = U \quad . \tag{7}$$

$\bar{E}_s/N_0$ is defined as an average value over all users in order to get an appropriate criterion for fair comparison with the equal power case. The criterion for convergence is still reaching the point of η=1 after a finite number of iterations. For the PIC this is fulfilled if $f(\eta)\geqslant\eta$, $\eta\in[0, 1]$. The number of iterations needed depends on the width of the tunnel. Whether the tunnel is open or not depends also on the power distribution. For PIC it turns out that equal power for all users is not the best choice, as presented e.g. in Schlegel and Shi (2004) and Caire et al. (2004). The optimization problem for PIC can be described by Caire et al. (2004)

$$\min_{P_1,\ldots,P_U} \sum_u P_u \quad \text{s. t.} \quad \begin{cases} f(\eta)\geqslant\eta+\varepsilon\ ,\ \eta\in[0;1] \\ P_u>0 \quad , \quad u=1,\ldots U \end{cases} \tag{8}$$

if the sum of all received powers should be minimized and single user performance for all users is claimed. These conditions can be adapted to the particular need. For example in this work it will be

$$\min_{P_1,\ldots,P_U} \max_u P_u \text{ s. t.} \begin{cases} P_b(u)\leqslant P_{b\ \max}\ \forall\ u=1,\ldots U \\ N_{it}\leqslant N_{it\ \max}\ \forall\ u=1,\ldots U \\ P_u>0\ \forall\ u=1,\ldots U \end{cases} \tag{9}$$

For taking a maximum number of iterations into account, not the transfer function but the trajectory is used for the first constraint. After each iteration up to the maximum the MUE for all users is calculated and either this value or the corresponding BER is considered. For the SIC a similar expression is used although the calculation of η is different (Eq. 6).

5.2 Differential evolution

The cost functions of the examined optimization problems contain many local optima (multimodal functions), especially for SIC. Furthermore, the search space is highly constrained. Therefore a starting point in the vicinity of the global optimum would be required for the employment of local optimization techniques. As the global optimum is not known in advance global optimization methods have to be used. In this paper the Differential Evolution (DE) algorithm is chosen that belongs to the class of evolutionary algorithms (Price, 1999). As the name indicates evolutionary algorithms are motivated by the natural evolution process. In contrast to local search algorithms not only one but several points (population) in the search space are regarded at a given time instance. The advantage is that evolutionary algorithms do not need any knowledge of the search space. Differential Evolution furthermore possesses the property of fast convergence

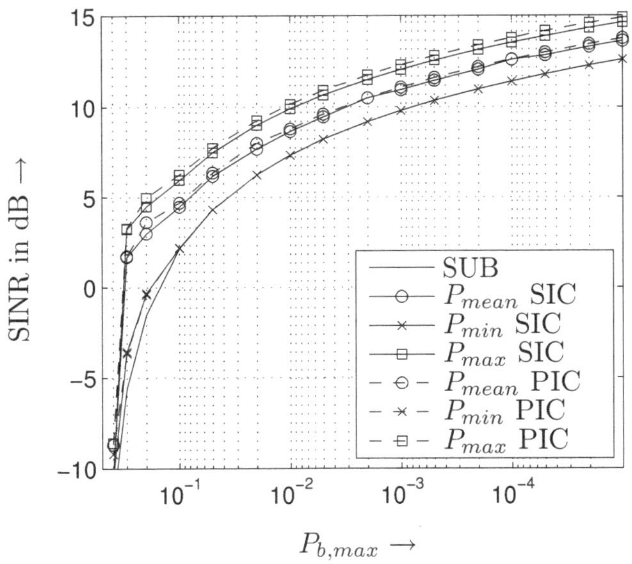

Fig. 7. Required receive power of all users dependent on the required BER for SIC and PIC, β=4, 20 iterations, mimimized according to Eq. (9).

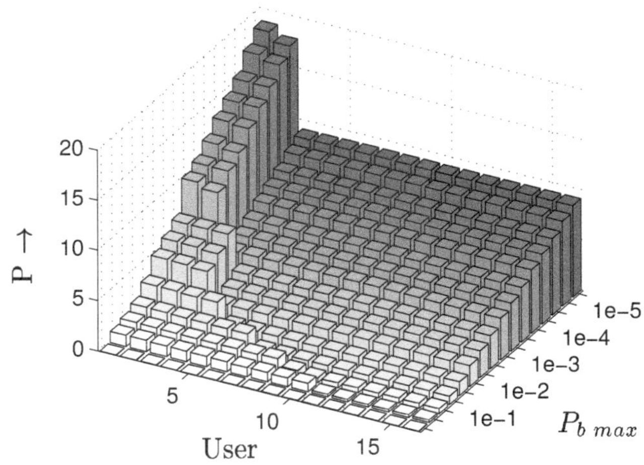

Fig. 8. Power profiles for different $P_{b\ \max}$ assuming Eq. (8).

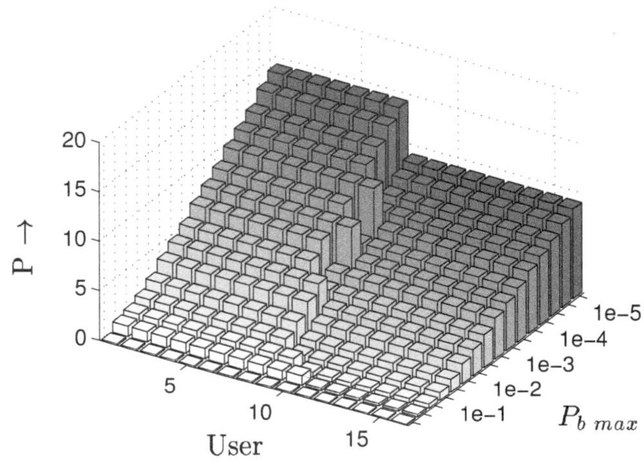

Fig. 9. Power profiles for different $P_{b\ \max}$ assuming Eq. (9).

due to an adaptive step size. Additionally, it is easy to use because of the few control parameters.

The starting population is generated randomly and should be sufficiently large for a high diversity. In this case a population size of NP=80 is used. Standard settings are used for other control parameters (Zielinski et al., 2005a). Population members of the current population are combined using the evolutionary operators mutation, recombination and selection to generate the next generation until a stopping condition is fulfilled. In this case the distribution of population members is monitored and the algorithm terminates when the maximum distance of every individual to the best population member is beneath a threshold (Zielinski et al., 2005b). In Section 6 the number of generations is restricted to 1000 due to limited computational resources.

6 Optimization results

As noted before in this paper not the sum of all powers is minimized but the maximum power. This seems to be more sufficient for cellular networks. Consequently, the overall power maybe increased but in the uplink the individual power constraint is the limiting point. In Fig. 5 are minimum, maximum and mean powers of the optimized profile depicted, which were optimized for different maximum uncoded BER. The lower bound for the receive power is the single user case. In this figure the difference between the optimization criteria (8) and (9) can be observed. The mean power is lower in the case of Eq. (8) which was the aim of this approach. But the maximum required power is up to 1 dB higher than for approach (9) over the whole BER range.

Results for the equal power case are not depicted, because the considered bit error rates will not be reached in this case. Just up to a load of ≈3 equal power systems may reach SUB.

The minimum power is in most cases equal to the minimum power for the single user case. Only for very low BERs this is no longer the case. This means on the other hand, for a given receive power the single user performance is not reached. This is due to the more general constraint only assuming the BER not the convergence to SUB. In Fig. 6 this case is illustrated with the corresponding transfer function. The iterative scheme gets stuck at $\eta \approx 0.8$. This plot corresponds to a uncoded target errror rate of 0.2. The results for PIC and SIC are compared in Fig. 7. As already noted in Weitkemper et al. (2005) SIC converges faster than PIC. With the additional constraint of a limited number of iterations which is also applied in this paper, this fact results in a decreased power. This effect gets smaller the more iterations are assumed. The resulting power profiles for different values of $P_{b\ \max}$ are depicted in Fig. 8 for the case of minimizing the sum of all powers. It can be seen that with decreasing error rate the number of users having a higher power than for the single user case gets smaller. But the ammount of power required for these users grows disproportionate to the minimum power. This behavior fulfills the requirement of minimum overall receive power, but in practice these high powers may not be possible. In Fig. 9 the power distributions with minimizing the maximum power can be seen. The number of users having an increased power is much higher than in Fig. 8 but the absolute maximum is smaller. In general the distribution gets more homogeneous. This distribution is more appropriate to practical cellular networks. In consideration of near-far effects the powers can be allocated to the users depending on their individual power constraint. For PIC the allocation order has no influence on the system performance and for SIC the detection order can be adapted if necessary. But as already observed in Weitkemper et al. (2005) the order of detection has minor influence at these high loads.

Any combination of individual and total power constraint can be implemented very easy. Even specific maximum supportable power by certain users can be taken into account.

7 Conclusions

In this paper some aspects of power profile optimization were investigated. The maximum bit error rate was considered as main constraint. At low BER this constraint gives different results than claiming convergence to the SUB. Additionally the expression to be minimized was adapted to the scenario of uplink communiaction in cellular networks. Not the overall power but the individual power is strictly limited and should be saved. So the optimization focusses on minimizing the maximum power. The results differ as well in general shape of the profiles as in the required maximum and mean power. Furthermore limited number of iterations was taken into account due to the limited time and resources for decoding in the base station. All these aspects deliver an optimization suitable for practicle system requirements.

References

Boutros, J. and Caire, G.: Iterative Multiuser Joint Decoding: Unified Framework and Asymptotic Analysis, IEEE Transactions on Information Theory, 48, 1772–1793, 2002.

Caire, G., Mueller, R., and Tanaka, T.: Iterative Multiuser Joint Decoding: Optimal Power Allocation and Low-Complexity Implementation, IEEE Transactions on Information Theory, 50, 1950–1973, 2004.

Kuehn, V.: Analysis of Iterative Multi-User Detection Schemes with EXIT Charts, in: IEEE International Symposium on Spread Spectrum Techniques and Applications (ISSSTA), Sydney, Australia, 2004.

Price, K.: An Introduction to Differential Evolution, in: New Ideas in Optimization, edited by: Corne, D., Dorigo, M., and Glover, F., pp. 79–108, McGraw-Hill, London, 1999.

Schlegel, C. and Shi, Z.: Optimal Power Allocation and Code Selection in Iterative Detection of Random CDMA, in: Zurich Seminar on Digital Communications, 2004.

Verdú, S.: Multiuser Detection, Cambridge University Press, 1998.

Weitkemper, P., Kühn, V., and Kammeyer, K.-D.: Analysis of Iterative Successive Interference Cancellation in SC-CDMA Systems, in: Fifth International Workshop on Multi-Carrier Spread-Spectrum (MC-SS 2005), Oberpfaffenhofen, Germany, 2005.

Weitkemper, P., Zielinski, K., Kammeyer, K.-D., and Laur, R.: Optimization of Interference Cancellation in Coded CDMA Systems by Means of Differential Evolution, in: 4th International Symposium on Turbo Codes & Related Topics, Munich, Germany, 2006.

Zielinski, K., Peters, D., and Laur, R.: Constrained Multi-Objective Optimization Using Differential Evolution, in: Third International Conference on Computational Intelligence, Robotics and Autonomous Systems, Singapore, 2005a.

Zielinski, K., Peters, D., and Laur, R.: Stopping Criteria for Single-Objective Optimization, in: Third International Conference on Computational Intelligence, Robotics and Autonomous Systems, Singapore, 2005b.

Calibration method for direct conversion receiver front-ends

R. Müller and H.-J. Jentschel

Institute of Traffic Communications Engineering, Dresden University of Technology, Germany

Abstract. Technology induced process tolerances in analog circuits cause device characteristics different from specification. For direct conversion receiver front-ends a system level calibration method is presented. The malfunctions of the devices are compensated by tuning dominant circuit parameters. Thereto optimization techniques are applied which use measurement values and special evaluation functions.

1 Introduction

Modern communication circuits are realized as integrated circuits. Because of process tolerances there are both defective and non-defective components produced. Test and calibration intend the separation of defective from non-defective modules as well as increasing yield. This paper presents a calibration method for direct conversion receiver (DCR) front-ends as device under test (DUT). The DCR front-end consists of a number of subsystems. These subsystems, e.g. mixer, LNA, filters, are called blocks and realize different functions. Basically, the DUT has a chain structure. The signal paths from the antenna to the A/D-converter corresponds to cascades of blocks. Therefore, the effects of the blocks on the signal superpose because the signal has to pass all blocks arranged in the signal path. From this follows, that conclusions can be given about the overall characteristic of the whole system using the output signal in relation to the input signal. Such analysis of the characteristic of an entire system is called system level test.

In reality, the implementation of such a system level test in combination with a system calibration is associated with some problems. There are evaluable characteristic system quantities needed to calibrate the system. These values can be obtained using special test signals. However, the nature of the measured information is restricted. There are only output signals available. Measuring points inside the DUT are not assumed. The analysis of the output signal should result in system values of the DUT, which can be compared to the specification or to the characteristic system quantities of a golden device. The problem in this context is as follows: it is possible to characterize enough the functionality of the DUT in respect of the requirements of the specifications using measured information of restricted nature. The integration of calibration mechanisms into the DUT requires adjustability of the block parameters. So the characteristic quantities of the blocks can be changed. Practically, block parameters correspond to circuit elements, e.g. resistors, capacitors.

It is clear from different reason, that it is impossible to make changeable all circuit elements in all blocks. Only a limited number of elements, the so called dominant parameters, can be prepared for tuning. Moreover, these parameters can be changed in a limited range only. So there are two problems in this context. First, the dominant parameters parameters must be selected taking in consideration practicability and costs. Second, it is possible to get desired changes of the DUT's output behavior by tuning a limited number of predetermined circuit elements.

At last there is the problem, that tuning of one parameter is affecting all system quantities on the output (Rühle, 2002). Because of this reason the use of optimization algorithms is proposed to correct the characteristic system quantities of the DUT. The dominant parameters are realized as variable parameters with a limited range. The optimization results in values for the dominant parameters for the correction of the DUT. In entirety the signal generator, signal analysis and optimization form a calibration at system level.

Instead of real circuits mathematical block models are considered to simplify the investigations. The feasibility of the calibration method can be verified with these models in principle. Malfunctions of the DUT are assigned to specific blocks and are included in the model.

Correspondence to: R. Müller
(reik.mueller@tu-dresden.de)

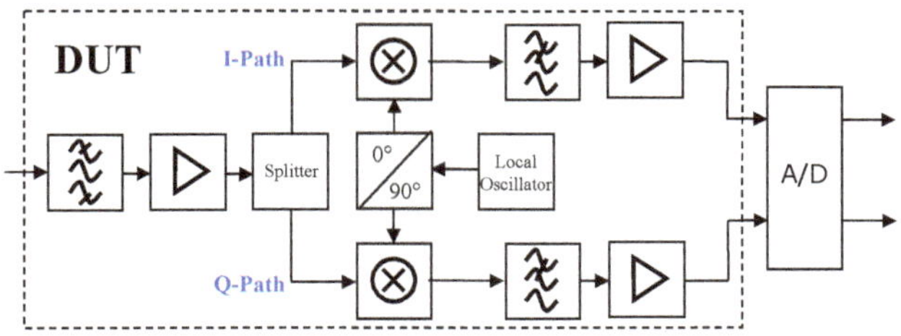

Fig. 1. The direct conversion receiver as test object.

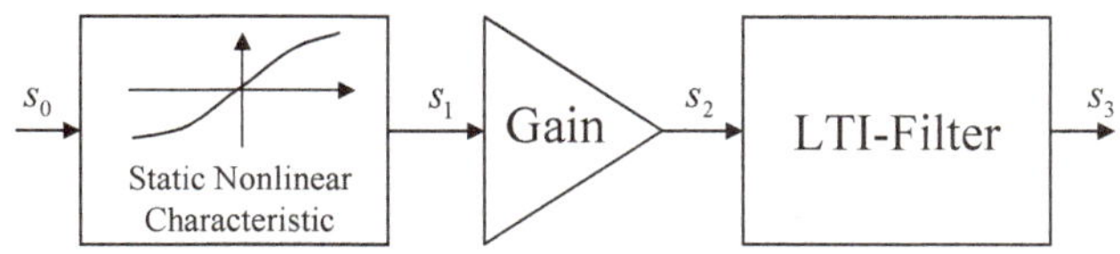

Fig. 2. Hammerstein model of the DUT.

2 The DCR as DUT

In our case a direct conversion receiver front-end is used as a DUT. Its block structure is shown in Fig. 1. The input block is the system filter (band pass filter). It is followed by a LNA. After pre-amplification the signal is split and fed to the I- and Q-path. The following mixers are driven by 90° phase shifted oscillator signals. The subsequent blocks in the chain are a low pass filter and a baseband amplifier in every path. The low pass filter and the amplifier often are applied as an active filter. These blocks determine the system characteristics, superpose in their influence on these characteristics and are responsible for functionality, properties and possible malfunctions of the DUT. But the influence on the functionality of the DUT is different per block (see Sect. 3). From this reason variable block properties can be used to compensate process tolerances.

3 The model of the DCR

A DCR front-end is a complex structure with a large number of devices. It is not possible to correct the value of all devices, because design effort and the need of chip area would be too high. Additionally a highly dimensional parametric space would result, which forms the basis of the correction of characteristic system values and can be applied only with high CPU-intensive computations and high effort of time. The latter is a problem of circuit design and the associated search for optimal variants of circuits (Müller et al., 2006). Because significant values have to be estimated by measurements and the measurements are time consuming, it is necessary to consider only dominant parameters for changing the DUT-properties. With such parameters the properties of the device can be changed efficiently, because there is a high sensitivity of the target values (Sect. 5) with reference to the variable device values. Such relations have to be analyzed by an sensitivity analysis of the whole circuit (Pursche, 2005; Hamida and Kaminska, 1993).
In the following it is assumed, that a limited number of tunable parameters has been chosen. The task is to correct the degradation of system performance caused by deviations in all parameters by this limited number of variable parameters, until the measurements of the DUT are possibly close to the measurements of the reference. The accuracy of the measurements is not regarded. The assumption is, that the measurement values are estimated faultlessly.
To achieve an easier verification of the proposed method the model has been kept simple consciously. The model only describes one path of the receiver front-end. One signal path of the DUT consists of the system filter (band pass filter, wide band), LNA, splitter, mixer, baseband filter (low pass filter) and a baseband amplifier (Fig. 1). The frequency shift of the mixing process has been considered as ideal, so it is not regarded in the model. The filter characteristics of all blocks have been summarized in a passive 2nd order Butterworth filter. Because the bandwidth of the system filter is much greater than that of the baseband filter, the cut-off frequency of the band pass filter has not to be considered. The splitter is only intended to affect the signal amplitude. The amplification of the active blocks superposes with the attenuation of the other blocks. The resulting amplification/attenuation is combined in an ideal amplification block with the amplification v. The nonlinearities of mixer, LNA and baseband amplifier are modeled by a static nonlinear characteristic. The analog to digital converter is considered as ideal. The final model is a Hammerstein model as shown in Fig. 2.
To show the relation between values of devices (e.g. capacitances) and the parameters of the model, the low pass filter has been realized as transfer function which includes device values. The transfer function of the low pass filter of the model is:

$$G(s) = \left[2 + K_1 s + K_2 s^2 + K_3 s^3 + K_4 s^4\right]^{-1} \quad (1)$$

with

$$K_1 = RC_2 + RC_1 + (L_1 + L_2)/R \quad (2)$$

$$K_2 = (C_1 + C_2)(L_1 + L_2) \quad (3)$$

$$K_3 = C_1 L_2 (RC_2 + L1/R) \quad (4)$$

$$K_4 = C_1 C_2 L_1 L_2 \quad (5)$$

The static nonlinear characteristic is modeled by:

$$s_1 = \theta(s_0) = c \times \tanh\left(\frac{s_0 + AP}{c}\right) - c \times \tanh\left(\frac{AP}{c}\right) \quad (6)$$

and realizes the operation point dependent characteristic of a differential amplifier. The parameter AP represents the operation point and is a variable parameter, $s_0{=}s_0(t)$ is the input signal in time domain and the parameter c stretches the characteristic to regulate the "strength" of the nonlinearity.

With the model in Fig. 2 the output signal $s_3(t)$ in dependence of the input signal $s_0(t)$ is:

$$s_3(t) = \int_{-\infty}^{\infty} g(\tau) \times s_2(t-\tau)\mathrm{d}\tau \qquad (7)$$

with $g(\tau)$ as impulse response according to Eq. 1. The filter input signal $s_2(t)$ results from Eq. 6 and the linear amplification v:

$$s_2(t) = v \times \theta(s_0(t)) \qquad (8)$$

4 Measurement and correction

The correction structure is shown in Fig. 3. For correction the present state of the DUT has to be estimated whether the specification is matched or not. A spectrum analysis is applied to the sampled output signal of the receiver. From the spectrum the following DUT parameters are estimated: cut-off frequency, over-all attenuation, IP3, I/Q-imbalance and frequency deviation of the oscillator signal. The extraction of these parameters is described in Lupea et al. (2003); Müller and Jentschel (2006a); Müller and Jentschel (2006b); DETAILS (2005). Using conventional measurement methods requires multiple measurements to estimate all these parameters (Kundert, 2002; Schaub and Kelly, 2004).
The evaluation of the measurement values includes a comparison with the specification and an optimization, described in the following sections. The evaluation results in parameters x_ν, which are numerical values and represent the new state of the variable circuit elements/block parameters. The control unit decodes the numerical parameters x_ν and transforms them into physical circuit values or operation point currents or voltages. Technical realizations are VGAs, varactors, fractional-N-PLLs, switched capacitors and switchable resistor and capacitor networks (Csipkes, 2006).
Subsequent the state of the DUT with the new parameter settings is estimated in relation to the specification again.

5 The objective function

The chain of DUT, measurement facilities and application of the control values realizes a function which function values are represented by measurement values. Every measurement is a calculation of a value of a function. A change in circuit device values changes the function. According to Fig. 3 the arguments of this function are represented by the parameters x_ν. It is convenient to summarize these parameters in a parameters vector.

$$\mathbf{x} = [x_1 \ldots x_k]^{\mathrm{T}} \qquad (9)$$

Every parameter x_ν can be set on discrete values between 0 and maximum register value $x_{\nu,\max}$ (control word for a device parameter). Dependent on the controlled device value

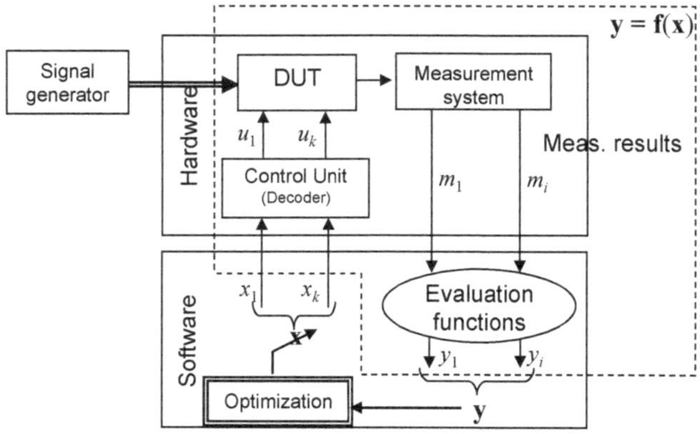

Fig. 3. Structure of the adjustment arrangement and scheme of the functional relation between parameter vector and target vector.

like capacitances or resistances the parameters x_ν are converted in different physical values. The relationship between the parameters and the physical values is represented by functions g_ν.

$$\mathbf{u} = [g_1(x) \ldots g_k(x)]^{\mathrm{T}} = \mathbf{g}(\mathbf{x}) \qquad (10)$$

The linear relation between u_ν and x_ν is:

$$u_\nu = u_{\nu,\min} + \left(u_{\nu,\max} - u_{\nu,\min}\right) \times x_\nu / x_{\nu,\max} \qquad (11)$$

and is for example valid for switchable resistor and capacitor networks or linear adjustable gain. There are also nonlinear relations like the capacitance control of varactors. The linear change of the control voltage of a varactor in conformity to Eq. 11 the capacitance changes in nonlinear way (Tietze and Schenk, 1990, p. 27).
With different settings u_ν of the circuit device values the input signal is transformed into an output signal $s_3(t)$=$s_3(s_0(t), \mathbf{u})$ in different way. The measurement of the output signal $s_3(t)$ results in measurement values m_μ. Hence, there is a relation between output signal $s_3(t)$ and measurement values. Therewith there is also a relation between circuit device values u_ν and the measurement values. Because of the manifold and complicated functional relations between output signal and measurement values, only the formal context is given:

$$\mathbf{m} = [m_1 \ldots m_i]^{\mathrm{T}} = [h_1(\mathbf{u}) \ldots h_i(\mathbf{u})]^{\mathrm{T}} = \mathbf{h}(\mathbf{u}) \qquad (12)$$

Measurement values are for example IP3, amplitude characteristic, NF, power loss, etc. The measurement values have to be adapted to the expected values of the optimization algorithm. The optimization goal is the maximum or the minimum of a function. As a result the measurement values m_i have to be transformed in an appropriate manner by evaluation functions:

$$\mathbf{y} = [y_1 \ldots y_i]^{\mathrm{T}} = [\zeta_1(\mathbf{m}) \ldots \zeta_i(\mathbf{m})]^{\mathrm{T}} = \zeta(\mathbf{m}) \qquad (13)$$

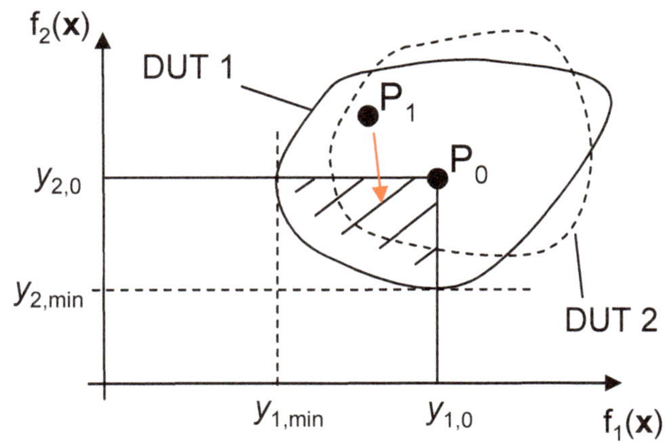

Fig. 4. Co-domain of two different DUT.

The transformation in target values has to be applied to all measurement values and is examined in Sect. 7 more detailed. Because there are several measurement values, a multi-criteria optimization problem occurs. With Eq. 9 to Eq. 12 and Eq. 13 the vectorial objective function follows formally (cp. Fig. 3):

$$\mathbf{y} = \mathbf{f}(\mathbf{x}) \tag{14}$$

Therewith the general relation between parameters of the DUT and objective function is shown. As it can be seen from Eq. 7 and from the description of Eq. 12, the objective functions $f_\mu(\mathbf{x})$ consists of complex analytical expressions and are difficult to present. That is why these relations are not specified in this paper.

6 Range of values

Equation 14 has a limited co-domain, because every scalar parameter x_ν is variable only in an limited range. There is no point accessible outside the limited range, even with arbitrary settings of $\mathbf{x}$. Because of process tolerances in circuit device values different DUT have co-domains which are differently limited. Using Eq. 14 the general formulation of the optimization task is:

$$\min_{\mathbf{x}} \mathbf{f}(\mathbf{x}) \tag{15}$$

Assuming the DUT is conform to the specification in any valid parameter combination, P_0 marks that point in the co-domain, where the minimum requirements of specification are sufficiently reached (Fig. 4). The evaluation functions ζ_i (cp. Sect. 7) ensure the minimum of a target value y_i is reached, when the appropriate property of the DUT is best possible. Hence, all points left and below the point P_0 represents a DUT that meets the specification. The hatched area in Fig. 4 shows that case for a two-dimensional co-domain. Consequently the calibration of the DUT is simplified to the task to shift the current point of state into the hatched area. If that condition has been achieved, the optimization can be finished. Thus, a search for the best point is not necessary and the requirements for optimization are simplified.
The indicator for compliance with the specification is examined in the following. The vector $\mathbf{r}_0$ from origin of ordinates to the point P_0 is calculated from specification and evaluation functions ζ_i. The point P_1 represents the current point of state with the position vector $\mathbf{r}_1$. From the point P_0 to the current point of state points the vector:

$$\mathbf{r}_\mathrm{D} = [d_1 \dots d_i]^\mathrm{T} = \mathbf{r}_1 - \mathbf{r}_0 = \begin{bmatrix} y_{1,1} - y_{1,0} \\ \vdots \\ y_{i,1} - y_{i,0} \end{bmatrix} \tag{16}$$

in the i-dimensional co-domain of the objective function. Where $y_{\mu,\nu}$ is the μ-th component of the vector ν. The exit condition is

$$d_\nu \le 0 \quad \text{with} \quad \nu = 1 \dots i \tag{17}$$

If this condition is complied, the value of the penalty function is 0, else it is

$$f_\mathrm{straf} = \sum d_\nu \quad \text{with} \quad \forall d_\nu > 0 \tag{18}$$

Hence, the goal of calibration is, that the value of the penalty function reaches 0. Then the DUT meets the specification.

7 Evaluation functions

In the following examples are given how to calculate values of the objective function from measurement values.
The intermodulation point (IP) is a value to rate nonlinear distortions. Mostly in communications there is an interest for linearity. I.e. an IP which is going towards infinity is wanted. To formulate a task of minimization with the 3rd order intermodulation point (IP3) the conversion from IP3 to a value of the objective function is:

$$y_{\mathrm{IIP_3}} = \zeta_1(\mathrm{IIP_3}) = -c_1 \times \mathrm{IIP_3} \tag{19}$$

From the value of the specification accordingly follows:

$$y_{\mathrm{IIP_3},0} = \zeta_1\left(\mathrm{IIP_{3,min}}\right) = -c_1 \times \mathrm{IIP_{3,min}} \tag{20}$$

where $\mathrm{IIP_{3min}}$ is the minimum acceptable IP3 according to the specification.
For rating the accuracy of the frequency of the local oscillator on chip, the frequency f_LO has to be measured and compared with the target frequency $f_{\mathrm{LO},0}$.

$$y_\mathrm{LO} = \zeta_2(f_\mathrm{LO}) = -c_2 \times \left(f_\mathrm{LO} - f_{\mathrm{LO},0}\right)^2 \tag{21}$$

The bound of y_LO pursuant to the specification is

$$\begin{aligned} y_{\mathrm{LO},0} &= -c_2 \times \left[\left(f_{\mathrm{LO},0} \pm \Delta f\right) - f_{\mathrm{LO},0}\right]^2 \\ &= -c_2 \times \Delta f^2 \end{aligned} \tag{22}$$

where Δf is the acceptable frequency deviation from the target frequency. For rating the constancy of the amplitude characteristic in the pass band of the DUT the spectrum of the output signal has to be estimated. At this, samples in frequency domain with an determined fundamental frequency occurs. With that samples and the input signal the transfer characteristic $|H(f_\nu)|$ can be estimated approximately. The sum of the square error of the transfer factors with reference to a constant H_{ref} (mean value or a reference) is a figure of merit of the waviness of the amplitude characteristic:

$$y_{\mathrm{Ampl}} = \zeta_3(H) = \sum_{\nu=\nu_{\min}}^{\nu_{\max}} \left(|H(f_\nu)| - |H_{\mathrm{ref}}|\right)^2 \tag{23}$$

with the bound $y_{\mathrm{Ampl},0} = \mathrm{const}$

The presented options for calculating values of the objective function from measurement values are also applicable for conversion of a number of further measurement values: I/Q-mismatch, cut-off frequency, signal power, power consumption, noise figure, etc.

So, the vector of the objective function can be written as:

$$\mathbf{r}_1 = \begin{bmatrix} y_{1,1} \\ y_{2,1} \\ y_{3,1} \\ \vdots \end{bmatrix} = \begin{bmatrix} y_{\mathrm{IIP3}} \\ y_{\mathrm{LO}} \\ y_{\mathrm{Ampl}} \\ \vdots \end{bmatrix} \tag{24}$$

The vector $\mathbf{r}_0$ has to be created from the acceptable bounds in equivalence to $\mathbf{r}_1$. Thereupon the value of the penalty function is calculated like shown in Sect. 6.

8 Verification of the calibration method

For verification of the feasibility of the calibration method, simulations with the model from Sect. 3 have been realized exemplarily. As reference model/golden device the following model parameters has been chosen: C_1=58.781 μF, C_2=24.348 μF, L_1=60.087 μH, L_2=146.95 μH, reference resistance R=50 Ω, v=1, AP=0 and c=10. The input signal is a multi-tone signal with 10 harmonics with the lowest frequency at 10 kHz, the highest frequency at 110 kHz and spectral gap between the lower and the higher harmonics. The amplitude of every harmonic is 0 dB. The frequency distance (fundamental frequency) between the harmonics is 10 kHz. The spectrum of the output signal is calculated. The complex amplitudes of the output signal at these frequency bins, which are occupied by the harmonics in the input signal, are written as a vector $\mathbf{v}$. From this vector the IIP_3 is calculated using the multi-tone test method described in DETAILS (2005). Additionally the power level $p_{3\mathrm{dB}}$ at the intended cut-off frequency and the mean signal power level $\bar{p}$ are calculated and compared with the specification. Hence, the values of the objective function are f_{IIP_3}, $f_{\bar{p}}$ and $f_{p_{3\mathrm{dB}}}$. Because of using a Butterworth filter, it can be assumed that there is linearity in the pass band. Because the oscillator frequency is not considered in the model, the application of Eq. 21 and Eq. 22 has been dropped.

The variable parameters of the DUT are u_1=C_1, u_2=C_2, u_3=v and u_4=AP. The remaining parameters of the DUT are invariant values. For simplification the parameter vector is considered a continuously variable. The transformation of the parameters x_ν in circuit device values u_ν linear functions g_ν are constructed according to Eq. 11. Two cases has been simulated:

- Case 1: the invariant values of the DUT are equal to that of the reference model.
- Case 2: the invariant values of the DUT have a deviation of +10% of the values of the reference model.

The initial values of the variable parameters have a deviation of −20% of the reference values in both cases. The initial value of the parameter AP has been set to 9.5. The DUT is stimulated with the same input signal like the reference model. The measurement values are estimated from the output signal of the DUT. The values of the objective functions are calculated according to Sect. 7 and the value of the penalty function according to Eq. 18.

For simplification of the optimization the optimization task has been reduced using a scalar objective function. The point of state of the golden device is inside the area of matched specification (cp. Fig. 4). Because the values of the objective function are calculated from measurement values and the measurement values are calculated from the output signal, the approximation of the output signal of the DUT to that of the golden device potentially leads to an approximation of the point of state in co-domain. The mean square error f_q of the deviation of the output spectrum $\mathbf{v}_{\mathrm{DUT}}$ of the DUT and the output spectrum $\mathbf{v}_{\mathrm{ref}}$ of the reference model is calculated.

$$f_q(\mathbf{x}) = |\mathbf{v}_{\mathrm{DUT}}(\mathbf{x}) - \mathbf{v}_{\mathrm{ref}}(\mathbf{x})|^2 \tag{25}$$

The mean square error is a scalar value that is optimized using Quasi-Newton-method (Matlab, 2007) in terms of the parameter vector $\mathbf{x}$. Doing so, the optimization task according to Eq. 15 is reduced to:

$$\min_{\mathbf{x}} f_q(\mathbf{x}) \tag{26}$$

While running the optimization the exit condition according to Eq. 17 or Eq. 18 respective is checked after every iteration step. The optimization is canceled if the specification is matched. The values of the specification has been chosen exemplarily like the parameters of the test object. So there is no direct applicatory significance. The values of specifications are IIP_3>25.4 dB, $p_{3\mathrm{dB}} \leq 27.87$ dB and $\bar{p}$=29.7 dB. The associated values of the DUT and the reference model are listed in Table 1.

Without consideration of the exit condition in case 1 the optimization of the DUT results in the same values of the parameter vector like the values of the reference model after 36 iterations and 210 measurements. With regard to the

Table 1. Measurement values.

CASE 1	IIP_3/dB	p_{3dB}/dB	$\bar{p}$/dB
Reference model	25.711	27.870	29.777
Unoptimized DUT	25.095	22.107	22.858
Spec. reached (exit)	25.710	27.839	29.757
Opt. without aborting	25.711	27.870	29.777
CASE 2	IIP_3/dB	p_{3dB}/dB	$\bar{p}$/dB
Reference model	25.711	27.870	29.777
Unoptimized DUT	24.296	22.402	23.523
Spec. reached (exit)	26.606	27.808	29.305
Opt. without aborting	26.610	28.597	29.910

specification the goal of correction is already reached after 31 iterations and 185 measurements. In this case the estimated parameters are C_1=58.870 μF, C_2=24.337 μF, v=0.998 and $AP = -0.068$.
In case 2 the optimization converges after 36 iterations and 195 measurements. Because the invariant parameters of the DUT are different to the parameters of the reference model, the variable parameters are estimated with C_1=53.933 μF, C_2=21.617 μF, v=0.990 and AP=0.040. The values of the specification are already achieved after 5 iterations and 30 measurements. At this the variable parameters are estimated with C_1=50.708 μF, C_2=22.365 μF, v=0.959 and AP=2.311.
The specification is matched after optimization and applying exit condition. Without applying the exit condition in case 2 the specification is not always matched during the progressive iteration process, i.e. the point of state temporarily leaves the region of matched specification in the co-domain.

9 Conclusions

A method to correct static deviations of parameters of DCR front-ends has been presented. Values of selected circuit devices of a DUT which does not match the specification are corrected applying an optimization algorithm until the specification is met. To enable the variability of the values of the selected circuit devices the circuit design has to be extended. Deviations of the values of the circuit devices caused by process tolerances can be compensated that way. Measurements of the DUT provide objective function values needed for optimization. The parameter alignment is done at the DUT directly. The optimization is an iterative method with several measurements at every iteration step. That is why the major disadvantage of this method concerns the measurement time, because it is possible that more than one hundred measurements are needed for correction.

In particular the noise increases the number of iteration steps which are needed. Additionally the duration of measurement has to be increased to average over time. With noise the accuracy of the measurement values decreases, which cause convergence problems.

Acknowledgements. This work has been supported by the German Government (BMBF). Additionally this work has been supported by Infineon Technologies AG.

References

Csipkes, G.-L.: Integrated realizations of reconfigurable low pass and band pass filters for wide band multi-mode receivers, Dissertation TU-Dresden, 2006.

DETAILS: Milestone Report M15-AP1.3 – Calibration and Test Friendly Design, BMBF project “DETAILS” Grant 01M3071, Milestone report 2005, 43–61, 2005.

Hamida, N. B. and Kaminska, B.: Multiple Fault Analog Circuit Testing by Sensitivity Analysis, J. Electron. Test., 4, 331–343, 1993.

Kundert, K. S.: Accurate and Rapid Measurement of IP_2 and IP_3, Designer’s Guide Consulting, Inc., http://www.designers-guide.org, 2002.

Lupea, D., Pursche, U., and Jentschel, H.-J.: RF-BIST: Loopback Spectral Signature Analysis. Proc. of Design, Automation and Test in Europe Conference - DATE 03, Munich, March 3–7 2003, 478–483, 2003.

Müller, R. and Jentschel, H.-J.: Testmethode für Direct Conversion Receiver, Analog’06, ITG Fachbericht 196, 95–100, 2006a.

Müller, R. and Jentschel, H.-J.: A Built-In-Self-Test view on estimation of mixer-parameters of direct conversion receiver front-ends, 18. ITG/GI/GMM Workshop “Testmethoden und Zuverlässigkeit von Schaltungen und Systemen”, Titisee, 12-14 März 2006, 71–75, 2006b.

Müller, D., Gräb, H., and Schlichtmann, U.: Optimierung analoger Schaltungsblöcke mittels Pareto-Wellenfront-Optimierung, Analog’06, ITG Fachbericht 196, 33–38, 2006.

Pursche, U.: Spektrale Signalflussmodellierung durch Harmonischen-Transfer-Matrizen für den Selbsttest und die Selbstkorrektur von Hochfrequenzschaltungen. Dissertation TU-Dresden, ISBN 3-8325-0996-8, 162–168, 2005.

Rühle, Th.: Entwurfsmethodik für Funkempfänger, Dissertation TU-Dresden, 46–56, 2002.

Schaub, K. B. and Kelly, J.: Production Testing of RF and System-on-a-Chip Devices for Wireless Communications, ARTECH HOUSE, Inc., ISBN 1-58053-692-1, 2004.

The MathWorks, Inc.: Matlab Optimization Toolbox User’s Guide, Sep. 2007, www.mathworks.com, 2007.

Tietze, U. and Schenk, Ch.: Halbleiter-Schaltungstechnik, 9. Auflage, Springer-Verlag Berlin, ISBN 3-540-19475-4, 1990.

Considering even-order terms in stochastic nonlinear system modeling with respect to broadband data communication

O. Schmitz[1], S.K. Hampel[1], H. Eul[1], and D. Schwingshackl[2]

[1]Leibniz Universität Hannover, Institut für Hochfrequenztechnik und Funksysteme, Appelstr. 9A, 30167 Hannover, Germany
[2]Infineon Austria AG, Siemensstr. 2, 9500 Villach, Austria

Abstract. As a tradeoff between efficiency and costs modern communication systems contain a variety of components that can at least be considered weakly nonlinear. A critical element in evaluating the degree of nonlinearity of any underlying nonlinear system is the amount of undesired signal strength or signal power this system is introducing outside the transmission bandwidth. This phenomenon called spectral regrowth or spectral broadening is subject to stringent restrictions mainly imposed by the given specifications of the particular communication standard. Consequently, achieving the highest possible efficiency without exceeding the linearity requirements is one of the main tasks in system design. Starting from this challenging engineering problem there grows a certain need for specialized tools that are capable of predicting linearity and efficiency of the underlying design. Besides a multitude of methods aiming at the prediction of spectral regrowth a statistical approach in modeling and analyzing nonlinear systems offers the advantage of short processing times due to closed form mathematical expressions in terms of input and output power spectra and is therefore further examined throughout this article.

1 Introduction

This article focuses on the derivation and analysis of a frequency domain model capable of predicting and estimating the degree of nonlinearity for a wide class of nonlinear systems by means of its self-induced nonlinear distortion. Besides the variety of already well established approaches, like the transient analysis or the harmonic balance analysis, the model presented in this article bears the advantage of a direct mathematical relation between input and output quantities in terms of closed form expressions. Therefore it is basically applicable in system-level simulations.

Historically this modeling approach has grown by the need for evaluating the spectral regrowth or spectral broadening caused by power amplifiers in classical wireless RF (Radio Frequency) transmission (see Gard, 2003; Raich, 2004). Driven at highest possible output power levels especially the power amplifier reveals nonlinear transmission behaviour, which is crucial regarding vicinal communication systems in RF transmission. Due to the fact that initial results in this area of research have been obtained by modeling classical RF transmission modules, fundamental contributions mainly deal with the modeling of nonlinear systems assuming a narrowband excitation stimulus, i.e., narrowband signalling, furthermore supposing the nonlinear system to be memoryless or static (see Gard et al., 1999).

Starting from these results the intention of this article is to adapt the present narrowband modeling approach and to modify it in a way, so that it is applicable for wideband communication systems as well. This effort not only requires the pure extension of the existing mathematical approach but also the consideration of a certain system dynamics, often called system memory, that is immediately associated with the given bandwidth the system is being probed with. The article is structured as follows: In Sect. 2 we introduce the classical narrowband nonlinear system modeling approach, also known as the bandpass approach, mainly based on the work of Gard (2003) and Raich (2004). Assuming the nonlinear system to occupy static temporal behaviour, we derive the systems time-domain output signal in terms of the narrowband input stimulus. Exploiting given inherent signal statistics it is then possible to derive the systems output autocorrelation function by means of the analytical expansion of so-called statistical higher order moments (see Davenport and Root, 1997). A subsequent Fourier transform finally yields the systems output PSD (Power spectral density) in terms of the input PSD allowing for an accurate estimation

Correspondence to: O. Schmitz
(schmitz@hft.uni-hannover.de)

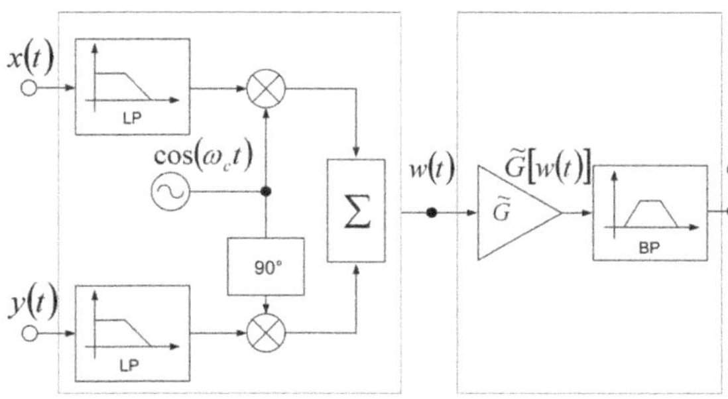

Fig. 1. Quadrature modulator and bandpass nonlinearity.

of the systems nonlinearity.

In Sect. 3 we modify the mathematical approach given in Sect. 2. Contrary to the classical narrowband assumption, we presume a broadband input signal requiring the expansion of additional higher order moments for the derivation of the autocorrelation function. As before in Sect. 2 the output PSD is obtained by applying the Fourier transform to the output autocorrelation. With regard to the input signal bandwidth we furthermore present the so-called generalized Hammerstein modeling approach, accounting for the temporal characteristics of the underlying broadband nonlinear system.

In Sect. 4 the derived frequency-domain modeling approaches are verified by time-domain simulations. For this purpose we compare mean distortion spectra computed according to the derived mathematical expressions as well as averaged distortion spectra taken from subsequent time-domain iterations for subset static and dynamic nonlinear systems.

Finally, in Sect. 5 we resume the presented results and draw essential conclusions.

2 Narrowband modeling of nonlinear systems

In this section we introduce the so-called bandpass modeling approach, mainly established by the work of Gard et al. (1999), Gard et al. (2001) and Zhou and Raich (2004). Fig. 1 therefore depicts a general digital transmission system followed by a static nonlinearity and a bandpass filter, whereas the combination of nonlinear system and bandpass filter is called bandpass nonlinearity for the following considerations.

At the output of the quadrature modulator the carrier signal modulated in ampltitude and phase can be stated as

$$w(t) = A(t)\cos(\omega_c t + \Theta(t)) \tag{1}$$

where $A(t)$ denotes the time varying amplitude, $\Theta(t)$ the time varying phase angle and ω_c the carrier frequency of the modulated signal in the bandpass domain. In terms of its complex envelope $\tilde{z}(t)$ we can rewrite Eq. (1)

$$w(t) = \frac{1}{2}\tilde{z}(t)e^{j\omega_c t} + \frac{1}{2}\tilde{z}^*(t)e^{-j\omega_c t} \tag{2}$$

with the complex envelope expressed by its in-phase component $x(t)$ and quadratue component $y(t)$ according to

$$\tilde{z}(t) = A(t)e^{j\Theta(t)} = x(t) + jy(t). \tag{3}$$

The modulated carrier is applied to a nonlinear circuit with nonlinear gain characteristic $\tilde{G}[w(t)]$ which is assumed to be a static nonlinearity containing no significant memory within the bandwidth of modulation. Generally, a complex power series is used to model the input output relations:

$$\begin{aligned}\tilde{G}[w(t)] &= \tilde{a}_1 w(t) + \tilde{a}_2 w^2(t) + \cdots + \tilde{a}_N w^N(t) \\ &= \sum_{n=1}^{N} \tilde{a}_n w^n(t),\end{aligned} \tag{4}$$

where the factors $\tilde{a}_1$ to $\tilde{a}_N$ represent the generally complex coefficients taken from a N-th degree taylor series expansion of a nonlinear input-output characteristic for instance. Due to the subsequent bandpass zonal filtering depicted in Fig. 1 only those nonlinear terms have to be considered, that contribute to distortion components at the fundamental frequency. Since even order nonlinear terms induce distortion at baseband as well as at even multiples of the fundamental frequencies, they do not have to be considered when applying the classical narrowband bandpass analysis. Thus Eq. (4) turns into

$$\begin{aligned}\tilde{G}[w(t)] &= \tilde{a}_1 w(t) + \tilde{a}_3 w^3(t) + \cdots + \tilde{a}_N w^N(t) \\ &= \sum_{n=0}^{(N-1)/2} \tilde{a}_{2n+1} w^{2n+1}(t),\end{aligned} \tag{5}$$

where N in this case denotes the highest odd-order power contained in the polynomial description. In order to simplify the analysis a binomial expansion is applied to compute the m-th power of $w(t)$, yielding

$$w^m(t) = \frac{1}{2^m}\sum_{k=0}^{m}\binom{m}{k}\tilde{z}(t)^k\tilde{z}^*(t)^{m-k}e^{j\omega_c(2k-m)t}. \tag{6}$$

According to the bandpass filtering depicted in Fig. 1, only the terms centered at the carrier frequency have to be taken into account. This implies $2k-m=\pm 1$ for odd m only. Substituting $k=(m+1)/2$, the terms centered around the carrier frequency are

$$w_{\omega_c}^m(t) = \left\{\frac{1}{2^{2m-1}}\binom{m}{\frac{m+1}{2}}\tilde{z}(t)^{\frac{m+1}{2}}\tilde{z}^*(t)^{\frac{m-1}{2}}\right\}e^{\pm j\omega_c t}. \tag{7}$$

It is convenient to express Eq. (7) in terms of odd-order powers $m=2n+1$ with $n=0\ldots(N-1)/2$:

$$w_{\omega_c}^{2n+1}(t) = \left\{\frac{1}{2^{2n}}\binom{2n+1}{n+1}\tilde{z}(t)^{n+1}\tilde{z}^*(t)^n\right\}e^{\pm j\omega_c t}. \tag{8}$$

The time-domain signal at the output of the bandpass nonlinearity can now be written as

$$\tilde{G}_{\omega_c}[w(t)] = \left[\sum_{n=0}^{(N-1)/2}\tilde{a}_{2n+1}w_{\omega_c}^{2n+1}(t)\right]e^{\pm j\omega_c t}. \tag{9}$$

Furthermore, in terms of the input signals complex envelope $\tilde{z}(t)$ Eq. (9) can be stated as

$$\tilde{G}_{\omega_c}[\tilde{z}(t)] = \sum_{n=0}^{(N-1)/2} \frac{\tilde{a}_{2n+1}}{2^{2n}} \binom{2n+1}{n+1} \tilde{z}(t)^{n+1} \tilde{z}^*(t)^n. \quad (10)$$

The expression in Eq. (10) describes the complex envelope of the first harmonic of a modulated carrier signal passed through a bandpass nonlinear circuit described by a complex power series. Originally this kind of description is based on a certain type of characterization of nonlinear systems. Within a so called AM-AM and AM-PM measurement campaign the amplitude of an unmodulated carrier signal is swept and the corresponding change in gain and phase at the output port of the circuit are measured at the input frequency. The amplitude gain response is known as the amplitude modulation to amplitude modulation transfer characteristic (AM-AM) and the amplitude phase response is the amplitude modulation to phase modulation characteristic (AM-PM). The fact that this way of modeling considers the determination of a complex large signal input output response clarifies the need for integrating a bandpass filter in the formal model depicted in Fig. 1.

In order to transfer the time-domain relation between input and output quantities according to Eq. (10) into a frequency-domain representation combining input and output PSDs, it is initially necessary to compute the output autocorrelation function in terms of the input autocorrelation. However, the straightforward approach we are aiming at in this article requires the assumption of certain signal statistics, that are inherently occupied by the input signal, assuming certain transmission schemes. By means of the the central limit theorem, which states that the sum of identically distributed, zero mean, independent random processes tends towards a zero mean Gaussian distribution (see Davenport and Root, 1987), it is possible to consider the input signal to be Gaussian distributed in case of multitone transmission. Hence the assumption of Gaussian statistics is quite common when focusing on multitone transmission schemes such as OFDM (Othogonal Frequency Division Multiplex) or DMT (Discrete Multitone Transmission) (see Zhou and Raich, 2004).

For further investigations the input signals complex envelope is therefore assumed to be accurately modeled by a real gaussian random process $x(t)$. The bandpass stimulus can then be expressed as

$$w(t) = \frac{1}{2} x(t) e^{j\omega_c t} + \frac{1}{2} x(t) e^{-j\omega_c t}. \quad (11)$$

The modulated carrier is applied to the input of a nonlinear system represented by the complex power series given by Eq. (5), finally yielding the time-domain output signal

$$\tilde{G}_{\omega_c}[\tilde{z}(t)] = \sum_{n=0}^{(N-1)/2} \frac{\tilde{a}_{2n+1}}{2^{2n}} \binom{2n+1}{n+1} x(t)^{2n+1}. \quad (12)$$

The output autocorrelation is then found by the expectation of the output signal

$$\tilde{R}_{gg}(\tau) = E[\tilde{G}_{\omega_c}(\tilde{z}_1)\tilde{G}^*_{\omega_c}(\tilde{z}_2)], \quad (13)$$

where $\tilde{z}_1 = \tilde{z}(t_1)$ and $\tilde{z}_2 = \tilde{z}(t_2) = \tilde{z}(t_1 - \tau)$. After combining Eq. (12) and Eq. (13) we obtain the expression given by Eq. (14) for the output autocorrelation.The expectation in Eq. (14) can only be evaluated if the moments of the random variable $x(t)$ are known.

$$\tilde{R}_{gg}(\tau) = \sum_{n=0}^{\frac{N-1}{2}} \sum_{m=0}^{\frac{N-1}{2}} \frac{\tilde{a}_{2n+1}\tilde{a}^*_{2m+1}}{2^{2(n+m)}} \binom{2n+1}{n+1} \cdots \\ \cdots \times \binom{2m+1}{m+1} E[x_1^{2n+1} x_2^{2m+1}]. \quad (14)$$

In case of a zero mean real Gaussian process the moments are given by

$$E[x_1 x_2 \ldots x_s] = \begin{cases} 0 & s\ odd \\ \sum \prod E[x_i x_j] & s\ even \end{cases} \quad (15)$$

whereas the summation given in Eq. (15) has to be performed over all distinct pairs of subscripts i and j of x.

For many cases it is sufficient to consider the nonlinear terms of an underlying nonlinear system up to the order of $N=7$, allowing for an accurate approximation of the nonlinear transmission behaviour. However, before the relevant moments evaluated for a seventh-order power series expansion are presented, we start by denoting the statistical assumptions and having a closer look on the applied nomenclature. Assume the random variables x_1 and x_2 (whereas x_2 is a time shifted instance of x_1) to be Gaussian distributed with zero mean, which implies

$$E[x_1] = E[x_2] = 0 \\ E[x_1^2] = E[x_2^2] = \sigma_{x_1}^2 = \sigma_{x_2}^2 = R_{xx}(0) = R_{x0}. \quad (16)$$

The expectation of the product of these two random variables is defined by the autocorrelation function

$$E[x_1 x_2] = R_{xx}(\tau). \quad (17)$$

Using these abbreviations Table 1 lists the moments evaluated for a seventh-order power series expansion, where only the odd-order powers corresponding to the presented bandpass approach are considered. If the Fourier transform is finally applied to the terms listed in Table 1, we obtain the output PSD in terms of its input PSD, remembering that the n-fold potentiation of the autocorrelation function in the time-domain turns out to be the n-fold convolution of the PSD in the frequency-domain.

A verification of the derived relations in terms of simulation results is given in Sect. 4.

Table 1. Odd order moments evaluated for a seventh order power series expansion.

Subscr. Eq. (14)	Expanded moment
$n = 0;\ m = 0$	$R_{xx}(\tau)$
$n = 0;\ m = 1$	$3R_{x0}R_{xx}(\tau)$
$n = 1;\ m = 1$	$9R_{x0}^3 R_{xx}(\tau) + 6R_{xx}^3(\tau)$
$n = 0;\ m = 2$	$15R_{x0}^2 R_{xx}(\tau)$
$n = 1;\ m = 2$	$45R_{x0}^3 R_{xx}(\tau) + 60R_{x0}R_{xx}^3(\tau)$
$n = 2;\ m = 2$	$225R_{x0}^4 R_{xx}(\tau) + 600R_{x0}^2 R_{xx}^3(\tau)+$ $120R_{xx}^5(\tau)$
$n = 0;\ m = 3$	$105R_{x0}^3 R_{xx}(\tau)$
$n = 1;\ m = 3$	$315R_{x0}^4 R_{xx}(\tau) + 630R_{x0}R_{xx}^3(\tau)$
$n = 2;\ m = 3$	$1575R_{x0}^5 R_{xx}(\tau) + 6300R_{x0}^3 R_{xx}^3(\tau)+$ $2520R_{x0}R_{xx}^5(\tau)$
$n = 3;\ m = 3$	$11025R_{x0}^6 R_{xx}(\tau) + 66150R_{x0}^4 R_{xx}^3(\tau)+$ $52920R_{x0}^2 R_{xx}^5(\tau) + 5040R_{xx}^7(\tau)$

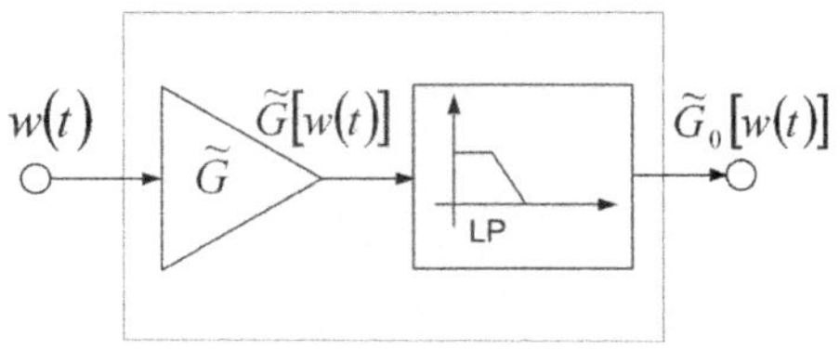

Fig. 2. Static Nonlinear system followed by an ideal low-pass filter.

3 The broadband extension

In case of a broadband input stimulus it is no longer sufficient to consider only the odd-order terms of a given nonlinear system, as indicated by Eq. (5). For considering even-order terms of a power series nonlinearity as well as odd-order terms, again, we have to pick up the term denoted by Eq. (6). In contrast to the previous section we now only focus on contributions that mix into the baseband. This assumption corresponds to the idea that an ideal lowpass filter is placed right behind the nonlinear system as indicated by Fig. 2.

That means instead of setting $2k-m=\pm 1$ we set $2k-m=0$ in Eq. (6), which yields

$$w_0^m(t) = \frac{1}{2^m}\binom{m}{\frac{m}{2}}\tilde{z}(t)^{\frac{m}{2}}\tilde{z}^*(t)^{\frac{m}{2}}. \tag{18}$$

Due to the assumption that m can only represent even values, we can state $m=2n$ where $n=1, 2, 3, \ldots, \frac{N}{2}$, provided N representing the highest even-order power within the power series polynomial. Thus Eq. (18) turns into

$$w_0^m(t) = \frac{1}{2^{2n}}\binom{2n}{n}\tilde{z}(t)^n\tilde{z}^*(t)^n. \tag{19}$$

After passing through the nonlinear circuit the even-order contributions in terms of the complex envelope can be stated

Table 2. Even order moments evaluated for a sixth order power series expansion.

Subscr. Eq. (21)	Expanded moment
$n = 1;\ m = 1$	$2R_{xx}^2(\tau) + R_{x0}^2$
$n = 1;\ m = 2$	$12R_{xx}^2(\tau)R_{x0} + 30R_{x0}^3$
$n = 1;\ m = 3$	$90R_{x0}^2 R_{xx}^2(\tau) + 15R_{x0}^4$
$n = 2;\ m = 2$	$24R_{xx}^4(\tau) + 72R_{xx}^2(\tau)R_{x0}^2 + 9R_{x0}^4$
$n = 2;\ m = 3$	$360R_{x0}R_{xx}^4(\tau) + 540R_{x0}^3 R_{xx}^2(\tau)+$ $45R_{x0}^5$
$n = 3;\ m = 3$	$720R_{xx}^6(\tau) + 5400R_{x0}^2 R_{xx}^4(\tau)+$ $4050R_{xx}^2(\tau)R_{x0}^4 + 225R_{x0}^6$

as

$$\tilde{G}_0[\tilde{z}(t)] = \sum_{n=1}^{\frac{N}{2}} \frac{\tilde{a}_{2n}}{2^{2n}}\binom{2n}{n}\tilde{z}(t)^{2n}\tilde{z}^*(t)^{2n}. \tag{20}$$

Again assuming a real gaussian random process $x(t)$ modulating the carrier amplitude, the output autocorrelation function can be written as

$$\tilde{R}_{gg}(\tau) = \sum_{n=1}^{\frac{N}{2}}\sum_{m=1}^{\frac{N}{2}} \frac{\tilde{a}_{2n}\tilde{a}_{2m}^*}{2^{n+m}}\binom{2n}{n}\binom{2m}{m}E[x_1^{2n}x_2^{2m}]. \tag{21}$$

As can easily be seen from Eq. (21), corresponding to the expectation value a lot of additional statistical moments have to be evaluated when accounting for even-order distortion. As previously done in Sect. 2 the additionally expanded moments are listed in Table 2. However, in contrast to the odd order considerations Table 2 indicates that the even order terms cause a remarkable amount of distortion power at DC (Direct Current) represented by the time-domain constants given by arbitrary powers of R_{x0}. In most practical applications these terms are not of considerable concern, as most transmission systems reveal an inherent highpass characteristic, therefore automatically rejecting these kind of distortion terms.

Again, as explained in the section before, the application of the Fourier transform to the specified moments given by Table 2 in combination with Eq. (21) yields the output PSD in terms of its input PSD. The employment of both mathematical approaches, therefore accounting for even and odd-order nonlinear terms, is finally capable of estimating the self-induced distortion of any given input signal, presuming the nonlinear system to occupy static temporal behaviour insofar as to be approximated according to a polynomial description.

However, this assumption can not be made in case of a broadband stimulus. In general the use of frequency selective circuitry by means of inductors or capacitors within the nonlinear device and furthermore thermal effects neccesitate a dynamic consideration of the underlying nonlinear system.

Since the so called generalized Hammerstein model is capable of approximating a wide class of nonlinear systems with an adequate degree of accuracy, while the complexity in terms of its description parameters is relatively low, this approach is well suited for an extension of the derived model regarding the temporal characteristics (see Mathews and Sicuranza, 2000).

Figure 3 depicts the typical structure of a generalized Hammerstein system in comparison to the static polynomial model. For the extension from the static polynomial model to the generalized Hammerstein model the power series polynomial has to be split up into parallel branches, each branch representing a homogeneous nonlinearity. The dynamic behaviour is introduced by linear FIR (Finite Impulse Response) filters that follow the particular homogeneous polynomials. Finally, all branches are merged at a summing element.

Concerning the structure depicted in Fig. 3 the mathematical expressions derived previously do not change essentially. Since every branch of the parallel structure represents a static homogeneous nonlinear system, the frequency- domain model for even and odd-order nonlinear polynomials can be applied directly to the dynamic model. The subsequent filtering introduced by the linear FIR filters in every branch is then implemented in the frequency-domain by the multiplication of the filter input power spectrum (output PSD of homogenous nonlinearity) with the squared magnitude of the filter transfer function. Consequently, if $S_{n,in}(\omega)$ denotes the output PSD of a nth order homogenous nonlinear system in one branch of the generalized Hammerstein system thus at the same time representing the filter input PSD, the output PSD $S_{n,out}(\omega)$ of the subsequent linear FIR filter defined by its transfer function $H_n(e^{j\omega})$ can be obtained by

$$S_{n,out}(\omega) = \left|H_n(e^{j\omega})\right|^2 S_{n,in}(\omega). \tag{22}$$

Thus the extension to the dynamic model does not require additional considerable effort regarding the computation of the output power specral density. Together with the derived approaches for static nonlinear systems this modeling approach is verfied in the following section by means of time-domain simulations.

4 Simulation results

The following section verifies the presented approaches by comparing output PSDs computed according to the derived mathematical expressions for given input PSDs with those taken from time domain-simulations for certain representative nonlinear systems. Within a time-domain simulation the given spectral mask in terms of a certain multicarrier allocation is adopted, while a random phase angle is assigned to every single tone. The time-domain signal is then obtained by performing an IFFT (Inverse Fast Fourier Transform) on

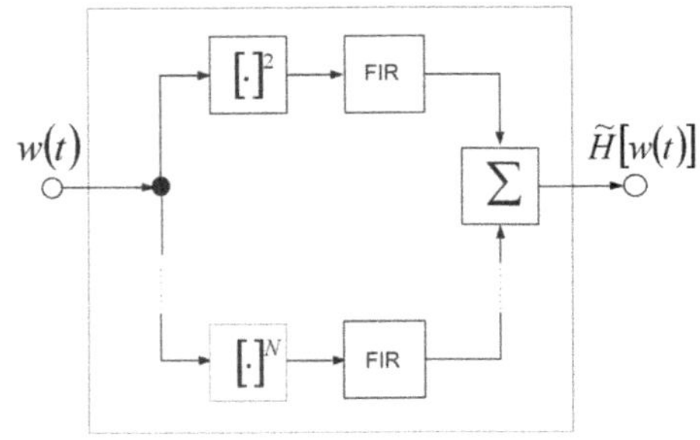

Fig. 3. Generalized Hammerstein system.

the given spectrum. The nonlinear system is finally passed by exponentiation of the time-domain signal according to the given polynomial description. The output PSD is obtainded by peforming the FFT (Fast Fourier Transform) operation. In case of dynamic systems the linear filtering is achieved by convolution with the filter impulse response.

This procedure is repeated for different choices of carrier phase sets in order to approximate a Gaussian "noise-like" stimulus. Subsequent averaging of the obtained distortion PSDs therefore yields an average output distortion PSD, comparable with those computed by the derived stochastical expressions.

First we start with a static polynomial model according to the classical bandpass approach explained in Sect. 2, hence totally described by its odd power series coefficients $\tilde{a}_n$. As in many practical applications, nonlinear effects shall be considered up to a degree of N=7. Therefore the polynomial description is given by

$$\tilde{G}[w(t)] = 10w^3(t) + 1000w^5(t) + 100000w^7(t) \tag{23}$$

whereas the odd coefficients are all chosen as to induce a comparable amount of distortion power. The simulation results obtained for this system are depicted in Fig. 4. Herein the frequency-domain representation of the input signal as well as the distortion PSDs obtained by time and frequency-domain simulations are given in a normalized representation. In case of multitone transmission, single tones are separated by each other corresponding to a certain carrierspacing. The normalization of the frequency axis by means of this carrierspacing yields the representation of the abscissa in terms of the depicted carrier index. The input signal (black graph) is given by two discrete "flat" frequency bands inducing the distortion PSDs computed by time-domain simulations as already described previously (blue graph) and the derived frequency-domain formula presented in Sect. 2 (red graph).

As can easily be seen, the distortion PSD obtained by means of the presented stochastical frequency-domain approach shows a good matching to the time-domain solution. The time-domain solution exhibits certain "ripples" that can be ascribed to the averaging process necessary for the approximation of a noise-like stimulus. In order to verify the mathematical modifications carried out in Sect. 3, we specify

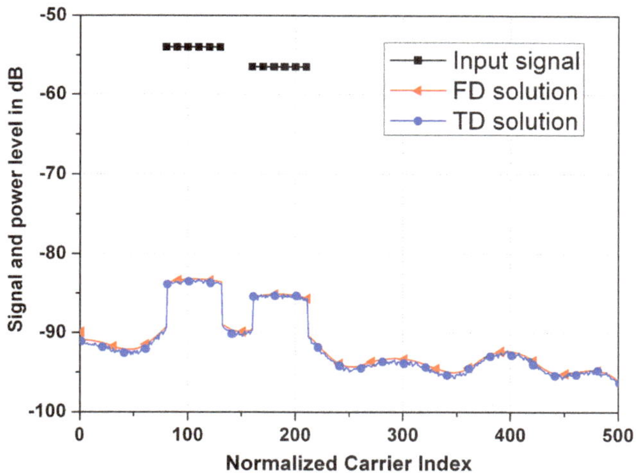

Fig. 4. Input signal and distortion PSDs computed in time and frequency domain for static odd-order model.

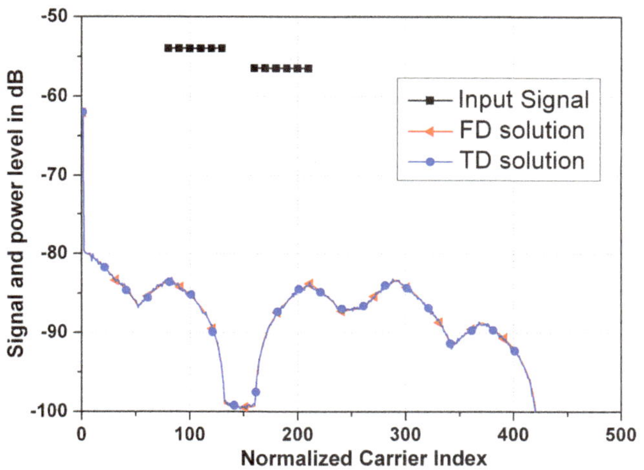

Fig. 5. Input signal and distortion PSDs computed in time and frequency domain for static even-order model.

a nonlinear polynomial model solely containing even-order terms. The chosen polynomial description is therefore given by

$$\tilde{G}[w(t)] = w^2(t) + 100w^4(t) + 10000w^6(t). \tag{24}$$

Figure 5 depicts the results obtained for the given even-order polynomial.

Again, we find a fairly accurate matching to the time domain iteration data. Whereas in case of the odd-order polynomial the induced in-band distortion is dominant, we recognize a remarkable amount of distortion power at the second harmonic of both input frequency bands as well as at DC for the even-order polynomial.

Simulation results in case of the dynamic Hammerstein approach are depicted in Fig. 6.

The filter coefficients needed for calculating the filter response in the frequency-domain were chosen incidentally,

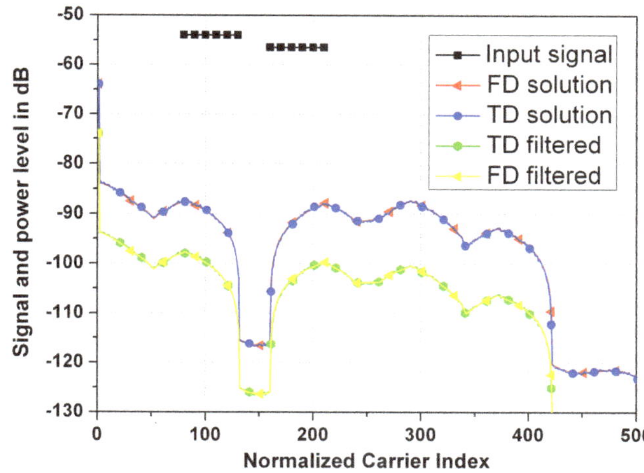

Fig. 6. Input signal and distortion PSDs computed in time and frequency domain for dynamic model corresponding to a seventh order generalized Hammerstein system.

while assuming a tap-length of L_{tap}=5 for every filter branch. The underlying nonlinear system was assumed to be of maximum order N=7. In addition to the results obtained without the filtering operation, thus reducing the model to a pure static nonlinearity, Fig. 6 depicts the filtered output PSDs computed in time and frequency-domain. Again, even in the dynamic case, time and frequency-domain solutions show a good matching. This is not surprising at all, since the dynamic model is just adding a frequency dependent scaling corresponding to the filter response for every single nonlinear order. Beyond the accuracy achievable with the presented approaches, the derived stochastical models bear the advantage of enormously reduced processing times compared to the time domain iteration procedure. In case of the dynamic model the computation time according to the stochastic frequency-domain approach reduces by a factor of 185 opposed to time-domain simulations, as the necessity for averaging the obtained spectra vanishes.

5 Conclusions

In this article we introduced a stochastical frequency-domain model for the time-efficient analysis and estimation of average distortion spectra induced by nonlinear systems. In a first step we therefore presented the classical narrowband RF modeling approach and derived the fundamental formulas, necessary for the closed form analysis of output distortion PSDs in terms of given input PSDs. Furthermore, we presented a mathematical modification enabling the already established narrowband approach to cope with broadband input stimuli, accounting for even-order terms as well as for odd-order terms. Considerations on the temporal characteristics of nonlinear systems excited by broadband stimuli finally led to the introduction of the generalized Hammerstein system. The verification of the derived mathematical expressions was given for three representative nonlinear systems, static as well as dynamic, in terms of a comparison to iterated time-domain solutions. Herein all systems showed an accurate matching between time and frequency-domain solutions. In case of the dynamic system the required computation time could be reduced considerably. Thus the presented stochastic frequency-domain approach is a well suited candidate for the time-efficient distortion analysis of a wide class of nonlinear systems, therefore applicable in system top level simulations.

Acknowledgements. The authors are grateful with R. Stolle, Infineon Germany, for his support on the research.

References

Davenport, W. B. and Root, W. L.: An Introduction to the Theory of Random Signals and Noise, John Wiley & Sons, Inc., 1987.

Gard, K. G.: Autocorrelation Analysis of Spectral Regrowth Generated by Nonlinear Circuits in Wireless Communication Systems, Dissertation, University of California, 2003.

Gard, K. G., Gutierrez, H. M., and Steer, M. B.: Characterization of Spectral Regrowth in Microwave Amplifiers Based on the Nonlinear Transformation of a Complex Gaussian Process, IEEE Transactions on Microwave Theory and Techniques, 47, 1999.

Gard, K. G., Steer, M. B., Larson, L. E.: Generalized Autocorrelation Analysis of Spectral Regrowth from Bandpass Nonlinear Circuits, IEEE MTT-S Digest, 2001.

Mathews, V. J. and Sicuranza, G. L.: Polynomial Signal Processing, John Wiley & Sons, Inc., 2000.

Raich, R.: Nonlinear system Identification and Analysis with Applications to Power Amplifier Modeling and Power Amplifier Predistortion, Dissertation, Georgia Institute of Technology, 2004.

Zhou, G. T. and Raich, R.: Spectral Analysis of Polynomial Nonlinearity with Applications to RF Power Amplifiers, EURASIP Journal on Applied Signal Processing, 2004.

Quantitative design space exploration of routing-switches for Network-on-Chip

M. C. Neuenhahn, H. Blume, and T. G. Noll

Chair of Electrical Engineering and Computer Systems, RWTH Aachen University, 52062 Aachen, Germany

Abstract. Future Systems-on-Chip (SoC) will consist of many embedded functional units like e.g. embedded processor cores, memories or FPGA like structures. These SoCs will have huge communication demands, which can not be fulfilled by bus-based communication systems. Possible solutions to this problem are so called Networks-on-Chip (NoC).

These NoCs basically consist of network-interfaces which integrate functional units into the NoC and routing-switches which connect the network-interfaces. Here, VLSI-based routing-switch implementations are presented. The characteristics of these NoCs like performance and costs (e.g. silicon area respectively logic elements, power dissipation) depend on a variety of parameters. As a routing-switch is a key component of a NoC, the costs and performance of routing-switches are compared for different parameter combinations. Evaluated parameters are for example data word length, architecture of the routing-switch (parallel vs. centralized implementation) and routing-algorithm.

The performance and costs of routing-switches were evaluated using an FPGA-based NoC-emulator. In addition different routing-switches were implemented using a 90 nm standard-cell library to determine the maximum clock frequency, power-dissipation and area of a VLSI-implementation. The power consumption was determined by simulating the extracted layout of the routing-switches. Finally, these results are benchmarked to other routing-switch implementations like Aetheral and xpipes.

1 Introduction

The complexity of VLSI-chips is increasing more and more due to recent improvements in VLSI-technology (ITRS, 2005). As the requirements given by applications concerning e.g. computing power and/or power-dissipation are increasing, VLSI-chips more and more turn to so called Systems-on-Chip. These SoCs consist of functional units like for example processor-cores, memories, DSPs or embedded FPGAs. The number of functional units in these SoCs is constantly increasing (Blume et al., 2005) and is supposed to reach the mark of hundred in the near future (Borkar et al., 2005).

Correspondence to: M. C. Neuenhahn
(neuenh@eecs.rwth-aachen.de)

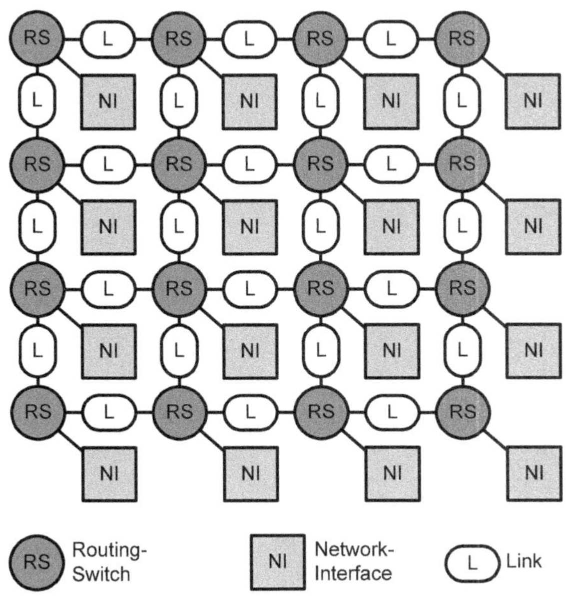

Fig. 1. NoC with mesh topology.

The communication demands of applications mapped onto these complex SoCs will be very high. For example, high throughput shall be provided at minimal latency, with guaranteed quality of service (QoS) and as much flexibility as possible. Furthermore, the communication should cause minimal costs, in terms of area and power consumption. As the number of functional units in SoCs will increase in the future the communication-structure should although be scalable. Traditional communication structures like on-Chip busses or point-to-point connections cannot fulfill these demands. Often quoted solutions to this problem are Network-on-Chip architectures (Benini and De Micheli, 2002).

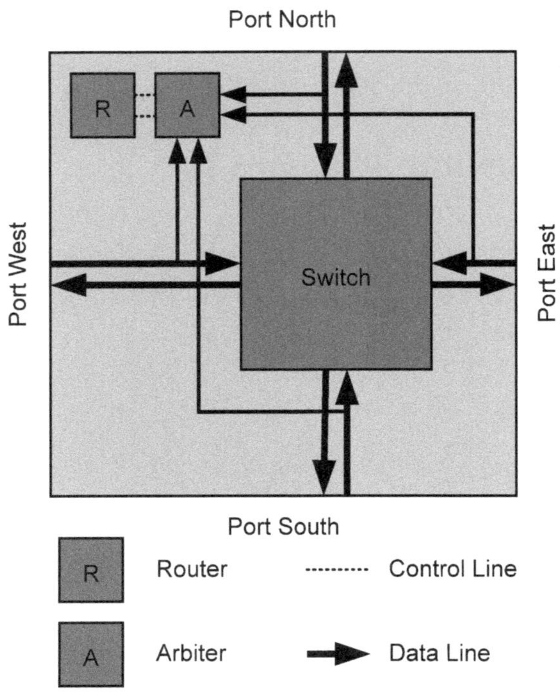

Fig. 2. Centralized architecture of a routing-switch with 4 ports.

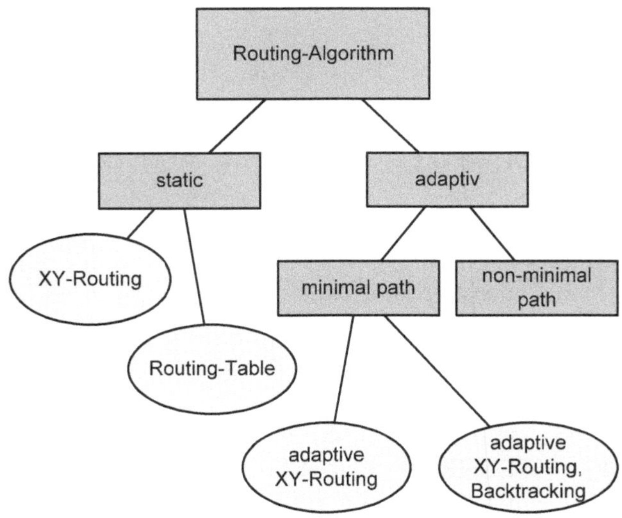

Fig. 3. Overview of routing-algorithms for Network-on-Chip.

The performance and costs caused by these NoCs strongly depend on NoC-specific parameters like topology (Neuenhahn et al., 2006), data word length, routing-algorithm and many more (Bjerregaard and Mahadevan, 2006). These NoC-parameters span a huge design-space for NoCs. As each application or application-class has different communication requirements the appropriate NoC-parameter combination that fulfills these requirements at minimal costs has to be found (Ahonen et al., 2005; Benini, 2006).

The implementation effort of NoCs with an altered parameter set, e.g. modified topology or data word length, should be low. Therefore, a modular and generic NoC-architecture like e.g. xpipes (Bertozzi and Benini, 2004) or the NoC-architecture presented in Sect. 2 is needed. Such generic architectures allow a modification of single NoC-parameters at minimum design effort.

A key-component of NoCs concerning area, energy dissipation and performance is the routing-switch. Hence, a detailed analysis of implementation costs and influence on performance of routing-switches featuring different NoC-specific parameters is needed.

The NoC-parameter routing-algorithm has great impact on the performance of NoCs. Many different algorithms for computer networks and NoCs have been proposed and evaluated (Duato et al., 2003; Moraes et al., 2004). But these evaluations did not regard the impact on implementation costs and focus only on performance aspects.

The architecture of routing-switches and the functionality of its building blocks are presented in Sect. 2. An evaluation of the performance of these routing-switches and implementation results concerning area and power are presented in Sect. 3. The NoC-parameters data word length, routing-algorithm and number of ports have been varied in cause of the experiments. Finally, an implemented routing-switch is compared to routing-switches recently published.

In Sect. 4 the design space of routing-switches is discussed, before in Sect. 5 a conclusion is given.

2 Modular and generic Network-on-Chip

The modular and generic NoC framework which was introduced in Neuenhahn et al. (2006) features three basic building blocks: network-interface, routing-switch and link. The network-interface is the gateway between NoC and a functional unit, the routing-switch routes data through the NoC and the link represents the connections between these building blocks. The arrangement of these blocks is defined by the NoC-parameter topology. The options of possible topologies range from classic topologies like mesh (see Fig. 1), torus or star topologies to application specific hierarchical topologies.

Every NoC component, network-interface, routing-switch and link, consists of different modules. Each of these modules can be implemented in different ways to realize a defined functionality. For example the module which is used for applying error-correcting or error-detecting codes in network-interfaces or routing-switches can be implemented using no coding, parity code or Huffman-codes (Neuenhahn et al., 2007).

This modular and generic approach enables the implementation of NoC with a variety of NoC-specific parameters at minimal design effort. In Neuenhahn et al. (2007) a NoC design-flow based on this framework was presented.

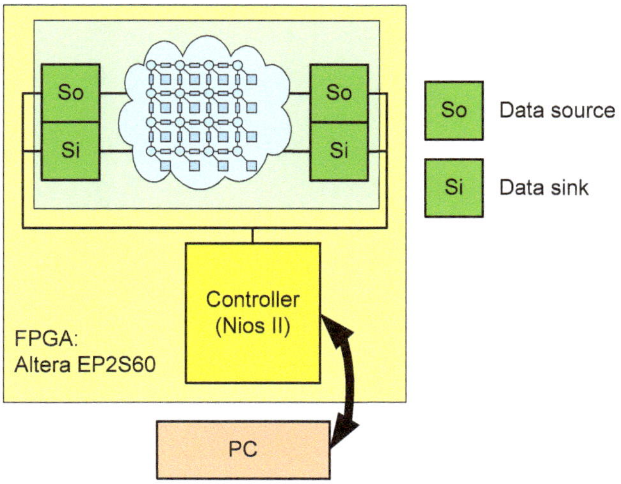

Fig. 4. FPGA-based emulation environment.

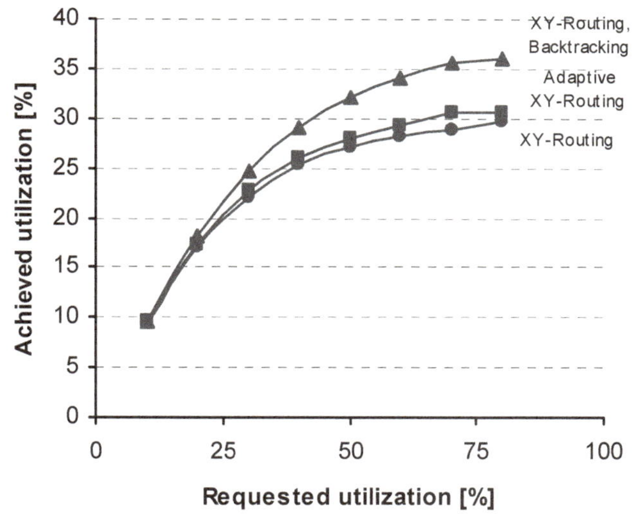

Fig. 5. Performance of different routing-algorithms for a NoC with Mesh topology and 16 functional units.

A routing-switch for NoCs consists of a switch, which connects the input ports with corresponding output ports. This switch is controlled by an arbiter, which grants or refuses the access to an output port. The routing-decision, to which output port data should be routed, is made in the router. This decision is based on the routing-algorithm, the position of the routing-switch in the network and the target address of the data. In case of adaptive routing-algorithms, the state of the routing-switch, e.g. used ports, is regarded as well. Here, circuit switching is used as switching technique and therefore no buffers are needed inside a routing-switch.

Different routing-switch architectures have been analyzed. The centralized architecture presented in Fig. 2 features the smallest area at negligible performance decrease compared to a distributed architecture. Depending on the timing-requirements, additional registers at the input ports can be applied to perform pipelining.

Routing-algorithms for NoCs can be distinguished in static and adaptive routing-algorithms, as depicted in Fig. 3. Static algorithms route the data always on the same path from the source to the destination. A widely used class of static routing-algorithms is dimension-order routing. In a NoC with mesh topology, as depicted in Fig. 1, such an algorithm is XY-routing. The data is at first routed in X-direction and afterwards in Y-direction over the NoC. The state of a routing-switch, e.g. if an output port is already used, is not regarded. In this case data transmission is aborted and reinitiated by the network-interface at a later point of time. The routing-algorithm can be implemented as a state-machine or as a routing-table.

Adaptive routing-algorithms can be distinguished in algorithms supporting only minimal paths and algorithms supporting non-minimal paths. The latter algorithms are not regarded in the following evaluation.

The adaptive XY-routing-algorithm routes data at first in X-direction. If an output port determined by the XY-routing is already used, an alternative output port is chosen, which leads the data directly towards the target. If no feasible output port can be determined, the data transmission is cancelled and reinitiated by the sending network-interface at a later point of time.

Using adaptive XY-routing with backtracking the data transmission is not cancelled when all feasible output ports of a routing-switch are blocked. Instead of that, another routing attempt at the previous routing-switch, if present, is performed.

3 Performance and implementation results

To evaluate the influence of the routing-algorithm on the performance of NoC an FPGA-based emulator (Neuenhahn et al., 2006) was used. This FPGA-based NoC-Emulator emulates a NoC on an FPGA. Even on modern FPGAs it is difficult to handle the complexity which is required for emulating a complex SoC. To solve this problem, the functional units of a SoC e.g. processor kernels or memories are replaced with abstract data sources and data sinks. With this abstracted functional units the NoC can be emulated.

The FPGA-based generic emulation environment depicted in Fig. 4 consists of an FPGA and a PC. The PC is used to synthesize the FPGA-configuration, initiate experiments and evaluate the results of the experiment. The FPGA contains the NoC to be emulated (see Sect. 2), traffic-sources and -sinks which emulate functional units of a SoC and a soft-core processor (Altera Corporation, 2007) which is used to control the data-sources and evaluate the raw results generated by the data-sinks.

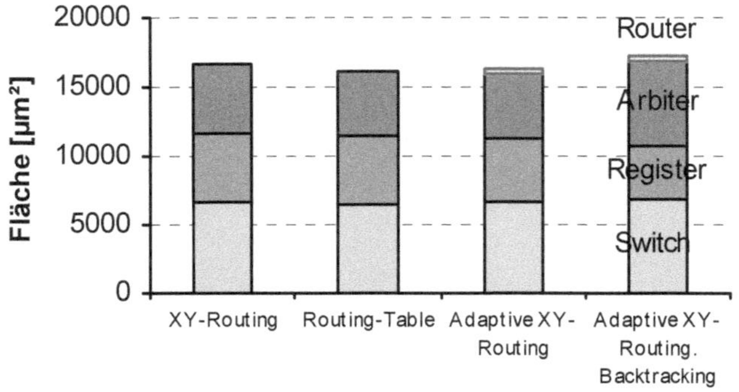

Fig. 6. Area distribution of routing-switches with 5 ports, 32 bit data word length, registers at the input port and varied routing-algorithms.

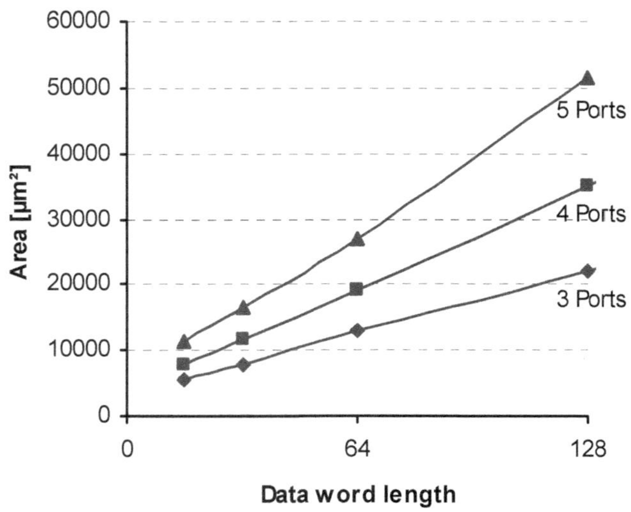

Fig. 7. Total area of a routing-switch with XY-routing and input port registers for varied data word length and different numbers of ports.

As depicted in Fig. 5, for a requested utilization smaller than 20% no influence on the performance of the NoC depending on the routing-algorithm occurs. Applying a higher utilization, adaptive algorithms result in a higher performance than the static routing-algorithms. The adaptive XY-routing with backtracking has obviously the highest performance. For a requested utilization the achieved utilization is about 15% higher than the achieved utilization obtained by the other algorithms.

Further experiments using other topologies and different NoC-sizes show that the differences in performance depending on these routing-algorithms increase for bigger and decrease for smaller NoCs.

The area needed for a VLSI-implementation of a routing-switch was obtained using the following design flow. At first, the VHDL-code describing the routing-switch was synthesized using Synopsys Design Compiler (Synopsys, 2007). The resulting netlist was used in the place-and-route tool Encounter (Mentor Graphics, 2006). For verification and power evaluation the resulting layout has been extracted into a SPICE-netlist (Cadence, 2006) which has been simulated using NanoSim (Synopsys, 2006). A 90 nm technology from TSMC and a standard-cell library from Artisan Components (Artisan Components Inc., 2005) has been used. The links, connected to each routing-switch are assumed to have a length of 1 mm. The resulting capacity on each output port is calculated to 0.148 pF (5 $\times$ minimum distance to neighbors, toggling against each other) and the operating frequency is 1 GHz. All results are computed using slow operating conditions.

As depicted in Fig. 6 the differences in area dissipation for routing-switches with 5 ports, 32 bit data word length and additional registers at the input ports featuring different routing-algorithms is less than 5%. The total area needed for control, router and arbiter, is about 20% of the total routing-switch area.

The correlation of routing-switch area, data word length and number of ports is presented in Fig. 7. There is a linear dependency of the area and the data word length. The gradient of the linear area function is determined by the number of ports. The area consumed by registers and the switch block increases with higher data word length whereas the area needed for router and arbiter remains constant.

The power consumption depends on the state of the routing-switch. As in this work circuit switching is performed, the routing-switch/output ports can be in the following states: "idle", "determine connection target", "wait for connection to establish", "active connection" and "destroy connection".

In the state "idle" no action is performed, while in the state "determine connection target" the arbiter and router determine the target port of the connection. As this state lasts only three cycles, the contribution to the overall energy-consumption is negligible. The state "wait for connection to establish" has approximately the same power consumption than the "idle" state. The highest power consumption and the dominant part in energy needed for a data transmission is used in the state "active connection". As the state "destroy connection" lasts only two clock-cycles, its share to the overall energy of a data transmission is negligible, too.

The power consumption of a single routing-switch is shown in Fig. 8. It is in the state "active connection" with one active connection. It depends nearly linear on the data word length, whereas the gradient is defined by the number of ports. A routing-switch can feature several connections simultaneously. Hence, these states can be superposed by other active connections.

To compare this routing-switch to implementations published, the area and maximum clock-frequency have been scaled to a 90 nm technology. In addition to that, the area has been normalized to one bit data word length, as

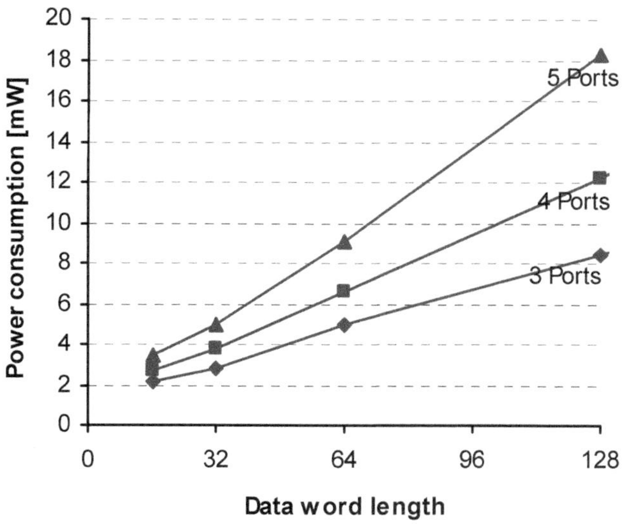

Fig. 8. Total power consumption of a routing-switch with XY-routing and input port registers for varied data word length and different numbers of ports with one active connection and a toggle-rate of approximately 25%.

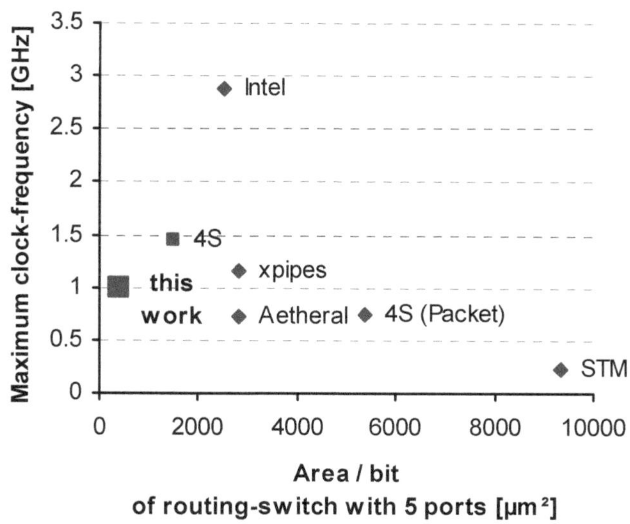

Fig. 9. Comparison of routing-switch implementations concerning area and maximal clock frequency. The area and maximum clock-frequency was scaled to a 90 nm technology.

the implementations presented feature data word lengths from 16 to 32 bit. The values are taken from Rijpkema et al. (2003)(Aetheral), Vangal et al. (2007) (Intel), Lattard et al. (2007)(STM), Stergiou et al. (2005) xpipes and Wolkotte et al. (2005)(4S and 4s(Packet)). These values are enlisted in Fig. 9.

The comparison in Fig. 9 shows that the implementation described here has the smallest area consumption. These results are based on the fact that in most implementations, besides the implementation named 4S, packet switching is performed, while in this work circuit switching is performed. Hence, no buffers are needed in this routing-switches resulting in smaller area.

The implementations 4S, xpipes and Intel outperform the implementation presented in this work concerning the maximum clock-frequency, but this implementation was optimized for the clock-frequency of 1 GHz. A further enhancement can easily be achieved by modifying the synthesis parameter at the cost of a slightly increased area.

4 Design space of routing-switches for NoC

In Fig. 10 different routing-switch implementations have been evaluated concerning the area used and the performance a NoC featuring these routing-switches achieves. The NoC has a mesh topology with 16 functional units attached. It features circuit switching and no error protection at data word level.

The data word length was varied from 16 to 256 bit; the routing-algorithms presented were used and input ports with and without registers were applied.

To achieve performance figures in terms of average latency, several experiments have been carried out using the FPGA-based emulator. Benchmarks based on random data traffic with a requested data rate of 204.8 GBit/s were applied to the various NoC instantiations. The NoC implemented with standard cells should operate at a clock-frequency of 1 GHz.

The dots in Fig. 10 representing routing-switch implementations result in a pareto-front. For each of these implementations laying on the pareto-front no other implementation exists featuring a higher performance (lower average latency) at lower or equal implementation costs, expressed in area. For a given combination of requested data rate and allowed average latency the NoC-parameter set fulfilling these requirements at minimal routing-switch area can be identified using this diagram.

5 Conclusion

In this work routing-switch implementations for different NoC-specific parameter combinations have been presented. It has been shown that the routing-algorithm has a significant impact on the performance and only a small impact on the resulting costs, especially for NoCs with a high data word length. Hence, the adaptive XY-routing with backtracking is the favorable routing-algorithm in most cases.

Furthermore, the dependency of area and power of routing-switches on the parameters data word length and port number has been presented. Compared to routing-switch implementations published, these implementations feature a much smaller area. The presented design space exploration is essential for determining an efficient NoC-specific parameter combination for given communication demands.

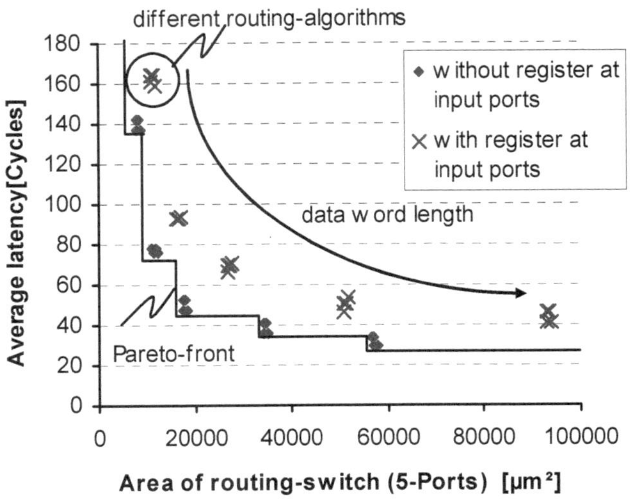

Fig. 10. Pareto-plot of routing-switch implementations.

References

Ahonen, T., Virtanen, S., Kylliäainen, J., Truscan, D., Kasanko, T., Sigäuenza-Tortosa, D., Ristimäaki, T., Paakkulainen, J., Nurmi, T., Saastamoinen, I., Isäannäainen, H., Lilius, J., Nurmi, J., and Isoaho, J.: A Brunch from the Coffee Table-Case Study in NoC Platform Design, Interconnect-Centric Design for Advanced SoC and NoC, 425–453, 2005.

Altera Corporation: Nios II Processor Reference Handbook, 2007.

Artisan Components, Inc.: TSMC 90nm CLN90G Process SAGE-X v3.0 Standard Cell Library Databook, 2005.

Benini, L.: Application Specific NoC Design, Design, Automation and Test in Europe, DATE '06, 2006.

Benini, L. and De Micheli, G.: Networks on chips: a new SoC paradigm, Computer, 35(1), 70–78, 2002.

Bertozzi, D. and Benini, L.: Xpipes: a network-on-chip architecture for gigascale systems-on-chip, Circuits and Systems Magazine, IEEE 4(2), 18–31, 2004.

Bjerregaard, T. and Mahadevan, S.: A survey of research and practices of Network-on-chip, ACM Comput. Surv., 38(1), 1–51, 2006.

Blume, H., Feldkaemper, H. T., and Noll, T. G.: Model-Based Exploration of the Design Space for Heterogeneous Systems on Chip, The Journal of VLSI Signal Processing, 40(1), 19–34, 2005.

Borkar, S. Y., Mulder, H., Dubey, P., Pawlowski, S. S., Kahn, K. C., Rattner, J. R., and Kuck, D. J.: Intel: Platform 2015: Intel® Processor and Platform Evolution for the Next Decade, 2005.

Cadence: Design Framework, 2006.

Duato, J., Yalamanchili, S., et al.: Interconnection Networks – An Engineering Approach, San Francisco, USA, Morgan Kaufmann Publishers, 2003.

ITRS: International Technology Roadmap for Semiconductors, 2005 Edition, Executive Summary, 2005.

Lattard, D., Beigne, E., Bernard, C., Bour, C., Clermidy, F., Durand, Y., Durupt, J., Varreau, D., Vivet, P., Penard, P., Bouttier, A., and Berens, F.: A Telecom Baseband Circuit based on an Asynchronous Network-on-Chip, Solid-State Circuits Conference, 2007, ISSCC 2007, Digest of Technical Papers, IEEE International, 2007.

Mentor Graphics: Encounter, 2006.

Moraes, F., Calazans, N., Mello, A., Moller, L., and Ost, L.: HERMES: an infrastructure for low area overhead packet-switching networks on chip, Integration, the VLSI Journal, 38(1), 69–93, 2004.

Neuenhahn, M. C., Blume, H., and Noll, T. G.: Quantitative analysis of network topologies for NoC-architectures on an FPGA-based emulator, Presentation, Kleinheubacher Tagung, 2006.

Neuenhahn, M. C., Lemmer, D., Blume, H., and Noll, T. G.: Quantitative Cost Modeling of Error Protection for Network-on-Chip, ProRISC Workshop, Veldhoven, 2007.

Rijpkema, E., Goossens, K. G. W., Radulescu, A., Dielissen, J., Meerbergen, J. v., Wielage, P., and Waterlander, E.: Trade Offs in the Design of a Router with Both Guaranteed and Best-Effort Services for Networks on Chip, Proceedings of the conference on Design, Automation and Test in Europe – Volume 1, 2003.

Stergiou, S., Angiolini, F., Carta, S., Raffo, L., Bertozzi, D., and De Micheli, G.: Xpipes Lite: a synthesis oriented design library for networks on chips, Design, Automation and Test in Europe, Proceedings, 2005.

Synopsys: NanoSim, 2006.

Synopsys: Design Compiler, 2007.

Vangal, S., Howard, J., Ruhl, G., Dighe, S., Wilson, H., Tschanz, J., Finan, D., Iyer, P., Singh, A., Jacob, T., Jain, S., Venkataraman, S., Hoskote, Y., and Borkar, N.: An 80-Tile 1.28TFLOPS Network-on-Chip in 65nm CMOS, Solid-State Circuits Conference, 2007, ISSCC 2007, Digest of Technical Papers, IEEE International, 2007.

Wolkotte, P. T., Smit, G. J. M., Rauwerda, G. K., and Smit, L. T.: An Energy-Efficient Reconfigurable Circuit-Switched Network-on-Chip, Parallel and Distributed Processing Symposium, Proceedings 19th IEEE International, 2005.

22

Simulating circuits with impasse points

T. Thiessen and W. Mathis

Institute of Theoretical Electrical Engineering, Leibniz University of Hannover, Germany

Correspondence to: T. Thiessen (thiessen@tet.uni-hannover.de)

Abstract. In this paper circuits with impasse points, i.e. with jumps in their configuration space will be analyzed. These non-regularized circuits exhibit a fold in their configuration space, which can lead to difficulties during the simulation with standard circuit simulators like SPICE. The former developed geometric approach to simulate these circuits without regularization will be extended by a detailed discussion of which coordinate system has to be chosen. Furthermore, two new approaches for a numerically efficient calculation of the hit points will be shown.

1 Introduction

There is a class of electronic circuits whose configuration space manifold $\mathcal{S}$ is folded within the embedding space $\mathcal{E}$ of currents and voltages. This fold can be related to so-called impasse points and leads under certain conditions to jumps from one stable part of $\mathcal{S}$ to another. A classical transient solution of such non-regularized circuits exhibiting impasse points is not possible (see e.g. Chua (1980), Reissig (1996), Chua and Deng (1989a)). However, a common method to overcome these simulation problems is to regularize the circuit by adding suitable located parasitic inductors L's or capacitors C's considering Tikhonov's Theorem (for further literature see Reissig (1996)). In previous works, a geometric concept was developed to simulate those circuits without regularization (see e.g. Thiessen and Mathis (2011a), Thiessen et al. (2013)). There, several electronic circuits exhibiting these behavior were studied and different approaches to calculate $\mathcal{S}$, jump and hit points were analyzed (e.g. Thiessen et al. (2012b), Thiessen et al. (2012a), Thiessen et al. (2011), Thiessen and Mathis (2011b), Thiessen and Mathis (2011a)). In this work improvements in the topic of hit point calculation will be shown. Especially the difficulty of multiple hit points will be studied in detail. Furthermore, a detailed description of the geometric system analysis will be given. There, the question of which coordinate system, i.e. system of equations, is best placed to deal with those non generic circuits will be answered.

2 Geometric system analysis

Common circuit simulators (e.g. SPICE) are based on MNA, which leads in the description of electronic circuits to a quasilinear differential-algebraic system of equations (DAE) (cf. Riaza (2008)):

$$\mathbf{A}_c\mathbf{C}(\mathbf{A}_c^T\mathbf{e})\mathbf{A}_c^T\dot{\mathbf{e}} = -\mathbf{A}_r\boldsymbol{\gamma}_r(\mathbf{A}_r^T\mathbf{e}) - \dots$$
$$\dots - \mathbf{A}_l\mathbf{i}_l - \mathbf{A}_u\mathbf{i}_u - \mathbf{A}_j\mathbf{i}_s(t) \quad (1a)$$
$$\mathbf{L}(\mathbf{i}_l)\dot{\mathbf{i}}_l = \mathbf{A}_l^T\mathbf{e} \quad (1b)$$
$$\mathbf{0} = \mathbf{v}_s(t) - \mathbf{A}_u^T\mathbf{e} \quad (1c)$$

This system of equations can also be formulated as

$$\mathbf{B}(\mathbf{c})\dot{\mathbf{c}} = \mathbf{h}(\mathbf{c}(t)) \ . \quad (2)$$

In the sense of Reich Reich (1990) eq. (2) is a triple $(\mathbb{R}^k, \mathbf{B}, \mathbf{h})$, where $\mathbf{B} : \mathbb{R}^k \to \mathbb{R}^{k\times k}$ is a linear operator and $\mathbf{h} : \mathbb{R}^k \to \mathbb{R}^k$ a diffeomorphism. Reich had proven that the DAE (2) is regular, if there is a differentiable manifold $\mathcal{S} \subset \mathbb{R}^k$ and a vector field $\mathbf{v} : \mathcal{S} \to \mathcal{TS}$, such that a differential mapping $\mathbf{w} : I \to \mathcal{S}$ $(I \subset \mathbb{R})$ is a solution of the vector field for all $t \in I$, if and only if the mapping $\mathbf{c} : \mathbf{j} \circ \mathbf{w} : I \to \mathbb{R}^k$ is a solution of the DAE, where $\mathbf{j} : \mathcal{S} \to \mathbb{R}^k$ is the natural injection Reich (1990).

This work focuses on the investigation of non-regular DAEs where impasse points exists. These impasse points will be called "jump points", because the transients will be continued by jumps in $\mathcal{E}$ from one point on $\mathcal{S}$ to another.

The systems of equations of the considered nonlinear dynamical circuits can be characterized by a semi explicit DAE:

$$\dot{\mathbf{x}} = \mathbf{g}(\mathbf{x}, \mathbf{y}, t) \quad (3a)$$
$$\mathbf{0} = \mathbf{f}(\mathbf{x}, \mathbf{y}, t) \quad (3b)$$

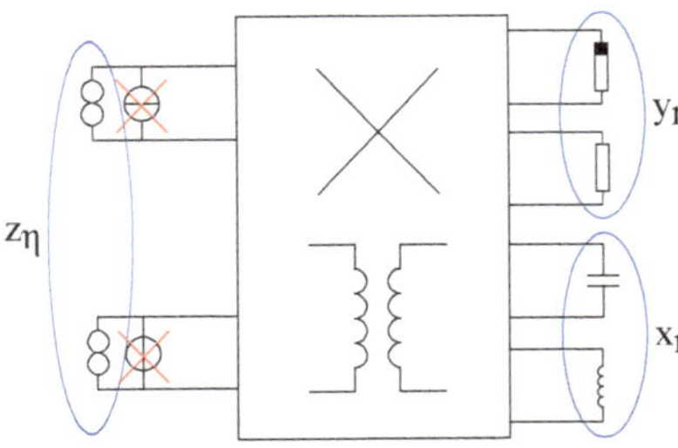

Fig. 1. Network for determining $\mathcal{S}$

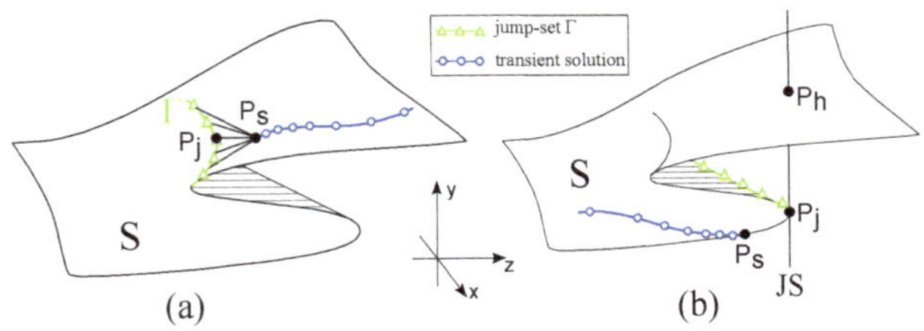

Fig. 2. Calculation principle of (a) P_j and (b) P_h

The vector $\mathbf{x} \in \mathbb{R}^n$ corresponds to the capacitor voltages and inductor currents and the vector $\mathbf{y} \in \mathbb{R}^m$ to additional voltages and currents in an electronic circuit.

Taken into account that $\mathcal{S}$ can exhibit a fold e.g. respectively an input voltage $\mathbf{v}_s(t)$ and not respectively t, the input sources have to be treated differently. For describing the behavior of the circuit for any input values, the independent and time dependent input sources will be replaced by norators and treated as further vector $\mathbf{z}$ fo unknowns (cf. Fig. 1). Therefore, an additional vector $\mathbf{z} \in \mathbb{R}^\eta$ has to be considered in the describing system of equations. The resulting autonomous, semi explicit system of equations then is described by:

$$\dot{\mathbf{x}} = \mathbf{g}(\mathbf{x}, \mathbf{y}, \mathbf{z}) \qquad \mathbf{g} : \mathbb{R}^k \to \mathbb{R}^n \tag{4a}$$

$$\mathbf{0} = \mathbf{f}(\mathbf{x}, \mathbf{y}, \mathbf{z}) \qquad \mathbf{f} : \mathbb{R}^k \to \mathbb{R}^m \tag{4b}$$

These system of Eq. (4) forms the basis of the further investigations. The distinction in $\mathbf{x}$, $\mathbf{y}$ and $\mathbf{z}$ is essential for calculating $\mathcal{S}$ and the jump and hit points. $\mathcal{S}$ can be defined as a subspace of $\mathcal{E} = \mathbb{R}^k$ (where $k = n+m+\eta$) and is represented by the solution set of the independent algebraic Eq. (4b). The calculation of the jump and hit points will be shown in Section 3 and 4.

Remark: As can be seen, the system of equations (1) do not distinct between $\mathbf{x}$, $\mathbf{y}$ and $\mathbf{z}$. One method to modify the system of equations (1) to (4) is shown in Thiessen et al. (2012a) and Thiessen et al. (2013). Another method for deriving a system of equations of type (4) is by applying the Augmented Nodal Analysis (ANA) shown in Riaza (2008). The resulting system of equations derived by the Augmented Nodal Analysis (ANA) is

$$\mathbf{C}(\mathbf{v}_c)\dot{\mathbf{v}}_c = \mathbf{i}_c \tag{5a}$$

$$\mathbf{L}(\mathbf{i}_l)\dot{\mathbf{i}}_l = \mathbf{A}_l^T\mathbf{e} \tag{5b}$$

$$\mathbf{0} = \mathbf{A}_r\boldsymbol{\gamma}_r(\mathbf{A}_r^T\mathbf{e}) + \mathbf{A}_l\mathbf{i}_l + \mathbf{A}_c\mathbf{i}_c + \mathbf{A}_u\mathbf{i}_u + \mathbf{A}_j\mathbf{i}_s(t) \tag{5c}$$

$$\mathbf{0} = \mathbf{v}_c(t) - \mathbf{A}_c^T\mathbf{e} \tag{5d}$$

$$\mathbf{0} = \mathbf{v}_s(t) - \mathbf{A}_u^T\mathbf{e} \tag{5e}$$

Here, the vectors $\mathbf{v}_c$ and $\mathbf{i}_l$ are counted among the vector $\mathbf{x}$. The vector $\mathbf{y}$ is composed of $\mathbf{i}_c$, $\mathbf{i}_u$ and $\mathbf{e}$ and the input vectors $\mathbf{i}_s$ and $\mathbf{v}_s$ are counted among the vector $\mathbf{z}$.

3 Jump points

It was shown by Chua and Deng (1989b), that almost all singular points of an autonomous semi explicit DAE are in fact impasse points, i.e. jump points. However, e.g. in Chua and Deng (1989a), Chua and Deng (1989b) and Andronov et al. (1966), the fact that $\mathcal{S}$ could contain a fold respectively z was neglected. Taken this into account and knowing that singular points are points, where the local solvability to y is not guaranteed (e.g. Thiessen et al. (2011), Chua and Deng (1989a)), the necessary jump condition can be specified by (cf. Chua and Deng (1989a), Andronov et al. (1966)):

$$\det\left(\partial_{\mathbf{y}}\mathbf{f}(\mathbf{x}, \mathbf{y}, \mathbf{z})\right) = 0 \text{ where } \mathbf{f}(\mathbf{x}, \mathbf{y}, \mathbf{z}) = \mathbf{0} \tag{6}$$

A point that is specified by eq. (6) and whose neighborhood includes each a Lyapunov-stable and -unstable point, is defined as proper jump point P_j. This sufficient jump condition can be verified by calculating the eigenvalues λ_i of the characteristic equation $\det\left(\partial_{\mathbf{y}}\mathbf{f}(\mathbf{x}, \mathbf{y}, \mathbf{z}) - \lambda \cdot \mathbf{E}\right) = 0$, where $\mathbf{E}$ is the identity matrix (cf. theory of discontinuous oscillators e.g. Andronov et al. (1966), Mishchenko and Rozov (1980)). The set of all points fulfilling these two conditions is called jump-set Γ, which represents a $l-1$-dimensional subset of $\mathcal{S}$. Of course, the calculation of the zero set of all points fulfilling the $m+1$ algebraic equations specified by Eq. (6) is difficult. However, not all zeros of Eq. (6) are of interest, but only in the actual chosen jump point during a simulation. Hence, the dynamics on $\mathcal{S}$ will be traced till reaching a stopping point P_s. This stopping point is defined as a point, where the step size of the numerical solver reaches a lower boundary (which is related to the machine constant of the simulating computer). In the next step, the "nearest" point on Γ will be calculated by choosing a suitable norm and defined as the actual jump point P_j (cf. Fig. 2 (a)).

4 Hit points

The dynamics on $\mathcal{S}$ is specified by Eq. (3a) and (3b). Since the dynamics is not defined at points on Γ, the trajectory increases very fast nearby Γ while tracing it. Therefore the numerical integration is stopped at P_s when the step size reaches the lower boundary. From there, the jump point P_j is calculated as described in Sect. 3. To trace the dynamics, we use a variable order solver based on the numerical differentiation formulas (NDFs) Shampine and Reichelt (1997).

Considering that the voltage across a capacitance and the current through an inductance is preserved, both have inertia through a jump process and do not change (i.e.$\mathbf{x}_j = \mathbf{x}_h$). Assuming that a jump happens instantaneous, i.e. $t_j = t_h$, a further restriction is the fixed value of $\mathbf{z}$ (i.e. $\mathbf{v}_s(t_j) = \mathbf{v}_s(t_h)$, $\mathbf{i}_s(t_j) = \mathbf{i}_s(t_h)$) during a jump. Consequently, a jump takes place in a tangential space of $\mathbb{R}^m$, which corresponds to the coordinate space of $\mathbf{y}$. In the following, the jump space will be denoted by $\mathcal{JS}$. This corresponds to the jump postulate of Chua and Alexander (1971).

Because we introduced $\mathcal{E}$, a hit point P_h can be calculated by the intersection of $\mathcal{JS}$ defined in P_j and $\mathcal{S}$ excluding the jump point itself ($P_h \in (\mathcal{JS}_{P_j} \cap \mathcal{S}) \setminus \Gamma$) (see Fig. 2 (b)).

Thus, the problem of the hit point calculation can be defined as

$$\mathbf{0} = \mathbf{f}(\mathbf{x}_j, \mathbf{y}, \mathbf{z}_j) =: \mathbf{h}(\mathbf{y}), \tag{7}$$

with the constraint

$$\mathbf{y}_h \neq \mathbf{y}_j \,. \tag{8}$$

In Thiessen and Mathis (2011a), Sarangapani et al. and Thiessen et al. (2012a) the hit point calculation was done in two steps: First a point $P_{j'}$ outside $\mathcal{S}$ was chosen so that $P_{h'} \in (\mathcal{JS}'_{P_{j'}} \cap \mathcal{S})$. In the next step, the actual hit point $P_h \in (\mathcal{JS}_{P_j} \cap \mathcal{S}) \setminus \Gamma$ with the initial condition $P_{h'}$ was calculated. There, the difficulty was to chose a suitable point $P_{j'}$ outside $\mathcal{S}$, so that the numerical solver is able to find $P_{h'}$.

Another approach (cf. Thiessen et al. (2011)) to calculate a hit point was to use a bisection method. This provided a set of possible points from which the closest to the corresponding jump point was chosen.

Now, an efficient approach to calculate the hit points of circuits with only one fold in $\mathcal{S}$ based on the penalty function will be shown.

4.1 Penalty function

The basic idea of a penalty function is to convert the zero problem of Eq. (7) with a constraint (8) in a zero problem without constraint (cf. Florian Jarre (2003)).

Thus, the optimization problem

$$\mathbf{h}(\mathbf{y}) \to \min \quad \mathit{with} \quad \mathbf{y} \neq \mathbf{y}_j \,, \tag{9}$$

where $\mathbf{h} : \mathbb{R}^m \to \mathbb{R}^m$ can be transformed to

$$\mathbf{p}(\mathbf{y}, \mathbf{r}) := [\mathbf{h}(\mathbf{y}) + \mathbf{r} \cdot l(\mathbf{y})] \to \min \,. \tag{10}$$

The penalty function $\mathbf{p}(\mathbf{y}, \mathbf{r})$ consists of the weighted sum of the objective function $\mathbf{h}(\mathbf{y})$ and the penalty function $l(\mathbf{y})$. The penalty function itself can be weighted by the vector $\mathbf{r} = (r_1, r_2, \ldots, r_m)^T$ which is chosen to be $\mathbf{r} = r \cdot (1, 1, \ldots, 1)^T$ in this work.

Here, an easy penalty function for the constraint of eq. (8) was chosen:

$$l(\mathbf{y}) = \frac{1}{\|\mathbf{y} - \mathbf{y}_j\|_2} \tag{11}$$

It follows that $l(\mathbf{y}) \to 0$ for $\mathbf{y} \neq \mathbf{y}_j$ and $l(\mathbf{y}) \to \infty$ for $\mathbf{y} \approx \mathbf{y}_j$.

To achieve a more robust solution, suitable initial conditions $\mathbf{y}_{h,0}$ outside $\mathcal{S}$ has to be chosen. It became apparent, that the addition of the inverse of the last step $\Delta\mathbf{y}$ to $\mathbf{y}_j$, yields a more robust numerical solution than by choosing e.g. $\mathbf{y}_{h,0} = \mathbf{0}$. This can be explained by the switching process of the considered circuits.

From P_h the dynamics can be traced, till reaching Γ again.

5 Multiple hit points

Another problem appears if the configuration space $\mathcal{S}$ of the electronic circuit is multiple folded, so that there are multiple possible hit points (cf. Fig. 3).

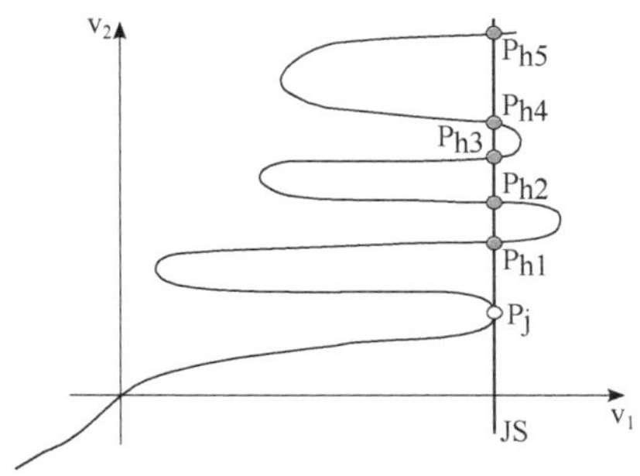

Fig. 3. Multiple hit points

In these cases, the penalty function approach is not suitable and a new approach based on the homotopy method is needed.

5.1 Homotopy method

The homotopy method is used to solve nonlinear algebraic equations of the form (7). The advantage is that the convergence region is much larger than the one by applying the penalty approach. Starting from an easy zero problem, the system of equations will be deformed by λ till reaching the original zero problem. This continuous deformation process is achieved by solving $\mathbf{H}(\mathbf{y}, \lambda) = \mathbf{0}$. A general homotopy can be given as follows:

$$\mathbf{H}(\mathbf{y}, \lambda) = \lambda \mathbf{h}(\mathbf{y}) + (1 - \lambda)\mathbf{B}(\mathbf{y}) = 0 \,. \tag{12}$$

This homotopy consists of a linear combination of two real functions: $\mathbf{h}(\mathbf{y})$, whose zeros are sought and $\mathbf{B}(\mathbf{y})$, a function for which a zero is known. The difficulty is to choose the proper function $\mathbf{B}(\mathbf{y})$ for the corresponding zero problem Trajkovic and Mathis (1995).

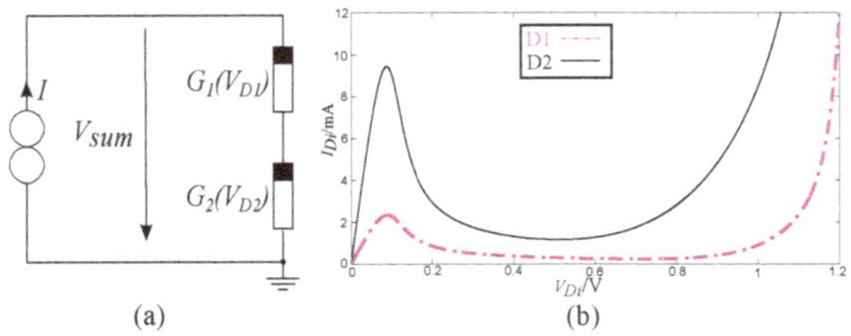

Fig. 4. (a) Analyzed TD circuit; **(b)** two TD characteristics

The zero curve of the homotopy map $\mathbf{H}(\mathbf{y}, \lambda)$ (homotopy path) can be tracked by different techniques. One technique, which is used in this work, is to use an ODE-based algorithm (cf. Watson (1990)). For this, the parametrization is not based on λ but on the arc length s of the homotopy path. Therefor, the equation $\mathbf{H}(\mathbf{y}(s), \lambda(s)) = \mathbf{0}$ has to be differentiated with respect to s and solved for λ and $\mathbf{y}$. The main advantage for using the arc length method is because therewith regressive homotopy paths can be traced. A disadvantage of such a multiple-step method is, that the error of the approximation of $\mathbf{y}$ increases with every step. By using a predictor corrector method, the error accumulation can be counteracted.

In the following the problem of multiple hit points will be displayed by the example of two series connected resonance tunneling diodes (cf. Fig. 4 (a)) Thiessen et al. (2012b). The chosen V-I characteristics of the tunnel diodes is shown in Fig. 4 (b). A good approximation of these V-I characteristics can be achieved by using the following equation (cf. Chang et al. (1993)):

$$G_j(V_{Dj}) = e \cdot \Big(c_1 \cdot V_{Dj} \cdot \Big[tan^{-1}(c_2 \cdot V_{Dj} + c_3) \\ - tan^{-1}(c_2 \cdot V_{Dj} + c_4)\Big] + c_5 \cdot V_{Dj}^m + c_6 \cdot V_{Dj}^l\Big) , \tag{13}$$

where c_i are constants related to the peak and valley currents and voltages, e is a scaling factor and m and l are integer fitting factors.

By using a norator, as explained in Section 2, the configuration space $\mathcal{S}$ of the series connection can be determined by solving the system of equations $I = G_1(V_{D1})$, $I = G_2(V_{D2})$, $V_{sum} = V_{D1} + V_{D2}$ (cf. Fig. 5).

It is noteworthy that the configuration space of the series connection consists of two separated manifolds: one main part proceeding through the origin and a second separated single loop which can only be reached by choosing suitable initial conditions (cf. Thiessen et al. (2012b)). If the current I is increased from zero, the first jump point of the non-regularized circuit of Fig. 4 (a) will be reached at the point P_j. From there, there are five possible hit points as shown in Fig. 5).

For the sake of completeness, the transients of the corresponding regularized circuit are also shown in Fig. 5) by the green lines with arrows. The regularized circuit can be achieved by adding small capacitances parallel to the diodes (cf. Thiessen et al. (2012b)).

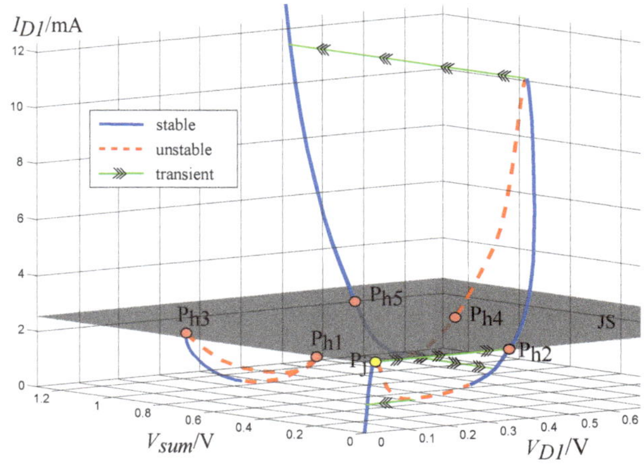

Fig. 5. $\mathcal{S}$ of the series connection of $D1$ and $D2$

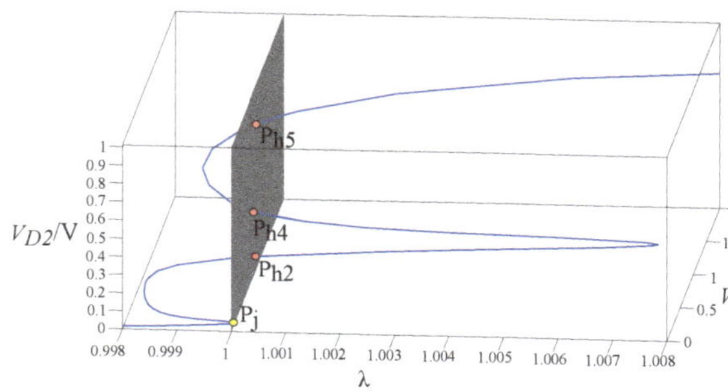

Fig. 6. Homotopy path

To calculate the hit points, the FPN homotopy shown in Rahimian et al. (2011) is used. The FPN homotopy consists of a combination of the fixed-point (FP) and Newton (N) function approach and can be formulated in two steps: (1) The function $\mathbf{h}(\mathbf{y})$ to be solved is multiplied by the fixed-point function giving

$$\mathbf{h}_{fp}(y) = \mathbf{h}(\mathbf{y})(\mathbf{y} - \mathbf{y}_0) . \tag{14}$$

(2) The function $\mathbf{B}(\mathbf{y})$ is formed by a combination of the fixed-point and Newton function (cf. Rahimian et al. (2011)) yielding

$$\mathbf{B}(\mathbf{y}) = (\mathbf{y} - \mathbf{y}_0) + (\mathbf{h}_{fp}(\mathbf{y}) - \mathbf{h}_{fp}(\mathbf{y}_0)) = \mathbf{0} . \tag{15}$$

Inserting Eq. (14) and eq.(15) in Eq. (12) gives

$$\mathbf{H}(\mathbf{y}, \lambda) = \lambda \mathbf{h}_{fp}(\mathbf{y}) + \\ (1 - \lambda) \cdot \big((\mathbf{y} - \mathbf{y}_0) + (\mathbf{h}_{fp}(\mathbf{y}) - \mathbf{h}_{fp}(\mathbf{y}_0))\big) = \mathbf{0} . \tag{16}$$

After some rearrangements, eq.(16) simplifies to

$$\mathbf{H}(\mathbf{y}, \lambda) = (1 + \mathbf{h}(\mathbf{y}) - \lambda) \cdot (\mathbf{y} - \mathbf{y}_0) . \tag{17}$$

With this homotopy, the corresponding homotopy path for calculating the hit points of the series connection can be seen in Fig. 6. The jump point P_j was chosen as starting point $\mathbf{y}_0$, which is also a solution of $\mathbf{h}(\mathbf{y})$. To calculate further zeros except P_j, the homotopy path was progressed beyond $\lambda_{\text{start}} = 1$. Therefore, a bifurcation point (BP) was inserted at P_j (cf. Dhooge et al. (2006)).

The numerical calculation of the homotopy path was done by a *numerical continuation method*, which includes a predictor corrector method. Therefore, the *MATLAB* toolbox *CL_MATCONT* was used Dhooge et al. (2006).

As can be seen from Fig. 5 and Fig. 6 not all hit points can be calculated with this method. The hit points on the separated single loop cannot be calculated, but all hit points on the main part of $\mathcal{S}$. This results due to the fact, that $\mathcal{S}$ consists of two separated manifolds.

The hit points on the separated single loop are in fact instable points and could be calculated by choosing suitable initial conditions. But, for tracking the transients, only the stable hit points are of interest, which are P_{h2} and P_{h4}. Now the question is, which of these both points is the right hit point ?

In the work of Chua and Alexander (1971) there is an inertia postulate which says: trajectories on a stable branch will continue on a stable part till reaching a jump point and than jump to the "nearest" stable part of $\mathcal{S}$. But a complete proof of this postulate is still pending. One possibility to verify this postulate is to analyze the catchment area of the dynamic of the corresponding regularized circuit, but this will be the focus on further studies.

6 Conclusions

In this work, a detailed description of the geometric system analysis was given. Thereby it was explained why the system of equations yielding from the ANA is the most suitable for applying the geometric approach. Furthermore, the penalty and the homotopy approach for a numerically efficient calculation of the hit points were shown. The penalty function method turned out to be suitable for circuits with only one fold in $\mathcal{S}$, but not for several folded configuration spaces. By using a homotopy method with a proper homotopy function, all hit points can be calculated assuming $\mathcal{S}$ consists only of one manifold. In this work a non generic case, where $\mathcal{S}$ consists of two separated manifolds was shown. In those cases, not all hit points can be calculated without further arrangements. Furthermore, the question of choosing the right hit points appeared, but shall be studied in further investigations.

Acknowledgements. The authors would like to thank the German Research Foundation (DFG) for the financial support and Serdar Ediz for his valuable contributions to the topic of hit point calculation.

References

Andronov, A., Vitt, A., and Khaĭkin, S.: Theory of Oscilators, International Series of Monographs in Physics, Vol. 4., Pergamon Press Ltd., 1966.

Chang, C., Asbeck, P., Wang, K.-C., and Brown, E.: Analysis of Heterojunction Bipolar Transistor/Resonant Tunneling Diode Logic for Low-Power and High-Speed Digital Applications, IEEE T. Electr. Dev., 40, 685–691, 1993.

Chua, L.: Dynamic nonlinear networks: State-of-the-art, IEEE T. Circuits Syst., 27, 1059–1087, 1980.

Chua, L. and Alexander, G.: The Effects of Parasitic Reactances on Nonlinear Networks, IEEE T. Circuits Syst., 18, 520–532, 1971.

Chua, L. O. and Deng, A.-C.: Impasse Points. Part 1: Numerical Aspects, International Journal of Circuit Theory and Applications, 17, 213–235, 1989a.

Chua, L. O. and Deng, A.-C.: Impasse Points, Part 2: Analytical Aspects, Int. J. Circ. Theor. App., 17, 271–282, 1989b.

Dhooge, A., Govaerts, W., Kuznetsov, Y. A., Mestrom, W., Riet, A. M., and Sautois, B.: MATCONT and CL_MATCONT: Continuation toolboxes in matlab, 2006.

Florian Jarre, J. S.: Optimierung, Springer-Verlag, 2003.

Mishchenko, E. F. and Rozov, N. K.: Differential Equations with Small Parameters and Relaxation Oscillators, Plenum Press, 1980.

Rahimian, S. K., Jalali, F., Seaderc, J., and White, R.: A new homotopy for seeking all real roots of a nonlinear equation, Computers and Chemical Engineering, 35, 403–411, 2011.

Reich, S.: On a Geometrical Interpretation of Differential-Algebraic Equations, Circ. Syst. Sign. Pr., 9, 367–382, 1990.

Reissig, G.: Differential-Algebraic Equations and Impasse Points, IEEE T. Circuits Syst., 43, 122–133, 1996.

Riaza, R.: Differential-Algebraic Systems: Analytical Aspects and Circuit Applications, World Scientific, 2008.

Sarangapani, P., Thiessen, T., and Mathis, W.: Differential Algebraic Equations of MOS Circuits and Jump Behavior (accepted), Advances in Radio Science.

Shampine, L. F. and Reichelt, M. W.: The MATLAB ODE Suite, SIAM Journal on Scientific Computing, 18, 1–22, 1997.

Thiessen, T. and Mathis, W.: Geometric Dynamics of Nonlinear Circuits and Jump Effects, International Journal of Computations & Mathematics in Electrical & Electronic Engineering (Compel 2011), 30, 1307–1318, 2011a.

Thiessen, T. and Mathis, W.: Geometrical Interpretation of Jump Phenomena in Nonlinear Dynamical Circuits, in: Joint 3rd Int'l Workshop on Nonlinear Dynamics and Synchronization (INDS 2011) & 16th Int'l Symposium on Theoretical Electrical Engineering (ISTET 2011), 1 –5, 2011b.

Thiessen, T., Gutschke, M., Blanke, P., Mathis, W., and Wolter, F.-E.: A Numerical Approach for Nonlinear Dynamical Circuits with Jumps, in: 20th European Conference on Circuit Theory and Design (ECCTD 2011), 461–464, 2011.

Thiessen, T., Plönnigs, S., and Mathis, W.: Transient Solution of Fast Switching Systems without Regularization, in: IEEE 55th International Midwest Symposium on Circuits and Systems (MWSCAS 2012), pp. 578 –581, 2012a.

Thiessen, T., Zorn, C., and Mathis, W.: On the Analysis of Series Connected Resonant Tunneling Diodes, Nonlinear Dynamics of Electronic Systems, Proceedings of NDES 2012, pp. 1 –4, 2012b.

Thiessen, T., Plönnigs, S., and Mathis, W.: Fast Switching Behavior in Nonlinear Electronic Circuits: A Geometric Approach, in: Selected Topics in Nonlinear Dynamics and Theoretical Electrical Engineering, edited by Kyamakya, K., Halang, W. A., Mathis, W., Chedjou, J. C., and Li, Z., Vol. 459 of Studies in Computational Intelligence, 99–116, Springer Berlin Heidelberg, 2013.

Trajkovic, L. and Mathis, W.: Parameter Embedding Methods for Finding DC Operating Points: Formulation and Implementation, in: International Symposium on Nonlinear Theory and its Applications (NOLTA 1995), 1159–1164, 1995.

Watson, L. T.: Globally convergent homotopy algorithms for nonlinear systems of equations, Nonlinear Dynamics, 1, 143–191, 1990.

23

Automatic focus algorithms for TDI X-Ray image reconstruction

J. Dörr, M. Rosenbaum, W. Sauer-Greff, and R. Urbansky

TU Kaiserslautern, Institute of Communications Engineering, 67653 Kaiserslautern, Germany

Correspondence to: M. Rosenbaum (rosenbaum@eit.uni-kl.de)

Abstract. In food industry, most products are checked by X-rays for contaminations. These X-ray machines continuously scan the product passing through. To minimize the required X-ray power, a Time, Delay and Integration (TDI) CCD-sensor is used to capture the image. While the product moves across the sensor area, the X-ray angle changes during the pass. As a countermeasure, adjusting the sensor shift speed on a single focal plane of the product can be selected. However, the changing angle result in a blurred image in dependance to the thickness of the product. This so-called "laminographic effect" can be compensated individually for one plane by inverse filtering. As the plane of contamination is unknown, the blurred image will be inversely filtered for different planes, but only one of these images shows the correctly focussed object. If the correct image can be found, the plane containing the contamination is identified. In this contribution we demonstrate how the correctly focussed images can be found by analyzing the images of all planes. Different characteristics for correctly and incorrectly focussed planes like sharpness, number of objects and edges are investigated by using image processing algorithms.

1 Introduction

For quality assurance in food industry more and more X-ray-scanners are used instead of simple metal detectors, because they are also able to detect non-metallic contaminations, to check filling levels or to check that a product is sealed properly. For such scanning applications continuous processing is desirable, because it fits in most production processes without the need of further adaptations. To minimize the required X-ray power, a Time, Delay and Integration (TDI) (Wong et al., 1992) CCD sensor can be used to capture images while the objects are moving across the sensor area. One way to obtain depth information is the use of laminographic techniques (Gondrom and Schröpfer, 1999; Rooks and Sack, 1995) in conjunction with the Radon Transform (Popularikas, 1996). This requires a complex mechanical setup not suitable for continuous scanning or requires two or more X-ray sources and detectors. This approach is not only expensive in terms of investment, in addition the second X-ray tube also needs power, cooling and better shielding against X-rays leaving the machine.

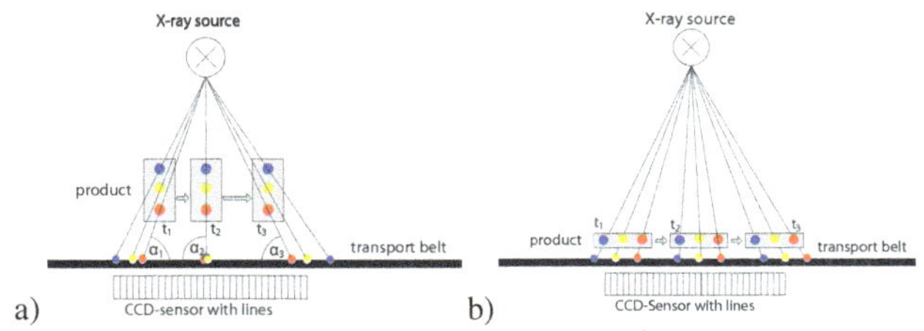

Fig. 1. The laminographic effect: result of changing angle on **(a)** high objects **(b)** no impact on flat objects

An alternative way is the use of a single sensor followed by inverse filtering. However, this technique causes the problem that the angle of the X-rays changes during the transport, as shown in Fig. 1.

This effect can be modeled as a convolution of the real image $a(x,y)$ with the point spread function (PSF) of the imaging system $h(x,y)$, yielding an image $s(x,y)$.

$$s(x,y) = a(x,y) * h(x,y) \quad (1)$$

In general, this PSF is two-dimensional, but in our case only the contribution in direction of the belt movement is significant, which can be modeled as a one-dimensional impulse response. This impulse response is height variant, which means it is only valid for a given height of the object above the sensor plane. Further information can be found in (Rosenbaum et al., 2011).

Images of thin objects can be deblurred by inverse filtering during post processing using a Wiener filter approach. This deblurring process performs well for thin objects, if the PSF is known. The PSF can be estimated for one specific plane or layer. By resizing this PSF one can obtain PSFs of other layers. However, it is unknown which layer contains the contamination. So the idea is to deblur the image with respect to the PSFs of different layers and to search for a correctly focussed image a posteriori. If no correctly focussed image can be identified, the object is not flat. If a correct image is found, the object is flat and the height of the object over the transport belt is known.

Humans are well able to decide for correctly focussed images. However, for practical industrial usage the decision has to be made by computer algorithms. Unfortunately, standard autofocus algorithms (Subbarao et al., 1993) can not be applied because of potential artifacts from inverse filtering. In this article a solution for these problems is proposed and demonstrated by simulation results.

The remainder of this paper is organized as follows: After an overview on inverse filtering which is visualized by resulting images from our simulation setup, three different autofocus algorithms are discussed. A combination of these algorithms is proposed and demonstrated by simulation results. Finally, conclusions are drawn.

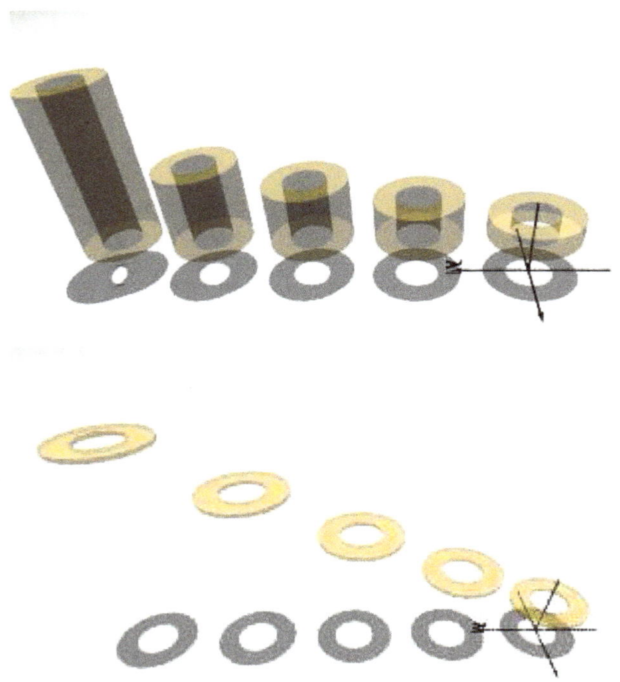

Fig. 2. Ray tracing model to test algorithms.

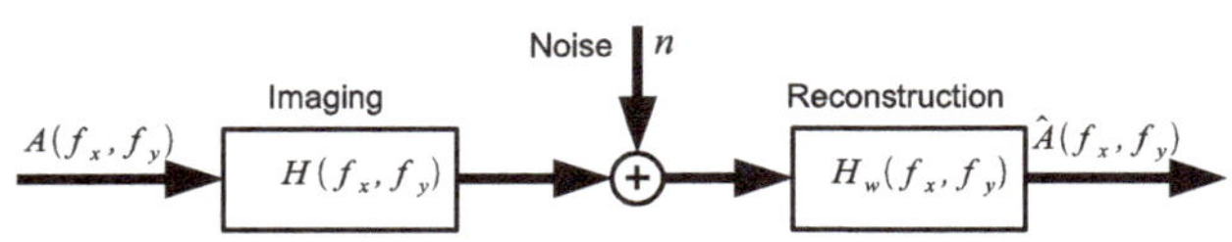

Fig. 3. Block diagram of the whole imaging process.

2 Simulation setup

To demonstrate and test the proposed algorithms, the ray tracing software POV-Ray (see www.povray.org) was used to build test images, where a visual light source models the X-ray tube. Figure 2 shows the model. Different washers and cylinders with different thicknesses in different layers were simulated. To model the conveyor belt, the light source is moved and the changing shadows are captured. Finally, the images of different light positions are merged to one image by calculating the mean value. This simulates the laminographic effect resulting from the use of an TDI X-ray system.

The results of the mean value calculations are inversely filtered for a specific number of layers (here 160). Finally we use these 160 inversely filtered layer-images for every washer to test the algorithms.

3 Inverse filtering

Blurring caused by the laminographic effect can be effectively addressed by inverse filtering since it can be regarded as a convolution of the real image with the point-spread-function of the imaging system additionally corrupted with noise n.

$$s(x,y) = a(x,y) * h(x,y) + n \tag{2}$$

The inverse filtering takes advantage of the frequency domain. A pure inversion of the transfer function $H(f_x, f_y)$ requires to take the reciprocal of it yielding $H_w(f_x, f_y) = H^{-1}(f_x, f_y)$, which poses problems because of spectral zeros and amplification of noise contained in the X-ray image. A well-known solution to this problem is Wiener filtering (Madissettti and Williams, 1998). Equation (3) shows the parametric form of the Wiener filter. K accounts for noise influence, which is ideally chosen as the reciprocal of the input signal-to-noise-ratio.

$$H_w(f_x, f_y) = \frac{H^*(f_x, f_y)}{|H(f_x, f_y)|^2 + K} \tag{3}$$

Figure 3 summarizes the imaging and reconstruction process, which produces an estimate $\hat{A}$ of the unknown original image A.

In the reconstruction process, an impulse response is only valid for a single focal plane. The characteristics of the imaging process allow to calculate an impulse response for every focal plane out of one known impulse response by using algorithms also known in image processing as bilinear or bicubic scaling (details can be found in Keys, 1981).

Fig. 4. Blurred image with compressed PSF.

Fig. 5. Inversely filtered with lightly spread PSF.

To estimate the height of an object above the belt, images have to be reconstructed with scaled PSFs. In our case, 160 images corresponding to 160 focal planes with a distance of 0.5 mm were calculated, yielding a range of 8 cm. As described in Sect. 2, we used simulated X-ray scenarios of a thin metallic washer, diameter 2 cm. Some examples of these images are shown in Figs. 4–7. The characteristics of these images and especially the differences between the correctly and incorrectly focussed images will be useful for understanding suitable algorithms.

If the inverse filtering PSF is too compressed, it nearly corresponds to a dirac-impulse and therefore the result of the convolution with this PSF resembles the original blurred image after the scanning process (Fig. 4).

The result is a very blurry image, but it shows only one object and not separated parts of the washer. Figure 5 depicts the result of an inverse convolution with a wider spread PSF.

If the PSF is spread farther, artifacts with sharp edges and many separated areas are the result as shown in Fig. 6.

Figure 7 depicts an example for images filtered with the correctly scaled PSF and which should be chosen as correctly focussed. They are characterized by sharp edges, only one object and not many changes of intensity values.

4 Algorithms

In general, the contaminants cannot be specified and the product may also vary. This means, the general properties which are different between correctly and incorrectly focussed images need to be used to detect the focal plane.

Fig. 6. Many separated objects and edges are the result of inverse filtering with widely spread PSF.

a) b)

Fig. 7. Results of inversion with correctly scaled PSF **(a)** washer on a low layer, **(b)** washer on a high layer.

In this contribution we propose to use the set of following features: (1) peaks of image autocorrelation function, (2) number of objects and (3) pixels outside objects.

4.1 Autocorrelation

Incorrectly focussed images with a too wide PSF suffer from periodic artifacts, which need to be detected. The idea is to measure these changes by autocorrelation. The more periodic artifacts appear in an image, the more peaks occur in the result of the image autocorrelation function.

The autocorrelation function is calculated for every image line. Figure 8 shows the resulting matrix after normalization.

After the summation of all lines the result is a „correlation line". To deal with varying image intensities, the maximum of the autocorrelation function is normalized to 1. Figure 9 shows an example for an autocorrelation function of a correctly focussed image and Fig. 10 for an incorrectly focussed image.

Obviously, the incorrectly focussed image produces more autocorrelation peaks than the correctly focussed. We propose to sum up the number of peaks of the correlation line. These sums for all images over the image height index are shown in Fig. 11. The value near the best image is very low in comparison to incorrectly focussed images. As can be seen, this criterion is suitable expect for low-index (compressed

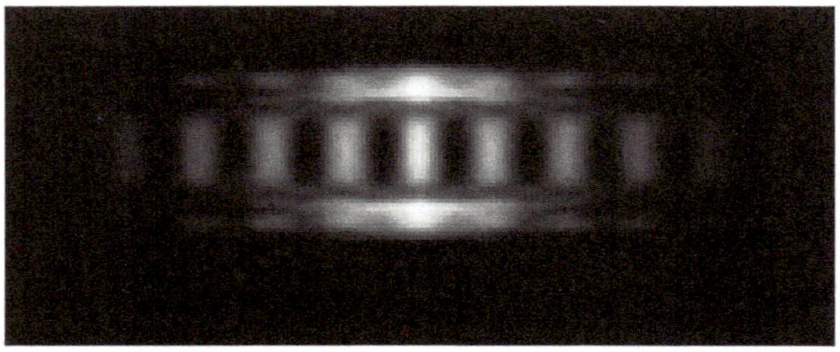

Fig. 8. Example of correlation matrix.

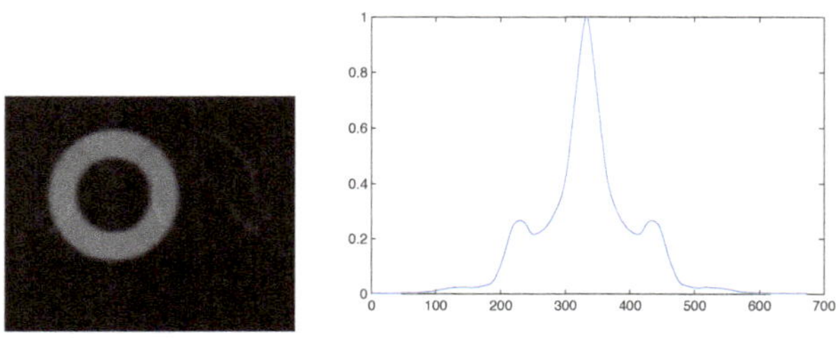

Fig. 9. Correlation line with one object (correctly focussed).

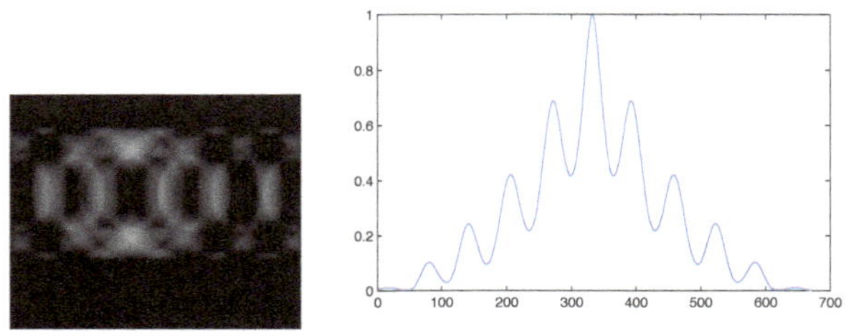

Fig. 10. Correlation line with multiple objects (PSF too wide, incorrectly focussed).

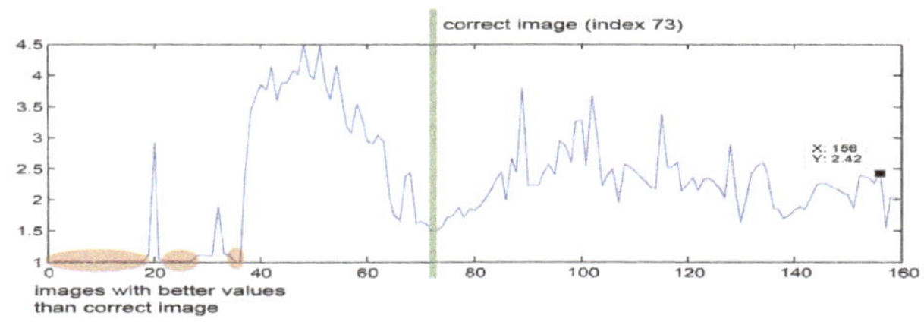

Fig. 11. Sum of maxima over image index, correctly focussed image leads to low values, but there are also images with lower values.

PSF) images. So, for these images another algorithm has to be employed.

4.2 Object count

Another difference between a correctly and most of the incorrectly focussed images is the number of separated objects. To count the objects, an area has to be defined and divided into seperate objects. This is done by a binarisation and a following object segmentation. The binarisation threshold has to be determined adaptively depending on the background brightness.

All connected white pixels are counted as one object. To find out which pixel groups belong together, well known segmentation algorithms (Ho, 2011) can be used.

Figure 12 shows two binarized images. Obviously, the incorrectly focussed image contains more objects (white areas) than the correctly focussed image.

Figure 13 shows the counted number of objects over the image height index.

Obviously many images have only one object, especially the low index images and the images around the correctly focussed image.

This algorithm is more likely usable for an image preselection because it cannot determine the correctly focussed image, but it can detect images which are clearly incorrectly focussed.

4.3 Remaining pixel analysis

The two previous algorithms work well if the PSF is widely spread. They fail for blurred images, which result from inverse filtering with a narrow PSF. A possible solution for this problem is to count the pixels which remain outside the objects after binarisation. This can be implemented very efficiently: The filtered image with white background is binarized after applying a noise suppressing Gaussian filter. This binarized image is multiplied with the inverted image. Since the color black corresponds to the value zero all pixels in object areas are suppressed. The number of remaining pixels will be summed up.

Figure 14 demonstrates the process, and Fig. 15 shows the sums of remaining pixels over the image hight index. As can be seen the process works well, therefore the correctly focussed image can be found easily by a minimum-search.

4.4 Combination of algorithms

Other test images showed that summing up the pixels out of objects could lead to a few wrong decisions. Therefore, a combination of the presented methods is proposed. The flowchart in Fig. 16 illustrates the concept of a full image detection process.

First, the remaining pixels and the number of objects are calculated. In addition, the number of edges are counted. This acts like counting the objects, but with a preceding edge-finding algorithm.

To identify that an image indicates a plane with a flat object, the following conditions are proposed: A correctly focussed image has to exhibit only one object, not more than a specific number of edges and a sum of remaining pixels lower than a specific threshold, which depends on the edge length. The thresholds have to be adapted to the specific

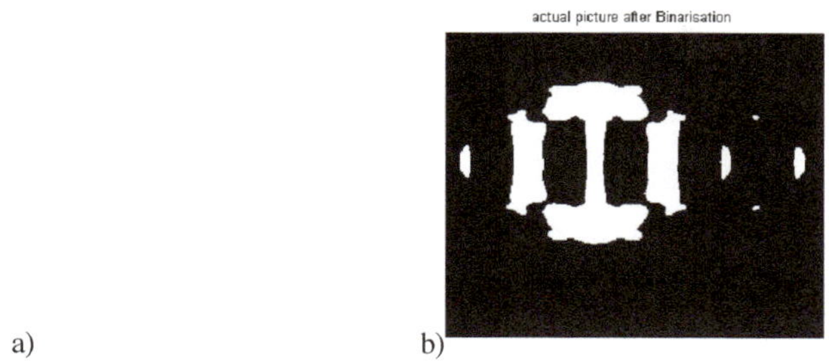

Fig. 12. After binarisation only one object in correctly focussed image **(a)** many objects in incorrectly focussed image.

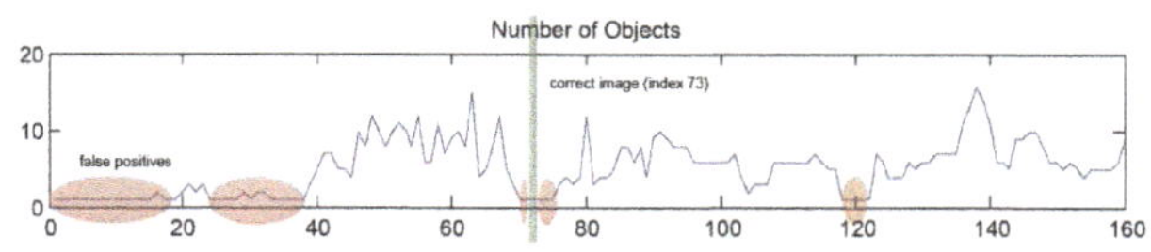

Fig. 13. Number of objects over image index for a washer on a low layer.

system. If no image satisfying the conditions can be identified, it is assumed that there is no distinguishable flat object.

To save computing time, the number of edges and the autocorrelations are calculated for images with one object only. The results of remaining pixels and of autocorrelations are combined to deliver a final result. In our example, the algorithm detects three images, which satisfy the conditions and the correct image (index 35 in this example) has been found.

5 Results

Figure 17 shows the results of the steps of the combined algorithm for one of the washers. The parameters of the algorithms, like threshold values or Gaussian filter values were set up in such a way that they operate successfully for all test images. Objects with a height below 2 centimeters are considered flat, yielding a distinguished sharp plane. For objects higher than that, which are considered non-flat, no erroneous images satisfying the conditions have been found.

6 Conclusions

It has been demonstrated that the proposed technique is able to identify distinguishable flat objects in a set of images resulting from inverse filtering of a laminographic image for different focal planes. The algorithm is based on combining methods from autocorrelation techniques, object counts and pixels-outside-objects analysis.

Together with the idea of height variant inverse filtering this paves the way to extracting 3-D-Information out of 2-D scanned objects by using the laminographic effect of TDI-X-ray systems. The concept works well in our simulation setup, the next step will be to verify these results for a real system.

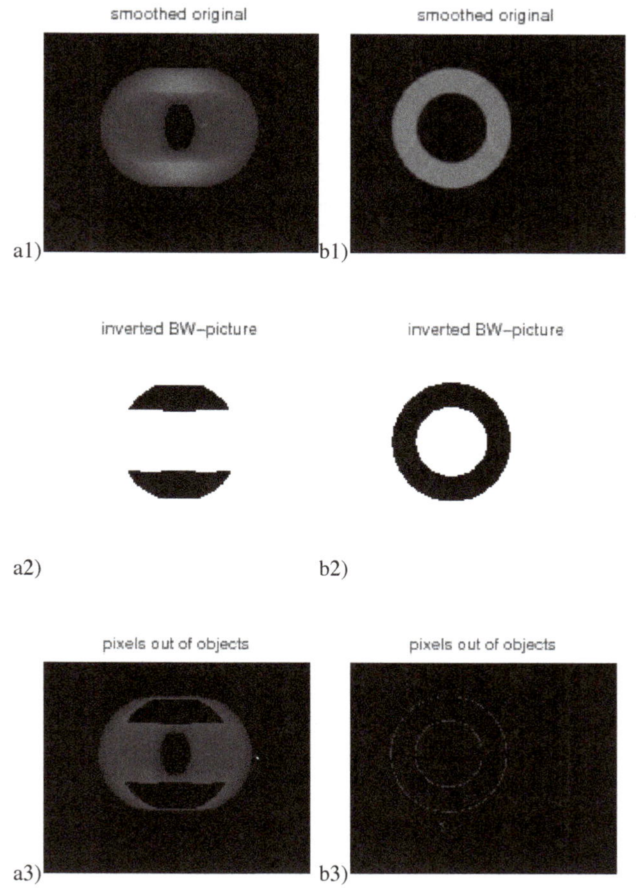

Fig. 14. After binarisation (a2, b2) and multiplication with original (a1,b1) there are many pixels remaining in the result of the blurred image (a1) and only a few pixels in the result of the sharp image (b3).

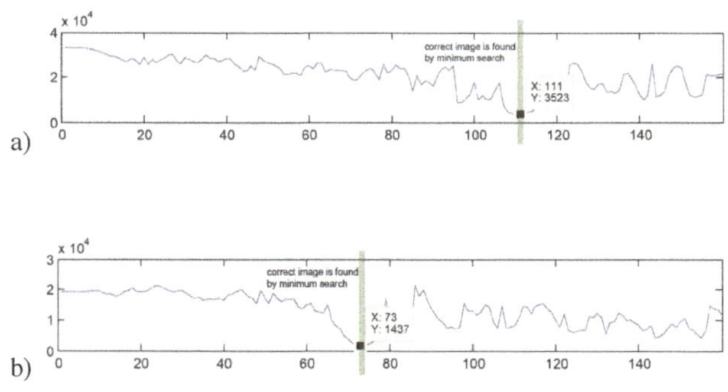

Fig. 15. Sum of non-black pixels out-of objects **(a)** washer on a low layer, **(b)** washer on a higher layer.

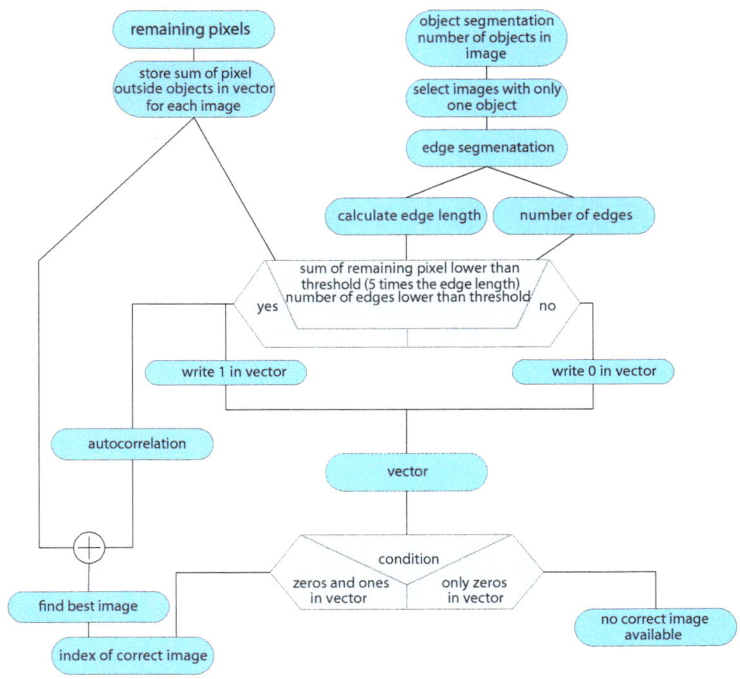

Fig. 16. Structure of combined process.

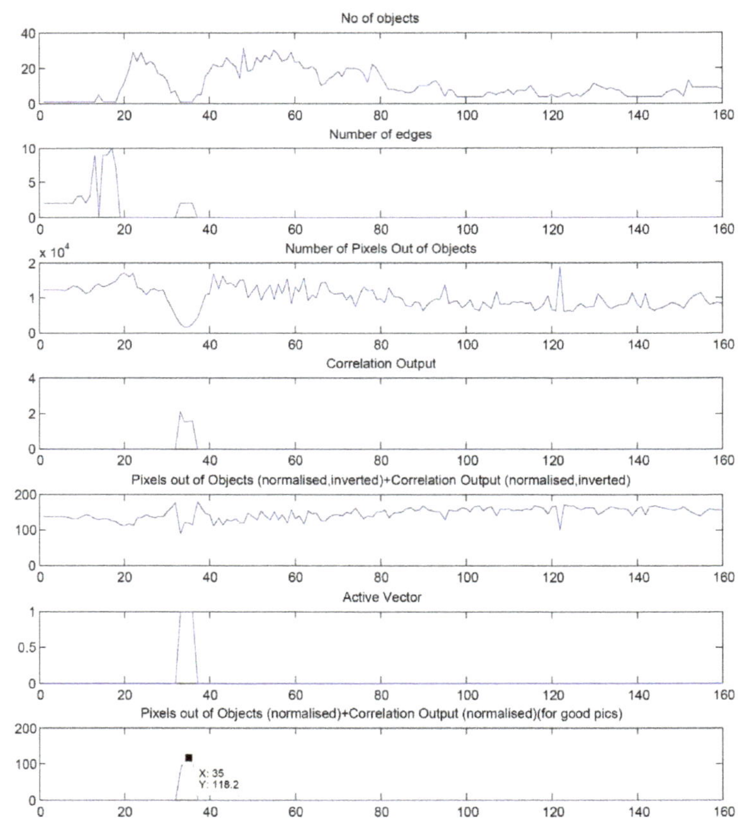

Fig. 17. Results of final system

References

Gondrom, S. and Schröpfer, S.: Digital computed laminography and tomosynthesis – functional principles and industrial applications, Proceedings of International Symposium on Computed Tomography and Application, Berlin, Germany, 1999.

Ho, P.-G. (ed): Image Segmentation, InTech, 2011.

Keys, R.: Cubic convolution interpolation for digital image processing, Int. Conf. Acoust. Spee., 29, 1153–1160, 1981.

Madisetti, V. K. and Williams, D.B. (eds): The Digital Signal Processing Handbook, CRC Press, 1998.

Popularikas, A. D. (ed.): The Transforms and Applications Handbook, CRC Press, 1996.

Rosenbaum, M., Sauer-Greff, W., and Urbansky, R.: Inverse filtering for time, delay and integration X-ray imaging, Adv. Radio Sci., 9, 135–138, doi:10.5194/ars-9-135-2011, 2011.

S. Rooks and T. Sack: X-ray Inspection of Flip Chip Attach Using Digital Tomosynthesis, Circuit World, 21, 51–55, 1995.

Subbarao, M., Choi, T. S., and Nikzad, A.: Focusing Techniques, Journal of Optical Engineering, 2824–2836, 1993.

Wong, H.-S., Yao, Y. L., and Schlig, E. S.: TDI charge-coupled devices: Design and applications, IBM J. Res. Develop., 36, 83–106, 1992.

24

Auxiliary branch method and modified nodal voltage equations

A. Reibiger

Dresden University of Technology, Department of Electrical Engineering, Mommsenstr. 13, 01062 Dresden, Germany

Abstract. A theorem is presented describing a transformation by means of which it is possible to assign to an elementary multiport with fairly general constitutive equations (including all kinds of controlled sources, nullors, ideal transformers, etc.) a modified multiport with the same all-pole terminal behavior. The branch set of this modified multiport is augmented with so called auxiliary branches whereas its constitutive equations are always in conductance form. Therefore an interconnection of a family of multiports transformed in this manner can always be analyzed by means of a system of nodal voltage equations. It will be shown that this system of equations is equivalent to a system of modified nodal voltage equations set up for the network that is an interconnection of the elementary multiports originally given.

1 Introduction

For an arbitrary elementary multipole with semi-implicite constitutive equations we consider an algorithm for the construction of a multipole having the same terminal behavior as the given elementary multipole but with constitutive equations in conductance form.

The idea for the construction of this algorithm goes back to a conference paper of Reinschke and Schwarz (1977). Former related results go back to papers of Klein (1958) and his collaborators Engelhardt and Heinz (1970); Thielmann (1973); Kremer (1978).

As an application of the theory of terminal behavior of networks developed in Reibiger (1985, 1986, 2003a) we present a reformulation of this algorithm which makes its network theoretic assumptions and consequences more transparent.

Correspondence to: A. Reibiger
(reibiger@iee.et.tu-dresden.de)

2 Network theoretical background

In this section we give a short review of results of our research on the foundations of network theory Reibiger (1985) to Reibiger (2003b).

As standing notations we use the symbols $\mathcal{U}$, $\mathcal{I}$, and $\mathcal{T}$ to denote the *space of branch voltage values*, the *space of branch current values*, and the *time axis*, respectively. We assume that all these spaces are provided with the structure of one-dimensional normed oriented real vector spaces. Int $\mathcal{T}$ denotes the set of all intervals of the time axis.

The assumptions $\mathcal{U}=\mathbb{R}$V, $\mathcal{I}=\mathbb{R}$A, and $\mathcal{T}=\mathbb{R}$s lead to the class of *electrical networks* and $\mathcal{U}=\mathcal{I}=\mathcal{T}=\mathbb{R}$ to that of *normalized networks*. In the same manner it is possible to introduce network classes which can be used for modeling mechanical or thermal systems, etc., cf. Koenig et al. (1967); Reinschke and Schwarz (1976).

For each nonvoid finite set $\mathcal{Z}$ and each interval $T \in \text{Int}\,\mathcal{T}$ we assume that the sets $\mathcal{U}^{\mathcal{Z}}$, $(\mathcal{U}^{\mathcal{Z}})^T$, $\mathcal{U}^{\mathcal{Z}} \times \mathcal{I}^{\mathcal{Z}}$,..., $(\mathcal{U}^{\mathcal{Z}})^T \times (\mathcal{I}^{\mathcal{Z}})^T$ are provided with the structure of normed linear spaces induced by the structures of the spaces $\mathcal{U}$ and $\mathcal{I}$[1]. The set $\mathcal{S} := \bigcup_{T \in \text{Int}\,\mathcal{T}} (\mathcal{U}^{\mathcal{Z}})^T \times (\mathcal{I}^{\mathcal{Z}})^T$ is denoted as the *universal signal set on* $\mathcal{Z}$.

Let $\mathcal{W} \subseteq \mathcal{S}$, where $\mathcal{S}$ is the universal signal set on $\mathcal{Z}$. Let $\text{sd}\mathcal{W} := \{T \mid \exists_{(u,i) \in \mathcal{W}}\ T = \text{dom}\, u\}$, and let $\dot{\vee}$ denote the exclusive or. The set $\mathcal{W}$ is denoted as *restriction compatible* if the following conditions are fulfilled:

(i) $\bigcup_{T' \in \text{sd}\mathcal{W}} T' \in \text{sd}\,\mathcal{W}$,

(ii) $|\,\text{sd}\,\mathcal{W}| = 1 \,\dot{\vee}\, (\forall_{T,T' \in \text{Int}\,\mathcal{T}}\ T \in \text{sd}\,\mathcal{W} \wedge T' \subseteq T$
$\Rightarrow T' \in \text{sd}\,\mathcal{W})$,

(iii) $\forall_{(u,i) \in \mathcal{W}} \forall_{T \in \text{sd}\mathcal{W}}\ T \subseteq \text{dom}\, u \Rightarrow (u|T, i|T) \in \mathcal{W}$.

The set $\mathcal{W} \subseteq \mathcal{S}$ is called *properly restriction compatible* if $|\,\text{sd}\,\mathcal{W}| > 1$ otherwise $\mathcal{W}$ is referred to as *trivially restriction compatible*.

[1] As usual in set theory we denote for any two sets X and Y the set of all mappings from X to Y by Y^X. For each $f \in X$ and $X' \subset X$ we denote with $f|X'$ the restriction of f to X'.

For each $(u,i)\in\mathcal{W}$, $b\in\mathcal{Z}$, and $\mathcal{Z}'\subset\mathcal{Z}$, $(\mathcal{Z}'\neq\emptyset)$ we denote with u_b, i_b, and $u_{\mathcal{Z}'}$, $i_{\mathcal{Z}'}$ the time functions defined for all $t\in\mathrm{dom}u=\mathrm{dom}i$ by $u_b(t):=u(t)(b)$ and $i_b(t):=i(t)(b)$, or $u_{\mathcal{Z}'}(t):=u(t)|\mathcal{Z}'$, and $i_{\mathcal{Z}'}(t):=i(t)|\mathcal{Z}'$, respectively. Under the same assumptions the set $\mathcal{W}_{\mathcal{Z}'}:=\{(u_{\mathcal{Z}'},i_{\mathcal{Z}'})|\ (u,i)\in\mathcal{W}\}$ is denoted as the *projection of* $\mathcal{W}$ *generated by* $\mathcal{Z}'$.

An *oriented graph* is defined as an ordered triple $(\mathcal{Z},\mathcal{K},\mathcal{A})$ of two finite disjoint sets $\mathcal{Z}$ and $\mathcal{K}$ with $(\mathcal{K}=\emptyset\Rightarrow\mathcal{Z}=\emptyset)$ and a map $\mathcal{A}:\mathcal{Z}\to\mathcal{K}\times\mathcal{K}$. $\mathcal{Z}$ is the *branch set*, $\mathcal{K}$ the *node set*, and $\mathcal{A}$ the *incidence mapping* of this graph. The incidence map assigns to each branch b the ordered pair $(v,w):=\mathcal{A}(b)$ of its *initial* and *terminal node*. Each branch is *oriented* from its initial to its terminal node.

Let $\mathcal{G}=(\mathcal{Z},\mathcal{K},\mathcal{A})$ and $\bar{\mathcal{G}}=(\bar{\mathcal{Z}},\bar{\mathcal{K}},\bar{\mathcal{A}})$ be oriented graphs. $\bar{\mathcal{G}}$ is a *subgraph* of $\mathcal{G}$ if $\bar{\mathcal{Z}}\subseteq\mathcal{Z}$, $\bar{\mathcal{K}}\subseteq\mathcal{K}$, and $\bar{\mathcal{A}}=\mathcal{A}|\bar{\mathcal{Z}}$. If $\mathcal{Z}'\subseteq\mathcal{Z}$, $\mathcal{A}'=\mathcal{A}|\mathcal{Z}'$ and $\mathcal{K}'=\{v|\ \exists b\in\mathcal{Z}\ v$ initial or terminal node of $b\}$ then $\mathcal{G}_{\mathcal{Z}'}:=(\mathcal{Z}',\mathcal{K}',\mathcal{A}')$ is denoted as the subgraph of $\mathcal{G}$ *generated* by $\mathcal{Z}'$.

As in Reibiger (1997); Reibiger et al. (1999); Reibiger (2003a,b) we define a *network* $\mathcal{N}$ as an ordered pair $\mathcal{N}=(\mathcal{C},\mathcal{V})$ of a skeleton and a constitutive relation. The *skeleton* $\mathcal{C}$ is an ordered pair $\mathcal{C}=(\mathcal{G}_{\mathrm{v}},\mathcal{G}_{\mathrm{c}})$ of two oriented graphs with the same branch and node set differing at most in their orientations. $\mathcal{G}_{\mathrm{v}}$ and $\mathcal{G}_{\mathrm{c}}$ are denoted as the *voltage* or *current graph* of $\mathcal{N}$, respectively. The *constitutive relation* $\mathcal{V}$ is a restriction compatible nonvoid subset $\mathcal{V}$ of the universal signal set on the branch set of $\mathcal{G}_{\mathrm{v}}$ and $\mathcal{G}_{\mathrm{c}}$. For simplicity this signal set is denoted as the *universal signal set of* $\mathcal{N}$. Analogously, the branch and node sets of the graphs $\mathcal{G}_{\mathrm{v}}$ and $\mathcal{G}_{\mathrm{c}}$ are denoted as those of $\mathcal{N}$.

The skeleton describes the topological structure of a network. The constitutive relation characterizes the physical properties assigned to the branch set of a network. The constitutive relation is a set of ordered pairs of time functions and it is therefore a binary relation in the sense of set theory.

In the standard case the constitutive relation of a network is properly restriction compatible. The treatment of linear time-invariant networks by means of Laplace transform, where only time functions defined on the intervall $[0,+\infty]$ are admitted, delivers typical examples of networks with trivially restriction compatible constitutive relations.

Let $\mathcal{N}$ be a network with branch set $\mathcal{Z}$. The elements of the universal signal set of $\mathcal{N}$ are denoted as *signals*. If (u,i) is a signal then u is called the corresponding *voltage on the branch set of* $\mathcal{N}$ and i is called the corresponding *current in the branch set of* $\mathcal{N}$. For each signal (u,i) and each $b\in\mathcal{Z}$, $\mathcal{Z}'\subset\mathcal{Z}$ $(\mathcal{Z}'\neq\emptyset)$ the time functions $u_b, i_b, u_{\mathcal{Z}'}$, and $i_{\mathcal{Z}'}$ are denoted as the corresponding *branch voltage*, *branch current*, *partial voltage on* $\mathcal{Z}'$, and *partial current in* $\mathcal{Z}'$, respectively.

The orientations of the branches of the voltage and the current graph determine the *reference directions* for branch voltages and branch currents, respectively. The constitutive relation and both of the Kirchhoff's laws have reference to these orientations.

For the computer-aided analysis of networks it is indispensable to separate off the constitutive relation of a network from the corresponding universal signal set, or an appropriate subset of this set, by means of a constitutive equation. In analogy to Willems (1991); Poldermann and Willems (1998) the notion of a constitutive equation can be introduced as follows.

Let $\mathcal{N}=:(\mathcal{C},\mathcal{V})$ be a network with universal signal set $\mathcal{S}$. If there exist a subset $\mathcal{S}'\subseteq\mathcal{S}$ with $\mathcal{V}\subseteq\mathcal{S}'$, an *equating set* $\mathcal{E}$, and two mappings $f,g:\mathcal{S}'\to\mathcal{E}$ such that the relationship

$$\mathcal{V}=\{(u,i)\in\mathcal{S}'|f(u,i)=g(u,i)\}\tag{1}$$

holds, then the equation $f(u,i)=g(u,i)$ is denoted as a *constitutive equation* of $\mathcal{N}$. The right hand side of Eq. (1) is denoted as a *representation* of the constitutive relation of $\mathcal{N}$ by means of a constitutive equation.

Let $\mathcal{N}$ be a network with universal signal set $\mathcal{S}$ and constitutive relation $\mathcal{V}$. Because the universal signal set contains a lot of very complicated signals, e.g. signals with nowhere differentiable voltages and currents, etc., it is useful to introduce in addition to the universal signal set some more specific signal sets. Of course, such a signal set $\mathcal{S}'$ of $\mathcal{N}$ must be a subset of its universal signal set and it must contain its constitutive relation, $\mathcal{V}\subseteq\mathcal{S}'\subset\mathcal{S}$. An important special case of such a signal set is a signal set with a prescribed type of branch signals.

Let $R:\mathcal{I}\to\mathcal{U}$ be a linear bijection then a set $\mathcal{S}_*\subseteq\bigcup_{T\in\mathrm{Int}\,\mathcal{T}}\mathcal{U}^T\times\mathcal{I}^T$ is denoted as a *branch signal type* if it fulfills the conditions:

(i) $\mathcal{S}_*=\{(u,i)|u\in\mathcal{S}_{*\mathrm{v}}\wedge i\in\mathcal{S}_{*\mathrm{c}}\wedge\mathrm{dom}\,u=\mathrm{dom}i\}$, where $\mathcal{S}_{*\mathrm{v}}:=\{u|\exists_i(u,i)\in\mathcal{S}_*\}$, $\mathcal{S}_{*\mathrm{c}}:=\{i|\exists_u(u,i)\in\mathcal{S}_*\}$.

(ii) $\mathcal{S}_*=\{(R\circ i,R^{-1}\circ u)|(u,i)\in\mathcal{S}_*\}$,

(iii) $\mathcal{T}\in\mathrm{sd}\mathcal{S}_*$, where $\mathrm{sd}\mathcal{S}_*:=\{T|\exists_{u,i}(u,i)\in\mathcal{S}_*\wedge T=\mathrm{dom}u\}$.

(iv) $\forall_{(u,i)\in\mathcal{S}_*}\forall_{T\in\mathrm{Int}\,\mathcal{T}}\ T\subseteq\mathrm{dom}u\Rightarrow(u|T,i|T)\in\mathcal{S}_*$.

Typical examples of branch signal types are sets of ordered pairs of continously differentiable, piecewise continously differentiable functions, locally Riemann-integrable functions, etc.

Let $\mathcal{Z}$ be a finite set, $\mathcal{S}$ the universal signal set on $\mathcal{Z}$, and $\mathcal{S}_*$ a branch signal type. A set $\mathcal{S}'\subseteq\mathcal{S}$ is called the $\mathcal{S}_*$*-signal set* on $\mathcal{Z}$ if $\mathcal{S}'=\{(u,i)\in\mathcal{S}|\ \forall_{b\in\mathcal{Z}}(u_b,i_b)\in\mathcal{S}_*\}$.

Let $\mathcal{N}$ be a network whose constitutive relation is represented by means of an equating set $\mathcal{E}$ and a constitutive equation $f(u,i)=g(u,i)$ as a subset of some signal set $\mathcal{S}'$. This constitutive equation is denoted as a constitutive equation in *Belevitch form* if the following additional conditions are fulfilled:

(i) The equating set $\mathcal{E}$ is a subset of $\bigcup_{T\in\mathrm{Int}\,\mathcal{T}}(\mathbb{R}^m)^T$, where m is some natural number.

(ii) The function g is defined for all $(u,i)\in\mathcal{S}'$ by $g(u,i):=0_{\mathrm{dom}\,u}$, where $0_{\mathrm{dom}\,u}$ denotes the zero function of $(\mathbb{R}^m)^{\mathrm{dom}\,u}$.

The resulting constitutive equation $f(u,i)=0_{\mathrm{dom}\,u}$ is usually simply written as $f(u,i)=0$.

In the standard case, for *nondegenerated* networks, the value of m is equal to the number of branches of the network.

Constitutive equations in resistance, conductance, hybrid, or chain form are special cases of constitutive equations in Belevitch form.

The set $\mathcal{H}$ of all signals (u,i) of a network where u and i fulfills Kirchhoff's voltage or current law, respectively, is denoted as the *Kirchhoff part* of its universal signal set. The intersection $\mathcal{L}:=\mathcal{H}\cap\mathcal{V}$ of Kirchhoff part and constitutive relation is denoted as the *behavior* or the *solution set* of that network.

Obviously, both Kirchhoff's laws lead for each network $\mathcal{N}$ to a finite system of homogeneous linear algebraic equations. Therefore, if the constitutive relation of $\mathcal{N}$ is represented by means of some system of constitutive equations, the problem of the determination of the solution set of $\mathcal{N}$ can be reduced to the determination of the solution set of the resulting system of equations. Such a resulting system of equations is denoted as a system of *behavioral equations* of $\mathcal{N}$.

Traditionally, matrix calculus is used for the formulation of constitutive and behavioral equations of networks. Using a numbering of the branches of the network under consideration it is always possible to transfer the representation of the elements of the universal signal set of a network into the language of matrix calculus. Let $\mathcal{Z}$ be the branch set of this network, $z:=|\mathcal{Z}|$, and $\zeta:\{1,\ldots,z\}\to\mathcal{Z}$ a bijection. The definitions $\boldsymbol{u}^{\zeta}:=(u_{\zeta(1)},\ldots,u_{\zeta(z)})^{\mathsf{T}}$ and $\boldsymbol{i}^{\zeta}:=(i_{\zeta(1)},\ldots,i_{\zeta(z)})^{\mathsf{T}}$ assign to each signal (u,i) of the universal signal set on $\mathcal{Z}$ its *matrix representation* $(\boldsymbol{u}^{\zeta},\boldsymbol{i}^{\zeta})$ with respect to ζ.

Above we have introduced the notion of a projection of a subset of the universal signal set of a network generated by a subset of the branch set of this network. The following definition makes an interesting use of this concept.

Let $\mathcal{N}$ and $\tilde{\mathcal{N}}$ be networks with branch sets $\mathcal{Z}$ and $\tilde{\mathcal{Z}}$, respectively, and solution sets $\mathcal{L}$ and $\tilde{\mathcal{L}}$, respectively, where $\mathcal{Z}\subset\tilde{\mathcal{Z}}$. We say that $\tilde{\mathcal{N}}$ *generates the solution set of* $\mathcal{N}$ if $\mathcal{L}=\tilde{\mathcal{L}}_{\mathcal{Z}}$.

A *multipole* is a network with a distingushed subset of its node set whose elements are used as accessible *terminals* for the interconnection with other networks.

A *multiport* is a multipole whose terminal set is partitioned into *terminal pairs*.

Obviously, each n-port can be identified with a $2n$-pole.

A network with n nodes is called an *elementary n-pole* if its graphs are forests whose branches connect these nodes. A network with $2n$ nodes and n branches is called an *elementary n-port* if its graphs are forests with one-branch components only.

On this basis it is possible to introduce special kinds of networks such as *independent voltage* and *current sources*, *open* and *short circuits*, *nullators* and *norators*, *resistors*, *controlled sources*, *gyrators*, *inductors*, *coupled inductors*, *capacitors*, etc.

Elementary n-poles deliver in some sense minimal models of technical devices with n terminals (Reibiger (2003a); Koenig et al. (1967), too). But it should be noted that in the general case it is only possible to construct a minimal model for some real technical device if networks with sufficiently general constitutive relations are admitted. Otherwise one has to try to synthesize an approximate realization by means of an interconnection of networks with more restricted constitutive relations.

Let $(\mathcal{N}^l)_{l\in L}$ be a family of networks $\mathcal{N}^l=(\mathcal{C}^l,\mathcal{V}^l)$ with branch sets $\mathcal{Z}^l$ and node sets $\mathcal{K}^l$. This family of networks is denoted as *interconnectable* if $\forall_{k,l\in L,k\neq l}(\mathcal{Z}^k\cap\mathcal{Z}^l=\emptyset\wedge\mathcal{K}^k\cap\mathcal{Z}^l=\emptyset\wedge\mathrm{sd}\mathcal{V}^k=\mathrm{sd}\mathcal{V}^l)$.

An *interconnection* of such a family of networks can be determined with the help of an equivalence relation $\simeq_{\mathrm{equ}}$ on the union $\bar{\mathcal{K}}:=\bigcup_{l\in L}\mathcal{K}^l$ of their node sets. Its skeleton is then defined by an identifiction of the elements of each of the equivalence classes $\mathcal{K}'\in\bar{\mathcal{K}}/\simeq_{\mathrm{equ}}$ and its constitutive relation is equal to $\{(u,i)\in\mathcal{S}|\,\forall_{l\in L}\,(u_{\mathcal{Z}^l},i_{\mathcal{Z}^l})\in\mathcal{V}^l\}$ where $\mathcal{S}$ denotes the universal signal sets on $\bigcup_{l\in L}\mathcal{Z}^l$. For details see Reibiger (2003b).

Sometimes it is appropriate to use a *formal interconnection* of a family of interconnectable networks. It is defined by means of the identical relation on the union of their node sets.

Obviously, each network can be represented both as an interconnection of elementary multipols and as an interconnection of elementary multiports, too.

Elementary multipoles and multiports are usually denoted as network elements in the literature. But in our formalization of network theory they are itself networks (obviously, networks with very simple skeletons) and they can be subnetworks of other, larger networks. Therefore we avoid to denote them as *elements*.

The notion of the terminal behavior of a network is of central importance as well for the application of network theory as for the further development of this theory, too. Indeed, a lot of theorems of network theory deliver propositions about the terminal behavior of some networks.

Let $\mathcal{N}$ be a given network with a prescribed subset of its node set whose elements are denoted as terminals. The terminal behavior of $\mathcal{N}$ has to characterize the behavior of this network with respect to all interconnections of its terminals with arbitrary admissible external networks $\tilde{\mathcal{N}}$ in such a manner that after an exchange of the given network $\mathcal{N}$ by another network $\hat{\mathcal{N}}$ with the same terminal behavior this exchange can not be observed in all the external networks $\tilde{\mathcal{N}}$ of these interconnections.

For an introduction let us first assume that $\mathcal{N}$ is a network with at least two nodes where exactly two of its nodes are distingushed as *terminals* at which $\mathcal{N}$ can be interconnected with external networks $\tilde{\mathcal{N}}$. The *terminal behavior* of $\mathcal{N}$ can then be defined as the set of all ordered pairs of terminal pair voltages and terminal currents corresponding to the solution sets of all interconnections of the given network $\mathcal{N}$ with arbitrary external networks $\tilde{\mathcal{N}}$. Obviously, each such ordered pair of terminal pair voltages and terminal currents is also compatible with the interconnection of $\mathcal{N}$ with a norator. Because this norator is an element of the set of all admissible interconnections of $\mathcal{N}$ with external networks, it suffices to consider for the determination of the terminal behavior of $\mathcal{N}$ merely its interconnection with only one norator. The terminal behavior of $\mathcal{N}$ is then essentially equal to the projection of the solution set of this interconnection generated by the branch set of the external norator.

By means of this idea we have given in Reibiger (1997) a geometrical proof of the Thévenin-Norton theorem.

These considerations can be immediately generalized to the case of the determination of the terminal behavior of networks which are connected to external networks at more than two terminals. As it is shown in Reibiger (1985, 1986, 2003a); Haase (1983) it suffices in this case to analyse the interconnection of a given network with a tree (and in more general cases with a forest) of norators connecting the terminals of the given network. These ideas are also applied in Clauß et al. (1995); Enge-Rosenblatt et al. (2007).

3 Central results

In this section we consider the relationship between four networks denoted as $\mathcal{N}, \mathcal{N}^{\mathrm{aug}}, \bar{\mathcal{N}}$, and $\bar{\mathcal{N}}^{\mathrm{aug}}$.

For this purpose we assume that there are given a branch signal type $\mathcal{S}_*$ and four finite pairwise disjoint nonvoid sets $\mathcal{Z}_1, \mathcal{Z}_2, \mathcal{Z}_3$, and $\mathcal{K}$ with $|\mathcal{Z}_2|=|\mathcal{Z}_3|$ and $|\mathcal{K}|=2|\mathcal{Z}_1\cup\mathcal{Z}_2|$. The $\mathcal{S}_*$-signal sets on $\mathcal{Z}:=\mathcal{Z}_1\cup\mathcal{Z}_2$ and $\mathcal{Z}^{\mathrm{aug}}:=\mathcal{Z}_1\cup\mathcal{Z}_2\cup\mathcal{Z}_3$ are denoted with $\mathcal{S}$ and $\mathcal{S}^{\mathrm{aug}}$, respectively.

The network $\mathcal{N}$ is defined as an elementary multiport with branch set $\mathcal{Z}$, node set $\mathcal{K}$, and a constitutive relation which can be represented by means of the $\mathcal{S}_*$-signal set $\mathcal{S}$ as the set of all signals $(u, i)\in\mathcal{S}$ obeying the semi-implicit constitutive equations

$$i_{\mathcal{Z}_1} = g(u_{\mathcal{Z}_1}, u_{\mathcal{Z}_2}, i_{\mathcal{Z}_2}), \tag{2}$$
$$0 = h(u_{\mathcal{Z}_1}, u_{\mathcal{Z}_2}, i_{\mathcal{Z}_2}), \tag{3}$$

where

$$g:\{(u_{\mathcal{Z}_1}, u_{\mathcal{Z}_2}, i_{\mathcal{Z}_2})\,|\,(u, i)\in\mathcal{S}\}\to\{i_{\mathcal{Z}_1}\,|\,(u, i)\in\mathcal{S}\}$$

and

$$h:\{(u_{\mathcal{Z}_1}, u_{\mathcal{Z}_2}, i_{\mathcal{Z}_2})|(u, i)\in\mathcal{S}\}\to\{i_{\mathcal{Z}_3}|(u, i)\in\mathcal{S}^{\mathrm{aug}}\}.$$

The network $\mathcal{N}^{\mathrm{aug}}$ is defined as a meshless multipole with branch set $\mathcal{Z}^{\mathrm{aug}}$ and terminal set $\mathcal{K}$ satisfying the following conditions: The voltage and current graph of $\mathcal{N}^{\mathrm{aug}}$ consist of at least $|\mathcal{Z}|$ components and and each of these components includes at most one branch of $\mathcal{Z}$. The subgraphs of these graphs generated by $\mathcal{Z}$ are equal to the corresponding graphs of $\mathcal{N}$. Its constitutive relation can be represented by means of the $\mathcal{S}_*$-signal set $\mathcal{S}^{\mathrm{aug}}$ together with a linear bijection $G:\mathcal{U}^{\mathcal{Z}_3}\to\mathcal{I}^{\mathcal{Z}_2}$ as the set of all signals $(u, i)\in\mathcal{S}^{\mathrm{aug}}$ obeying the following constitutive equations in conductance form

$$i_{\mathcal{Z}_1} = g(u_{\mathcal{Z}_1}, u_{\mathcal{Z}_2}, G\circ u_{\mathcal{Z}_3}), \tag{4}$$
$$i_{\mathcal{Z}_2} = G\circ u_{\mathcal{Z}_3}, \tag{5}$$
$$i_{\mathcal{Z}_3} = h(u_{\mathcal{Z}_1}, u_{\mathcal{Z}_2}, G\circ u_{\mathcal{Z}_3}). \tag{6}$$

Theorem 1: The networks $\mathcal{N}$ and $\mathcal{N}^{\mathrm{aug}}$ have the same terminal behavior with respect to $\mathcal{K}$.

PROOF: For a proof of this theorem we connect $\mathcal{N}$ and $\mathcal{N}^{\mathrm{aug}}$ with a norator tree consisting of $|\mathcal{K}|-1$ branches such that in the resulting network exactly one branch of this norator network is parallel to a branch of $\mathcal{Z}$ and all these parallel connections are pairwise connected by means of one of the remaining norators.

Then both of these interconnections are analyzed. It is easy to see that the projections of the solution sets of these interconnections generated by the branch set of the norator-tree are equal. That means, $\mathcal{N}$ and $\mathcal{N}^{\mathrm{aug}}$ have the same terminal behavior. □

Remark 1: Theorem 1 and both of the subsequent theorems, too, can be generalized to the case $\mathcal{Z}_1=\emptyset$. Then the Eqs. (2) and (4) are to delete and in the remaining equations the term $u_{\mathcal{Z}_1}$ must be omitted. □

Remark 2: Let $(\mathcal{N}^l)_{l\in L}$ be a family of interconnectable elementary multiports whose constitutive relations satisfy the assumptions of Theorem 1. Then the formal interconnection of these multiports is an elementary multiport, too, fulfilling these assumptions likewise. □

Let $\simeq_{\mathrm{equ}}$ denote an equivalence relation defined on the terminal set of $\mathcal{N}$ and $\mathcal{N}^{\mathrm{aug}}$. The networks $\bar{\mathcal{N}}$ and $\bar{\mathcal{N}}^{\mathrm{aug}}$ are then defined by an identification of the elements of each of the equivalence classes $\mathcal{K}'\in\mathcal{K}/\simeq_{\mathrm{equ}}$.

Theorem 2: The network $\bar{\mathcal{N}}^{\mathrm{aug}}$ generates the solution set of $\bar{\mathcal{N}}$. □

Our proof of this theorem is based on the following lemma. Its proof follows directly from the fact that the branches of $\mathcal{Z}_3$ are not included in any mesh of $\bar{\mathcal{N}}^{\mathrm{aug}}$. For the formulation of this lemma we need some additional notations.

We set $z_1{:=}|\mathcal{Z}_1|$, $z_2{:=}|\mathcal{Z}_2|$, $z_3{:=}|\mathcal{Z}_3|$, $z{:=}|\mathcal{Z}|{=}z_1{+}z_2$. Nullity and rank of the graphs of $\bar{\mathcal{N}}$ are denoted by n and r, respectively. M denotes a reduced mesh-branch incidence matrix of the voltage graph of $\bar{\mathcal{N}}$ and K a reduced node-branch incidence matrix of the current graph of $\bar{\mathcal{N}}$. By definition, M is an $n{\times}z$ and K an $r{\times}z$ matrix. The assignment of the columns of these matrices to the branch set of $\bar{\mathcal{N}}$ is given by a bijective numbering $\zeta{:}\{1,\dots,z\}{\to}\mathcal{Z}$ satisfying $\zeta(\{1,\dots,z_1\}){=}\mathcal{Z}_1$ and $\zeta(\{z_1{+}1,\dots,z\}){=}\mathcal{Z}_2$.

Lemma: Suppose $(M_1 \quad M_2)$ denote the natural partition of the matrix M corresponding to the partition $(\mathcal{Z}_1,\ \mathcal{Z}_2)$ of $\mathcal{Z}$, then there exists an $n{\times}z_3$ zero matrix 0 such that

$$(M_1 \quad M_2 \quad 0)$$

is a reduced mesh-branch incidence matrix of the voltage graph of $\bar{\mathcal{N}}^{\text{aug}}$.

Analogously, if $(K_1 \quad K_2)$ denotes the natural partion of K corresponding to the partition $(\mathcal{Z}_1,\ \mathcal{Z}_2)$ of $\mathcal{Z}$, then there exist appropriate zero matrices and a regular $z_3{\times}z_3$ matrix K_3 such that

$$\begin{pmatrix} K_1 & K_2 & 0 \\ 0 & 0 & K_3 \end{pmatrix}$$

is a conformably partitioned reduced cut-branch incidence matrix of the current graph of $\mathcal{N}^{\text{aug}}$. □

Proof of Theorem 2: For simplicity we identify in the following the partial voltages and currents of each signal (u,i) of the universal signal set of $\mathcal{N}^{\text{aug}}$ corresponding to the branch sets $\mathcal{Z}_1$, $\mathcal{Z}_2$, and $\mathcal{Z}_3$ with column matrices of length z_1, z_2, and z_3, resp.

Obviously, the constitutive Eqs. (4), (5), and (6) deliver together with

$$\begin{pmatrix} M_1 & M_2 & 0 \end{pmatrix} \begin{pmatrix} u_{\mathcal{Z}_1} \\ u_{\mathcal{Z}_2} \\ u_{\mathcal{Z}_3} \end{pmatrix} = 0 \qquad (7)$$

and

$$\begin{pmatrix} K_1 & K_2 & 0 \\ 0 & 0 & K_3 \end{pmatrix} \begin{pmatrix} i_{\mathcal{Z}_1} \\ i_{\mathcal{Z}_2} \\ i_{\mathcal{Z}_3} \end{pmatrix} = 0 \qquad (8)$$

a system of behavioral equations for $\bar{\mathcal{N}}^{\text{aug}}$. Let (u,i) be a solution of $\bar{\mathcal{N}}^{\text{aug}}$, i.e., a solution of the system of the Eqs. (4) to (8). Then the Eqs. (7) and (8) imply

$$\begin{pmatrix} M_1 & M_2 \end{pmatrix} \begin{pmatrix} u_{\mathcal{Z}_1} \\ u_{\mathcal{Z}_2} \end{pmatrix} = 0 \qquad (9)$$

and

$$\begin{pmatrix} K_1 & K_2 \end{pmatrix} \begin{pmatrix} i_{\mathcal{Z}_1} \\ i_{\mathcal{Z}_2} \end{pmatrix} = 0\,. \qquad (10)$$

Since K_3 is regular, it follows from Eq. (8) that

$$i_{\mathcal{Z}_3} = 0$$

and therefore, together with Eqs. (5) and (6), that $u_{\mathcal{Z}_1}, u_{\mathcal{Z}_2}$, $i_{\mathcal{Z}_1}$, and $i_{\mathcal{Z}_2}$ satisfy the Eqs. (2) and (3). But the Eqs. (2), (3), (9) and (10) are a system of behavioral equations of for $\bar{\mathcal{N}}$. That means, $(u_{\mathcal{Z}}, i_{\mathcal{Z}})$ is a solution of $\bar{\mathcal{N}}$. □

Let $\mathcal{K}_{\text{dat.}}$ be an arbitrary chosen equivalence class of the quotient set $\mathcal{K}/{\simeq_{\text{equ}}}$ and let $n_{\text{dat.}}$ be that node of $\bar{\mathcal{N}}$ and $\bar{\mathcal{N}}^{\text{aug}}$ corresponding to this equivalence class.

Theorem 3: If the branches of $\mathcal{Z}_3$ generate starlike subtrees of the voltage and current graph of $\bar{\mathcal{N}}^{\text{aug}}$ with $n_{\text{dat.}}$ as center node, then the system of nodal voltage equations for $\bar{\mathcal{N}}^{\text{aug}}$ with respect to $n_{\text{dat.}}$ as datum node is equivalent to a system of modified nodal voltages for $\bar{\mathcal{N}}$.

Proof: Our proof consists of two parts.

Part 1: Analysis of $\bar{\mathcal{N}}$ by means of a system of u-i-$u^{\emptyset}$ equations:

$$\begin{pmatrix} u_{\mathcal{Z}_1} \\ u_{\mathcal{Z}_2} \end{pmatrix} = \begin{pmatrix} K_1^{\mathsf{T}} \\ K_2^{\mathsf{T}} \end{pmatrix} u^{\emptyset}\,,$$

$$\begin{aligned} i_{\mathcal{Z}_1} &= g(u_{\mathcal{Z}_1}, u_{\mathcal{Z}_2}, i_{\mathcal{Z}_2})\,, \\ 0 &= h(u_{\mathcal{Z}_1}, u_{\mathcal{Z}_2}, i_{\mathcal{Z}_2})\,, \end{aligned}$$

$$\begin{pmatrix} K_1 & K_2 \end{pmatrix} \begin{pmatrix} i_{\mathcal{Z}_1} \\ i_{\mathcal{Z}_2} \end{pmatrix} = 0\,.$$

Elimination of $u_{\mathcal{Z}_1}, u_{\mathcal{Z}_2}$, and $i_{\mathcal{Z}_1}$ delivers a system of modified nodal voltage equations for $\bar{\mathcal{N}}$:

$$K_1\, g(K_1^{\mathsf{T}} u^{\emptyset},\, K_2^{\mathsf{T}} u^{\emptyset},\, i_{\mathcal{Z}_2}) + K_2\, i_{\mathcal{Z}_2} = 0\,, \qquad (11)$$

$$h(K_1^{\mathsf{T}} u^{\emptyset},\, K_2^{\mathsf{T}} u^{\emptyset},\, i_{\mathcal{Z}_2}) = 0\,. \qquad (12)$$

Part 2: Analysis of $\bar{\mathcal{N}}^{\text{aug}}$ by means of nodal voltage equations:

Under the assumptions of Theorem 3 the branches of $\mathcal{Z}_3$ generate starlike subtrees with the datum node as center node. Therefore the submatrix K_3 of the node-branch incidence matrix of $\bar{\mathcal{N}}^{\text{aug}}$ is equal to a $z_3{\times}z_3$ unit matrix E_3 and it follows

$$\begin{pmatrix} u_{\mathcal{Z}_1} \\ u_{\mathcal{Z}_2} \\ u_{\mathcal{Z}_3} \end{pmatrix} = \begin{pmatrix} K_1^{\mathsf{T}} & 0 \\ K_2^{\mathsf{T}} & 0 \\ 0 & E_3 \end{pmatrix} \begin{pmatrix} u_1^{\emptyset} \\ u_2^{\emptyset} \end{pmatrix},$$

$$\begin{aligned} i_{\mathcal{Z}_1} &= g(u_{\mathcal{Z}_1}, u_{\mathcal{Z}_2}, i_{\mathcal{Z}_2})\,, \\ i_{\mathcal{Z}_2} &= G \circ u_{\mathcal{Z}_3}\,, \\ i_{\mathcal{Z}_3} &= h(u_{\mathcal{Z}_1}, u_{\mathcal{Z}_2}, i_{\mathcal{Z}_2})\,, \end{aligned}$$

$$\begin{pmatrix} K_1 & K_2 & 0 \\ 0 & 0 & E_3 \end{pmatrix} \begin{pmatrix} i_{\mathcal{Z}_1} \\ i_{\mathcal{Z}_2} \\ i_{\mathcal{Z}_3} \end{pmatrix} = 0\,.$$

Obviously, $i_{\mathcal{Z}_3}=0$. Elimination of $i_{\mathcal{Z}_2}$ and $u_{\mathcal{Z}_3}$ yields

$$K_1 g(K_1^{\mathsf{T}} u_1^{\emptyset}, K_2^{\mathsf{T}} u_2^{\emptyset}, G \circ u_2^{\emptyset}) + K_2 G \circ u_2^{\emptyset} = 0, \quad (13)$$

$$h(K_1^{\mathsf{T}} u_1^{\emptyset}, K_2^{\mathsf{T}} u_2^{\emptyset}, G \circ u_2^{\emptyset}) = 0. \quad (14)$$

It is easy to see that the assignment $(u^{\emptyset}, i_{\mathcal{Z}_2}) \mapsto (u_1^{\emptyset}, u_2^{\emptyset})$ defined by $u_1^{\emptyset}:=u^{\emptyset}$ and $u_2^{\emptyset}:=G^{-1} \circ i_{\mathcal{Z}_2}$ delivers a bijection between the solution sets of the equation systems Eqs. (11) and (12), and Eqs. (13) and (14). □

The modified nodal voltage Eqs. (11) and (12) are set up here under the assumption that the constitutive equations of $\bar{\mathcal{N}}$ are the semi-implicit Eqs. (2) and (3). However, in the literature it is for this purpose usually assumed that the constitutive equations are given in hybrid form. Observe, the constitutive equations $u_1=0$, $i_1=0$ of a nullor are not in hybrid form.

Examples

(1°) *Nullor*: A nullor is an elementary two-port. If $\mathcal{Z}:=\{1,2\}$ denotes its branch set, then its constitutive relation can be characterized by the equations

$$u_1=0, \quad i_1=0$$

or equivalently

$$i_1=0, \quad 0=G \circ u_1 ,$$

where $G:\mathcal{U}\to\mathcal{I}$ linear bijection. Therefore the associated network $\mathcal{N}^{\text{aug}}$ can defined by

$$\mathcal{Z}_1=\{1\}, \quad \mathcal{Z}_2=\{2\}, \quad \mathcal{Z}_3=\{3\},$$

$$i_1=0, \quad i_2=G\circ u_3, \quad i_3=G\circ u_1 .$$

(2°) *Independent Voltage Source*: An independent voltage source is an elementary one-port. If $\mathcal{Z}:=\{1\}$ denotes its branch set, then its constitutive relation can be together with a prescribed voltage $u_1^{\mathrm{e}} \in \mathcal{S}_*$ characterized by the equation

$$u_1=u_1^{\mathrm{e}}|\mathrm{dom}\, u_1,$$

or equivalently

$$0 = G\circ(u_1-u_1^{\mathrm{e}}|\mathrm{dom}\, u_1) ,$$

where $G:\mathcal{U}\to\mathcal{I}$ linear bijection. Therefore the associated network $\mathcal{N}^{\text{aug}}$ can be defined by

$$\mathcal{Z}_1=\emptyset, \quad \mathcal{Z}_2=\{1\}, \quad \mathcal{Z}_3=\{2\},$$

$$i_1=G\circ u_2 , \quad i_2=G\circ(u_1-u_1^{\mathrm{e}}|\mathrm{dom}\, u_1) .$$

(3°) *Current Controlled Voltage Source*: A current controlled voltage source is an elementary two-port. If $\mathcal{Z}:=\{1,2\}$ denotes its branch set, then its constitutive relation can be characterized by the equations

$$u_1=0, \quad u_2=R_{\mathrm{trf}}\circ i_1$$

or equivalently

$$0=G\circ u_1, \quad 0=G\circ u_2-G\circ R_{\mathrm{trf}}\circ i_1 ,$$

where $R_{\mathrm{trf}}:\mathcal{I}\to\mathcal{U}$ is an arbitrary map and $G:\mathcal{U}\to\mathcal{I}$ is a linear bijection. Therefore the associated network $\mathcal{N}^{\text{aug}}$ can be defined by

$$\mathcal{Z}_1=\emptyset, \quad \mathcal{Z}_2=\{1,2\}, \quad \mathcal{Z}_3=\{3,4\},$$

$$i_1=G\circ u_3 , \quad i_2=G\circ u_4 , \quad i_3=G\circ u_1$$

$$i_4=G\circ u_2-G\circ R_{\mathrm{trf}}\circ i_1 .$$

If R_{trf} is a bijection, too, then it is possible to start from the equations

$$i_1=R_{\mathrm{trf}}^{-1}\circ u_2 , \quad 0=G\circ u_1 .$$

The associated network $\mathcal{N}^{\text{aug}}$ can now be defined by

$$\mathcal{Z}_1=\{1\}, \quad \mathcal{Z}_2=\{2\}, \quad \mathcal{Z}_3=\{3\},$$

$$i_1=R_{\mathrm{trf}}^{-1}\circ u_2 , \quad i_2=G\circ u_3 , \quad i_3=G\circ u_1 .$$

4 Concluding remarks

On the basis of some recent results on the foundations of network theory we have presented an algorithm for the construction of a multipole having the same all-pole terminal behavior as an given elementary multiport whose constitutive relation is defined by means of a system of semi-implicit constitutive equations such that the constitutive relation of the modified multipole can be represented by means of constitutive equations in conductance form.

A row of other applications of the auxiliary branch method to circumvent restrictions of the input languages of circuit simulators can you find in Haase (1983); Clauß et al. (1995); Reibiger and Elst (1983).

References

Clauß, C., Haase, J., Kurth, G., and Schwarz, P.: Extended Admittance Description of Nonlinear n-Poles, AEU-Arch. Elektron. Ub., 49, 91–97, 1995.

Enge-Rosenblatt, O., Haase, J., and Clauß, C.: Important Characteristics of VHDL-AMS and Modelica with Respect to Model Exchange, EOOLT Workshop, Berlin, Juli 2007, 2007.

Engelhardt, R. and Heinz, M.: Behandlung idealer gesteuerter Quellen bei der Netzwerkanalyse mit Digitalrechnern, Nachrichtentech. Z., 23, 8–10, 1970.

Haase, J.: Verfahren zur Beschreibung und Berechnung des Klemmenverhaltens resistiver Netzwerke, Diss. TU Dresden, 1983.

Klein, W.: Der Übertrager als Zweipolnetz, AEU-Arch. Elektron. Ub., 12, 133–137, 1958.

Koenig, H. E., Tokad, Y., Kesavan, H. K., and Hedges, H. G.: Analysis of Discrete Physical Systems, McGraw-Hill, 1967.

Kremer, H.: Numerische Berechnung linearer Netzwerke und Systeme, Springer-Verlag, 1978.

Poldermann, J. W. and Willems, J. C.: Introduction to Mathematical Systems Theory – A Behavioral Approach, Springer, 1998.

Reibiger, A.: On the Terminal Behaviour of Networks, Proc. ECCTD'85, Prague, 224–227, 1985.

Reibiger, A.: Über das Klemmenverhalten von Netzwerken, Wiss. Z. TU Dresden, 35, 165–173, 1986.

Reibiger, A.: Geometrical Proof of the Thévenin-Norton Theorem", Proc. ECCTD'97, Budapest, 1, 24–28, 1997.

Reibiger, A.: Terminal Behaviour of Networks, Multipoles and Multiports, Proc. 4th MATHMOD, Vienna, (CD), 2003a.

Reibiger, A.: Networks, Decomposition and Interconnection of Networks, Proc. ISTET'03, Warsaw, 245–248, 2003b.

Reibiger, A. and Elst, G.: Network Theoretic Interpretation of Multibody Systems, Acta Polytechnica Práce ČVUT v Praze, 18, (III,4), Práce, 109–129, 1983.

Reibiger, A. and Loose, H.: Bond Graphs and Matroids (cf. also Proc. 3rd MATHMOD, Vienna, 2000), Math. Comput. Simulat., 53, 323–332, 2000.

Reibiger, A., Loose, H., and Nähring, T.: Multidimensional Networks, Proc. X. Int. Symp. Theor. Electrical Engineering, ISTET'99, Magdeburg, 293–298, 1999.

Reibiger, A., Mathis, W., Nähring, T., and Trajkovi'c, Lj.: Mathematical Foundations of the TC-Method for Computing Multiple DC-Operating Points, Int. J. Appl. Electrom., 17, 169–191, 2003.

Reinschke, K. and Schwarz, P.: Verfahren zur rechnergestützten Analyse linearer Netzwerke, Akademie-Verlag, 1976.

Reinschke, K. and Schwarz, P.: Computer-Aided Analysis of Linear Networks with Ideal Elements, Proc. SSCT 77, Kladno, 506–511, 1977.

Thielmann, H.: Analysis of Linear Active (n+1)-Poles by Simple Transformations of the Nodal Admittance Matrix, IEEE Trans. Circ. Theory CT-20, 575–577, 1973.

Willems, J. C.: Paradigms and Puzzles in the Theory of Dynamical Systems, IEEE T. Automat. Contr., AC-36, 259–294, 1991.

25

An energy efficient weakly programmable MIMO detector architecture

C. Gimmler-Dumont and N. Wehn

University of Kaiserslautern, Germany

Correspondence to: C. Gimmler-Dumont (gimmler@eit.uni-kl.de)

Abstract. Energy efficient processing is mandatory in todays' mobile devices. For the upcoming multiple-antenna systems, algorithmic flexibility enables the dynamic reaction to changing channel conditions. We show that most of the tree search based MIMO detection algorithms are based on the same algorithmic kernels and present a weakly-programmable architecture based on these observations. In this way, the detection algorithm can be chosen and parameterized during runtime according to the current channel conditions and QoS requirements leading to a highly energy efficient implementation. The architecture has been implemented and synthesized on a 65 nm technology, resulting in an area of 0.26 mm^2 and a power consumption of only 15 mW.

1 Introduction

Multiple-antenna or MIMO systems have the potential to increase the data rate of wireless communication systems and, thus, they have been adopted by recent communication standards like WiMax, WiFi, or LTE. The best communications performance is achieved for an iterative processing between MIMO detector and channel decoder. As for channel codes, the encoding on the transmitter side is simple but the detection complexity of the MIMO signals at the receiver increases exponentially with the number of antennas.

There exist a wide range of sub-optimal algorithms which allow trade-offs of communications performance versus implementation complexity and energy consumption. Linear detection methods, e.g. successive interference cancellation (SIC) introduced by Foschini et al. (1999), have the lowest complexity but also the lowest communications performance. Optimal soft-input soft-output sphere detection is the most demanding detection algorithm but it offers the best communications performance and the ability for iterative detection and decoding (Vikalo et al. (2004)). A fixed target error rate can be achieved with different detection algorithms which cover a signal-to-noise ratio (SNR) range of more than 10 dB.

In current standards, like LTE, the quality of service (QoS) is dynamically adjusted at runtime, e.g. higher data throughput rates are specified for higher SNR. This is due to the fact that for higher SNR, the required communications performance can be achieved by lower complexity algorithms, which enables higher throughput. Energy consumption is one of the most critical implementation metrics for mobile receivers. Thus, complexity adaptive processing is key for enabling energy efficient architectures (Yu et al. (2012)). Therefore, flexible energy efficient MIMO detection architectures are mandatory to fulfill the requirements of modern communication standards like LTE.

Visionary implementation styles are based on libraries of flexible algorithmic kernels, e.g. the Nucleus concept proposed by Ramakrishnan et al. (2009). These kernels offer core functionalities which can be reused in different algorithms or scenarios. However, defining the best set of kernels is one of the major challenges of this approach. Extracting the common parts of different algorithms allows a high degree of shared and highly optimized components and, thus, a flexible implementation with a low overhead. Weakly programmable architectures based on such kernels can lead to highly energy efficient hardware implementations.

However, in the literature, mainly highly optimized architectures are presented which perform exactly one algorithm, e.g. Borlenghi et al. (2011); Studer et al. (2011). There exist only a few processor architectures which are able to perform different algorithms. In Jafri et al. (2009), the processor architecture is based on very small-grained operations, i.e. complex number operations. In Chen et al. (2012), a processor architecture based on matrix operations is presented which performs maximum ratio combining, linear

MMSE detection, and MMSE-SIC detection. However, none of the processor architectures supports tree search algorithms which offer the best communications performance. Thus, they only cover a limited range of the possible channel scenarios.

Almost all detection algorithms can be mapped to a search in a tree structure. We will show that all these algorithms can be constructed by only five coarse-grained algorithmic kernels: an enumeration unit which decides which child node is visited next, a computation unit for the interference reduction, a computation unit for the Euclidean distance, administration of nodes, and administration of computed metrics. We present an architecture based on these kernels which is able to perform most of the existing algorithms. The complexity of these algorithms covers all classes from linear up to almost exponential complexity, e.g. linear SIC, fixed effort detection (Wong et al. (2002); Barbero and Thompson (2008)), or sphere detection. As all algorithms are performed with the same algorithmic kernels, the overhead for flexibility is negligible.

In contrast to the existing approaches, we present the first weakly programmable MIMO detector architecture which offers just the necessary flexibility. In this way, the detection algorithm can be chosen and parameterized during runtime according to the current channel conditions and QoS requirements leading to a highly energy efficient implementation. The architecture has been implemented and synthesized on a 65 nm technology, resulting in an area of 0.26 mm^2, a power consumption of 15 mW and throughputs between 35 Mbit/s up to 720 Mbit/s.

In the following sections, we will introduce the employed system model and the basics of tree search based MIMO detection. In Sect. 4, we will derive the algorithmic kernels which are shared by most MIMO detection algorithms. Implementation results of the flexible detection architecture and the trade-off between throughput and communications performance resulting from this flexibility are presented in Sect. 5. Sect. 6 concludes the paper.

2 System Model

In this paper, we focus on a bit interleaved coded modulation (BICM) scheme like shown in Fig. 1. The source generates a random information word $\boldsymbol{u}$ of length K_c which is encoded by the channel encoder. The interleaved codeword $\boldsymbol{X}^N$ is mapped directly to complex symbols s chosen from a 2^Q-ary QAM modulation scheme. M_T symbols are combined in one transmission vector $\boldsymbol{s}_t$. The whole modulated sequence is represented by

$$\boldsymbol{S}^T = (\boldsymbol{s}_1, \boldsymbol{s}_2, \ldots, \boldsymbol{s}_t, \ldots, \boldsymbol{s}_T) \tag{1}$$

T time slots are needed to transmit all symbols of one codeword. The transmission of vector $\boldsymbol{s}_t$ in time step t is modeled

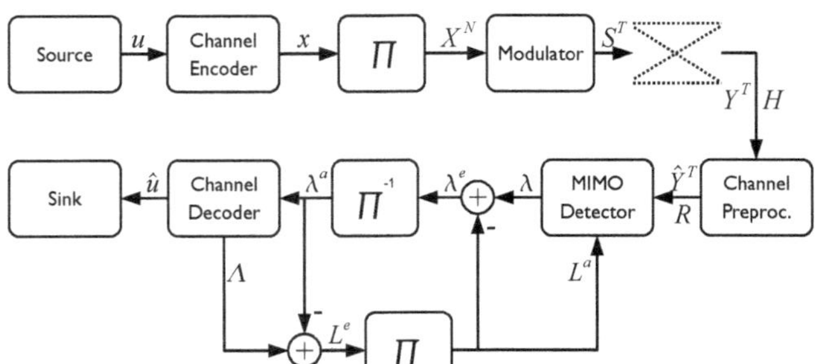

Fig. 1. MIMO-BICM system with feedback loop between MIMO Detector and Channel Decoder

by

$$\boldsymbol{y}_t = \boldsymbol{H}_t \cdot \boldsymbol{s}_t + \boldsymbol{n}_t \tag{2}$$

with $\boldsymbol{H}_t$ the channel matrix of dimension $M_T \times M_R$ and $\boldsymbol{n}_t$ the noise vector of dimension M_R whose entries are zero-mean and unit variance Gaussian variables. The elements of $\boldsymbol{H}_t$ are modeled as independent, complex, zero-mean, Gaussian random variables. Real and imaginary part are independent variables each with variance $\sigma^2 = N_0/2$. All received vectors $\boldsymbol{y}_t$ are gathered in the matrix $\boldsymbol{Y}^T$

$$\boldsymbol{Y}^T = (\boldsymbol{y}_1, \boldsymbol{y}_2, \ldots, \boldsymbol{y}_t, \ldots, \boldsymbol{y}_T) \tag{3}$$

Before the decoding starts, the channel preprocessing applies a QR decomposition on $\boldsymbol{Y}^T$ and $\boldsymbol{H}_t$. This results in the transformed received vectors $\hat{\boldsymbol{Y}}^T$ and updated channel matrices $\boldsymbol{R}_t$. The decoding process iterates over the MIMO detector and the channel decoder, which exchange probability information about the codeword. The soft-in-soft-out MIMO detector determines the likelihood of the bits for each received vector $\boldsymbol{y}_t$ using the a-priori information $\boldsymbol{L}^a$ from the channel decoder. Only the extrinsic information $\boldsymbol{\lambda}^e = \boldsymbol{\lambda} - \boldsymbol{L}^a$ is passed on to the channel decoder.

The channel decoder processes the whole codeword at a time. It uses the interleaved a-priori information $\boldsymbol{\lambda}^a$ from the MIMO detector for the calculation of the estimated information bit sequence $\hat{\boldsymbol{u}}$ and the a-posteriori logarithmic likelihood ratios (LLRs) $\boldsymbol{\Lambda}$ of the codeword. The extrinsic information $\boldsymbol{L}^e = \boldsymbol{\Lambda} - \boldsymbol{\lambda}^a$ is returned to the MIMO detector thus closing the iterative loop.

3 MIMO Detection

A received symbol vector $\boldsymbol{y}_t$ can be seen as a weighted superposition of the entries of $\boldsymbol{s}_t$, disturbed by Gaussian noise. The task of the MIMO detector is the equalization and separation of the originally sent sequence of symbols $\boldsymbol{s}_t$. It works on one received vector $\boldsymbol{y}_t$ at a time. For all detection related explanations, the time indexes of $\boldsymbol{y}$, $\boldsymbol{H}$, etc. are dropped for ease of notation. Even if not mentioned specifically for each equation, the vectors $\boldsymbol{s}$ and $\boldsymbol{x}$ are always the complex representation and the bit representation, respectively, of the same symbol vector. $x_{q,m}$ denotes the qth bit of the mth symbol in $\boldsymbol{s}$.

There exist a wide range of suboptimal algorithms for MIMO detection, which allow trade-offs of communications performance versus implementation complexity and energy consumption. In this paper, we restrict ourselves to the most common detection algorithms which are based on a tree search. It can be shown that the likeliness of having sent a certain vector $\boldsymbol{s}$ is related to the metric $d(\boldsymbol{s})$ (Vikalo et al. (2004)).

$$d(\boldsymbol{s}) = \|\boldsymbol{y} - \boldsymbol{H}\boldsymbol{s}\|^2 - N_0 \sum_{\forall q,m} \ln P(x_{q,m}) \tag{4}$$

Smaller values of $d(\boldsymbol{s})$ relate to a higher probability of $\boldsymbol{s}$ having been sent. The metrics $d(\boldsymbol{s})$ are used to compute LLR values for the channel decoder.

$$\lambda(x_{q,m}) \approx \min_{\forall \boldsymbol{s}|x_{q,m}=+1} d(\boldsymbol{s}) - \min_{\forall \boldsymbol{s}|x_{q,m}=-1} d(\boldsymbol{s}) \tag{5}$$

For the optimum calculation of Eq. (5), several minimum searches on $d(\boldsymbol{s})$ over all vectors $\boldsymbol{s}$ have to be performed. However, calculating all possible $d(\boldsymbol{s})$ quickly becomes infeasible for a larger number of receive- and transmit antennas and/or higher order modulations as the complexity grows with 2^{QM}. In order to simplify the metric calculations as well as the minimum search, the calculation of Eq. (4) is mapped on a tree structure. Therefore, the channel matrix $\boldsymbol{H}$ is decomposed into a unitary matrix $\boldsymbol{Q}$ and an upper-triangular matrix $\boldsymbol{R}$. The Euclidean distance is rewritten as

$$\|\boldsymbol{y} - \boldsymbol{H}\boldsymbol{s}\|^2 = \|\hat{\boldsymbol{y}} - \boldsymbol{R}\boldsymbol{s}\|^2 \tag{6}$$

with $\hat{\boldsymbol{y}} = \boldsymbol{Q}^H \boldsymbol{y}$. Equation Eq. (4) is replaced by the equivalent metric

$$d(\boldsymbol{s}) = \|\hat{\boldsymbol{y}} - \boldsymbol{R}\boldsymbol{s}\|^2 - N_0 \sum_{\forall q,m} \ln P(x_{q,m}) \tag{7}$$

The triangular structure of $\boldsymbol{R}$ allows the recursive calculation of $d(\boldsymbol{s})$

$$d_m = d_{m+1} + \gamma_m(s^{(m)}) \tag{8}$$

with the starting point $d_{M+1} = 0$ and $d(\boldsymbol{s}) = d_1$. The metric update $\gamma_m(s^{(m)})$ depends on the partial symbol vector $s^{(m)} = (s_m, s_{m+1}, \ldots, s_M)$.

$$\gamma_m(s^{(m)}) = |\hat{y}_m - \sum_{j=m}^{M} R_{m,j} s_j|^2 - N_0 \sum_{q=1}^{Q} \ln P(x_{q,m}) \tag{9}$$

This equation can be further simplified by introducing the interference-reduced symbol $\hat{y}'_m$, which is the same for all children of a node.

$$\hat{y}'_m = \hat{y}_m - \sum_{j=m+1}^{M} R_{m,j} s_j \tag{10}$$

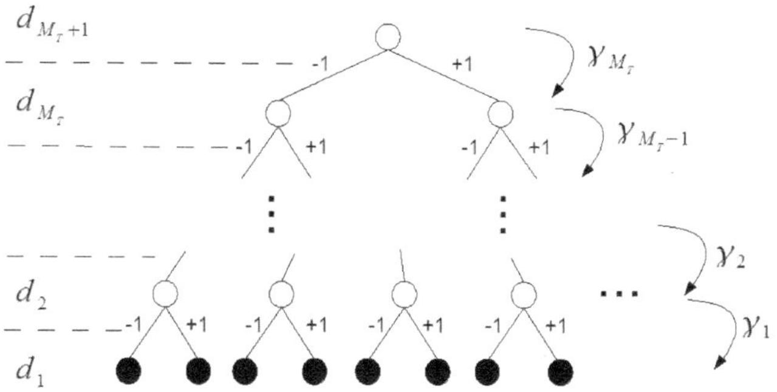

Fig. 2. Detection problem represented by a tree for the modulation alphabet $\{-1, +1\}$ (BPSK) and M_T antennas

$$\gamma_m(s^{(m)}) = |\hat{y}'_m - R_{m,m} s_m|^2 - N_0 \sum_{q=1}^{Q} \ln P(x_{q,m}) \tag{11}$$

The recursive calculation of Eq. (8) can be represented by a tree with $M+1$ levels, as shown for the modulation alphabet $\{-1, +1\}$ in Fig. 2. The root node corresponds to d_{M+1} and each leaf node corresponds to the metric $d(\boldsymbol{s})$ of one possible vector $\boldsymbol{s}$. Each level corresponds to the detection of one symbol s_m. Branches are labeled with an element of the modulation alphabet. When advancing from a parent node to a child node, the child node's metric d_m is calculated from the metric of its parent d_{m+1} and the branch metric γ_m.

Based on this tree search, many different MIMO detection algorithms exist which approximate Eq. (5). The main differences between the algorithms can be described by how they traverse the tree, e.g,. breadth-first, depth-first, or metric-first, and how branches of the tree are pruned. In general, these algorithms exhibit different communications performances and implementation complexities. All tree search based algorithms require the same channel preprocessing which performs the QR decomposition of the channel matrix $\boldsymbol{H}$. This preprocessing only has to be done when the channel is changing and not for every transmission vector. For these reasons, we concentrate only on the MIMO detection itself. QR decomposition implementations can be found in Nazar et al. (2010) for example.

4 Algorithmic kernels

Complexity adaptive processing enables energy efficient implementations. For different channel conditions, a variety of MIMO detection algorithms exist which trade-off communications performance and energy consumption. An energy efficient architecture, thus, needs the flexibility to support several detection algorithms. A weakly programmable architecture which is based on algorithmic kernels needed by all algorithms may offer this flexibility with a negligible overhead. In this section, we will derive the algorithmic kernels

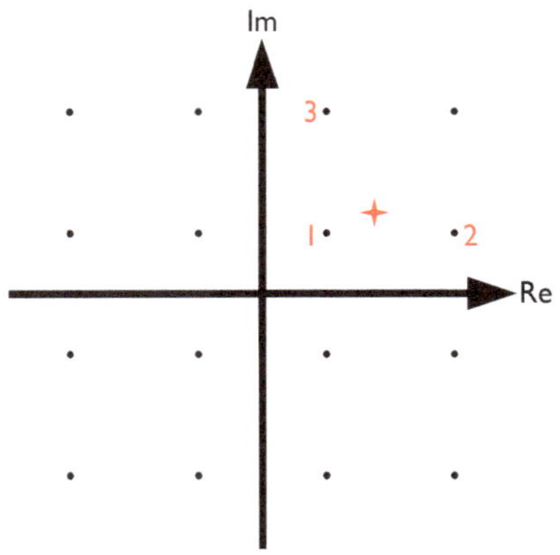

Fig. 3. SISO detection problem

of tree search based MIMO detection which are common to most detection algorithms.

We will start with some observations on the detection in single-input single-output (SISO) systems. The SISO transmission can be modeled by Eq. (2) and the detection is then performed according to Eq. (5) for $M_T = M_R = 1$. We will now depict the different steps involved in the SISO detection. Thereby, we will differentiate between maximum likelihood (ML) or hard-output and maximum a-posteriori (MAP) or soft-output detection. Figure 3 illustrates the SISO detection problem. A complex value y (represented by a red star) has been received. The ML solution is simply the modulation symbol which is closest to y (indicated by the number 1). The algorithm consists therefore of a single step: determining the closest modulation symbol. For the MAP solution, the closest symbols for each bit being 0 or 1 have to be found according to Eq. (5). This requires the enumeration of the best symbols starting with the closest one (indicated by the numbers 1, 2, 3), followed by the calculation of the Euclidean distance between y and those symbols. Each distance is used to update the minima for the calculation of Eq. (5).

The detection of MIMO signals cannot be mapped on a diagram as shown in Fig. 3 as the signal space is multi-dimensional and not rectangular. Therefore, we used the QR decomposition of the channel matrix $\boldsymbol{H}$ to map the detection problem on a tree search. The Euclidean distance is then computed as follows

$$\left\| \begin{pmatrix} \hat{y}_1 \\ \hat{y}_2 \\ \vdots \\ \hat{y}_M \end{pmatrix} - \begin{pmatrix} r_{1,1} & r_{1,2} & \cdots & r_{1,M} \\ 0 & r_{2,2} & \ddots & \vdots \\ \vdots & \ddots & \ddots & \vdots \\ 0 & \cdots & 0 & r_{M,M} \end{pmatrix} \begin{pmatrix} s_1 \\ s_2 \\ \vdots \\ s_M \end{pmatrix} \right\|^2 \tag{12}$$

The last row of Eq. (12) equals the SISO detection problem. After finding the most likely symbol for the last row, the result can be used to remove the interference of s_M from all other layers. The row $M-1$ is then interference free and can be treated accordingly. The repetition of these steps results in the successive interference cancellation (SIC) algorithm which solves the MIMO detection problem by splitting it into several SISO detection problems and by adding an interference cancellation step.

The SIC algorithm traverses the tree only once from top to bottom, thus, it has a low complexity. However, for a better communications performance, it is necessary to approximate Eq. (5) over many MIMO vectors $\boldsymbol{s}$ and not as a SISO detection problem. Real tree search algorithms like the depth-first sphere search or the K-best algorithm traverse the tree and compute the distances of several complete vectors. In addition to the detection steps introduced before, a node administration unit is required which stores the nodes which have been reached by the search but are not completed yet.

The main tree search based detection algorithms can thus be constructed by only five *coarse-grained algorithmic kernels*:

- The *enumeration unit EN* determines the visiting order for the children of one particular node.
- The *interference reduction unit IR* computes the interference-reduced symbols according to Eq. (10).
- The *metric computation unit MC* computes the recursive metric according to Eq. (11).
- The *node administration unit NA* stores all intermediate nodes with their results and chooses the node which is visited next in the tree.
- The *minima administration unit MA* stores the results of leaf nodes and updates the minima for the calculation of Eq. (5).

Table 1 is summarizing again which kernels are used by which algorithm. We chose soft-input soft-output sphere detection as tree search algorithm (Hochwald and ten Brink (2003)). In the computation of Eq. (5) it considers all symbol vectors $\boldsymbol{s}$, which lie inside a sphere of radius r around the received vector $\boldsymbol{y}$, i.e., for which $d(\boldsymbol{s}) < r$. Whenever a partial metric d_i exceeds the sphere radius, the corresponding part of the tree is excluded from the search. The number of processed nodes in the tree is dynamic and depends on the current channel realization. When a large radius is chosen, sphere detection offers near-optimal communications performance at the cost of a low throughput. The throughput can be increased by reducing the radius, which, however, will lead to a degradation of the communications performance.

5 Results

Based on these observations, we designed a MIMO detector architecture consisting of five algorithmic kernels. The configuration of these kernels allows the processing of different algorithms with negligible overhead. The design was implemented on a 65 nm low power bulk CMOS library from ST

Table 1. Algorithmic kernels which allow the construction of most tree search based detection algorithms.

Algorithm / Kernels	EN	IR	MC	NA	MA	Throughput
SISO hard-output	X					1200 Mbit s^{-1}
SISO soft-output	X		X		X	400 Mbit s^{-1}
MIMO SIC hard-output	X	X				480 Mbit s^{-1}
MIMO SIC soft-output	X	X	X		X	390 Mbit s^{-1}
MIMO sphere soft-output	X	X	X	X	X	40–120 Mbit s^{-1}

Table 2. Implementation Results: Area of the kernels after synthesis and the area of the whole detector after Place & Route.

Kernel	Area
EN	0.012 mm^2
MC	0.038 mm^2
IR	0.022 mm^2
MA	0.042 mm^2
NA	0.050 mm^2
Sum of the kernels	0.164 mm^2
After P&R, incl. I/O and control	0.254 mm^2

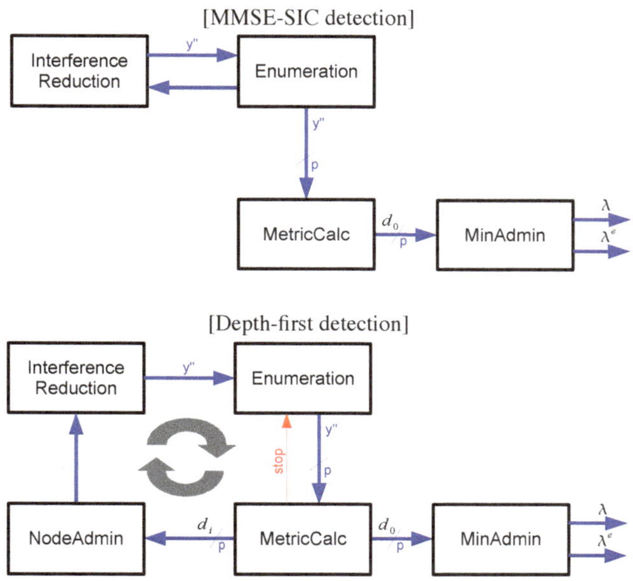

Fig. 4. The presented MIMO detector architecture consists of five algorithmic kernels, which can be configured for different detection algorithms.

Microelectronics. We considered the following PVT parameters: Worst Case (WC, 1.1 V, 125°C), Nominal Case (NOM, 1.2 V, 25 °C) and Best Case (BC, 1.3 V, –40 °C). Synthesis was performed with Synopsis Design Compiler in topographical mode, Placement & Routing (P&R) with Synopsys IC Compiler. Synthesis as well as P&R were performed with the Worst Case PVT settings of the 65nm library. The final design runs with a clock frequency of 300 MHz after P&R. The implementation results for the different parts of the detector are summarized in Table 2. Throughput results for the different algorithms are shown in Table 1. The power consumption after P&R is 14.4 mW.

Two examples for the configuration are illustrated in Fig. 4: MMSE-SIC detection and soft-input soft-output sphere detection. The scheduling of the operations for the SIC algorithm is always fixed. Therefore, we discuss only the configuration of the sphere search in Fig. 4. Especially for iterative receivers, the soft-input soft-output sphere detector offers the best communications performance. During runtime, throughput can be traded off against communications performance by adjusting the sphere radius. However, due to the nature of the depth-first search, the throughput is dynamic and varies with channel conditions and over the outer iterations. The configuration in Fig. 4 utilizes all five algorithmic kernels. All operations can be performed in parallel if data is available. The *enumeration* unit determines a sequence of the best children of a node. The *metric calculation* computes the recursive metric for each node. The enumeration is stopped when a maximum number is reached or when one of the child nodes violates the radius constraint. Intermediate nodes which fulfill the radius constraint are stored together with their metrics in the *node administration* until their processing is continued. Leaf nodes are stored in the *minima administration*. Whenever possible, the *interference reduction* unit processes a node from the *node administration* and passes the result to the *enumeration* unit. This recursive loop continues until all nodes within the sphere have been computed. In contrast to other depth-first sphere decoders (e.g. Burg et al. (2005), Witte et al. (2010)) which employ a one-node-per-cycle architecture, the presented architecture computes two nodes per cycle. This is a novel approach, which doubles the throughput compared to state-of-the-art implementations.

In order to show the throughput flexibility (and thus the energy efficiency) of the implemented design, we determined the maximum achievable throughput for different channel conditions. Therefore, we simulated a MIMO-BICM system including a 64-state convolutional code as channel code, using 4×4 antennas with 16-QAM modulation and a codeword length of 2304 bits. We tried to reach a frame error rate

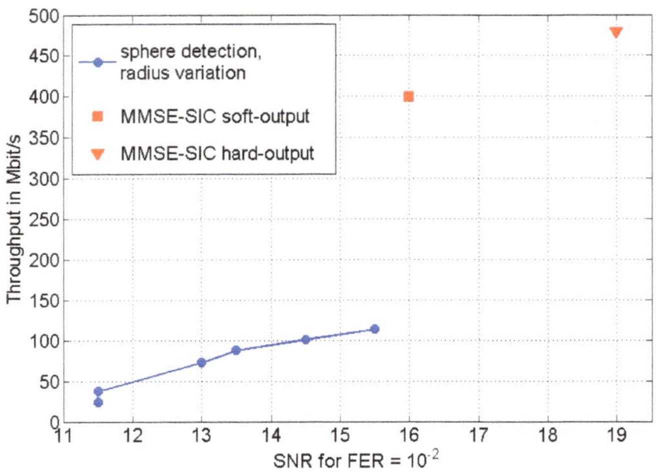

Fig. 5. Throughput vs. QoS flexibility of the presented detector architecture

(FER) of 10^{-2} for varying channel conditions, i.e. different signal-to-noise ratios (SNR). The results are shown in Fig. 5. The presented architecture covers a wide range of SNR values (almost 10 dB) and is thus able to react to different channel environments.

6 Conclusions

In this paper, we presented the first weakly-programmable MIMO detector architecture which offers just the necessary flexibility. We have demonstrated that the common tree search based MIMO detection algorithms are all constructed by the same algorithmic kernels allowing the design of a weakly-programmable architecture. In this way, the detection algorithm can be chosen and parameterized at run-time according to the current channel conditions and QoS requirements. This approach leads to a highly energy-efficient implementation.

Acknowledgements. This work was partly supported by the German Research Foundation (DFG) as part of the priority program "Dependable Embedded Systems" (SPP 1500).

References

Barbero, L. G. and Thompson, J. S.: Fixing the Complexity of the Sphere Decoder for MIMO Detection, IEEE T. Wirel. Commun., 7, 2131–2142, 2008.

Borlenghi, F., Witte, E. M., Ascheid, G., Meyr, H., and Burg, A.: A 772Mbit/s 8.81bit/nJ 90nm CMOS soft-input soft-output sphere decoder, in: Proc. IEEE Asian Solid State Circuits Conf. (A-SSCC), 297–300, 2011.

Burg, A., Borgmann, M., Wenk, M., Zellweger, M., Fichtner, W., and Bolcskei, H.: VLSI implementation of MIMO detection using the sphere decoding algorithm, IEEE J. Solid-St. Circ., 40, 1566–1577, doi:10.1109/JSSC.2005.847505, 2005.

Chen, X., Minwegen, A., Hassan, Y., Kammler, D., Li, S., Kempf, T., Chattopadhyay, A., and Ascheid, G.: FLEXDET: Flexible, Efficient Multi-Mode MIMO Detection Using Reconfigurable ASIP, in: Field-Programmable Custom Computing Machines (FCCM), 2012 IEEE 20th Annual International Symposium on, 69–76, doi:10.1109/FCCM.2012.22, 2012.

Foschini, G. J., Golden, G. D., Valenzuela, R. A., and Wolniansky, P. W.: Simplified processing for high spectral efficiency wireless communication employing multi-element arrays, IEEE J. Sel. Area. Comm., 17, 1841–1852, 1999.

Hochwald, B. and ten Brink, S.: Achieving Near-Capacity on a Multiple-Antenna Channel, Communications, IEEE Transactions on, 51, 389–399, doi:10.1109/TCOMM.2003.809789, 2003.

Jafri, A. R., Karakolah, D., Baghdadi, A., and Jezequel, M.: ASIP-based flexible MMSE-IC Linear Equalizer for MIMO turbo-equalization applications, in: Proc. DATE '09. Design, Automation & Test in Europe Conf. & Exhibition, 1620–1625, 2009.

Nazar, G. L., Gimmler, C., and Wehn, N.: Implementation comparisons of the QR decomposition for MIMO detection, in: Proceedings of the 23rd symposium on Integrated circuits and system design (SBCCI '10), 210–214, ACM, doi:10.1145/1854153.1854204, 2010.

Ramakrishnan, V., Witte, E., Kempf, T., Kammler, D., Ascheid, G., Leupers, R., Meyr, H., Adrat, M., and Antweiler, M.: Efficient and portable SDR waveform development: The Nucleus concept, in: Military Communications Conference, 2009. MILCOM 2009, IEEE, 1–7, doi:10.1109/MILCOM.2009.5379897, 2009.

Studer, C., Fateh, S., and Seethaler, D.: ASIC Implementation of Soft-Input Soft-Output MIMO Detection Using MMSE Parallel Interference Cancellation, IEEE J. Solid-St. Circ., 46, 1754–1765, doi:10.1109/JSSC.2011.2144470, 2011.

Vikalo, H., Hassibi, B., and Kailath, T.: Iterative Decoding for MIMO Channels Via Modified Sphere Decoding, Wireless Communications, IEEE Transactions on, 3, 2299–2311, 2004.

Witte, E. M., Borlenghi, F., Ascheid, G., Leupers, R., and Meyr, H.: A Scalable VLSI Architecture for Soft-Input Soft-Output Single Tree-Search Sphere Decoding, IEEE T. Circuits II, 57, 706–710, doi:10.1109/TCSII.2010.2056014, 2010.

Wong, K., Tsui, C., Cheng, R., and Mow, W.: A VLSI architecture of a K-best lattice decoding algorithm for MIMO channels, in: Proc. IEEE Int. Symp. Circuits and Systems ISCAS 2002, Vol. 3, doi:10.1109/ISCAS.2002.1010213, 2002.

Yu, Z., van Berkel, C., and Li, H.: A complexity adaptive channel estimator for low power, in: Design, Automation Test in Europe Conference Exhibition (DATE), 2012, 1531 –1536, 2012.

26

Best usage of free-space capacitors in ASIC regulators

T. Maier[1]**, D. Droste**[1]**, and M. Siegel**[2]

[1]Robert Bosch GmbH, Reutlingen, Germany
[2]Karlsruher Institut für Technologie (KIT), Karlsruhe, Germany

Correspondence to: T. Maier (tobias.maier2@de.bosch.com)

Abstract. In this work we examine how to improve the performance of voltage regulators in application specific integrated circuits (ASICs) by placing capacitors into free layout space. The problem arising after layout, when there are areas not covered by functional elements, is where to connect the free-space capacitors (FSCs), as they can be connected to the input or the output net of a voltage regulator. Therefore we designed a testbench for mathematical calculations and one for simulations to identify the influence of a capacitance connected at these certain positions. We mainly focused on PSR analysis while not losing sight of transient effects. The results of calculation and simulation illustrate that the best solution is to split the capacitance half by half to both possible nets if no output capacitance was installed during design. Otherwise a ratio of one to one for input capacitance to output capacitance has to be set up for best performance.

1 Introduction

In application specific integrated circuits (ASICs) regulators are used to create an output voltage from a certain input voltage range. The ideal regulator has a constant output voltage independent of a load change and without any influence from the input voltage. This influence not only includes DC voltage levels (line regulation), but also any AC distortion. The independence of the output voltage from the input is called power supply rejection (PSR) (Ong and Chan, 2010). In literature also often the term PSRR (power supply rejection ratio) (Baker, 2010; Holberg, 2010) is found, which is more true for amplifier as there a ratio between the signal amplification and the power distortion damping exists, while in regulators only the power distortion damping is of interest. Nonideal regulators, which are in fact all implemented regulators, suffers from both, AC distortions and transient load changes. The PSR is limited in its damping and also frequency dependent, i.e. in most cases the PSR gets worse with higher frequencies due to capacitive coupling. Load changes create a distortion at the output voltage as the regulation loop needs time to settle to the adopted IR drop. The investigations of improvement of PSR and load stability are important, especially if there are sensitive front-ends on the ASIC. They would be disturbed by these effects and then wrong data would be returned by precious analog to digital converters (ADCs) to signal processing devices like on- or off-chip microcontrollers or DSPs. Therefore a better ASIC performance should be reached by increasing the PSR of the regulator.

Many authors have been dealing with topics concerning regulators. New methods to increase PSR or faster load regulation loops are e.g. 2-stage regulation (Yang et al., 2009) or double regulation (Lim et al., 2011). Newest trends also tend to capacitor-less low dropout regulators, where a high PSRR could be reached by an additional fast output buffer, which drives the regulation transistor (Abbasi et al., 2010). Another method to increase PSRR is to add a left-hand zero by nested Miller compensation (Huang and Liu, 2011) or to use a negative feedback loop (Shirahatti et al., 2010). Capacitor-less regulators are voltage regulators, which don't need an external bypass or stabilization capacitor as the Texas Instruments TPS78915 (TI, 2000) would need. Nevertheless, capacitor-less does not exclude capacitors at all, as internal capacitors are still used for these regulators. An advantage of on-chip capacitors is that they have very low ESR(QuadTech, 2003) values due to the short metal connections. For on-chip capacitors, often the leakage is stated in the design documents. A typical leakage value is around 1e-8 to 1e-7 $\mathrm{A\,cm^{-2}}$ (TI, 2009) which would result in a minimum parallel resistance value for 1.8 V and 1 $\mathrm{mm^2}$ of around 1.8 $\mathrm{G\Omega}$. Smaller ESR value as typical off-chip capacitors and low leakage makes capacitor choice discussion for ESR (Morita, 2011) and leakage obsolete.

As this work researches PSR effects, noise reduction techniques like low-pass filtering of the Bandgap reference

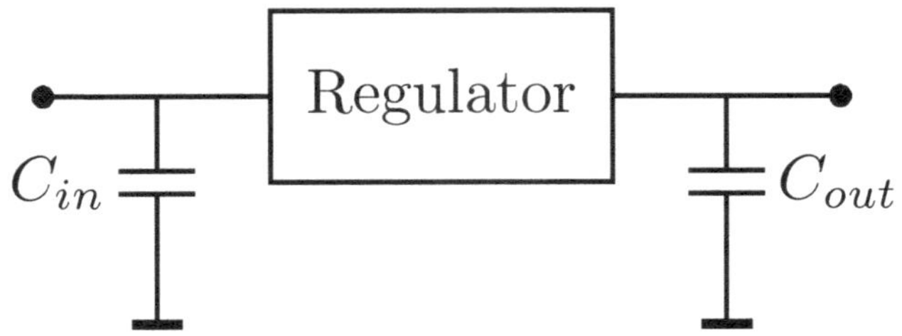

Fig. 1. Possibel positions for free-space capacitor

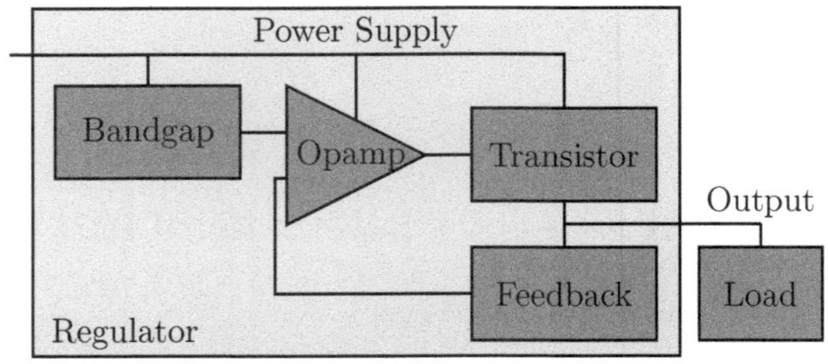

Fig. 2. Regulator under research

voltage with external (TI, 2000) or possible internal capacitors are not further investigated.

This work deals with the question, which net is the best to connect free-space capacitors (FSCs). A free-space capacitor is a capacitor, created in ASIC areas, which are not covered by functional elements after layout. These areas are often gaps between elements of the regulator. More free space therefore leads to more capacitance. The FSCs are placed into the finished layout and their capacitance values are derived from their size, which is then back-annotated from the finalized layout to the corresponding schematic. As a higher capacitance would increase the PSR performance more and the capacitor has no need to have an exact value over process variations and temperature, the most important parameter, when choosing an appropriate capacitor from the design kit, is the capacitance per area. As the functionality of the regulator was already proved through simulations before layout, the additional capacitors are used to further improve the regulator and not for stability reasons.

The free-space capacitors are connected to ground with one side in every case, but which is the best net to connect the other side of the capacitors?

Should FSCs be connected to the input or the output net of the regulator (see Fig. 1), or should it be split to both nets? Both nets can be seen as supply nets, which could be stabilized by an additional capacitance. There are the same arguments for both nets, e.g. the capacitance between a net and ground serves as a low pass, which rejects AC distortions when taking net resistances into account or that FSCs are a charge storage which helps with transient load changes.

It should be mentioned that the reliability of an ASIC should not be decreased by additional free-space capacitors as on-chip capacitors generally have a low leakage. Of course the designer should pay attention that the regulator is still stable with an additional output capacitance, what is normally the case. A capacitor at the input could only have an effect on the time, the regulator needs to start up, if the capacitance is very large. Nevertheless it is worth to re-simulate the regulator after all changes from layout are back annotated.

2 Regulator under research and testbench

2.1 Regulator under research

For this work, the regulator under research is a low-dropout regulator (LDO), on which the best usage of a free-space capacitor is investigated. The LDO consists of 4 major elements, shown in Fig.2. The first one is a bandgap that is used to create a reference voltage to which the output voltage is compared. Next, there is an operational amplifier (op-amp) which is used to compare the bandgap voltage to the feedback voltage and to set the regulation transistor's gate voltage corresponding to the result of the comparison. This regulation transistor is the LDO's third element and is used to regulate the voltage drop between supply net and output net. Generally the transistor can be a n-MOS or p-MOS transistor and here a n-MOS device is used.

The voltage drop thereby is generated through an IR drop, where the current I is the current of the load plus the current of the feedback network and the resistance R is the resistance of the transistor, R_{ds_on}. This resistance is regulated by the gate voltage, set by the op-amp. For the regulation transistor it is necessary that its area is big enough so that the maximum current flow density is not exceeded for any load. Large area also helps transporting away the generated heat in the transistor.

The 4th element used in the LDO is the feedback network. In most applications it exists of two resistors forming a voltage divider, dimensioned in a way that the divided voltage corresponds to the bandgap voltage, when the output voltage of the LDO has the nominal value (Baker, 2010). The load in this work is a constant resistive load plus a parallel variable current source. An additionally capacitor is also taken into account at the later part of this work.

2.2 Testbench

For a good prediction to which net the capacitance should be connected, a good testbench is necessary. Aside to the regulator itself and the load, the testbench also includes the pads, the bond wires and the metal resistances from the on-chip wires. For the calculations, the circuit was simplified to reduce the number of elements. The inner circuitry of the bandgap and the op-amp, connected between the supply wire *VDD3V* and ground, are included to the testbench by a impedance $Z(s)$. This impedance is assumed to have one pole, based on simplified simulation results. Including the PSR transfer characteristic of the bandgap $A_{\mathrm{BG}}(s)$ and the op-amp $A_{\mathrm{op}}(s)$ is done by the regulation transistor's gate voltage V_{G}. The Feedback influence from the resistive voltage divider (V_{FB}) is also included in $V_G = A_{\mathrm{op}}(s)\,(A_{\mathrm{BG}}(s)\,(V_{\mathrm{in}} - V_{\mathrm{drop}}) - V_{\mathrm{FB}})$ where $V_{\mathrm{drop}} = I_1\,(2s\,L_{\mathrm{bond}} + R_{\mathrm{met}})$ is the voltage drop of the package circuitry.

3 Calculation and simulation

3.1 Calculations

Based on the testbench, shown in Fig.3 calculations were effected to estimate the results and to verify that the diagrams gained by simulation show the correct behavior. The testbench has two input variables, V_{in} and I_{load}, but the calculations are done only for V_{in} as the point of interest is the PSR analysis. As the results depend on many parameter (e.g. $R_{\mathrm{FB1}}, R_{\mathrm{FB2}}, R_L, R_{\mathrm{met}}, L_{\mathrm{bond}}$) only the generalized zero pole forms are shown in Eqs. (1) and (4). The highest order coefficients of numerator denominator from Eqs. (1) and (4) are shown in Eqs. (2),(3) and (5) respectively, for the parameters directly affected by the FSCs, C_{in} and C_{out}. All other parameters are substituted by c_{ni} and c_{di}.

For the frequency behavior of *VDD3V*, Eqs. (1),(2) and (3) show that the number of poles and zeros keeps constant if either C_{in} or C_{out} or both are added. This is due to $Z(s)$ already having a pole, as the additional C_{in} would create. But Eqs. (2) and Eq. (3) show that the poles and zeros are shifted to lower frequencies especially at the change of C_{in}.

$$\frac{a_{2,VDD3V}\,s^2 + a_{1,VDD3V}\,s + a_{0,VDD3V}}{b_4 s^4 + b_3 s^3 + b_2 s^2 + b_1 s + b_0} \tag{1}$$

$$a_{2,VDD} = (c_{n1}\,C_{\mathrm{out}} + c_{n2})\,C_{\mathrm{in}} + c_{n3}\,C_{\mathrm{out}} + c_{n4} \tag{2}$$

$$\begin{aligned} b_4 = {} & (c_{d1}\,C_{\mathrm{out}} + c_{d2})\,C_{\mathrm{in}}^2 \\ & + (c_{d3}\,C_{\mathrm{out}} + c_{d4})\,C_{\mathrm{in}} + c_{d5}\,C_{\mathrm{out}} + c_{d6} \end{aligned} \tag{3}$$

The frequency behavior of *AVDD* to V_{in} (Eq. 4) is similar to that of *VDD3V* (Eq. 1). As both have the same denominator, the poles are shifted the same amount, but Eq. (5) shows that zeros are shifted otherwise.

$$\frac{AVDD}{V_{\mathrm{in}}} = \frac{a_{2,AVDD}\,s^2 + a_{1,AVDD}\,s + a_{0,AVDD}}{b_4 s^4 + b_3 s^3 + b_2 s^2 + b_1 s + b_0} \tag{4}$$

$$\begin{aligned} a_{2,AVDD} = {} & (c_{n21}\,C_{\mathrm{out}} + c_{n22} + c_{n23}\,C_{\mathrm{in}})\,s^2 \\ & + (c_{n24}\,C_{in} + c_{n25}\,C_{\mathrm{out}} + c_{n6})\,s + c_{n27} \end{aligned} \tag{5}$$

3.2 Simulation

After the calculations showed the principal behavior, simulations were executed. A simulation testbench was implemented, containing the elements added for package and the transistor level circuitry of the bandgap and the operational amplifier. To see the effects of a capacitor added to the input or the output clearly, the AC effects of the metal resistance and the bond inductances were simulated separately. *AVDD* is affected by a metal resistance change only by a pole movement towards lower frequencies. Adding the bond inductances adds a pole at about $f \propto \frac{1}{L_{\mathrm{bond}}}$ to the transfer functions, i.e. the higher the bond inductance the lower the frequency of the pole.

The complete regulator is supplied by a voltage of 3.3 V and has an output voltage of 1.8 V. The values for bond inductances were zero, 5 nH and 50 nH to see the effect of the inductance. With generally assumed value of $\approx 1\,\mathrm{nH\,mm^{-1}}$ (Lee, 2003) inductance per bondwire length, a value of 5 nH was used for the later simulations.

The resistance was varied between zero, 5 Ω and 50 Ω for the simulation of the wire effects. From earlier layout extractions a value of 5 Ω for thick supply wires is assumed for further simulations.

For output transistor a n-MOS transistor is chosen, while the opamp was implemented as folded-cascode. The total capacitance available for connection was extracted from layout with 114 unit capacitors, each with a value of 401 fF. So the total amount of capacitance is around 45.7 pF.

In Fig.4 the AC influence of the free-space capacitor towards *VDD3V* and *AVDD* is the point of interest. For the *VDD3V* net (upper 3 curves) the AC behavior is quite simple. Adding the capacitance to the output net has no effect to the input net in the frequency range of interest. The more capacitance is added to the input net the earlier *VDD3V* is damped and so has higher damping values at higher frequencies. I.e. the more capacitance is added to the net *VDD3V*, the lower is the frequency of the first pole, what corresponds to the calculation results. Of course the effects on *VDD3V* directly affect *AVDD*, as *AVDD* is derived from *VDD3V*.

For *AVDD* (lower curves in Fig.4) the influence is clearly visible, but very depending on the position of the capacitor. If the capacitor is at the input supply net (solid style), the curve is the worst of all three in middle frequencies. It profits from the shift of the pole of the input net from far right towards lower frequencies in the range of low gigahertz. Connecting the capacitor to the output or regulated net (dashed-dotted style) shows a high improvement in the mid frequency area. This improvement profits from the pole shift beginning from some tens of megahertz towards high frequencies. The third possible option, splitting the capacitor and connecting half the amount to the input and the other half to the output net, generates the dashed curve. As only half the amount is connected to the output net, the pole movement is quite good, but not so far towards lower frequencies as if all the capacitance

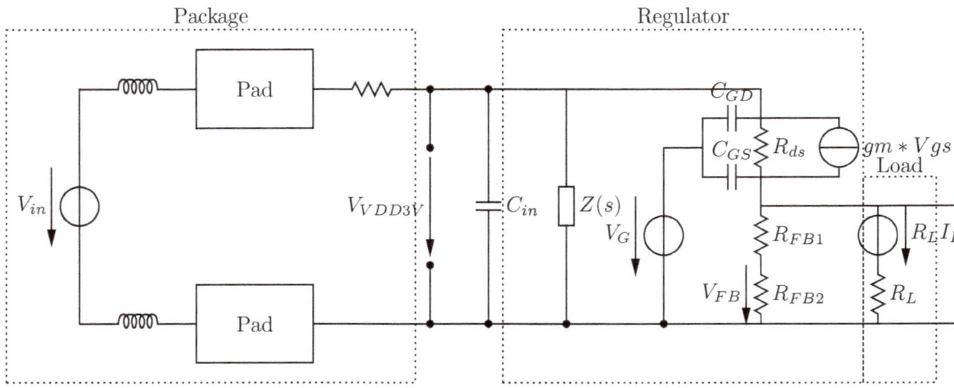

Fig. 3. Complete testbench for calculation

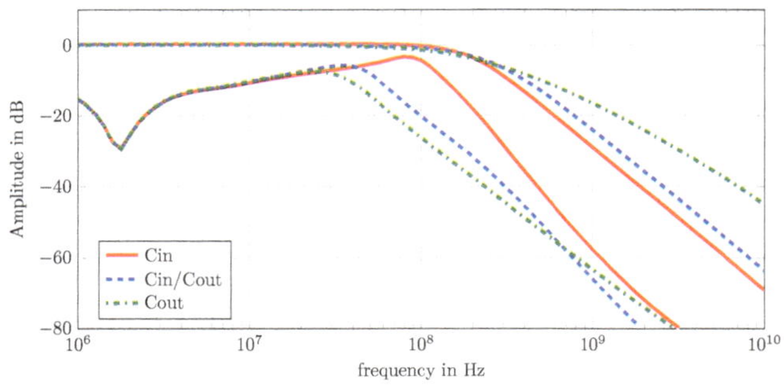

Fig. 4. Effect of FSC to VDD3V and AVDD

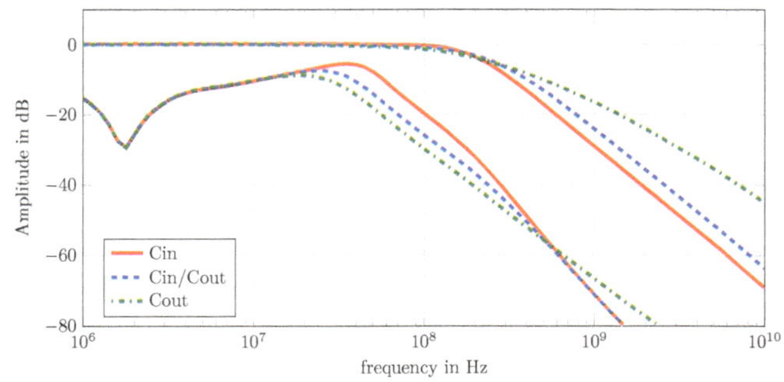

Fig. 5. Effect of FSC plus add. cap to VDD3V and AVDD

is connected to the output net. The additional shift of the input net pole towards lower frequencies by the connection of the 2nd half of the capacitance to the input net, helps to improve the frequency behavior of the regulator at higher frequencies. So splitting could be called the best solution as in mid-frequency range it already damps very well and at high frequencies shows the best damping.

Another point of interest is shown in Fig.5, where again the frequency behavior of *AVDD* and *VDD3V* is shown. The changed parameter now is again the capacity, in this case not only for the additional FSC, but rather for the FSC and a fixed capacitor at the output net. Often a capacitor is installed at the output net for e.g. better transient behavior or stability. There also the question arises where to best connect the additional FSCs.

Again the diagrams were plotted for three cases: All capacitance at the input net (solid curves), all at the output net (dashed-dotted curve) and a half-half split of the capacitance(dashed curve).

For the net *AVDD* (lower 3 curves) the best mid frequency behavior is again reached by connecting all capacitance to the output net. But then the high frequency behavior is very bad because of the missing early pole on the net *VDD3V* (upper 3 curves), which supplies the regulator. Connecting all the capacitance to the input net leads to very bad mid frequency behavior, but a better high frequency behavior as if connecting all capacitance to the output. This is due to the earlier pole in *VDD3V*. With splitting the capacitance, the mid frequency behavior is lying between the two other cases but is not much worse as if connecting all FSC to the output. But with split FSCs, AC behavior of net *AVDD* profits from the *VDD3V* pole and therefore shows the same high frequency behavior as if connecting all the FSC to the input. So splitting is again the best solution.

Simulations show that splitting is best as long as splitting can create an allover ratio of $1:1$ for all the capacitance at the input net to all the capacitance at the output net. If the output capacitance is greater than the free-space capacitance at all, the best solution is connecting all FSC to the input net.

In the diagrams in Figs. 4 and 5 the shape of AC response is only visible in the range between 1 MHz and 10 GHz. In Fig.6 the complete shape is visible for the configuration with a constant capacitance and the free-space capacitance.

What is investigated in the upper and lower curves of Fig.6 is the influence of the bandgap to the regulator. In the lower curves, the bandgap voltage and especially the bandgap currents are kept constant while in the upper curves, the bandgap voltage and currents are that of the simulated transistor circuit. The influence is clearly visible especially at very low frequencies, where the difference is about 20 dB.

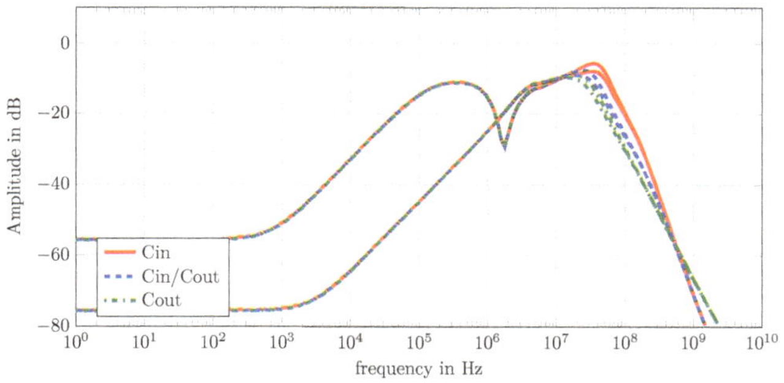

Fig. 6. Influence of bandgap to AVDD

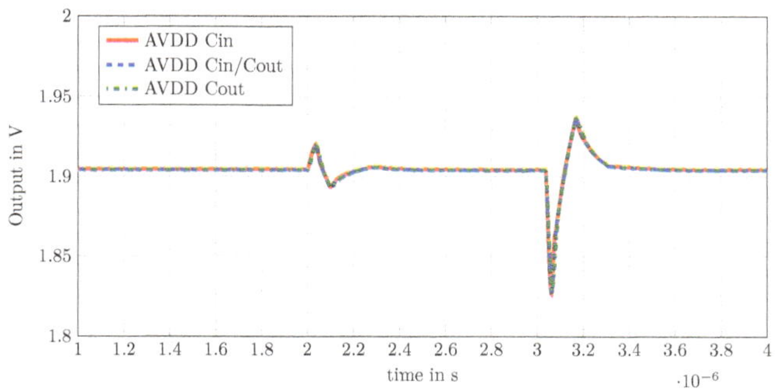

Fig. 7. Transient behavior of AVDD at load change

This simulation is done to prove the necessity to damp the input net, as the bandgap, the operational amplifier and the regulation transistor are connected to that net and so a damping on that net helps improving the output net very much. The diagram also shows that the values of the free-space capacitors are too small to damp the low frequencies where the PSR distortion from bandgap reaches the output net in the range from DC to about 1 MHz.

So far, all the simulations only cover AC behavior and point out how the FSCs influence the frequency behavior. When simulating the transient behavior of load changes, the effects are only very small on the nets *AVDD* (see Fig.7) and *VDD3V*. As it can be seen in the diagram, the effect of a load change at the output is almost the same, independent to which net the FSC is connected. The explanation for this effect is that the charge, stored on the capacitor, is very small as the capacitor itself is very small. Therefore the capacitor could only stabilize the voltage for a very short time before the voltage on the capacitor drops together with the output voltage.

4 Conclusions

In this work, we show that it is possible to improve PSR effects of LDOs by the application of FSC during post-layout optimization. We especially point out that FSCs placed half by half to the input and output net of a LDO improve AC performance more than just adding them to one of the two nets. If the LDO already has an output capacitor installed, the best solution is to add capacitance to the input net till the input and output nets have the same amount of capacitance and split the rest of the capacitance.

Splitting the capacitance shifts the main poles of input and output net towards lower frequencies, decreasing PSR effects in mid-frequency range with the output capacitance and in high-frequency range with the input capacitance. It is worth mentioning that due to the small capacitance, placed into the rare free space, the damping AC effects on the output net are also only very small. An additional free-space capacitor also does not have a significant effect on transient load changes.

This work emphasizes, that in contrast to other work, where the focus is on changes in error-amplifier or improvement of control loop, a simple, well understood design can still be improved if the cooperation between design and layout is excellent and all space on the expensive ASIC is utilized.

Therefore the analog designer should already pay attention to post-layout optimization opportunities during design of the circuit to advise layouter where to connect free-space capacitors. An important instrument to fulfill that job is the creation of a proper testbench, containing next to most of the parasitic elements also opportunity devices for post-layout optimization. These could be ideal devices with an estimated value of FSCs which are later replaced by PDK (process develepont kit) devices where the real value is back-annotated. This approach is legitimately if the regulator is stable with and without the FSCs and just improved by additional input or output capacitance.

Nevertheless a simulation with extracted parasitics from layout needs to be done in every case.

References

Abbasi, M., Bagnall, D., and Bn, V.: A high PSRR capacitor-less on -Chip low dropout voltage regulator, in: 8th International Symposium on Intelligent Systems and Informatics (SISY), 2010, 361–364, doi:10.1109/SISY.2010.5647405, 2010.

Baker, R. J.: CMOS Circuit Design, Layout, and Simulation (Third Edition), IEEE Press, doi:10.1002/9780470891179, 2010.

Holberg, P. E. A. D. R.: CMOS Analog Circuit Design, Oxfort University Press, 2010.

Huang, W.-J. and Liu, S.-I.: PSRR-enhanced low-dropout regulator, Electron. Lett., 47, 17–18, doi:10.1049/el.2010.2869, 2011.

Lee, T.: The Design of CMOS Radio-Frequency Integrated Circuits, Cambridge University Press, http://books.google.de/books?id=io1hL48OqBsC, 2003.

Lim, H. J., Singh, R., Chuan, K., Nuttman, D., and Je, M.: Double regulated voltage supply for high precision MEMS accelerometers, in: 13th International Symposium on Integrated Circuits (ISIC), 2011, 571–574, doi:10.1109/ISICir.2011.6131927, 2011.

Morita, G.: Low Dropout Regulators – Why the Choice of Bypass Capacitor Matters, Analog Dialogue 45-01 Back Burner, 2011.

Ong, G. and Chan, P.: A Low Quiescent Biased Regulator With High PSR Dedicated to Micropower Sensor Circuits, Sensors Journal, IEEE, 10, 1266–1275, doi:10.1109/JSEN.2009.2038793, 2010.

QuadTech: Equivalent Series Resistance (ESR) of Capacitors, Tech. rep., QuadTech, Incorporated, http://www.low-esr.com/QT_LowESR.pdf, 2003.

Shirahatti, S., Nandi, A., and Priyatamkumar: A Capacitor-less Low Drop out Voltage Regulator, in: IEEE Int. Conference on Communication Control and Computing Technologies (ICCCCT), 2010 , 177–182, doi:10.1109/ICCCCT.2010.5670547, 2010.

TI: TPS78915, TPS 78918, TPS78928, TPS78930 Ultralow-Power Low-Noise 100 mA Low-Dropout Linear Regulators Datasheet, Texas Instruments, 2000.

TI: Process Control Document and Design Parametrics for the LBC8 (LinBiCMOS-8) Technology, Texas Instruments, version 2.4, 23 June 2009, 2009.

Yang, S., Zou, X., Zou, Z., and Chen, X.: A Loop-Improved Capacitor-Less Low-Dropout Regulator for SoC Power Management Application, in: Asia-Pacific Power and Energy Engineering Conference, 2009, APPEEC 2009, 1–4, doi:10.1109/APPEEC.2009.4918052, 2009.

Comparison of reconfigurable structures for flexible word-length multiplication

O. A. Pfänder[1], **R. Nopper**[1], **H.-J. Pfleiderer**[1], **S. Zhou**[2], **and A. Bermak**[2]

[1]Microelectronics Department, University of Ulm, Germany

[2]Electrical and Computer Engineering Department, The Hong Kong University of Science and Technology, Hong Kong S.A.R.

Abstract. Binary multiplication continues to be one of the essential arithmetic operations in digital circuits. Even though field-programmable gate arrays (FPGAs) are becoming more and more powerful these days, the vendors cannot avoid implementing multiplications with high word-lengths using embedded blocks instead of configurable logic. But on the other hand, the circuit's efficiency decreases if the provided word-length of the hard-wired multipliers exceeds the precision requirements of the algorithm mapped into the FPGA. Thus it is beneficial to use multiplier blocks with configurable word-length, optimized for area, speed and power dissipation, e.g. regarding digital signal processing (DSP) applications.

In this contribution, we present different approaches and structures for the realization of a multiplication with variable precision and perform an objective comparison. This includes one approach based on a modified Baugh and Wooley algorithm and three structures using Booth's arithmetic operand recoding with different array structures. All modules have the option to compute signed two's complement fix-point numbers either as an individual computing unit or interconnected to a superior array. Therefore, a high throughput at low precision through parallelism, or a high precision through concatenation can be achieved.

1 Introduction

Reconfigurable hardware is becoming more and more popular today, thanks to the continuous evolution and improvement of FPGA devices. FPGA-based systems can be reprogrammed after production by overwriting memory contents and thus adapted to new requirements or even completely different application scenarios. One beneficial effect on the economic side is the potential to save the development and/or production costs of an application-specific integrated circuit (ASIC), specifically for low and medium volume production. FPGAs achieve their high flexibility using programmable lookup tables and reconfigurable routing. However, implementing algorithms with high word-lengths and complex arithmetics on FPGAs can consume large hardware resources and obtain comparatively slow clock frequencies.

On the other hand, arithmetic operations such as addition, multiplication and combinations such as multiply-accumulate (MAC) are required in almost every DSP algorithm, e.g. digital filter designs or Fast Fourier Transform (FFT) implementations (Parhami, 2000). Furthermore, special applications make use of multi-precision arithmetic, i.e. working at different levels of precision during their execution time. Examples for this type of applications are iterative approximation of non-linear functions (Lachowicz and Pfleiderer, 2008), neural networks with different precision in training phase and operation phase (Bermak and Martinez, 2003), and also adaptive filter designs with variable input or coefficient definitions. Since multiplication is crucial in all these situations, optimizing this particular operation can help to save hardware cost and power dissipation. In our approach, we utilize embedded fixed-wired multiplier structures in FPGAs as dedicated hardware. Above that, we upgrade these multipliers with data exchange interfaces and also a sign handling capability. Hence, not only the operand number system, but also the word-length of the multiplication can be directly controlled at run-time, substituting the re-programming step by setting a number of appropriate control signals.

Correspondence to: O. A. Pfänder
(oliver.pfaender@uni-ulm.de)

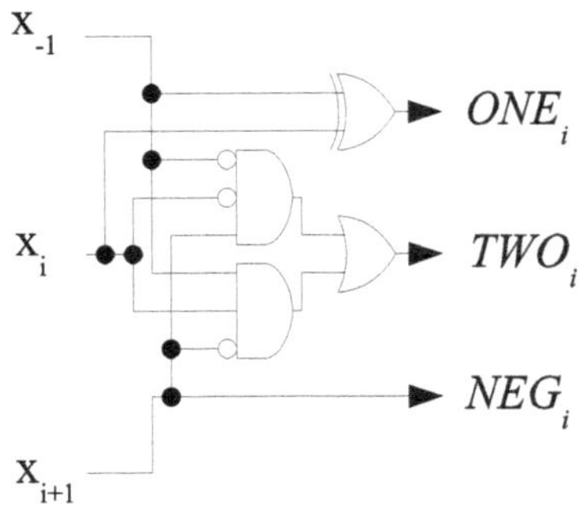

Fig. 1. Realization of a periodical radix-4 Booth encoder, overlapping 3 input bits.

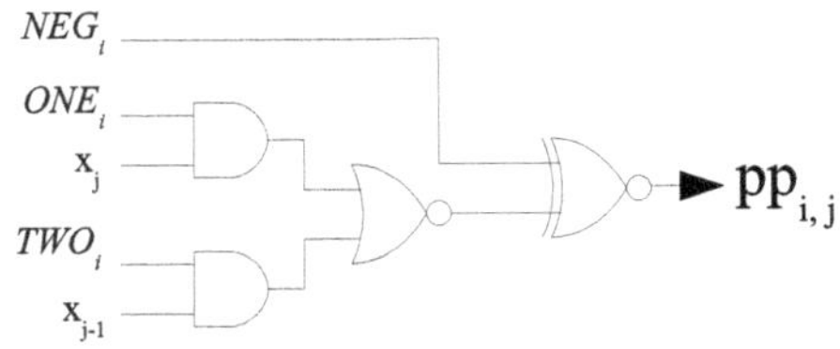

Fig. 2. Partial product generation for non-MSB bits from Booth-recoded signals.

This paper is organized as follows: Sect. 2 describes the basic concept of concatenating several individual multiplier blocks to form a superior multiplier, also covering the improvements over our previous designs, Sect. 3 compares the different approaches in terms of hardware usage and speed, and Sect. 4 presents a novel structure for a block-serial multiplier, followed by a conclusion.

2 Basic concept

In order to overcome the high area and routing expenses when realizing higher word-length multiplication in FPGAs, it is to the best advantage to embed multiplier blocks into the fabric and connect them to the routing network. This is already state-of-the-art in modern commercially available FPGA devices such as the Xilinx Virtex-II Pro and the Virtex-IV with its DSP48 blocks. Both examples feature 18×18-bit multipliers, while our multiplier elements are designed as fully functional $n\times n$-bit blocks that can be combined to form a superior multiplier array. For this work, the parameter n was set to 8 because the percental hardware overhead for reconfiguration would be too large for a lower number (Pfänder et al., 2004), and a larger number would constrain the word-length flexibility too much. By concatenating $m\times m$ uniform blocks, a superior $(m\cdot n)\times(m\cdot n)$-bit multiplier can be realized. Since every element is a fully functional multiplier itself and can work separately at its basic word-length, a high parallelism can be achieved.

There are several degrees of freedom for the multiplier elements' inner structure. Firstly, there is the choice of multiplication algorithm, i.e. parallel, serial, serial/parallel etc., and secondly, the number system must be assigned, in order to handle signed numbers. We have shown in Pfänder and Pfleiderer (2005) that a modified Baugh and Wooley algorithm can be efficiently applied to a parallel and a serial/parallel multiplier that accepts two's complement signed numbers. In both cases, this leads to a regular array structure – also permitting an asymmetric operand scheme – and it also shows a small overhead compared to an unsigned and stand-alone multiplier. Reconfigurable interfaces and configuration options include the exchange of carry and sum input signals with neighboring elements and also the inversion of partial products in certain array positions. In addition, introducing pipeline stages is also possible with the Baugh and Wooley scheme, as shown by Di and Yuan (2003). New improvements of our previous designs include two major updates, namely the the adoption of Booth's arithmetic recoding of input operands (Booth, 1951) and the use of a carry-save instead of a carry-ripple adder array. While the Booth recoding is aimed specifically at reducing the number of partial products to save hardware costs, while still retaining the full precision in contrast to e.g. truncation (Kidambi et al., 1996) both measures help to further accelerate the multiplication.

The radix-4 Booth algorithm encodes three overlapping digits of the multiplier into one new digit (Booth, 1951). The number of multiplier digits and thus the number of partial products is reduced by 50%, thus accelerating the multiplication significantly thanks to a shorter critical path. Although additional effort is needed for the encoding circuit, it only consists of a few logic gates as depicted in Fig. 1. Also, the partial products are one bit larger than the original multiplicand and some additional effort is needed for the partial product generation (PPG). But since their number is divided by two and the addition array is considerably smaller, the encoding can be compensated and the hardware usage is reduced. Lee (2005) presents a dedicated circuit to handle different word-lengths in this context, but its scalability is limited and this method results in a large overhead. Figure 2 shows the logic circuit for our PPG.

Our previous designs, e.g. the configurable parallel/parallel multipliers presented in Pfänder et al. (2004), were based on a carry-ripple adder array. This approach can be combined with a Booth recoding scheme as we have shown in Zhou et al. (2007) which already leads to the speed improvement shown in Table 1 over a standard array multiplier. The arbitrary array size is still possible, while the PPG is realized separately. Control signals manipulate the carry propagation and also the sign handling functionality in certain perimeter cells at run-time. For further acceleration thanks to a shorter critical path, we have re-shaped the inner structure of the partial product addition (PPA) stage and applied a carry-save addition technique instead. This enables us to design the PPA independently from the final adder, which

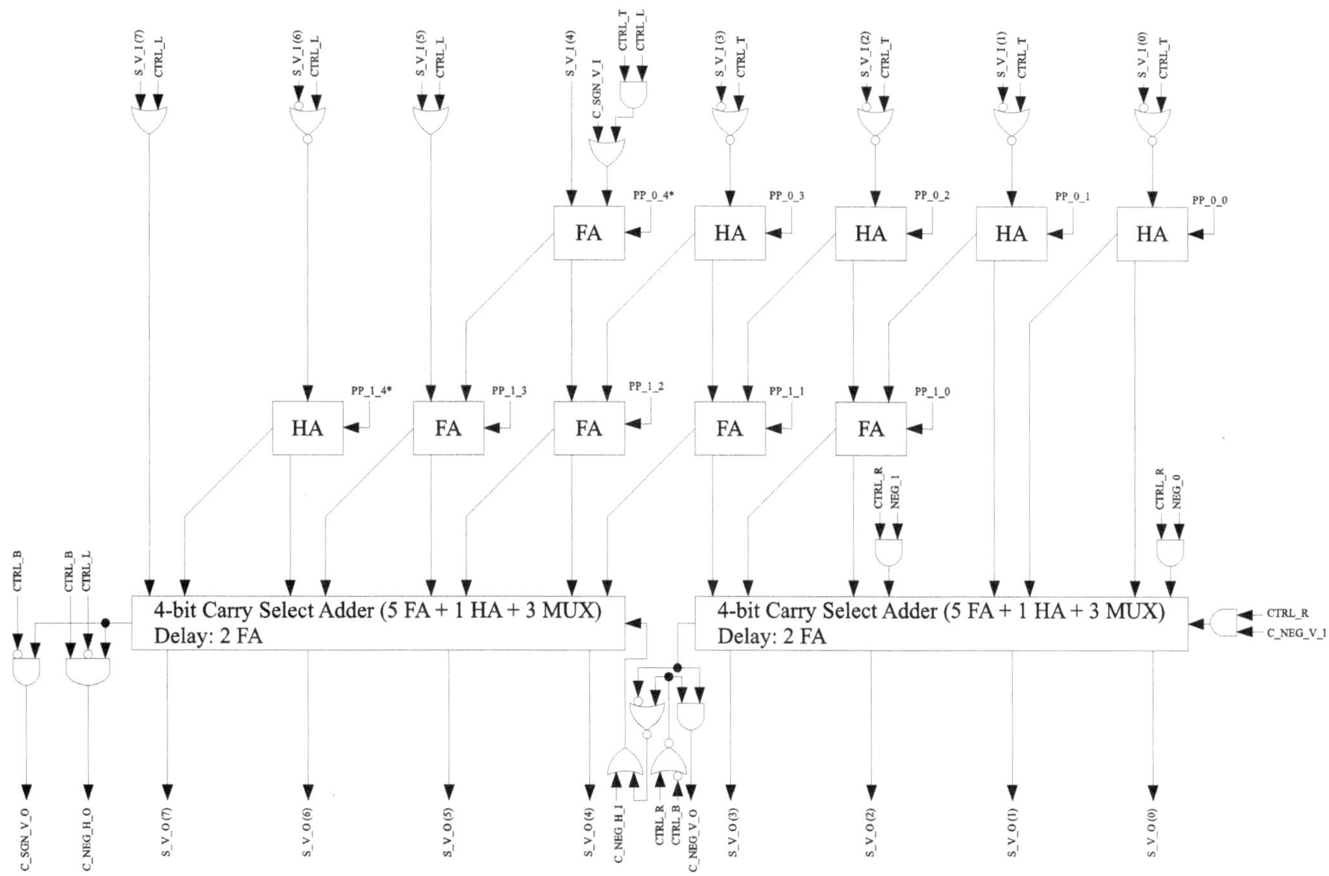

Fig. 3. Reconfigurable Booth multiplier with carry-save partial product generation and accelerated carry-select final adder.

can either be a standard carry-ripple adder row or an accelerated carry-select adder as pictured in Fig. 3.

3 Comparison

In order to perform a meaningful assessment in terms of hardware complexity and data throughput, the different structures for flexible word-length multiplication are compared with a selection of previous designs. We have adopted two benchmarks for comparison: The theoretical and hardware-independent approach is explained in Sect. 3.1 and the implementation results of an abstract hardware description in VHDL targeting a Xilinx Virtex-II Pro FPGA device is summarized in Sect. 3.2. Although this method does not lead to the same results as when using a semi-custom design flow toward an ASIC, it helps to quantify the designs and estimate the implementation costs on a common basis, i.e. the number of occupied slices in the FPGA device is assumed to scale up with the area usage. From this data, we derive a combined and normalized figure of merit in order to compare the multiplier concepts, display them as vertical-bar charts and discuss the results.

Table 1. Speed improvement of Booth recoding scheme over parallel array multiplier and percental overhead for reconfiguration (Zhou et al., 2007).

Word-length (bit)	Speed Advantage	Overhead for Reconfiguration
16	13.8%	2.0%
24	21.5%	1.7%
32	22.5%	1.6%

3.1 Theoretical hardware and speed comparison

Our theoretical hardware and speed comparison takes into account the number of required transistors and the number of full adder delays for a multiplier element. While the transistor numbers as depicted in Fig. 4 calculated for one exemplary implementation, e.g. assuming 24 transistors for one full adder, show no significant differences for the hardware usage, the multiplication times are more interesting, because they represent a hardware-independent figure for comparing the speed. Figure 5 shows the estimated delay in terms of full adders over the multiplier element's word-length.

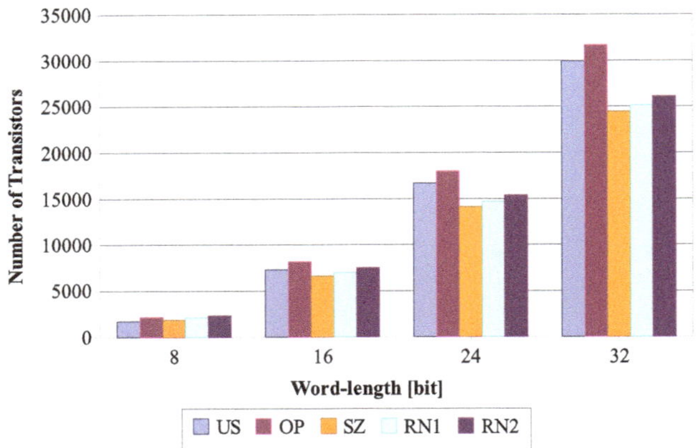

Fig. 4. Theoretical hardware usage for different word-lengths, expressed in terms of transistor numbers.

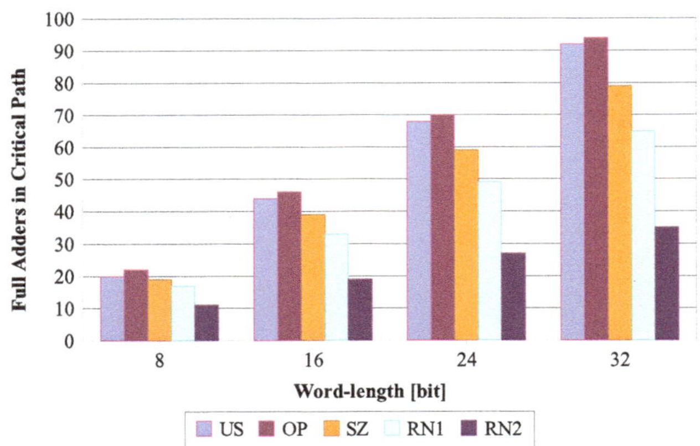

Fig. 5. Theoretical multiplication delay for different word-lengths, expressed in terms of the number of full adders in the critical path.

The diagram legends in Figs. 4–7 contain the following abbreviations: US stands for an unsigned parallel/parallel array. OP is one of our previous designs from Pfänder et al. (2004) using a modified Baugh and Wooley algorithm (Baugh and Wooley, 1973) and a parallel/parallel array. SZ is our first approach to introduce a Booth recoding scheme with a carry-ripple adder array, as covered in Zhou et al. (2007). RN1 incorporates a carry-save adder array and a carry-ripple final adder, while RN2 uses a carry-select final adder instead, as pointed out in the previous section. The numbers used for the diagrams in Figs. 4–5 represent the hardware usage of a basic multiplier element scaled to different word-lengths.

3.2 FPGA synthesis hardware and speed comparison

In order to get actual numbers for the hardware usage of our multiplier elements, we use the synthesis results from mapping the VHDL structural description onto a Xilinx Virtex-II Pro FPGA using the vendor's Integrated Synthesis Environment (ISE) software. More precisely, the number of occupied slices in the FPGA device is used as a measure for area usage,

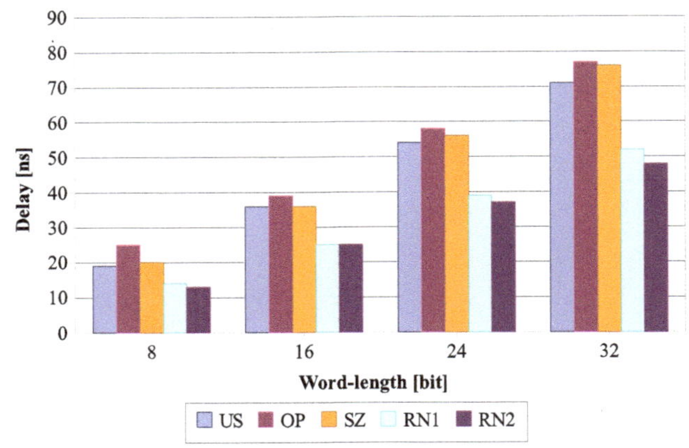

Fig. 6. Estimated delay for a cascaded superior multiplier assembled from 8×8-bit basic elements, without interconnect influence.

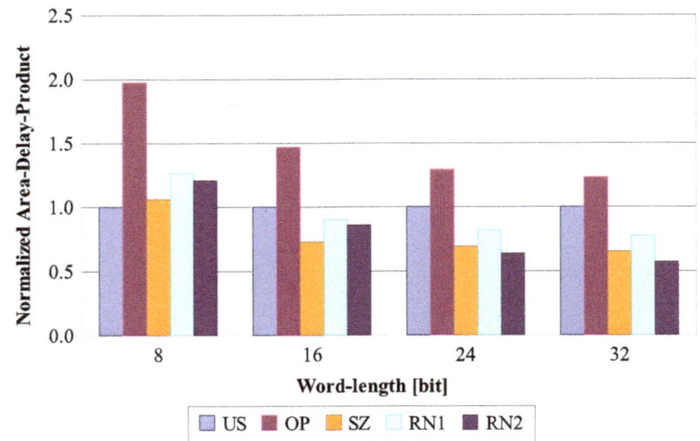

Fig. 7. Combined assessment criterion for scaled multiplier elements with different word-lengths, normalized with respect to an unsigned parallel multiplier.

and the critical path delay from the synthesis report is interpreted as a measure for multiplication time. Our experiments have shown that the synthesis results reflect the trends of the hardware-independent comparison and therefore confirm the theoretical prognosis.

The estimated multiplication time in nanoseconds for a cascaded multiplier array that was constructed from 8×8-bit basic elements is taken from the synthesis reports and shown in Fig. 6. Moreover, the influence of the interconnect and routing network has to be considered, but this should be done using a semi-custom ASIC design flow instead.

3.3 Combined comparison of area and delay

In addition to the comparisons summarized in the previous subsections, we have adopted the product of area and delay as a cost function normalized to the respective values of an n-bit unsigned parallel multiplier:

$$\text{Cost} = \frac{\text{No. of transistors} \times \text{No. of full adder delays}}{\text{Unsigned reference multiplier}} \tag{1}$$

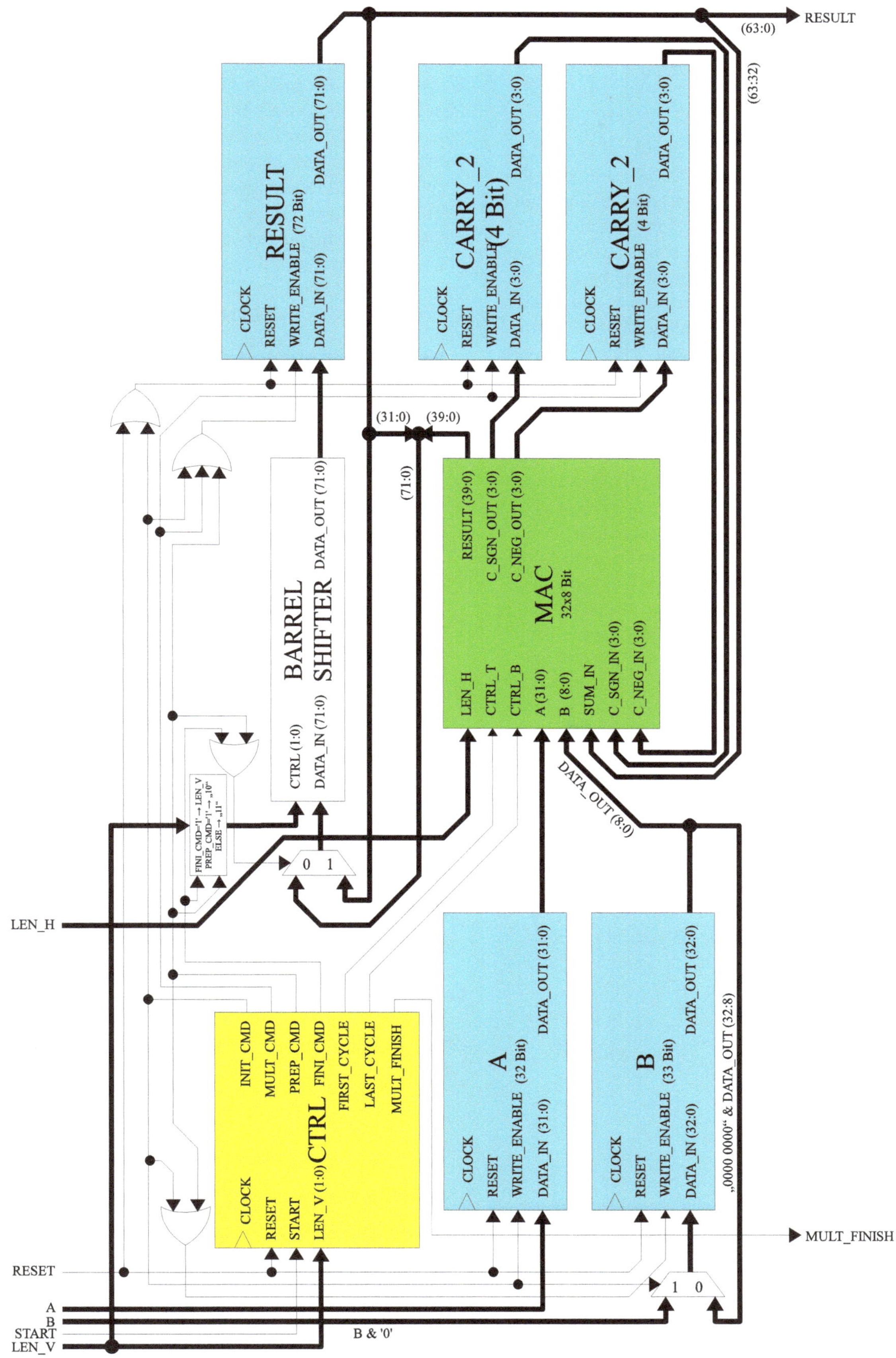

Fig. 8. Block-serial multiplication scheme with control circuit, registers, shifter and a configurable 32×8-bit MAC core.

Figure 7 plots the cost function over different word-lengths for scaled multiplier elements and clearly shows the significantly higher efficiency due to the Booth encoding scheme.

4 Novel block-serial approach

Assembling arbitrary large multiplier structures through the concatenation of configurable basic blocks leads to high routing efforts and a complex interface design communicating with the circuit periphery. In order to avoid this hardware overhead a priori, we have come up with an approach to provide a configurable precision multiplier with less operating expense by introducing a half-serial computation of the multiplier operand. This means that only one of the PPA rows is actually implemented in hardware – in the case of an 8-bit basic block this is equivalent to a 32×8-bit multiplier-accumulator (MAC). The other cells are replaced by buffered intermediate results that are fed back to the MAC sum input. After an certain number of computation steps, the multiplication result is valid at the output with the required precision that can be 64-bit maximum. With this arrangement, multiplications of 8×8-bit up to 32×32-bit can be realized, where both input operands are varied independently in steps of 8-bit. The major attribute of this block-serial structure is the fact that in contrast to a fully parallel multiplier, a number of clock cycles depending on the word-length n of the multiplier is required for computing the product. Since it is segmented in blocks of 8-bit, the number of computation steps including multiplication, shifting and addition is $n/8$. The multiplier's word-length is given by the 2-bit input `LEN_V` as a positive number in the range of $[0 \ldots 3]$ where $n=(\texttt{LEN\_V}+1)\cdot 8$. The principle design, composed of a control circuit, a shifter registers for the operands, a 32×8-bit MAC and feedback lines for the intermediate results to the sum input is shown in Fig. 8 on page 117. CTRL incorporates a finite state machine that receives and interprets `LEN_V` and addresses the registers and the MAC block accordingly. The outstanding feature of this design is its possibility to compute at a precision that is configurable at run-time, performing $(8\cdot x)\times(8\cdot y)$-bit up to 32×32-bit multiplications and using 60% less hardware than a fully parallel 32×32-bit array multiplier with sign handling capability and data exchange interfaces.

5 Conclusions

In this work, we have described and compared a number of four different run-time reconfigurable multiplier architectures in terms of hardware usage and speed. Starting from a previous design based on a parallel/parallel carry-ripple array, we have improved the architecture by a Booth recoding scheme, a carry-save adder array and an accelerated final adder. The overhead for a reconfigurable module over a static unsigned carry ripple multiplier varies between 24% and 37% at a module word length of 8-bit, but the critical path can be reduced by up to 45%, therefore accelerating the design significantly. For an 8-bit basic building block, our new approach requires 50% less hardware and decreases the delay by about 60% according to our synthesis experiments targetting a Xilinx Virtex-II Pro FPGA device. Moreover, we have applied a normalized area-delay product for fair comparison and verified the significant benefit of the Booth encoder, while still retaining the possibility to change the multiplier's precision at run-time. We have also presented a novel block-serial multiplication approach which turned out to be noticeably more area-efficient than a full 32×32-bit reconfigurable multiplier, since it requires 60% less hardware. We are currently working on the implementation of our multiplier architectures in standard-cell ASIC technology in order to avoid the FPGA synthesis limitations and make further improvements, and also developing a detailed routing and interface design.

Acknowledgements. The work described in this paper is supported by a grant from the Research Grant Council of Hong Kong S.A.R. and the German DAAD (Project No. G-HK019/05).

References

Baugh, C. R. and Wooley, B. A.: A Two's Complement Parallel Array Multiplication Algorithm, IEEE Trans. Computers, C-22, 1045–1047, 1973.

Bermak, A. and Martinez, D.: A compact 3D VLSI classifier using bagging threshold network ensembles, IEEE Trans. Neural Networks, 14, 1097–1109, 2003.

Booth, A. D.: A Signed Binary Multiplication Technique, Quart. J. Mechanics and Appl. Mathematics, 4, 236–240, 1951.

Di, J. and Yuan, J. S.: Run-time Reconfigurable Power-aware Pipelined Signed Array Multiplier Design, in: International Symposium on Signals, Circuits and Systems, 2, 405–408, 2003.

Kidambi, S. S., El-Guibaly, F., and Antoniou, A.: Area-efficient multipliers for digital signal processing applications, IEEE Trans. Circuits and Systems II, 43, 90–95, 1996.

Lachowicz, S. W. and Pfleiderer, H.-J.: Fast Evaluation of Nonlinear Functions using FPGAs, Adv. Radio Sci., 6, 2008.

Lee, H.: Power-Aware Scalable Pipelined Booth Multiplier, IEICE Trans. Fundamentals, E88-A, 3230–3234, 2005.

Parhami, B.: Computer Arithmetic: Algorithms and Hardware Designs, Oxford University Press, Oxford/New York, 2000.

Pfänder, O. A. and Pfleiderer, H.-J.: Dynamische Rekonfiguration von arithmetischen Einheiten auf Bitebene, Adv. Radio Sci., 3, 319–323, 2005, http://www.adv-radio-sci.net/3/319/2005/.

Pfänder, O. A., Hacker, R., and Pfleiderer, H.-J.: A Multiplexer-Based Concept for Reconfigurable Multiplier Arrays, in: International Conference on Field-Programmable Logic and Applications, pp. 938–942, Antwerp, Belgium, 2004.

Zhou, S., Pfänder, O. A., Pfleiderer, H.-J., and Bermak, A.: A VLSI Architecture for a Run-Time Multi-Precision Reconfigurable Booth Multiplier, in: International Conference on Electronics, Circuits and Systems, Marrakech, Morocco, 2007.

28

FPGA implementation of trellis decoders for linear block codes

S. Scholl[1], E. Leonardi[2], and N. Wehn[1]

[1]Microelectronic Systems Design Research Group, University of Kaiserslautern, 67663 Kaiserslautern, Germany
[2]Institute of Informatics, Federal University of Rio Grande do Sul, 91501970 Porto Alegre, Brazil

Correspondence to: S. Scholl (scholl@eit.uni-kl.de)

Abstract. Forward error correction based on trellises has been widely adopted for convolutional codes. Because of their efficiency, they have also gained a lot of interest from a theoretic and algorithm point of view for the decoding of block codes. In this paper we present for the first time hardware architectures and implementations for trellis decoding of block codes. A key feature is the use of a sophisticated permutation network, the Banyan network, to implement the time varying structure of the trellis. We have implemented the Viterbi and the max-log-MAP algorithm in different folded versions on a Xilinx Virtex 6 FPGA.

1 Introduction

Forward error correction is widely used in today's communication systems for the correction of transmission errors. In the last years and decades many different error correction schemes have been introduced and successfully adopted in various communication standards. Prominent examples for channel codes are convolutional codes, Reed-Solomon codes, turbo codes and LDPC codes.

The optimal correction strategy is called maximum likelihood (ML) decoding. Since ML decoding is very complex for many practically used codes, most of the decoding algorithms are suboptimal heuristics, e.g. the turbo decoding algorithm (Lin and Costello, 2004). However, in this paper we consider two algorithms, that can efficiently perform ML decoding of convolutional codes or small block codes: the Viterbi algorithm (VA) (Viterbi, 1967) and the BCJR algorithm (Bahl et al., 1974).

The efficiency of the VA and BCJR algorithm originate mainly in the exploitation of the code's structure, which is graphically represented as a trellis diagram. In the past a vast amount of research has been carried out on trellis based decoding for convolutional codes, including works from general theory to hardware implementations for real world applications.

However, not only convolutional codes can be described by trellis diagrams. Also block codes can be represented as trellis and thus efficient decoding algorithms for convolutional trellises can also be applied to block codes. Trellis decoders for block codes have many different applications as standalone ML decoder or as components of larger decoding heuristics. Here we want to point out just a few use cases:

- as a maximum likelihood decoder for small block codes
- as a component decoder for turbo product codes, Pyndiah et al. (1994)
- as a check node decoder for generalized LDPC codes, Tanner (1981)
- as a component for soft decision decoding of Reed-Solomon codes, e.g. in Vardy and Be'ery (1991), Liu and Lin (2004)

Trellises of block codes mostly have a special structure (called time varying trellis), that poses a major challenge for the hardware designer. So far hardware architectures and implementations have not been considered yet. In Kim et al. (2003) a trellis decoder was implemented on an FPGA, but its use is restricted to a small group of block codes, that do not have a time varying structure.

In this paper we propose an architecture, that is able to handle all block codes of reasonable size. We solve the challenge posed by the time varying structure by introducing a optimized Banyan permutation network, that is tailored to the application. We evaluate the architectures for the VA and BCJR as well as folded versions by implementing them on a Virtex 6. To our best knowledge, this is the first hardware implementation of trellis decoding for arbitrary block codes.

The paper is structured as follows: in Sect. 2 we first present the construction rules for a trellis, followed by a brief description of the algorithms in Sect. 3. Section 4 describes the proposed architectures and the implementation results can be found in Sect. 5.

2 Trellis construction for block codes

We consider a binary block code with block length N and K information bits. The code is defined by its parity check matrix (PCM) $\mathbf{H}$ of dimension $(N-K)\times N$. The columns of $\mathbf{H}$ are denoted by $\mathbf{h}_k$. A valid code word is denoted by $\mathbf{x}=(x_0,x_1,...,x_{N-1})$ and the received log-likelihood ratios (LLRs) by $\mathbf{y}=(y_0,y_1,...,y_{N-1})$

A trellis diagram is a graphical representation of the code word space of a channel code. Every path in a trellis connecting the starting and end point correspond to exactly one code word.

The trellis diagram for block codes can be constructed in two different ways: using the generator matrix or the parity check matrix (PCM) of the code.

The first construction method is based on the generator matrix (see Lin and Costello (2004) for more information). Since this construction method is quite complex and requires the generator matrix to be in a special form (trellis oriented generator matrix form), we apply the second construction method based on the PCM.

The PCM method provides full flexibility on the matrix, so that any PCM can be used. Furthermore, the trellis structure can easily be deduced from the PCM, which makes it suitable for a hardware implementation. Since the trellis describes a block code, we call it a bit-level trellis – in contrast to the convolutional trellis. In the following we describe the construction method of Wolf (1978).

The bit-level trellis has N trellis steps (sometimes called time steps), and $M=2^{N-K}$ states per step. The states are labelled by a binary $N-K$ tuple $\mathbf{s}_k^m$, where $m=0,1,...,M-1$ denotes the index of a state in step $k=0,1,...,N$.

To construct the trellis, the states between step k and $k+1$ are connected by branches. There are two different types of branches. One corresponds to code bit '0' ($x_k=0$) the other one to code bit '1' ($x_k=1$). The branches are established recursively as follows: From each state $\mathbf{s}_k^m$ of step k two branches depart to state

$$\mathbf{s}_{k+1}^m=\mathbf{s}_k^m \qquad \text{for } x_k=0$$

$$\mathbf{s}_{k+1}^l=\mathbf{s}_k^m+\mathbf{h}_k \qquad \text{for } x_k=1$$

Since only paths from $\mathbf{s}_0^0$ to $\mathbf{s}_N^0$ represent code words, all other paths can be expurgated.

There are some remarkable differences between a convolutional trellis and a bit-level trellis. A bit-level trellis is in general a time-varying trellis, i.e. state transitions change over time – in contrast to the convolutional trellis.

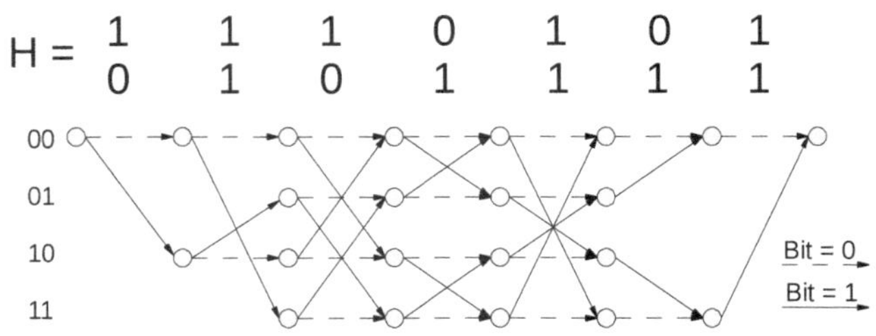

Figure 1. Example for a trellis construction using the PCM **H**

Besides the time varying property, in a bit-level trellis only one bit is associated with a time step. In a convolutional trellis usually two or more bits correspond to one time step.

Additionally, it should be pointed out that the "0" branches always connect states having the same label. This property can be exploited in hardware, as we will see later.

A small example for $N=7$ and $K=5$ with a $M=4$ state trellis is depicted in Fig. 1 to clarify the trellis construction.

3 Trellis decoding algorithms

Once the trellis representation of a code is obtained, it enables the use of very efficient decoding algorithms, like the VA and the BCJR algorithm. In this section we give a short summary of these two algorithms for bit-level trellises and their variants in the log domain.

3.1 Viterbi algorithm

The VA (Viterbi, 1967) performs ML decoding efficiently on a trellis. It looks for the most probable path in the trellis by recursively building up paths through the trellis and discarding unlikely paths in every step. We shortly repeat the VA in the logarithm domain, that is usually used for hardware implementation.

For the VA every trellis state $\mathbf{s}_k^m$ is assigned a state metric $\alpha(\mathbf{s}_k^m)$. The state metrics for step k are calculated recursively from those of step $k-1$.

1. initialize the state metrics at step 0: $\mathbf{s}_0^0$ with 0 and all other $\mathbf{s}_0^m$ ($m\neq 0$) with -infinity

2. For all states $k=1,..,N-1$ and all indices $m=0,1..,M-1$ calculate

$$\alpha\left(\mathbf{s}_k^l\right)=\max\left[\alpha\left(\mathbf{s}_{k-1}^l\right),\ \alpha\left(\mathbf{s}_{k-1}^m\right)+y_{k-1}\right] \tag{1}$$

where $\mathbf{s}_k^l=\mathbf{s}_{k-1}^m+\mathbf{h}_{k-1}$

3. output the path which led to the maximum state metric $\alpha\left(\mathbf{s}_N^0\right)$, called the traceback step.

More detailed information on the VA can easily be found in literature (Lin and Costello, 2004; Wolf, 1978).

3.2 BCJR and Max-log-MAP algorithm

A drawback of the VA is that it does not provide any soft output information, which is required by modern decoding heuristics. However, the BCJR algorithm (Bahl et al., 1974) provides such additional information.

For hardware implementations is it advantageous to use the BCJR algorithm in the logarithm domain, which is called log-MAP, or its low complexity version max-log-MAP (Robertson et al., 1995). In this paper we consider the max-log-MAP, because it provides low complexity without degrading the decoding performance significantly.

In the max-log-MAP (and also the BCJR algorithm) every state is assigned two state metrics: the forward state metrics $\alpha\left(\mathbf{s}_k^m\right)$ and the backward state metrics $\beta\left(\mathbf{s}_k^m\right)$.

Max-log-MAP decoding consists of three phases:

1. forward recursion (calculates state metrics $\alpha\left(\mathbf{s}_k^m\right)$)
2. backward recursion (calculates state metrics ($\beta\left(\mathbf{s}_k^m\right)$)
3. soft output calculation (using $\alpha\left(\mathbf{s}_k^m\right), \beta\left(\mathbf{s}_{k+1}^m\right)$ and y_k)

The forward recursion is equal to that of the VA in Eq. (1). For the backward recursion the states are processed in reversed order, i.e. from right to left. Details of the algorithm can be found e.g. in Lin and Costello (2004).

It can already be seen, that the recursion steps are important in both the VA and the max-log-MAP. It consists of add-compare-select (ACS) operations in Eq. (1) and requires permutations of state metrics according to the time varying branch structure. In the following section, we will focus on the hardware architecture for such a recursion unit for bit-level trellises.

4 Proposed architectures

In this section we present the hardware architectures for the VA and the max-log-MAP decoder for bit-level trellises. We propose the use of a Banyan network to implement the flexible state transitions. Finally we present a folded architecture, which reduces the decoder area to enable the implementation for trellises with a large number of states.

4.1 Architectures for Viterbi and Max-Log-MAP algorithm

The top level architectures for Viterbi and max-log-MAP decoder are depicted in Figs. 2 and 3. Although they look different, their core functionality is the recursion unit, which is identical in both cases.

The recursion unit for a simple case of a 4 state trellis is shown in Fig. 4. All state metrics $\alpha\left(\mathbf{s}_k^m\right)$ of one time step k are

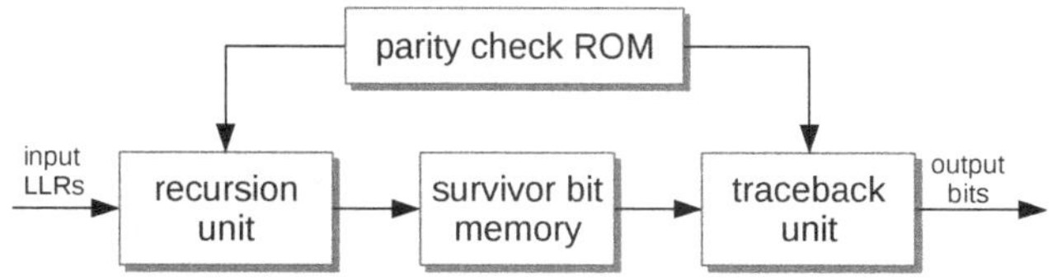

Figure 2. Top level architecture of the Viterbi decoder

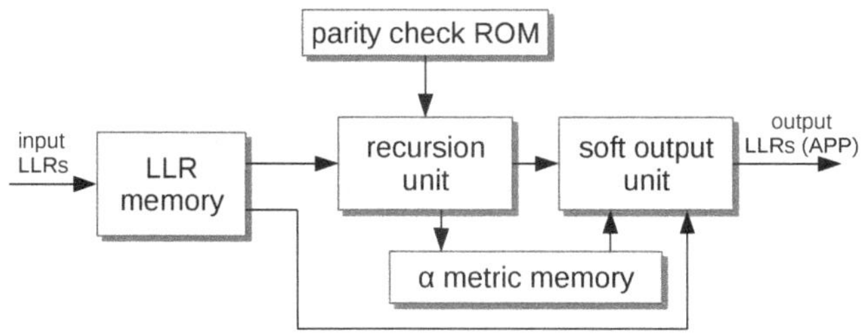

Figure 3. Top level architecture for the max-log-MAP decoder

calculated in parallel. The different trellis steps are calculated consecutively, i.e. one trellis step per clock cycle. The state metrics are temporarily stored in the registers.

One iteration consists of routing the state metrics through the permutation network first. The routing is done according to the branch structure of the trellis and is determined by the columns of the PCM $\mathbf{h}_k$. Then the ACS units calculate Eq. (1). Finally the state metrics are stored in the register again.

Note, that the state metrics are directly fed back to the ACS units. This maps the branches for code bit '0' to the hardware.

The state metrics and received LLRs $\mathbf{y}$ are represented as fixed point numbers. To keep the number of bits low and therefore to save resources without provoking catastrophic overflows, modulo arithmetic is used here (Hekstra, 1989).

4.2 Permutation network

An essential part of the trellis decoders is the permutation network (PN). It routes the state metrics along the Bit "1" branches during the recursion.

The PNs in the trellis decoders must have $M = 2^{N-K}$ inputs and outputs. The network is controlled by the column bits of the PCM $\mathbf{h}_k$ and consists of 2×2 butterfly switches, which can permute or route through its two inputs.

We first investigate the use of the well known Benes PN (Benes, 1964) and in a second step propose the use of a more sophisticated network, the Banyan PN (Goke and Lipovski, 1973).

The Benes PN is capable of performing all possible $M!$ permutations, although in this application only M are required. Therefore, the Benes PN uses more resources than necessary. A drawback of the Benes PN is its elevated number of $2 \cdot \log_2(M) - 1$ stages, which directly lengthens

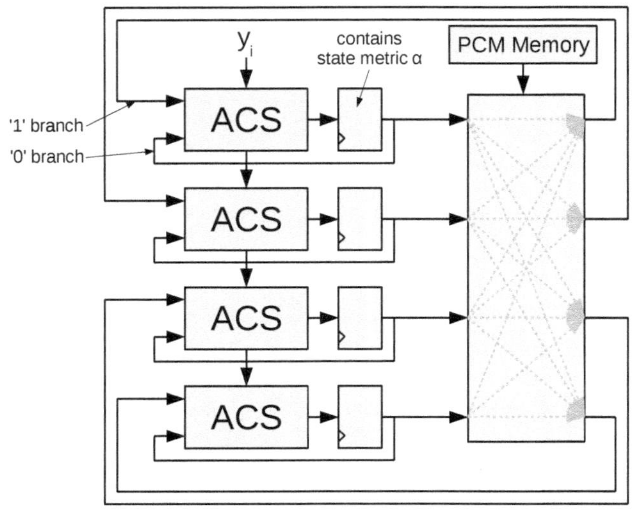

Figure 4. Architecture of the recursion unit

the maximum combinational path of the recursion unit and thus decreases the throughput. Furthermore the Benes PN requires a complex controlling logic, which maps the bits from the PCM to the control bits of the switches, see Fig. 7.

The second network studied is the Banyan PN. It performs the necessary permutation with approximately half the number of stages ($\log_2(M)$) in comparison to the Benes PN. This reduces the signal propagation time and improves throughput. Moreover the Banyan PN has approximately half the number of switches (see Fig. 5). Another advantage is that the bits of the columns of the PCM can directly be applied to control the switches in the network. No extra controlling logic is required. The small number of switches and the saving of the complex controlling logic makes the Banyan PN the network of choice for this application.

In the following, we provide our method of how to build up such a network. The Banyan PN is constructed from two Butterfly PNs and a final permutation stage. The smallest Butterfly PN is composed of a single switch and larger ones can be constructed recursively, i.e. a network with M inputs is obtained from two Butterfly PNs with $M/2$ inputs.

We place the second Butterfly PN below the first one and denote the outputs by $S(i)$ with $i = 0, 1, .., M/2 - 1$ and $i = M/2, .., M - 1$. The final stage is a column of $M/2$ switches that placed right of the two Butterfly PNs. Their inputs are denoted by $I(i)$ with $i = 0, 1, .., M - 1$ and their outputs by $B(i)$. The connections between the two Butterfly PNs and the final stage is done by the following algorithm:

Finally, we connect the outputs $B(i)$ of the final stage to the outputs $O(i)$ of the Banyan PN by using the following algorithm:

Each input of the network has to be able to reach all the M different output addresses and that is only possible with $N - K$ stages. Thus it is impossible to do all the required permutations with a network with less stages than the Banyan.

Algorithm 1. Intermediate connections of a Butterfly PN.

```
for (i = 0 to (M/4 -1)) do
    I(2*i) ← S(2*i)
    I(2*i + 1) ← S(2*i + M/2)
    I(2*i + M/2) ← S(2*i + 1)
    I(2*i + M/2 + 1) ← S(2*i + M/2 + 1)
end for
```

Algorithm 2. Output connections of a Banyan PN.

```
for (i = 0 to (M/2 -1)) do
    O(i) ∗←B(2 i)
    O(i + M/2) ∗←B(2 i + 1)
end for
```

Figure 6 shows the construction of an 8×8 Banyan PN using the algorithms from above.

4.3 Folding

For trellises with a large number of states M, the above presented architecture grows quickly and may become too large for an FPGA implementation. To counteract this problem, we propose folding to reduce the required resources of the decoder at the expense of a reduced throughput.

In the folded architecture not all M state metrics in one trellis step are calculated in one clock cycle. Instead the state metrics are calculated, e.g. in two clock cycles. In this case half of the ACS units can be reused.

Furthermore it reduces the size of the PN by more than half given that its construction is recursive (see above). Note, that the folded Banyan PN with $M/2$ exactly fulfils the permutation requirements of the folded architecture, and is thus suitable for folded architectures.

The number of clock cycles required to calculate the state metrics of one trellis step is called folding factor f and must be a power of 2.

Due to folding the size of the PN, the number of ACS units and the soft output unit (for the max-log-MAP decoder) is largely reduced. However, some additional hardware resources are needed to distribute the state metrics to the PN and ACS units and registers to store the partially calculated state metrics.

The area reduction of the soft output unit is directly proportional to the folding factor f. Therefore folding is especially advantageous for the max-log-MAP decoder.

5 FPGA implementation

To evaluate the architectures in detail we have implemented the architectures on a Xilinx Virtex 6 (XC6VLX75T-3) FPGA using ISE 14.1. The architectures for the VA, the max-log-MAP and their folded version have been evaluated for trellises with different number of states. We analyse the

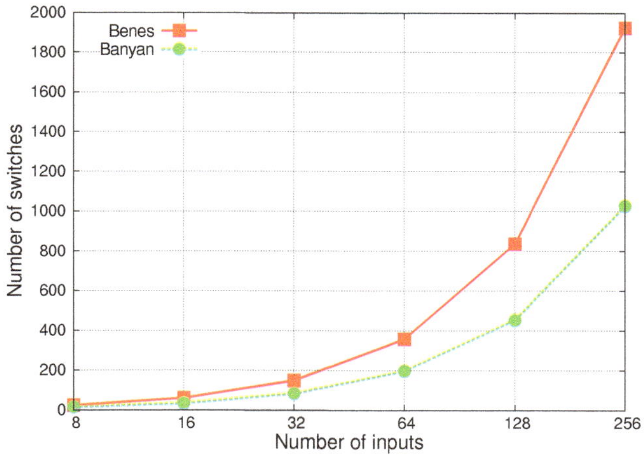

Figure 5. Number of switches used by the Benes and the Banyan PN

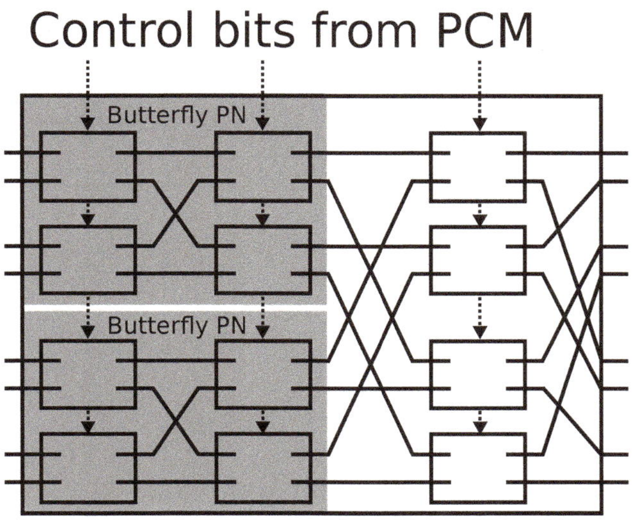

Figure 6. Eight input Banyan PN: it needs less switches than the Benes PN and no controlling is needed

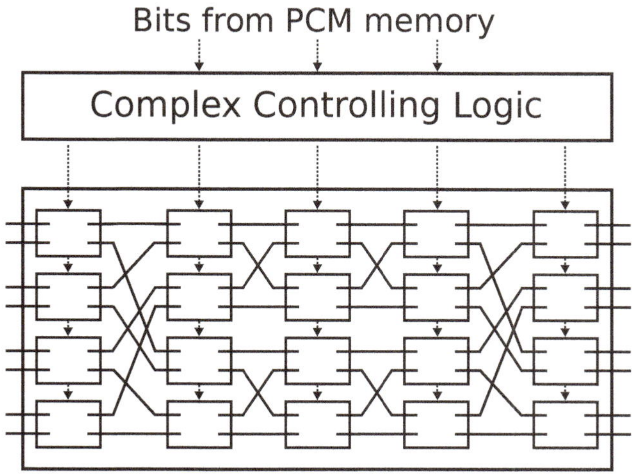

Figure 7. Eight input Benes PN: it needs more switches than the Banyan PN and additional controlling is needed

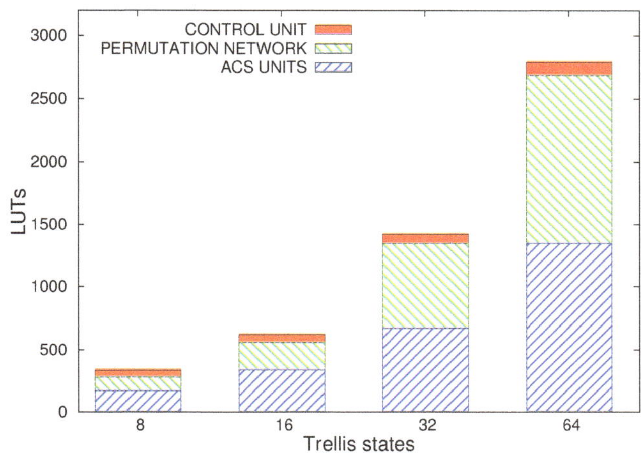

Figure 8. Viterbi decoder: look-up tables

resource consumption regarding the required look-up tables (LUTs) after place & route.

The numbers presented are only dependent on the number of trellis states M and not on the number of trellis steps N. The number of trellis steps N only influences the size of the memories. However this is not the critical resource in the design.

The quantization of the metrics is dependent on the context in which the decoders are used. However it has been experienced by simulations that a state metric quantization of 7 Bits for the VA and 8 bits for the max-log-MAP is reasonable.

In Figs. 8 and 9 the required LUTs for the unfolded architecture ($f = 1$) is shown, separated for each decoder part.

For the VA almost 50 % of the LUTs are occupied by the ACS units. The PN also consumes nearly 50 % of the LUTs. Therefore the ACS units and the PN are the dominating parts. The remaining fraction is occupied by the controlling, which also includes the traceback unit.

The max-log-MAP is dominated by the soft output unit, which needs approximately 50 % of all LUTs. The PN and the ACS units consume around 25 % each.

In Figs. 10 and 11 the number of required LUTs for different folded architectures are shown. It shows how the occupied resources reduce and allows for the implementation of larger trellises. The throughput decrease for the VA can be seen in Fig. 12.

6 Conclusions

In this paper we have investigated hardware architectures for bit-level trellises. We have selected the trellis construction based on the PCM, because it provides maximum flexibility. After a brief review of the VA and the BCJR decoding algorithms, we proposed efficient hardware architectures. A key

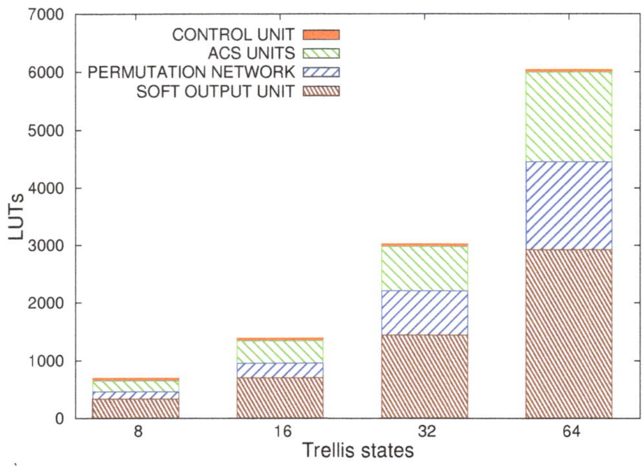

Figure 9. Max-log-MAP decoder : look-up tables

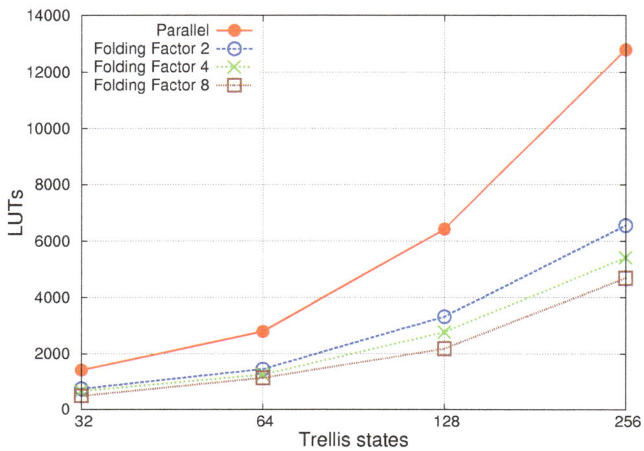

Figure 10. Viterbi decoder: LUTs dependent on folding factor

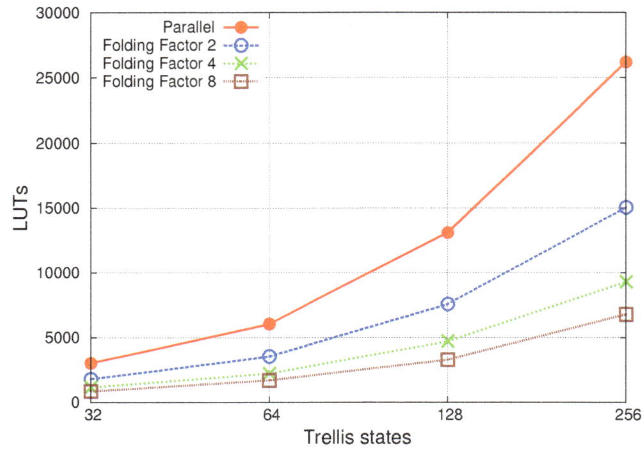

Figure 11. Max-log-MAP decoder: LUTs dependent on folding factor

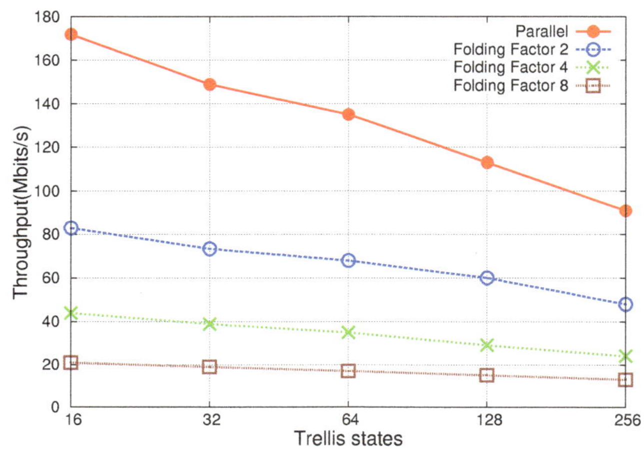

Figure 12. Viterbi decoder: throughput dependent on folding factor

feature is the Banyan PN, which maps the time variant property of the trellis efficiently to hardware and significantly outperforms standard solutions like the Benes PN. Furthermore we presented a folded version of the architecture to enable the implementation for large trellises. Finally the resource consumption and throughput of the architecture have been evaluated on a Xilinx Virtex 6 FPGA.

Acknowledgements. We gratefully acknowledge partially financial support by the DFG (project-ID: KI 1754/1-1) as well as by the Center of Mathematical and Computational Modelling of the University of Kaiserslautern. We thank F. Kienle for his valuable comments and suggestions.

Edited by: J. Götze

References

Bahl, L., Cocke, J., Jelinek, F., and Raviv, J.: Optimal Decoding of Linear Codes for Minimizing Symbol Error Rate, IEEE T. Inform. Theory, IT-20, 284–287, 1974.

Benes, V. E.: Optimal Rearrangeable Multistage Connecting Networks, The Bell System Technical Journal, 4, 1641–1656, 1964.

Goke, L. R. and Lipovski, G. J.: Banyan Networks for Partitioning Multiprocessor Systems, 1st Annual Symposium on Computer Architecture, 21–28, 1973.

Hekstra, A. P.: An Alternative to Metric Rescaling in Viterbi Decoders, IEEE T. Commun., 37, 1220–1222, 1989.

Kim, S., Ryoo, S., and Lee, S.: Block Turbo Codes Using Multiple Soft Outputs, in: Proceedings of the 3rd ISTC, Vol. 1, 247–250, Brest, 2003.

Lin, S. and Costello, D.: Error Control Coding 2nd., Prentice Hall PTR, Upper Saddle River, New Jersey, USA, 2004.

Liu, C. Y. and Lin, S.: Turbo encoding and decoding of Reed-Solomon codes through binary decomposition and self-concatenation, IEEE T. Commun., 52, 1484–1493, 2004.

Pyndiah, R., Glavieux, A., Picart, A., and Jacq, S.: Near optimum decoding of product codes, in: Proc. IEEE Global Telecommunications Conf. GLOBECOM '94. Communications: The Global Bridge, 339–343, 1994.

Robertson, P., Villebrun, E., and Hoeher, P.: A Comparison of Optimal and Sub-Optimal MAP decoding Algorithms Operating in the Log-Domain, in: Proc. 1995 International Conference on Communications (ICC '95), 1009–1013, Seattle, Washington, USA, 1995.

Tanner, R. M.: A Recursive Approach to Low Complexity Codes, IEEE T. Inform. Theory, IT-27, 533–547, 1981.

Vardy, A. and Be'ery, Y.: Bit-level soft-decision decoding of Reed-Solomon codes, IEEE T. Commun., 39, 440–444, 1991.

Viterbi, A. J.: Error Bounds for Convolutional Codes and an Asymptotically Optimum Decoding Algorithm, IEEE T. Inform. Theory, 13, 260–269, 1967.

Wolf, J.: Efficient maximum likelihood decoding of linear block codes using a trellis, IEEE T. Inform. Theory, 24, 76–80, 1978.

Systematic design of output filters for audio class-D amplifiers via Simplified Real Frequency Technique

E. Hintzen, T. Vennemann, and W. Mathis

Institute for Theoretical Electrical Engineering, Leibniz Universität Hannover, Hannover, Germany

Correspondence to: T. Vennemann (vennemann@tet.uni-hannover.de)

Abstract. In this paper a new filter design concept is proposed and implemented which takes into account the complex loudspeaker impedance. By means of techniques of broadband matching, that has been successfully applied in radio technology, we are able to optimize the reconstruction filter to achieve an overall linear frequency response. Here, a passive filter network is inserted between source and load that matches the complex load impedance to the complex source impedance within a desired frequency range. The design and calculation of the filter is usually done using numerical approximation methods which are known as Real Frequency Techniques (RFT). A first approach to systematic design of reconstruction filters for class-D amplifiers is proposed, using the Simplified Real Frequency Technique (SRFT). Some fundamental considerations are introduced as well as the benefits and challenges of impedance matching between class-D amplifiers and loudspeakers. Current simulation data using MATLAB is presented and supports some first conclusions.

1 Introduction and motivation

The development of audio technology in recent years is primarily characterized by more efficient and compact power components and systems. Due to the rapid growth and ongoing development of mobile devices, class-D amplifiers have become prevalent in modern applications.

The concept of class-D amplifiers can handle only switching signals, so the circuit complexity is much higher compared to linear power amplifier concepts to guarantee an overall linear signal transmission. The audio signal has to be initially encoded by a modulator (e.g. pulse width modulation). Then, the signal can be amplified by the switching stage and finally it has to be decoded by a demodulator (reconstruction filter). An alternative encoding method is called ZePoC (Zero Position Coding). ZePoC was initially used and implemented for audio coding by the Institute for Theoretical Electrical Engineering in Hannover. Much research was done over years resulting in a number of publications. A good overview could found in the white paper from Oliva et al. (2005).

The reconstruction filter is usually realized with conventional filter design techniques by defining pass and stop band and setting desired ripple and tolerance. Then, the transfer function is calculated with the help of well-known approximation procedures like Butterworth, Tschebyscheff or Cauer (Quek, 2006). A further work regarding the optimization of class-D output filter has been published by Doy (2002). In practice, a purely resistive load is often used to model the loudspeaker impedance at the output of the filter, thus neglecting complete frequency dependencies of real loudspeakers. In consequence of committing these effects, the frequency response of the filter will no longer be linear in the audio band and will distort the signal. This contribution deals with a new method of class-D output filter design which allows the designer to take into account the given impedance for an optimum overall frequency response. Therefore, some basic preliminaries and specifications have to be made first.

2 Specifications for class-D filter design

An existing class-D amplifier platform which has been introduced and successfully developed by the Institute for Theoretical Electrical Engineering in Hannover is used for a first filter design. Based on the sample rate frequency of $f_s = 48\,\text{kHz}$ for pulse code modulation systems (PCM) like CD or DVD players, and taking account of the Nyquist-Shannon

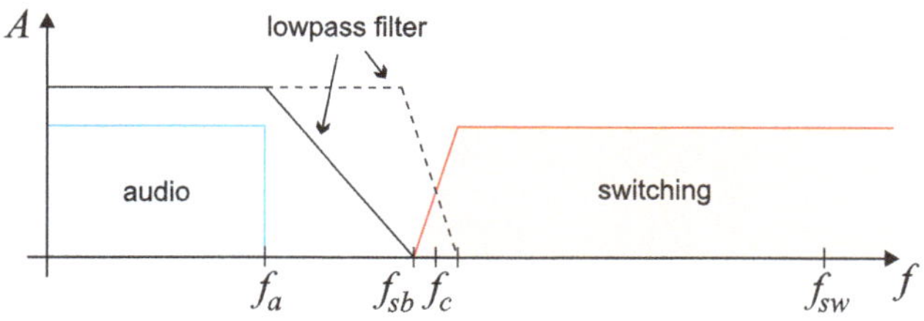

Figure 1. Frequency spectrum with audio and switching band.

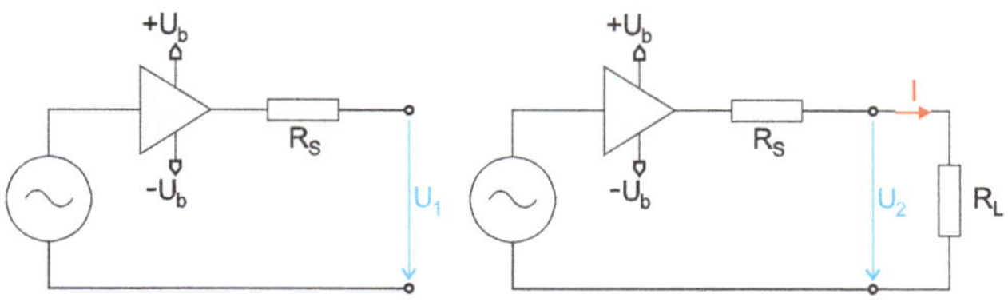

Figure 2. Output impedance measurement.

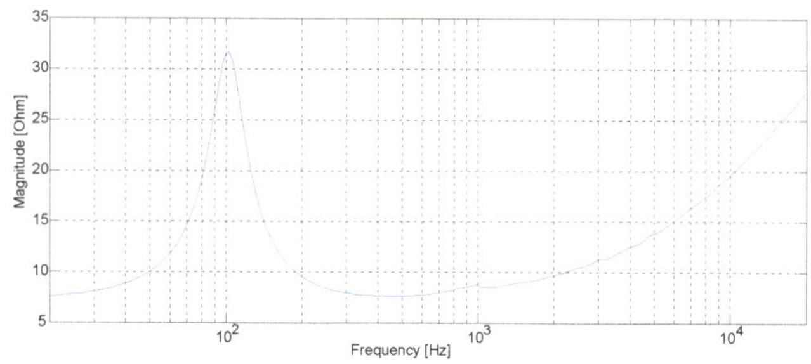

Figure 3. Measured impedance response of 8 Ω fullrange speaker.

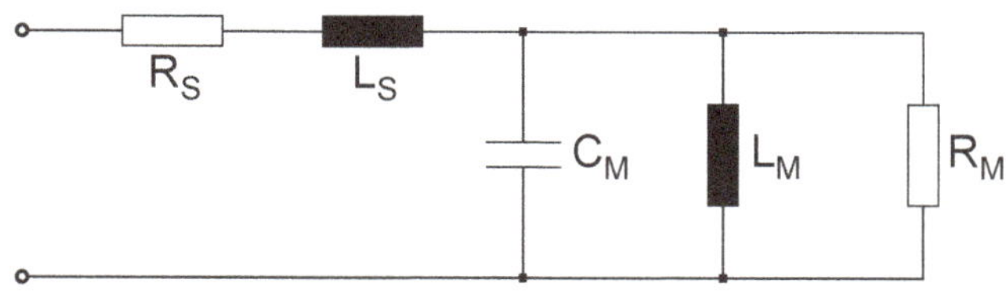

Figure 4. Electrical equivalent circuit of dynamic loudspeaker.

sampling theorem, the maximum signal frequency is limited to 24 kHz. However, the human ear is only capable of detecting frequencies up to 20 kHz, thus the audio frequency band can be defined from 20 Hz to $f_a = 20$ kHz. The frequency response of the overall circuit, which consists of class-D output impedance, reconstruction filter and loudspeaker impedance, has to be completely flat within this entire wave band. Further on, the carrier frequency of the single-sideband ZePoC modulation is set to $f_c = 50$ kHz and then the switching frequency of the class-D amplifier is given by $f_{sw} = 2f_c = 100$ kHz. The resulting frequency spectrum is shown in Fig. 1.

Since the radio frequency energy initially rises slowly, the transition band of the reconstruction filter can be arranged symmetrically to the carrier frequency. On the other hand, the frequency band between f_a and f_{sb} does not contain any information, so that the transition band can be moved towards the audio band. In both cases, a sufficiently high attenuation should be reached at f_{sb} to prevent distortion. Before introducing the SRFT algorithm, the impedances on the input and output side of the filter are described.

2.1 Class-D output impedance

The output impedance of a class-D switching amplifier is among other things dependent on the used power supply, but essentially it is determined by the resistance of the power transistors. Since the output voltage of the switching amplifier is a pulse width modulated signal with a fixed switching frequency, it is assumed that the output impedance of the amplifier can be accepted as a constant within the considered frequency band. Typically, the output resistance of a class-D switching amplifier is in the range of only a few milliohms.

A precise mathematical description or even the calculation of the output resistance of a switching amplifier is only possible with high effort. Therefore, the resistance is determined by a simple measurement procedure which is presented in Metzler (1993). In a first measurement, the amplifier is operated at no-load and a sine wave generator with fixed frequency is used as a signal source. The generator causes a defined voltage at the amplifiers output which is measured. In a second step, the amplifier is operated with a definded load and the voltage drop is measured. Then, knowing differential voltage $U_1 - U_2$ and current through load resistor $I = \frac{U_2}{R_L}$, the output resistance is given by $R_S = R_L \cdot (\frac{U_1}{U_2} - 1)$. The measurement setup is shown in Fig. 2. By applying this measurement method, the output resistance of the class-D amplifier can be computed approximately to be $R_0 = 100\,\mathrm{m\Omega}$.

2.2 Loudspeaker impedance

The SRFT is a numerical approximation procedure and can handle real frequency measurement and simulation data. This also means that the filter design has to be done specifically for one type of loudspeaker and the choice of a speaker has to be made during the preliminary stages. For the first design a 6.5" fullrange speaker with a nominal impedance of 8 Ω is used. Measuring the loudspeaker impedance is done by using a software which has to be extended by a simple measurement setup. The measured impedance response of the speaker is shown in Fig. 3.

The impedance starts with DC resistance of the voice coil at about 8 Ω. The magnitude increases with the self-resonance of the speaker cone around 100 Hz to more than 30 Ω, followed by a linear frequency range at around 4 kHz. Here, the impedance is purely resistive. Finally, the inductance of the voice coil dominates the loudspeaker impedance at around 10 kHz and the magnitude increases steadily. In addition, for simulation purposes and to verify later results, an electrical equivalent circuit of the loudspeaker is needed. A

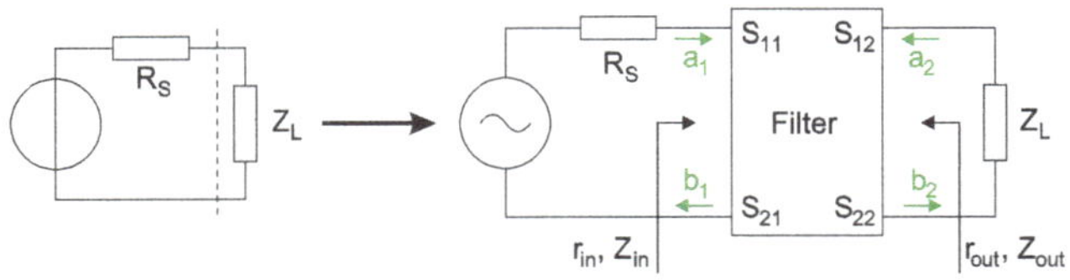

Figure 5. Single matching problem regarded as a two-port.

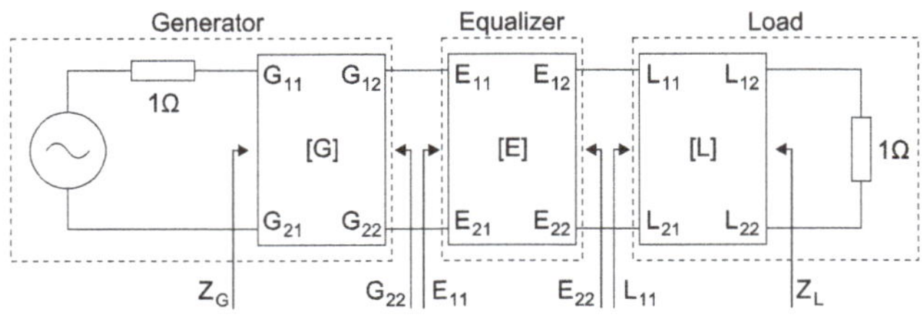

Figure 6. Equivalent network using Darlington's theorem.

very simple equivalent circuit of a dynamic speaker is shown in Fig. 4. It neglects skin and hysteresis effects which primarily contribute to the increase in impedance at high frequencies. Measured data or the technical data sheet of the loudspeaker can be used to calculate the equivalent circuit elements.

3 Simplified Real Frequency Technique

The Simplified Real Frequency Technique (SRFT) is an alternative approach to the Real Frequency Technique (Carlin, 1977) and was first introduced in the work of Yarman (1982). This technique was developed for complex matching problems between generator and load in radio frequency and describes a given network regarded as a two-port as shown in Fig. 5.

In general, a two-port is described by its two-port parameters. The SRFT uses scattering parameters (S-parameters) to describe the electrical behavior of a lossless network by

$$\begin{pmatrix} b_1 \\ b_2 \end{pmatrix} = \begin{pmatrix} S_{11}(p) & S_{12}(p) \\ S_{21}(p) & S_{22}(p) \end{pmatrix} \begin{pmatrix} a_1 \\ a_2 \end{pmatrix}. \quad (1)$$

Using Darlington-Cauer-Piloty's theorem (Darlington, 1939), the network can be described as a ladder network with multiple two-ports. The resulting network is shown in Fig. 6. The SRFT algorithm uses the transducer power gain (TPG) of a network for optimization which is definded by

$$T = \frac{P_\mathrm{L}}{P_\mathrm{S,max}}, \quad (2)$$

where P_L is power to the load and $P_\mathrm{S,max}$ is maximum available power from source. Now, the purpose of the SRFT is to achieve the best possible TPG over a desired frequency band. Further, TPG of the ladder network in Fig. 6 can be expressed as a function of its two-port parameters by

$$T(\omega) = |G_{21}|^2 \frac{|E_{21}|^2}{|1 - E_{11}G_{22}|^2 |1 - \hat{E}_{22}L_{11}|^2} |L_{21}|^2 \quad (3)$$

with

$$\hat{E}_{22} = E_{22} + \frac{E_{21}^2 G_{22}}{1 - E_{11}G_{22}} \quad (4)$$

being the output reflection coefficient of the filter. The two-port parameters of source G_ij and load L_ij are known since they can be simply determined by measurement (Yarman, 2010).

The crux of SRFT is to generate the unknown parameters E_ij of the equalizer using Belevitch's theorem (Belevitch, 1968). The theorem states that an ideal reciprocal lossless two-port network can only be realized if the **S**-matrix is given by

$$\mathbf{S}(p) = \frac{1}{g(p)} \begin{pmatrix} h(p) & f(p) \\ f(p) & -\frac{f(-p)h(-p)}{f(p)} \end{pmatrix}, \quad (5)$$

where $g(p)$ is a strict Hurwitz polynomial of degree n (minimum number of inductors and capacitors) and $h(p)$ a real polynomial with $\deg(h) \le \deg(g)$. The transmission zeros of the transfer function are represented in $f(p)$. Since the matching problem between amplifier and loudspeaker is considered to be a lowpass problem, we may choose all transmission zeros at infinity. This choice corresponds to setting $f(p) = 1$.

The coefficients h_n of the polynomial $h(p)$ are unknown and chosen as initial values. In order to determine the coefficients g_n of the Hurwitz polynomial $g(p)$ as a function of the initial values, a numercial procedure is used. Once, the polynomial $g(p)$ is constructed, the S-parameters and TPG can be computed using Eqs. (3) and (4). For optimization, an error function is defined by

$$\begin{aligned} T_\mathrm{E}(\omega) &= T(\omega) - T_0 \\ &= |G_{21}|^2 \frac{|E_{21}|^2}{|1 - E_{11}G_{22}|^2 |1 - \hat{E}_{22}L_{11}|^2} |L_{21}|^2 - T_0 \end{aligned} \quad (6)$$

with T_0 being the desired TPG. As a result of optimization, the coefficients h_n of $h(p)$ are known. Finally, the LC filter is constructed by synthesis of the driving-point impedance which is given by

$$Z_\mathrm{in}(p) = \frac{a(p)}{b(p)} = \frac{1 + E_{22}}{1 - E_{22}} = \frac{g(p) + h(p)}{g(p) - h(p)}. \quad (7)$$

As a consequence of the choice of all transmission zeros at infinity, the input impedance $Z_\mathrm{in}(p)$ may be realized by the well-known Cauer canonic realization using continued

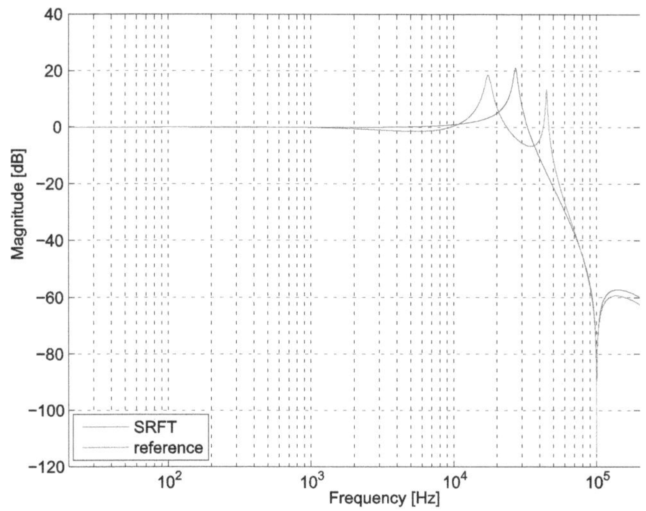

Figure 7. Frequency response of reference and computed SRFT filter.

fraction expansion about $p = \infty$:

$$Z_{\mathrm{in}}(p) = Z_1(p) + \cfrac{1}{Y_2(p) + \cfrac{1}{Z_3(p) + \cfrac{1}{\cdots}}}. \tag{8}$$

The terms $Z_i(p)$ are purely imaginary impedances (inductors), whereas $Y_i(p)$ are purely imaginary admittances (capacitors). Thus, a LC ladder network can be realized directly by use of this expression.

4 Simulation using MATLAB

In a series of simulations especially one type of filter has been found which is going be introduced briefly in the following. In Fig. 7 the computed frequency response of the SRFT filter is presented. A common Butterworth filter is used for reference. It can be seen that the response of the SRFT filter is maximum flat within the frequency band of 0 Hz to 10 kHz, whereas the response of the reference filter has some ripple at around 5 kHz. Further on, the resonance peak of the reference filter is at around 18 kHz which means that the resonance is within the audio frequency band. The peak of the SRFT filter is shifted towards higher frequencies, thus the response is longer flat. In both cases, the stop band attenuation is sufficiently high. The frequency response can be optimized even further by inserting an additional transmission zero to the transfer function nearby the switching frequency. This is done by connecting a further capacitor C_P in parallel with L_2 as shown in Fig. 8. The computed filter element values are: $L_1 = 392.5\,\mathrm{nH}$, $L_2 = 5.26\,\mu\mathrm{H}$, $C_1 = 81.2\,\mu\mathrm{F}$, $C_2 = 6.72\,\mu\mathrm{F}$ and $C_P = 481.56\,\mathrm{nF}$.

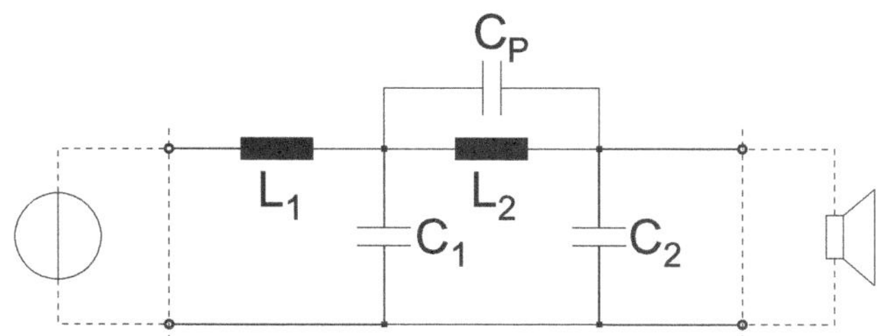

Figure 8. Realized LC ladder network of computed SRFT filter.

5 Conclusions

In this paper, the systematic design of output filters for audio class-D amplifiers via SRFT is presented. Therefore, a short introduction of specifications for filter design was given. After describing the class-D output impedance and the loudspeaker impedance, the SRFT algorithm was presented. Finally, some first simulation results were presented using MATLAB and the computed LC ladder network was illustrated. The simulations also revealed that the optimization criterion (TPG) is not chosen well, since there's usually no power adjustment in audio electronic but voltage adaption. Thus, the computed frequency responses are mostly independend of the desired TPG. Further work could deal with adaption of the optimization criterion.

Edited by: J. Götze

References

Belevitch, V.: Classical Network Theory, Holden Day, San Francisco, 276–279 1968.

Carlin, H. J.: A New Approach to Gain-Bandwidth Problems, IEEE T. Circuit Theory, CAS-24, 4, April 1977.

Darlington, S.: Synthesis of Reactance 4-poles, J. Math. Phys. Camb., 18, 257–353, 1939.

Doy, T.: Class D Audio Amplifier Output Filter Optimization, Application Note, Maxim Integrated Products, Sunnyvale, 8 pp., 2002.

Metzler, B.: Audio Measurement Handbook, Audio Precision, Beaverton, 59–60, 1993.

Quek, Y. B.: Class-D LC Filter Design, Application Report, Texas Instruments, 20 pp., April 2006.

Oliva, A., Paolini, E., and Ang, S. S.: A New Audio File Format for Low-Cost, High Fidelity, Portable Digital Audio Amplifiers, 15 pp., Texas Instruments, White Paper, October 2005.

Yarman, B. S.: A Simplified Real Frequency Technique for Broadband Matching Complex Generator to Complex Loads, RCA Review, 43, 529–541, September 1982.

Yarman, B. S.: Design of Ultra Wideband Power Transfer Networks, Wiley Chichester, 616–619, 2010.

Parameter estimation under non-stationary circumstances using extended signal model

M. Lechtenberg[1], **J. Götze**[1], **K. Görner**[2], **and C. Rehtanz**[2]

[1]Information Processing Lab, TU Dortmund, Otto-Hahn-Str. 4, 44227 Dortmund, Germany
[2]Institute of Energy Systems, Energy Efficiency and Energy Economics, TU Dortmund, Emil-Figge-Str. 70, 44227 Dortmund, Germany

Correspondence to: M. Lechtenberg (matthias.lechtenberg@tu-dortmund.de)

Abstract. In the context of transient processes in power system measurements, a sampled signal is analyzed with respect to electro-magnetic influences. These are most likely superposed sinusoids that can be found next to the fundamental system sinusoid. However, such signal is not stationary, i.e. has a varying model order which means that temporary and exponentially damped sinusoids appear eventually. In our previous work, the exponential damping was only considered indirectly but not explicitly incorporated in the signal model used. Regarding the amplitude estimation, this led to inaccuracies since the amplitude was incorrectly assumed constant within a window of samples. In this paper, we use ESPRIT's ability to also estimate a parameter of exponential damping, improve the corresponding signal model, advance the amplitude estimation algorithm and back projection algorithm. Especially short-term signal components can be tracked far more precise.

1 Introduction

Modern power system operation requires sophisticated monitoring and analysis since the power distribution policy changes from unidirectional central power generation to a highly meshed grid of distributed and varying generation. PMUs are widely recognized for their accurate synchrophasor measurement subject to dynamic conditions. However, only stationary signals and modulated signals with low frequency components (i.e. mainly electro-mechanic transients) are considered, complying with the requirements of the IEEE Standard for synchro-phasors C37.118-2011; reviewed in this context by Görner et al., 2013). Electromagnetic transients e.g. induced by switchings in the network are not considered.

When estimating these electro-magnetic signal components, it has to be considered that these components are not part of the signal for all time. They are induced by events like switchings or line faults and fade out exponentially. In terms of a model, the net can be seen as a circuit of inductances, capacitances and resistances forming meshes with individual resonance frequencies. Such mesh might be excited by an event and (in consequence) may start to oscillate. This oscillation is typically exponentially damped. The mentioned events become more frequent these days due to the evolution towards more dynamic operation of the net.

Waveform analysis is a central part of this monitoring and analysis process. Yet, waveform analysis can only be done using sampling windows (Banerjee and Srivastava, 2012). Focusing on the prior to this work not explicitly considered exponential damping, it is worth mentioning that subspace analysis and frequency estimation are not affected by the damping influencing the current window. Furthermore, the subspace-based frequency estimation algorithm ESPRIT can also estimate the parameter of exponential damping. However, the fact that a sample at the beginning of a sampling window is affected by exponential damping differently than the on at the end of the window has drawbacks on the amplitude estimation, naturally. This can be circumvented by incorporating the exponential fading parameter explicitly.

As a framework for this paper, the waveform analysis is done by subspace extraction in order to enable the detection and estimation of electro-magnetic parameters aside the system's fundamental frequency. The subspace-based parameter estimation algorithm ESPRIT (Paulraj et al., 1986) is

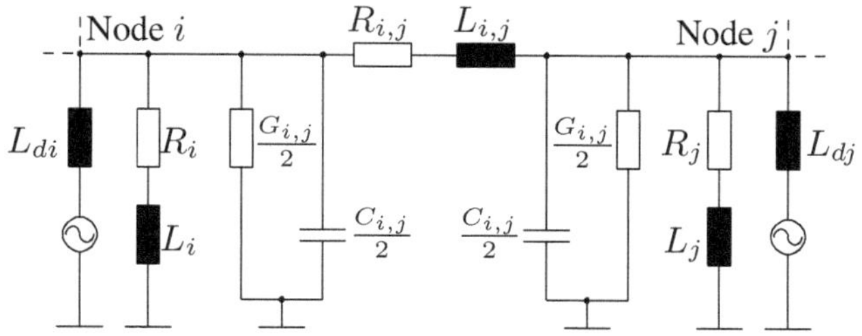

Figure 1. PI-modeled transmission line between nodes i and j.

supported by a subspace tracker of the PAST-family (Yang, 1995a). Further (complex) amplitude estimation is done with the help of a Least-Squares (Sorenson, 1970) approach. The extracted parameters are gathered and rated by the DaPT algorithm (Lechtenberg and Götze, 2010; Lechtenberg et al., 2012). The extension back projection introduced in Lechtenberg et al. (2013) will also be used to emphasize the benefit.

In the following subsection, the signal model with and without exponential damping will be explained. The framework including ESPRIT, PAST, DaPT, Back Projection and LS-based amplitude estimation are reviewed in Sect. 2. The paragraphs explaining LS-based amplitude estimation also introduce signal reconstruction and RMSE calculation for evaluation. A simulation demonstrates the application of these methods in Sect. 3. This paper is completed by a summary of the work and the results in a final section, also providing a glance at future work.

1.1 Line modeling and signal model

The transmission net can be modeled as a circuit of resistances, inductances and capacitances with partly overlaid meshes, see Fig. 1. Each mesh with inductances and capacitances taken from the circuit has its own resonance frequency and can be excited by a switching event. In such case, each participating mesh in a node provides its resonance frequency as a superposed sinusoid in the voltage/current signal of this node.

These sinusoids appear during a transient process when the transmission net merges from one steady-state condition into another. A resulting signal model for one segment of constant model order might be written as follows:

$$x(n) = \sum_{i=1}^{p} c_i(n) e^{jn\omega_i f_s^{-1}} + w_{\mathrm{awgn}}(n) \tag{1}$$

However, this signal model does not consider exponential damping. In consequence, such effect would indirectly be part of the complex amplitude $c_i(n)$; this indicates that the model is not sufficient. The following equation adds the missing exponential factor:

$$x(n) = \sum_{i=1}^{p} c_i(n) e^{-\gamma n f_s^{-1}} e^{jn\omega_i f_s^{-1}} + w_{\mathrm{awgn}}(n) \tag{2}$$

The input samples $x(n)$ may also be used in the form of a vector $\boldsymbol{x}(n)$ describing a set of subsequent samples. The number of sinusoids (rank) is p. The complex amplitude $c_i(n) = a_i(n)\exp(j\varphi_i)$ is built of the (initial) amplitude $a_i(n)$ and the phase φ_i (at time instant $n = 0$). $\omega_i = 2\pi f_i$ describes the angular frequency. The parameter γ is the exponential fading factor; since the argument is negative, a positive γ indicates damping. The component $\{c_1, \omega_1\}$ represents the fundamental frequency. The resonance frequency of an excited LC-oscillation mesh is embodied in $\omega_i|_{i>1}$. Note that to increase comprehensibility, the signal model is reduced to an only-exponential model although real-valued measurements would require a description with sines or cosines (or – via Euler's identity – doubled exponentials).

2 Process chain

2.1 Subspace estimation

The concept of the presented parameter estimation was already discussed before (e.g. Lechtenberg and Götze, 2011). The samples from the measurement are prepared with a sliding window before the respective data is fed to the subspace estimator.

The samples' space can be described by a set of vectors forming a basis. The obvious way to do this is to use windowed sample data to estimate the auto-correlation matrix (e.g. via exponentially weighted averaging) and to perform an eigenvalue/-vector decomposition of this matrix. These eigenvectors form one possible basis of the samples' space. Knowing the rank p, the signal's subspace can be extracted by selecting the vectors corresponding to the p greatest eigenvalues. Consequently, the other vectors span the noise subspace.

A less computational complex method to find a basis for the signal's subspace is depicted by the so-called subspace trackers. They do not produce eigenvectors; they provide vectors forming a basis describing the same space as a basis defined by eigenvectors does. Next to algorithms like PROTEUS (Champagne and Liu, 1998) or YAST (Badeau et al., 2008) the class of PAST-based algorithms (Yang, 1995b) is common in such context. In this work, the OPAST algorithm (Abed-Meraim et al., 2000) was identified to be a good choice.

PAST-based algorithms are based on the nearly unconstrained minimization of the cost function

$$J\left(\hat{\mathbf{W}}\right) = E\|\boldsymbol{x} - \hat{\mathbf{W}}\hat{\mathbf{W}}^H \boldsymbol{x}\|^2 \tag{3}$$

where $\hat{\mathbf{W}}$ is not constrained to hold eigenvectors. The minimization is based on the idea of gradient-descent methods and incorporates exponential weighting for updating $\hat{\mathbf{W}}$ which increasingly better approximates the basis vectors $\mathbf{W}$ (Yang, 1995b). In exponential weighting $b(n) = \beta b(n-1) +$

$(1-\beta)\hat{b}$, the factor β is also called forgetting factor. The pseudo window length $w_{\text{PAST}} = \frac{1}{1-\beta}$ gives an impression on the rage of samples affecting the current average.

Assuming valid results of OPAST, the following equations provide noise samples $\boldsymbol{w}$ and an estimation of the noise power σ^2:

$$\boldsymbol{w}(n) = \boldsymbol{x}(n) - \mathbf{W}\boldsymbol{y} \quad (4)$$

$$\hat{\sigma}^2_{\text{awgn}} = \|\boldsymbol{w}(n)\|_2^2 \quad (5)$$

with an internal PAST variable $\boldsymbol{y}$ only weakly related to eigenvalues.

2.2 Parameter estimation

The a priori selected basis vectors describing the desired signal are fed to the subspace-based parameter estimator ESPRIT (Paulraj et al., 1986). It is based on the analysis of the rotational invariance of two subsequent basis vectors. Due to the principle of rotational invariance, these vectors should only differ in a constant rotation $\exp(j \cdot n \cdot \Omega_i)\big|_{n=1}$ (with $\Omega_i = 2\pi \frac{f_i}{f_s}$) depending on the eigenfrequency f_i, which is the desired parameter.

Remembering the signal model in Eq. (2), the signal can also be described by subspaces like in the equation

$$E\left[\boldsymbol{x}\boldsymbol{x}^H\right] = \mathbf{S}(n) + \sigma^2_{\text{awgn}}\mathbf{I} \quad (6)$$

$$\text{with } \boldsymbol{x} = \mathbf{A}(n, \boldsymbol{\omega})\boldsymbol{c}(n) + \boldsymbol{w}_{\text{awgn}}(n). \quad (7)$$

The matrix $\mathbf{A}$ contains the exponential functions $\exp\left(j\omega_i \frac{n}{f_s}\right)$ (and $\exp\left(-\frac{\gamma_i}{f_s}\right)$) to form the superposition in the horizontal dimension and the burst description in the vertical dimension. $\mathbf{S}(n)$ describes the signal subspace, which can be written as a product of the eigenvectors $\mathbf{W}(n)$ and the eigenvalues (on a diagonal matrix) $\mathbf{D}(n)$:

$$\mathbf{S}(n) = \mathbf{W}(n)\mathbf{D}(n)\mathbf{W}(n)^H. \quad (8)$$

One time-step can be described by a multiplication with a diagonal matrix $\boldsymbol{\Phi}$ having diagonal entries $\exp\left(2\pi j \cdot \frac{f_i}{f_s} - \frac{\gamma_i}{f_s}\right)$:

$$\mathbf{W}(n+1) = \mathbf{W}(n)\boldsymbol{\Phi}. \quad (9)$$

The key idea of ESPRIT is to estimate the rotation $\boldsymbol{\Phi}$. Thinking of the burst dimension of $\mathbf{A}$, this rotation is also the factor between the upper $N-1$ lines of $\mathbf{W}$ and the lower. Due to noise, $\hat{\boldsymbol{\Phi}}$ will have off-diagonal elements. So an EVD is performed on $\hat{\boldsymbol{\Phi}}$. The frequencies can be extracted from these eigenvalues by calculating their angle. These frequencies are the input of DaPT.

Recapitulating the idea of ESPRIT, the to be estimated factor is an exponential function with a complex argument. However, this algorithm also works for factors like

$$\phi_{i,i} = e^{\left(j \cdot \frac{\omega_i}{f_s} - \frac{\gamma_i}{f_s}\right)}. \quad (10)$$

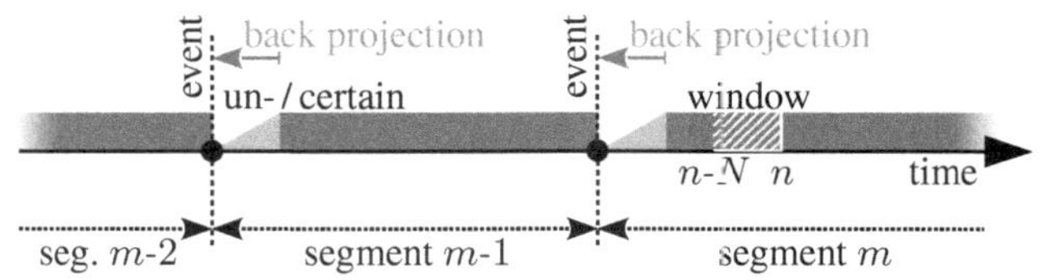

Figure 2. Quasi-stationary segments of measurement samples, uncertainty regions, sample window and back projection concept.

As can be seen, the idea of ESPRIT is not restricted to rotational poles but allows poles aside the unit circle (Badeau et al., 2003). The exponential factor can be retrieved by calculating the logarithm of the absolute value of a $\phi_{i,i}$.

2.3 Database-assisted Parameter estimation – DaPT

Assuming the feed of ESPRIT to contain a few more basis vectors than the rank of the signal, these additional vectors belong to the noise subspace and are named auxiliaries (Yang, 1995b). In contrast to the signal's vectors, theses will not result in constant parameter estimations within the ESPRIT algorithm. The idea of DaPT (Database-assisted Parameter estimation, Lechtenberg and Götze, 2011) is to rate the temporal presence of each frequency.

Entries with high rating can be hypothesized as part of the desired signal and entries with a low rating are assumedly noise and to be forgotten. By this, the algorithm recognizes a change in rank by simply counting entries with high ratings. This rank is advanced by a small number (the auxiliaries) and looped back to the block that selects the vectors to be fed to ESPRIT.

In every recursion, the incoming frequency estimation is mapped to the most suitable database entry. Ideally, both values match neglecting the measurement noise. For entries that have been successfully mapped, the quality is increased (up to a maximum value); for others, it is decreased (down to zero). A zero-quality-entry will be deleted. The frequency entry is corrected according to the drift measures' severity indexes exceeding a threshold. The measures' values are exponentially weighted over time (double-exponential smoothing) and its severity index is updated sign-depended.

This database has entries for each signal component containing frequency f_i, fading parameter γ_i and (complex) amplitude c_i. The complex amplitude is to be estimated separately (see Sect. 2.5). Amongst other fields, DaPT also has a field valuing the quality of an entry, which is the actual rating. It rates how often a frequency has been part of the last estimation recursions. Based on that, components can be marked reliable (i.e. present), and therefore are accounted for rank estimation and output.

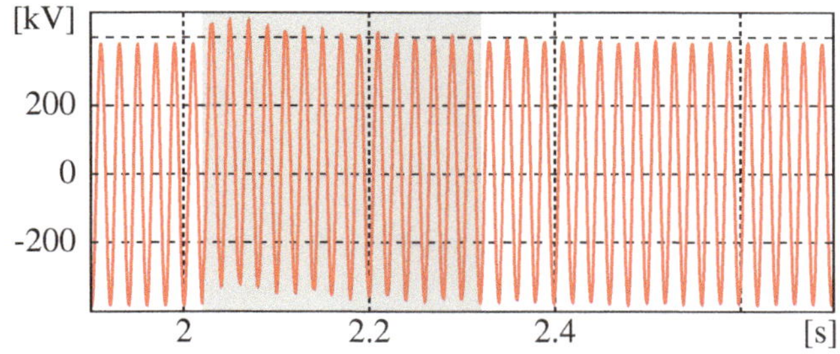

Figure 3. Input signal with an event shortly after 2 s producing decaying DC offset and additional sinusoids.

2.4 Segmentation and back projection

Anticipating, after the estimation of all signal parameters, a reconstruction of the signal is possible. The difference of this reconstruction to the input signal is an indicator for the fulfillment of the assumptions necessary for the algorithms. Calculating the MSE of this signal error and monitoring (by statistical means like $n \cdot \sigma$-rules) enables the robust detection of a violated stationarity assumption. These values can be expected to be fairly constant and may not contain significant steps. If a significant violation is detected, the time instant indicating a transient process from one stationary state to another may be found. This evaluation provides a so-called segmentation of the signal (Lechtenberg et al., 2013).

Furthermore, with the help of this segmentation information, it is possible to reconstruct the parameters without the uncertainties of the compromised windows. When an event is detected and the segmentation is triggered, a timer is set to wait a short while until the parameters can be expected to be valid/certain again (see Fig. 2). With the help of these valid parameters, the estimation for the time between the event and the time-out can be redone time-reversed. Of course, this requires the input samples to be memorized back to the time instant of segmentation.

Although this enables the reparation of the corrupted estimates, the delay resulting from the sliding window principle cannot be compensated, obviously. In consequence, the resulting parameter estimation provides clean estimates both prior to (causality principle) and after (back projection) the event.

2.5 Estimation of complex amplitude

Phase and amplitude estimation can be obtained by a simple LS approach referring to the signal model in Eq. (6). Since time index, frequency and exponential factor are known, the matrix **A** can be built. Together with the samples, the complex amplitude can be estimated. Using the same method, the synchrophasor can be determined by referencing a constant frequency – i.e. the system's fundamental frequency – and calibrating an offset-phase to adapt to the UTC-time (Martin et al., 2006). However, a real-time system like FPGA is recommended for meeting the temporal requirements of PMU measurements.

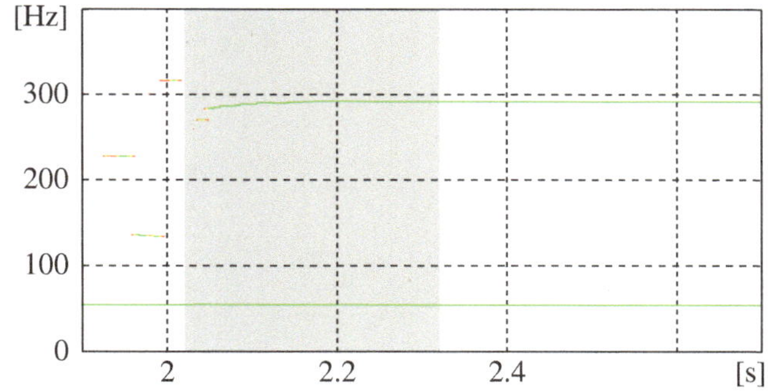

Figure 4. Estimation of frequency parameter without back projection.

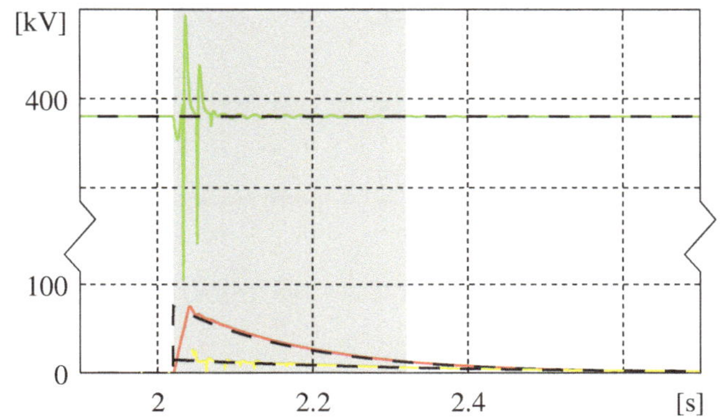

Figure 5. Amplitude estimation without back projection using signal model without damping (dashed: theoretical; with a break in scale).

Since the construction of the matrix **A** is not constrained to a fixed window length and window time offset but only needs to fulfill the LS requirement, i.e. an overdetermined system of equations, the application of back projection (Sect. 2.4) is straight forward.

The signal model and the matrix **A**, resp., can easily be used to reconstruct the input signal. This enables the calculation of the (R)MSE ((rooted) mean squared error) between this reconstruction and the original input samples:

$$\mathrm{RMSE}(n) = \sqrt{\frac{1}{N} \sum_{i=n-N+1}^{n} [x(i) - \mathbf{A}(i,\boldsymbol{\omega})\mathbf{c}(i)]^2}. \tag{11}$$

3 Simulation

The signal processing is done with The Mathworks MATLAB™. The sampling frequency is $f_s = 5\,\mathrm{kHz}$; the window length is $w_l = 100$ and the PAST forgetting factor is $\beta = 0.8$. The simulation is based on an electric power signal of 50 Hz powered to 380 kV that is constantly present. At the time of 2 s, two other signal components are added:

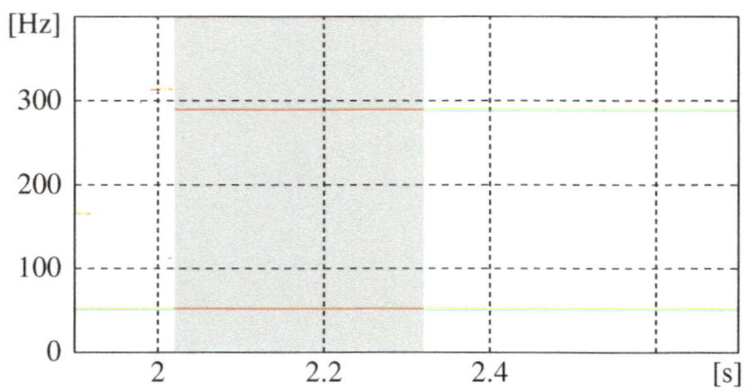

Figure 6. Estimation of frequency parameter with back projection.

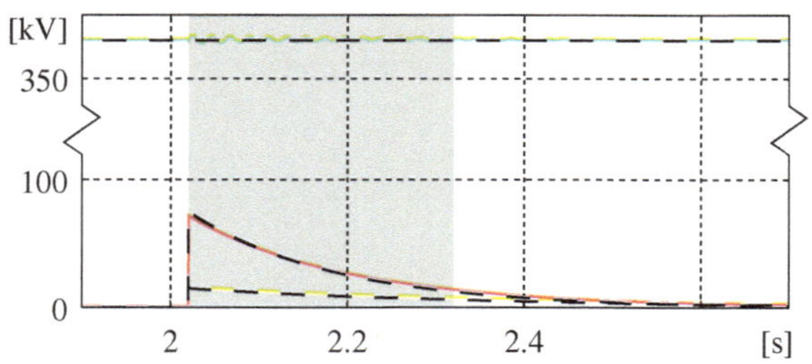

Figure 7. Amplitude estimation without back projection (dashed: theoretical; with a break in scale.)

another sinusoid of ≈ 290 Hz at ≈ 15 kV with an exponential damping factor of 3 and a direct component of ≈ 75 kV with an exponential factor of 6. The signal has additive white Gaussian noise resulting in an SNR of 80 dB. The following demonstration is based on one run of this simulation producing a signal like in Fig. 3.

When parameter estimation is done without respecting the constraint of stationarity, the results will be corrupted right after the model order changes. The less samples prior the model order change affect the estimation (due to exponential averaging) the less corrupted the estimates are. This is illustrated by Fig. 4 for frequency estimation and Fig. 5 for subsequent amplitude estimation.

In Lechtenberg et al. (2013), the concept of (segmentation and) back projection is introduced. Using back projection circumvents the corruption by recalculating the threatened region of samples. Figure 6 demonstrates this for the frequency estimation. However, in contrast to the frequency being correctly modeled as constant, the amplitude is also modeled as constant; this is not correct due to the not considered exponential damping, see Eq. (1). In Fig. 7, the estimates are clearly off the true values.

Using the signal model in Eq. (2) for the amplitude estimation requires more information; it utilizes a set of input samples and the ESPRIT-estimated parameters f_i and γ_i for frequency (Fig. 6) and damping (Fig. 8) to solve a set of equations for the only missing parameter c_i. As Fig. 9 visualizes, the signal model supports the complete signal. A comparison of the RMSE evaluating the estimation against the input signal also confirms this study contrasting both prior presented simulations (Fig. 10).

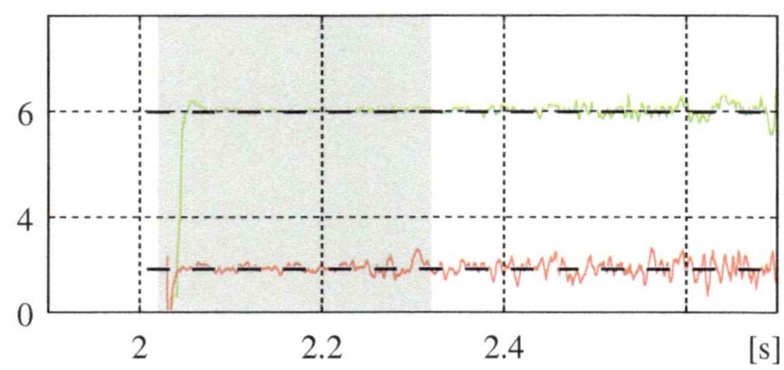

Figure 8. Estimation of exponential damping factor by ESPRIT (dashed: theoretical).

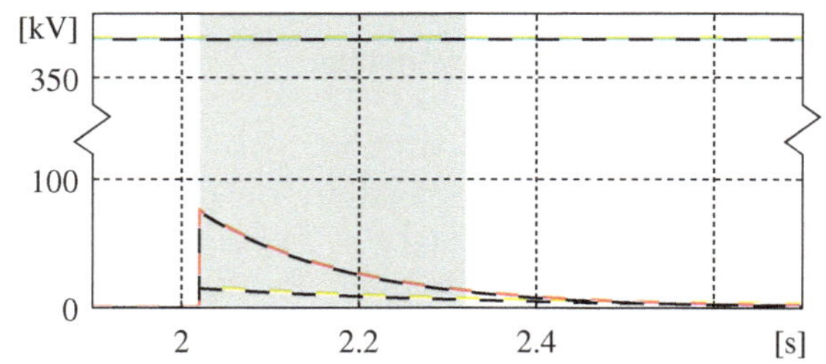

Figure 9. Fully enabled amplitude estimation (dashed: theoretical; with a break in scale).

Although the amplitudes of an estimation without support for damping inside the signal model will not match (see Fig. 7), these amplitudes can be used to estimate the damping factor γ_i as can be seen in Fig. 11. This estimation of γ_i is less corrupted than in Fig. 8 since the effect of back projection simulates stationarity. The segmentation information from the back projection and the information about the exponential damping can jointly be used to estimate the initial amplitude of such temporary sinusoid; Fig. 12 demonstrates this.

4 Conclusions

Parameter estimation in an environment of superposed sinusoids in which not all of them are continuously present but emerging step-like and fading in an exponentially damped fashion is not trivial. There are three main challenges:

1. the estimation of the number of currently present sinusoids (estimation of rank/model order),
2. the segmentation of the signal at the time instants of model order change (to ensure stationarity), and
3. the choice of the correct signal model.

Only if all of these items are considered, it is possible to extract all signal parameters (i.e. frequency, exponential

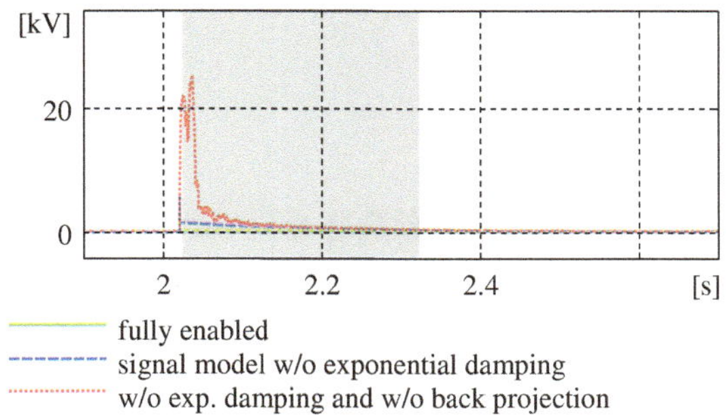

Figure 10. RMS of difference between reconstructed signal and input signal, RMSE.

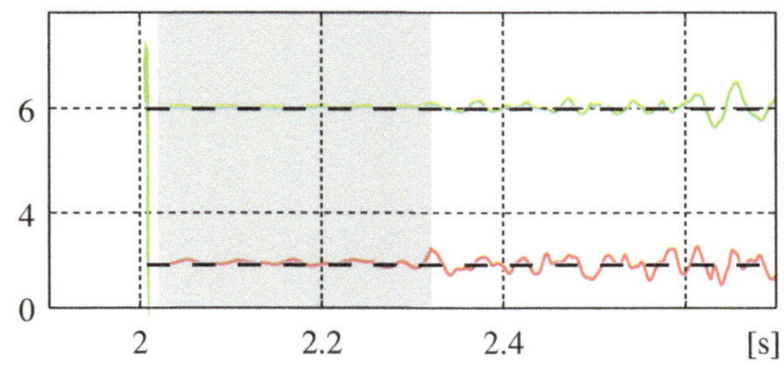

Figure 11. Estimation of exponential damping factor by logarithm of estimated amplitude (dashed: theoretical).

damping factor, (initial) amplitude, and segmentation information). The success may be evaluated by reconstructing the signal and comparing it to the input sequence.

In this paper, we advanced the signal model of superposed sinusoids by adding an exponential factor for damping. By exploiting ESPRIT's capability of not only estimating the frequency parameter but the complete complex pole, the subsequent least-squares estimation can match the signal model with the only on free parameter; the complex amplitude describes the phase offset and the initial amplitude of a sinusoid.

Considering the direct component to be a sinusoid of 0 Hz, it can jointly be estimated with the other signal components. In terms of complex computation, the only difference is that real sinusoids have to be considered with a positive and a negative frequency (Euler's identity) but the direct component only with one.

Future plug-ins to this work may use information from all phases of the 3-phase transmission system. Amplitude modulations (which can also be described as sinusoids very close in frequency) are still a challenge to be handled. A possible FPGA implementation may be considered for the fully enabled system.

Acknowledgements. This work was supported by DFG (German Research Foundation).

Edited by: W. Mathis

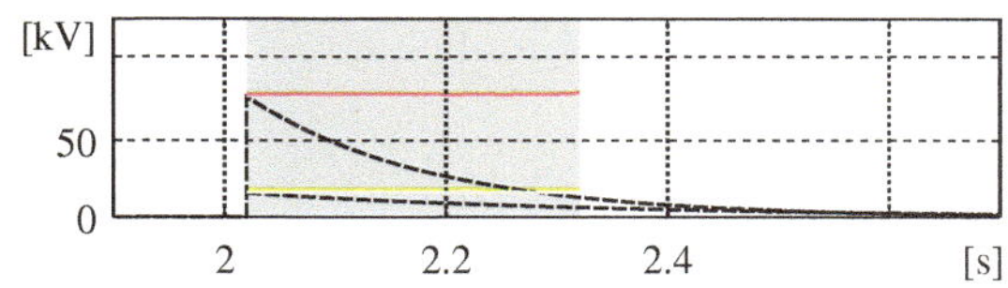

Figure 12. Estimation of initial amplitude and (dashed) theoretical amplitude.

References

Abed-Meraim, K., Chkeif, A., and Hua, Y.: Fast orthonormal PAST algorithm, IEEE Signal Proc. Lett., 7, 60–62, 2000.

Badeau, R., Richard, G., and David, B.: Adaptive ESPRIT algorithm based on the PAST subspace tracker, Int. Conf. Acoust. Spee., Vol. 6, 229–232, 2003.

Badeau, R., Richard, G., and David, B.: Fast and Stable YAST Algorithm for Principal and Minor Subspace Tracking, IEEE T. Signal Proces., 56, 3437–3446, 2008.

Banerjee, P. and Srivastava, S.: A Subspace-Based Dynamic Phasor Estimator for Synchrophasor Application, IEEE T. Instrum. Meas., 61, 2436–2445, 2012.

Champagne, B. and Liu, Q.-G.: Plane rotation-based EVD updating schemes for efficient subspace tracking, IEEE T. Signal Proces., 46, 1886–1900, 1998.

Görner, K., Lechtenberg, M., Rehtanz, C., and Götze, J.: A Method for Adaptive Time-Synchronized Measurement during Transients, in: 2013 International Conference on Power Systems Transient, 10th, 2013.

Lechtenberg, M. and Götze, J.: A Database Post-Processing Approach for Robust Frequency Estimation, in: Kleinheubacher Tagung, 2010.

Lechtenberg, M. and Götze, J.: Database Assisted Frequency Estimation, in: 4th IEEE International Conference on Computer Science and Information Technology, 2011.

Lechtenberg, M., Götze, J., Rehtanz, C., and Görner, K.: Database Assisted Frequency Estimation for Power System Measurement, in: 2012 IEEE Electrical Power and Energy Conference, 2012.

Lechtenberg, M., Götze, J., Görner, K., and Rehtanz, C.: Improved Parameter Estimation under Non-Stationary Circumstances using Segmentation and Back Projection, in: IEEE International Conference on Smart Instrumentation, Measurement and Applications (ICSIMA 2013), Kuala Lumpur, Malaysia, 2013.

Martin, K., Benmouyal, G., Adamiak, M., Begovic, M., Burnett, R. J., Carr, K., Cobb, A., Kusters, J., Horowitz, S., Jensen, G., Michel, G., Murphy, R., Phadke, A., Sachdev, M., and Thorp, J.: IEEE Standard for Synchrophasors for Power Systems, IEEE Std C37.118-2005 (Revision of IEEE Std 1344-1995), 1, 1–57, 2006.

Paulraj, A., Roy, R., and Kailath, T.: A subspace rotation approach to signal parameter estimation, P. IEEE, 74, 1044–1046, 1986.

Sorenson, H. W.: Least-squares estimation: from Gauss to Kalman, IEEE Spectrum, 7, 63–68, 1970.

Yang, B.: An extension of the PASTd algorithm to both rank and subspace tracking, IEEE Signal Proces. Lett., 2, 179–182, 1995a.

Yang, B.: Projection approximation subspace tracking, IEEE T. Signal Proces., 43, 95–107, 1995b.

31

Indoor radio channel modeling and mitigation of fading effects using linear and circular polarized antennas in combination for smart home system at 868 MHz

S. Wunderlich, M. Welpot, and I. Gaspard

Hochschule Darmstadt, Department of Electrical and Computer Engineering, Darmstadt, Germany

Correspondence to: I. Gaspard (ingo.gaspard@h-da.de)

Abstract. The markets for smart home products and services are expected to grow over the next years, driven by the increasing demands of homeowners considering energy monitoring, management, environmental controls and security.

Many of these new systems will be installed in existing homes and offices and therefore using radio based systems for cost reduction. A drawback of radio based systems in indoor environments are fading effects which lead to a high variance of the received signal strength and thereby to a difficult predictability of the encountered path loss of the various communication links.

For that reason it is necessary to derive a statistical path loss model which can be used to plan a reliable and cost effective radio network. This paper presents the results of a measurement campaign, which was performed in six buildings to deduce realistic radio channel models for a high variety of indoor radio propagation scenarios in the short range devices (SRD) band at 868 MHz.

Furthermore, a potential concept to reduce the variance of the received signal strength using a circular polarized (CP) patch antenna in combination with a linear polarized antenna in an one-to-one communication link is presented.

1 Introduction

Indoor radio channels suffer usually from slow fading. That means that the characteristics of the channel vary only slowly in time compared with the duration of a data symbol. This can become a very dramatic effect in the communication between two (or more) radio devices. Especially if these devices are non-mobile devices, like in a smart home system. The reason for that is, that if the receiver is located in a deep fade it will probably remain in this deep fade for a long (or unlimited) time period (Vireerackody, 1993). A deep fade can significantly decrease the signal-to-noise ratio (SNR) of a communication link, which results in a reduced data rate or in a complete drop out of the radio link.

Another effect which can also cause deep fades is fast fading. Fast fading occurs in indoor scenarios due to moving obstacles such as moving people (Horvat et al., 2013). In a fast fading channel, the transmitter may take advantage of the variations in the channel conditions using time diversity (which is often used by smart home systems). However, this paper does not cover the influences of moving people and objects on the radio channel, for further information on this topic, see e.g. Horvat et al. (2013).

The main reason for fading in indoor environments is multipath propagation. Transmitted radio waves are reflected and scattered on walls, ceilings, floors and other obstacles and can combine at the receiver in a destructive manner, due to different phase shifts and amplitudes of the arriving signal paths (Molisch, 2011). Another effect is depolarization. The scattered and reflected waves that contribute to multipath fading can also transfer energy from the transmitted polarization plane into the orthogonal polarization plane (called cross-polarization coupling) (Cox et al., 1986). Such coupling occurs as a result of oblique reflections from the walls as well as due to scattering from indoor clutter, such as furniture (Kyritsi and Cox, 2001). This effect can cause significant degradation in signal quality if the transmitting and receiving antennas are using the same polarization. This effect contributes also to fading (polarization fading).

Since the delay spread of indoor channels is in the range of nanoseconds (Hashemi and Tholl, 1992) and the bandwidth

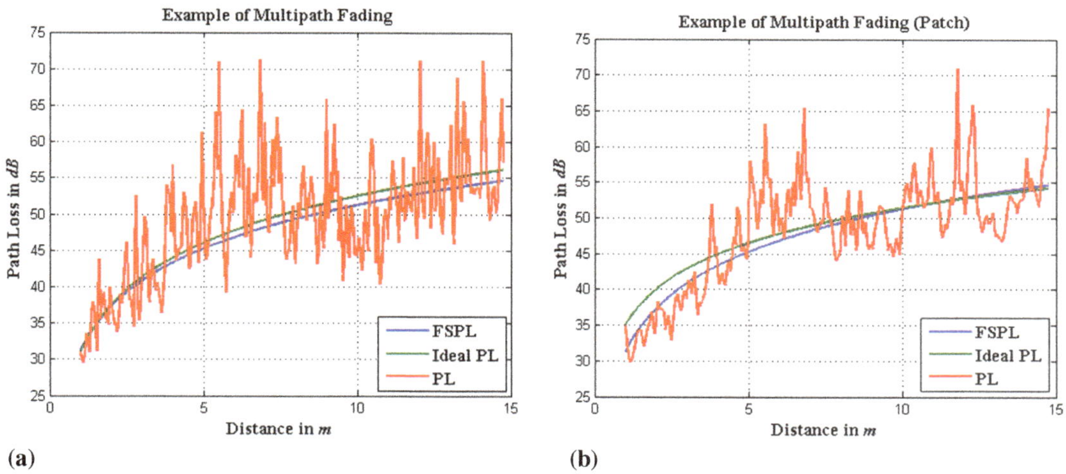

Figure 1. **(a)** A channel encounters strong fading using two vertical linear polarized antennas. The red graph (PL) represents the actual path loss the channel encounters (which was measured during the measurement campaign), the blue graph (FSPL) represents the theoretical free space path loss (for reference) and the green graph (Ideal PL) represents the actual path loss the channel encounters without fading effects. This graph was created by using formula 5 with a minimum mean square error (MMSE) approach to calculate the path loss exponent based on the original data set (red graph). **(b)** Mitigation of fading using a CP antenna in interaction with a vertically linear polarized antenna.

of the most indoor devices is relatively small (the SRD bandwidth at 868 MHz is below 600 kHz), the channel can be assumed to show flat fading, so the coherence bandwidth is much larger than the signal bandwidth and the channel does not show a frequency-selective behavior.

These fading effects can lead to a significant reduced signal quality between sender and receiver and represent a significant problem for radio communication in indoor transmission scenarios.

One way to mitigate fading introduced by cross-polarization coupling is to use a circular polarized antenna in interaction with a linear polarized antenna. The benefit of using a circular polarized antenna at one end of the communication link is the irrelevance of the orientation angle of the polarized wave, so the CP antenna can receive radio waves at an arbitrary polarization plane and the effect of cross-polarization coupling can be annihilated. The only drawback of this method is the additional 3 dB polarization loss introduced by using a circular polarized antenna in combination with a linear polarized antenna. One example of the reduction of fading using a circular polarized antenna in comparison to a vertical linear polarized antenna can be seen in Fig. 1. So this method is based on the idea that all values are decreased by 3 dB but the mean value of the received power level is much higher than without this method. This is due to the fact that the influence of the fading spikes is drastically reduced.

The variance of the received power in case of two linear polarized antennas is obviously much larger than in case of using one linear and one circular polarized antenna for the same radio link. Another positive effect of using a CP antenna is that the orientation of the linear polarized antenna does not influence the communication link anymore. This is particularly advantageous since the orientation of the antennas can seldom be guaranteed for the most indoor application scenarios. This approach is especially practical if a master-slave relationship between a central radio module and one or more radio subunits is given (like in centralized smart home systems). The central radio module can use the CP antenna while the minor radio devices still use low-cost and space-saving linear polarized antennas.

To verify this theoretical concept a measurement campaign in six buildings was performed (see Sect. 4).

So the two main objectives of this paper are:

1. Presentation of the results of a potential fading mitigation technique using a circular polarized patch antenna in combination with a linear polarized antenna. This antenna is subject to further improvements regarding size and cost reduction. The antenna itself is a well-known design (truncated-corner square patch), however the advantage of using a circular polarized antenna at one end of the link and a linear polarized antenna at the other end of the link in comparison to the conventional usage of linear polarized antennas at both ends should be proven on the basis of a comprehensive measurement campaign.

2. Development of a path loss prediction model which covers a versatile number of scenarios and can be used for a simple estimation of the expected path loss in an actual smart home installation. There does not exist many of these prediction models for this specific frequency domain which are based on an extensive measurement campaign.

2 Measurement setup and procedure

2.1 Measurement setup

The measurement setup is depicted in Fig. 2. The transmitting part of the setup is shown on the left side of the figure. The transmitting antenna is a cross polarized log-periodic antenna which is used to generate a horizontal and vertical polarized electromagnetic wave (denoted with XSLP9142, Schwarzbeck Mess-Elektronik, 2013). The circular polarized patch antenna can be used as an alternative transmitting antenna.

The two input ports of the cross polarized antenna are fed by two carrier wave (CW) generators at 873.8 MHz and 874 MHz (The prototype antenna was designed for a resonance frequency at $f = 873.9$ MHz and not for the actual SRD frequency at $f = 868.3$ MHz.) at an output power of $P_{\mathrm{CW_{dBm}}} = 14$ dBm.

These slightly different frequencies were chosen to distinguish between the two polarizations states in the spectrum which were received by a circular polarized (CP) patch antenna (or a dipole antenna as reference) and measured by a spectrum analyser (SA) which stored it on a PC for further data processing.

The height of the phase center of the transmitting and receiving antenna was $h = 1.16$ m. The transmitting antenna was located at a fixed position while the receiving antenna was moved along the measuring track during a single measurement run. A truncated-corner patch antenna design was used for the circular polarized antenna with a beamwidth of $\theta_{-3\,\mathrm{dB}} \approx 60°$, an antenna gain of $g = -1$ dBi and a return loss of RL = 30.8 dB. The spatial resolution of a single measurement path was $d = 0.05$ m (distance between two measurement points) which results in a spatial sampling rate (SSR) of

$$\begin{aligned} \lambda &= \frac{c_0}{f} = \frac{c_0}{874\,\mathrm{MHz}} = 0.343\,\mathrm{m} \\ \mathrm{SSR} &= \frac{\lambda}{d} = \frac{0.343\,\mathrm{m}}{0.05\mathrm{m}} = 6.86 \end{aligned} \qquad (1)$$

samples per wavelength which results in a sufficient detectability of fading spikes (where λ is the wavelength, c_0 is the speed of light and f is the operating frequency).

The path loss PL can be fairly simple measured by using the relationship:

$$\mathrm{PL} = P_{\mathrm{Tx_{dBm}}} - P_{\mathrm{Rx_{dBm}}} \qquad (2)$$

where $P_{\mathrm{Tx_{dBm}}}$ is the transmitted and $P_{\mathrm{Rx_{dBm}}}$ is the received power. Taking into account antenna gains and cable losses it follows:

$$\begin{aligned} \mathrm{PL} &= (P_{\mathrm{CW_{dBm}}} - a_{\mathrm{Tx}} + g_{\mathrm{Tx}}) - (P_{\mathrm{SA_{dBm}}} + a_{\mathrm{RX}} - g_{\mathrm{Rx}}) \\ \mathrm{PL} &= P_{\mathrm{CW_{dBm}}} - a_{\mathrm{Tx}} - a_{\mathrm{RX}} + g_{\mathrm{Tx}} + g_{\mathrm{Rx}} - P_{\mathrm{SA_{dBm}}} \end{aligned} \qquad (3)$$

where $P_{\mathrm{SA_{dBm}}}$ is the power at the spectrum analyzer and a is the cable loss and g is the respective antenna gain at Rx and Tx.

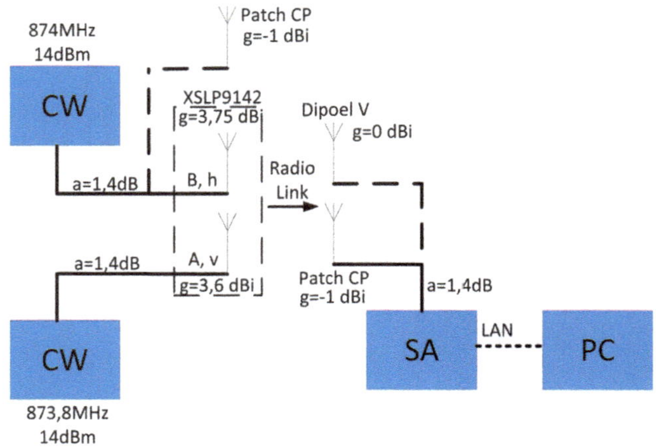

Figure 2. Measurement and test setup for the indoor radio channel modeling and the circular polarized patch antenna. (SA = spectrum analyser, CW = carrier wave generator).

2.2 Measurement procedure

The measurements were performed in six buildings on the campus of the Hochschule Darmstadt. All of these buildings differ in parameters like material, geometry, usage etc. So a huge variety of different fading scenarios was achieved (see Table 1).

Although all measurements were mostly taken in office-like environments there are some similarities between these environments and residential environments. For example the Building D1617 shows similar room geometry, building materials and furniture like many residential areas.

A section of a floor plan of one of the buildings can be seen in Fig. 3. The plan shows the entrance and the hallway of building B11. Basically two types of measurements were performed, line-of-sight (LOS) measurements within rooms and floors (denoted with yellow arrows) and non-line-of-sight (NLOS) measurements with walls, doors and furniture between transmitter and receiver (denoted with red arrows). The starting point and the direction of the single measurement runs were chosen randomly to guarantee an equal distribution of all possible propagation scenarios within the building.

3 Path loss prediction model

The log-distance path loss model was used for the path loss prediction model (Seidel and Rappaport, 1992). It is based on the assumption that the (mean) path loss (PL) is a function of distance d to the γ-th power

$$\mathrm{PL} \sim d^{\gamma} \qquad (4)$$

where γ is the path loss exponent which indicates how fast path loss increases with distance. A logarithmic distance is used in the actual model, so the exponent γ becomes a factor

Table 1. List of buildings which were used during the measurement campaign.

Building (name)	Type	Building materials	Designated usage
B11	One-story lightweight construction building	Wood	Lectures and seminars
B14	Two-storied shipping container architecture	Metal and wood	Lectures and seminars
C10	High-rise building (15 floors)	Reinforced concrete	Offices, lectures, seminars and labs
D11	Industrial building	Bricks and metal	Workshops, lectures and labs
D1617	Office building (5 floors)	Bricks and drywalls	Offices, seminars and labs

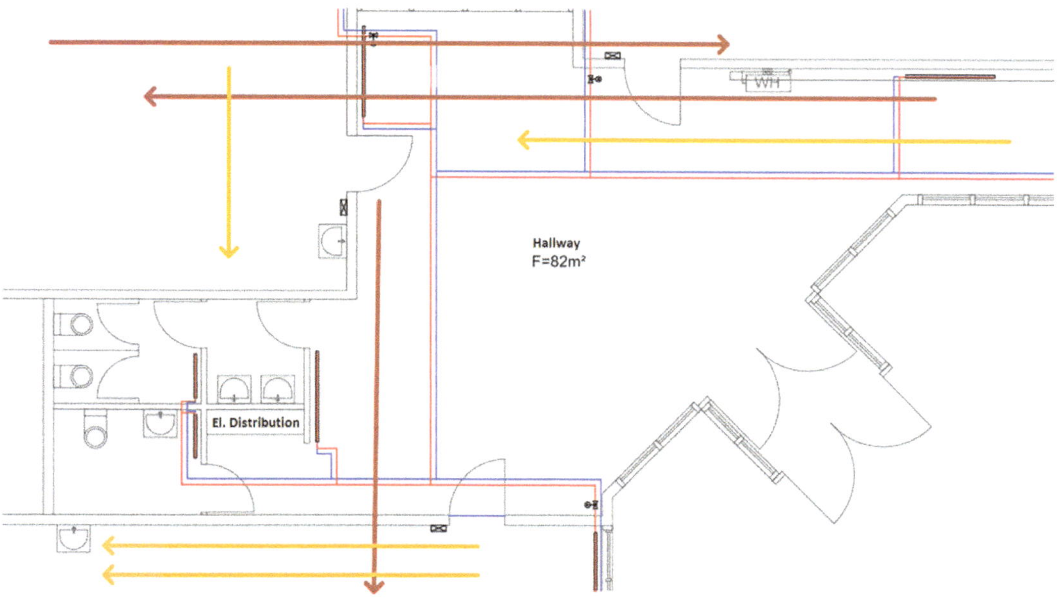

Figure 3. Section of a floor plan of one of the buildings (B11) used for the measurement campaign.

and the formula reads:

$$\mathrm{PL} = \mathrm{PL}_0 + 10 \cdot \gamma \cdot \log_{10} \frac{d}{d_0} + X_g \quad (5)$$

where PL_0 is the path loss at a certain distance d_0 and X_g is a normal distributed random variable which represents fading with zero mean and a standard deviation σ in dB. PL_0 and d_0 represent the path loss from the transmitter to the distance d_0 where the measurement starts. d_0 can be freely chosen and was determined to be one meter for all LOS measurements ($d_0 = 1$ m). The value of γ is dependent on the propagation scenario of the measurement, it is two ($\gamma = 2$) for free space path loss (FSPL), greater than two ($\gamma > 2$) when the path loss increases faster than in free space and smaller than two ($\gamma < 2$) when it increases slower than in free space.

For a homogenous environment γ can be assumed to remain constant for different distances between transmitter and receiver. However, if the propagation properties are changing (e.g. a wall separates transmitter and receiver) the path loss exponent γ is also changing. For this reason it is appropriate to divide the measuring track into several separate segments. γ can then be calculated separately for each of the segments. To compensate the additional path loss introduced by obstacles which influence the path loss exponent it is necessary to introduce new attenuation factors in Eq. (5) (Seidel and Rappaport, 1992):

$$\mathrm{PL} = \mathrm{PL}_0 + 10 \cdot \gamma \cdot \log_{10} \frac{d}{d_0} + X_g + m \cdot \mathrm{FAF}_{\mathrm{dB}} + n \cdot \mathrm{WAF}_{\mathrm{dB}} \quad (6)$$

where $\mathrm{FAF}_{\mathrm{dB}}$ represents a "Floor Attenuation Factor" and $\mathrm{WAF}_{\mathrm{dB}}$ means "Wall Attenuation Factor". m and n are representing the number of floors respectively walls which are penetrated by the radio wave (Seidel and Rappaport, 1992). This model is more flexible and more accurate than the

Table 2. Path loss exponent γ and standard deviation σ in dB of fading for every building.

Results	γ	γ	γ	σ	σ	σ	σ	σ	σ
Tx antenna	V	H	CP	V	H	CP	V	H	CP
Propagation				LOS	LOS	LOS	NLOS	NLOS	NLOS
B11	1.59	1.95	1.8	3.18	4.54	2.51	5.28	4.91	11.60
B14	1.94	1.70	1.71	4.59	4.11	4.58	5.45	5.19	7.25
C10	1.69	1.56	1.65	4.84	5.00	4.86	7.93	6.94	5.63
D11	1.99	2.07	2.03	3.86	4.28	3.3	5.50	5.50	4.15
D1617	1.95	2.3	2.17	4.3	5.02	3.67	5.33	5.12	6.01
Average Value	1.83	2.00	1.90	4.21	4.68	3.83	5.62	5.39	5.98

Table 3. Standard deviation σ in dB of fading in dependency of the used receiver antenna. Rx antenna: V = vertical polarized dipole, CP = circular polarized patch antenna.

Rx antenna	V	V	CP	CP	CP	V	V	CP	CP	CP
Tx antenna	V	H	V	H	CP	V	H	V	H	CP
Propagation	LOS	LOS	LOS	LOS	LOS	NLOS	NLOS	NLOS	NLOS	NLOS
σ/dB	4.90	5.40	3.52	3.96	3.83	6.00	5.67	5.52	5.12	5.98

conventional log-distance path loss model (Hernandez et al., 2012), since the hard-to-predict influences of walls and floors on the path loss is shifted in separate summands. So the path-loss exponent becomes only dependent from the geometry of the room. All later presented results show only the path-loss exponent and not the FAF and WAF values since there was not enough data to calculate these values for single walls and floors. However, since the path-loss exponent is calculated separately for every segment between walls (or floors) there is no need to calculate these values (d_0 was chosen according to the actual distance between sender and starting point for NLOS measurements).

4 Results

The results of over 13 000 measuring points and 270 000 single measurements were recorded and processed to obtain a general fading model of the investigated buildings and an adequate data set to study the effects of the optimized circular polarized antenna in combination with a linear polarized antenna on fading.

Table 2 shows the cumulated results of the measurement campaign separated for every building. The first three columns are representing the path loss exponents for a vertically (V), horizontally (H) and circular polarized (CP) transmitted wave.

The last six columns representing the standard deviation of fading for a LOS and NLOS connection between transmitter and receiver. All results are mean values of the results of the two available receiving antennas: a vertical polarized dipole and the circular polarized patch antenna.

Some observations in Table 2 are remarkable:

1. The path loss exponent is smallest if a vertically polarized wave is transmitted. One possible explanation for that is that the walls in a building can be treated as dielectric materials. Horizontal polarized waves can penetrate the walls if the angle of incidence fits to the permittivity of the wall material (Brewster angle phenomenon). Vertical waves do not suffer a similar effect and will be therefore reflected (according to Kyritsi and Cox, 2001).

2. The average path loss exponent is below two ($\gamma < 2$). This means that the path loss inside of the measured building increases slower than in free space. The reason for that is that the rooms and hallways of the buildings serve as a kind of wave guides due to reflections at floors and ceilings (see also the previously mentioned subitem).

3. For a NLOS scenario the standard deviation is smallest if a horizontal wave is transmitted. A possible explanation for this phenomenon is the following: horizontal waves suffer less from reflections on walls (see the first subitem), so multipath fading becomes less important.

The results of Table 2 are independent from the used receiver antenna. Table 3 shows the standard deviation in dependency of the used receiver antenna (vertically polarized dipole or circular polarized patch antenna). Two points are remarkable:

1. The standard deviation is the smallest for the transmitter/receiver antenna combination linear to circular (or vice versa). This outcome confirms the theoretical considerations introduced in Sect. 1.

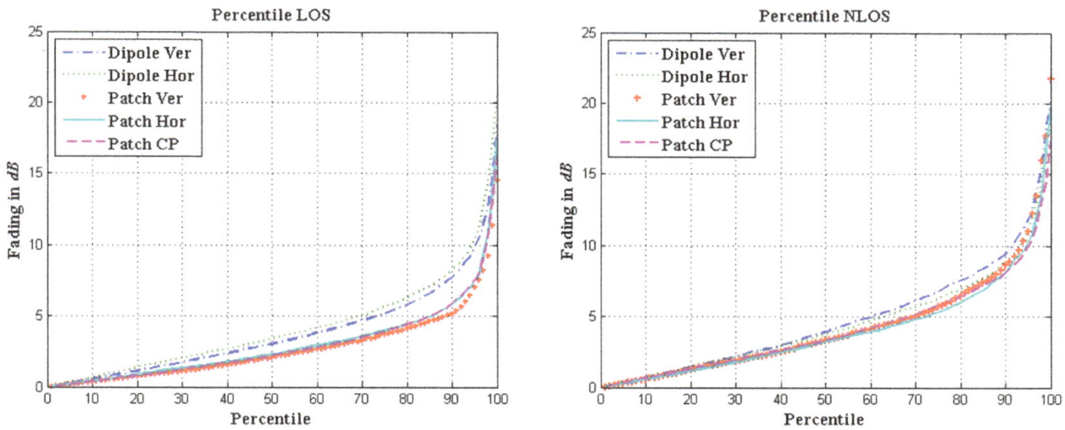

Figure 4. Percentile measure of fading (Dipole/Patch = Rx antenna; Ver/Hor/CP = Tx antenna).

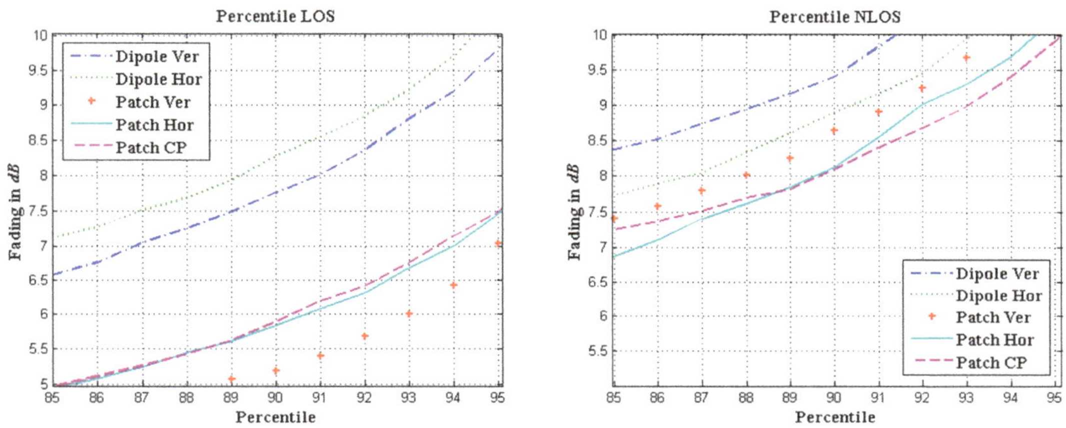

Figure 5. Percentile measure of fading, segment of the 90th percentile.

2. The combination circular/circular shows also good results, especially for LOS conditions. The explanation for this is that the energy of single-bounced reflected paths of the circular polarized transmitted wave disappears at the receiver since an odd number of reflections cause a reverse of the handedness of the circular polarized wave (Kajiwara, 1995).

A graphical representation of fading can be seen in Fig. 4. It shows a percentile measure of the encountered fading. Most measurement values of the patch antenna are below the according measurement values of the dipole. For example the 90th percentile (see Fig. 5) shows that 90 % of the measurement values of the patch (LOS) are below a value of 5–6 dB, while 90 % of the measurement values of the dipole are below a fading value of 7.5–8.5 dB. The same situation shows for a NLOS connection, the fading of 90 % of the measurement values of the patch are below 8–9 dB, while 90 % of the values of the dipole are below 9–9.5 dB.

Two other papers which describe similar topics like this paper are e.g. Seidel and Rappaport (1992) and Kajiwara (1995). Kajiwara (1995) discusses different anti-multipath fading schemes using circular polarization under LOS conditions at 10 GHz. The results show that the mean amplitude fade level is reduced by 7–11 dB by using CP transmission/reception. The results by using CP to linear transmission/reception are slightly less good.

Seidel and Rappaport (1992) introduce a 914 MHz path loss prediction model with a similar approach like in this paper. Different propagation scenarios were considered (e.g. grocery store, retail store and two office buildings). The path loss exponent and standard deviation was determined within a value range of $\gamma = 1.81 - 5.04$ and $\sigma = 4.3 - 16.3$ dB, these relatively high values occur due to fact that the knowledge of the number of floors and walls between sender and receiver were not taken into account (see also Sect. 3). In relatively open environments (retail/grocery store) the path loss exponent is much smaller $\gamma = 1.81 - 2.18$ which is similar to the values which were encountered in this paper. The standard deviation of the most scenarios $\sigma = 4.3 - 8.7$ dB is also similar to the values presented in this paper.

The paper Molisch et al. (2004) presents results for UWB (ultra wide band) channel models at 100–1000 MHz and

2–10 GHz. Even though UWB channel models are not fully comparable to the narrowband model which was used in this paper, the results show some similarities to the results presented in this paper. The path loss exponent and standard deviation was determined within a value range of $\gamma = 1.63 - 1.79$ and $\sigma = 0.83 - 2.22$ dB for LOS conditions and $\sigma = 2 - 3.51$ dB for NLOS conditions (in the 2–10 GHz band) for residential and office environments. The results for the 100–1000 MHz band are only given for an office-like environment under NLOS conditions with $\gamma = 2.4$ and $\sigma = 5.9$ dB.

5 Conclusion

The results of a measurement campaign were presented. This measurement campaign was performed to create a measurement based fading model and to evaluate the possibility to fight fading using a circular polarized and a linear polarized antenna in an one-to-one transmission link within an indoor scenario.

The path loss exponent was measured within a range of values $\gamma = 1.83 - 2$ and the standard deviation with mean value of $\sigma = 4.24$ dB for LOS conditions and $\sigma = 5.67$ dB for NLOS conditions. These values (especially the path loss exponents which is less or equal 2) seem to be in accordance with other studies presented in the previous chapter.

The method to combat fading using an optimized circular polarized truncated corner patch antenna has shown significant results. The results have shown that it is possible to mitigate fading by a notable amount, up to 3.5 dB for LOS and 1.5 dB for NLOS conditions (based on the 90th percentile), by replacing a linear polarized antenna (which is normally used in indoor SRD communication systems) with a circular polarized antenna. This is of practical importance since this is a very cost-efficient, space-saving and easy-to-realize method.

Acknowledgements. This project (HA-Project-Nr.: 344/12-34) is funded by LOEWE, Hesse's program of excellence in research.

Edited by: J. Götze

References

Cox, D. C., Murray, R. R., Arnold, H. W., Norris, A., and Wazowicz, M.: Cross-polarization coupling measured for 800 MHz radio trans-mission in and around houses and large buildings. Antennas and Propagation, IEEE T. Antenn. Propag., 34, doi:10.1109/TAP.1986.1143714, 1986.

Hashemi, H. and Tholl, D.: Analysis of the RMS delay spread of indoor radio propagation channels, Discovering a New World of Communications, IEEE International Conference, 2, 875–881, doi:10.1109/ICC.1992.268160, 1992.

Hernandez, M., Li, H.-B., Dotlic, I., Miura, R. (NICT), IEEE P802.15 Working Group for Wireless Personal Area Networks (WPANs): Channel models for TG8, IEEE P802.15-12-0459-00-0008, 2012.

Horvat, G., Rimac-Drlje, S., and Žagar, D.: Fade Depth Prediction Using Human Presence for Real Life WSN Deployment, Radioengineering, 22, 758–769, 2013.

Kajiwara, A.: Line-of-Sight Indoor Radio Communication Using Circular Polarized Waves, IEEE T. Veh. Technol., 44, 487–493, 1995.

Kyritsi, P. and Cox, C.: Propagation characteristics of horizontally and vertically polarizedelectric fields in an indoor environment: simple model and results, Vehicular Technology Conference, IEEE VTS 54th, 3, 1422–1426, doi:10.1109/VTC.2001.956431, 2001.

Molisch, A.: Wireless Communications, John Wiley & Sons Ltd., second edition, p. 27, 2011.

Molisch, A., Balakrishnan, K., Chong, C.-C., Emami, S., Fort, A., Karedal, J., Kunisch, J., Schantz, H., Schuster, U., and Siwiak, K.: IEEE P802.15.4, 662–703, IEEE 802.15.4a channel model – final report, http://ieee802.org/15/pub/04/15-04-0662-01-004a-channel-model-final-report-r1.pdf, last access: 2 September 2014, 2004.

Schwarzbeck Mess-Elektronik: XSLP 9142 Kreuzpolarisierte Breitband UHF-SHF Log.-Per. Messantenne Dual Polarized UHF-SHF Broadband Log.-Per. Test-Antenna 800 MHz ... 3 (5) GHz, http://www.schwarzbeck.com/Datenblatt/manx9142.pdf, last access: 22 July 2014, 2013.

Seidel, S. and Rappaport, T.: 914 MHz path loss prediction for indoor wireless communication in multifloored buildings, IEEE T. Antenn. Propag., 40, 207–217, 1992.

Vireerackody, K.: Characteristics of a simulated fast fading indoor radio channel, Vehicular Technology Conference, 43rd IEEE, 231–235, doi:10.1109/VETEC.1993.507051, 1993.

32

Investigation of 3G radio network topologies considering performance and exposure aspects

H. Buddendick[1], **P. Wertz**[1], **T. F. Eibert**[1], **F. M. Landstorfer**[1], **and G. Wölfle**[2]

[1]Institute of Radio Frequency Technology, Universität Stuttgart, 70550 Stuttgart, Germany
[2]AWE Communications GmbH, Otto-Lilienthal-Str. 36, 71034 Böblingen, Germany

Abstract. Five different radio network topologies, in the following referred to as "scenario", have been analyzed with regard to their capability to cover a medium size city with 3G mobile services. T-Mobile launched this study to assess different radio network topologies in terms of exposure. Thus, the main focus is to examine if, for exposure reasons, a medium size city can be covered with outer-city located sites.

The radio network topologies discussed in this paper can generally be divided into two categories: In category one, the sites are located in the city, and in the other, the sites are located outside of the built-up area. In both cases the number of sites per scenario varies, and additionally the antenna height and site distance to the built-up area vary for the second category.

The investigations are performed with a ray optical propagation model based on 3-D building data and a dynamic 3G network simulator. A mix of four services and a realistic spatial traffic distribution based on land usage is assumed.

The performance of the scenarios, assessed by the evaluation of coverage and capacity, is examined together with the exposure.

It is shown that network topologies with sites located only outside the city are not an adequate solution due to capacity constraints and coverage problems with high data services. A promising network topology is a reduced set of inner-city sites with higher order sectorization.

1 Introduction

In recent years, the number of people using mobile voice services increased significantly and the availability of mobile communications is taken for granted by most people in industrialized countries. In addition, mobile data services, like web browsing and email, become more and more popular.

In order to provide high capacity radio networks which also offer high bit rate data services for a large percentage of the population, network operators have to redesign their network structure. A straight-forward way, which is preferred by many operators, is to increase the number of base stations and to reduce their distance.

However, if new base stations are to be installed, the network operators are often faced with strong resistance of the local public, since people are afraid of an impact on their health. T-Mobile, like other network operators, takes the people's concerns seriously and is interested to roll-out and extend the network in a widely accepted way. Since it is expected that critical people will probably rather accept alternative network solutions suggested from a third, independent party, T-Mobile launched the study treated in this paper.

This study examines several network topologies to find promising candidates to cover a typical medium size middle European city with mobile services and to fulfill the requirements regarding network performance, economic efficiency and exposure issues. The focus is mainly on UMTS data services, as they are the most crucial ones regarding service coverage. Dynamic system simulations are used to evaluate and compare the above mentioned properties of different network topologies.

The simulation tools (propagation model and 3G system simulator), the considered environment and the investigated network topologies are described in Sect. 2. The results in terms of coverage, capacity and exposure are presented and compared in Sect. 3. Finally Sect. 3 sums up the results.

Correspondence to: H. Buddendick
(buddendick@ihf.uni-stuttgart.de)

2 Simulation environment and assumptions

2.1 Propagation model

For radio network planning, the prediction of the mobile propagation channel between the sector antennas and the user location in terms of path loss is the most important input parameter. Empirical propagation models only provide limited accuracy in urban environments. On the other hand, ray optical propagation models deliver much more accurate results (Rautiainen et al., 2002). However, ray tracing models have a high computational demand (Zhang et al., 2000). Therefore, a semi-deterministic propagation model based on dominant paths (Wahl et al. , 2005) was utilized in this study.

The main idea of this Urban Dominant Path model (UDP) is that only a few paths contribute a considerable amount of energy to the received power, and that only these paths have to be considered. The physical effects of reflections and diffractions resulting in the typical wave guiding and shadowing phenomena are taken into account.

In comparison to rigorous 3-D ray tracing models the UDP model is much faster but yields comparably prediction accuracy. For a carrier frequency close to the UMTS downlink frequency, a measurement campaign was carried out in the frame of the EU funded MAESTRO project Heyn et al. (2005). The UDP model was validated with these measurement results, yielding standard deviations of the prediction error from 3.5 dB to 5.6 dB for different measurement routes.

For mobile stations that are located inside buildings a semi-empirical penetration model is applied.

2.2 Dynamic 3G system simulator

It is well known that WCDMA based 3G mobile networks — in contrast to GSM networks – show some quite dynamic performance characteristics. In addition, coverage and capacity of the network are strongly correlated (e.g. cell breathing). A system simulation tool has to account for these characteristics. In the same way, a meaningful investigation of exposure aspects has to deal with realistic network loads and the resulting transmission power levels.

One possibility is to perform a dynamic 3G system simulation. The main drawback of dynamic simulations is the high computational effort. But for general studies and detailed simulations within limited areas, like for the investigations described in this paper, a 3G dynamic simulation tool is well suited and the computation times are in an acceptable range.

The tool used within these studies (Wertz et al., 2002) splits the simulation process into two parts: a dynamic simulation and a scan-mobile process. In the dynamic simulation the network evaluation is performed with mobile stations generated according to statistically defined arrival and serving processes. Thereby the spatial user distribution can be controlled by user densities that depend on the land usage data. Both indoor and outdoor users are generated.

For newly generated mobile stations a check is performed whether the available resources in terms of spreading codes and transmission power is sufficient to fulfill the respective service requirements. If not enough resources are available, or in case of bad pilot coverage, the mobile station is blocked. Fast power control is modeled on UMTS slot basis. If the power control is unable to ensure the required signal to noise ratio target for a given period of time, the mobile station is dropped. The UMTS soft/softer handover procedure is simulated in detail.

In the downlink the imperfect orthogonality due to multipath propagation effects is considered using an orthogonality factor. In accordance with T-Mobile an average orthogonality factor α=0.75 was used in this study.

Several statistical outputs are collected in the dynamic simulation. These are mainly the cell and network throughput and the blocking and dropping rates, as well as transmission power and code utilization statistics.

After a certain simulation time the dynamic simulation is interrupted and a probe-mobile scans the complete simulation area in order to provide coverage information (see Fig. 2 and Fig. 3 in Sect. 3.1). The main inputs to the probe-mobile process from the dynamic simulation are instantaneous uplink load and downlink transmission power levels. Both describe the interference conditions for uplink and downlink, respectively.

The coverage is determined for the complete simulation area for several points in time. Considering all time instances, the coverage probability is determined per pixel as the ratio between successful coverage tests and the total number of tests. These tests represent a quasi-static evaluation based on the fulfillment of the E_b/N_0-targets with the defined maximum link power and the evaluation of the CPICH C/I. However, they do not account for available spreading code resources. Therefore the coverage plots mainly show the area in which the given traffic density can be distributed.

2.3 Simulation area and network topologies

The selected simulation area should include a medium size city and fulfill two demands: first, the terrain should be flat since any effects caused by topographic undulations should be avoided for our investigations. Second, 3-D building data should be available in order to reach the best accuracy of the wave propagation model.

However, 3-D building data in suitable quality are only available for large cities. Thus, we selected a district of Bonn, Germany. This district, called Beuel, was formerly an individual city and kept its old city structure with a small city center, industrial area and residential area.

The land usage types of Beuel and the surrounding area is shown in Fig. 1. The urban area considered in the investigations, corresponding to the area of the former individual city of Beuel, is located within the black polygon. To simulate

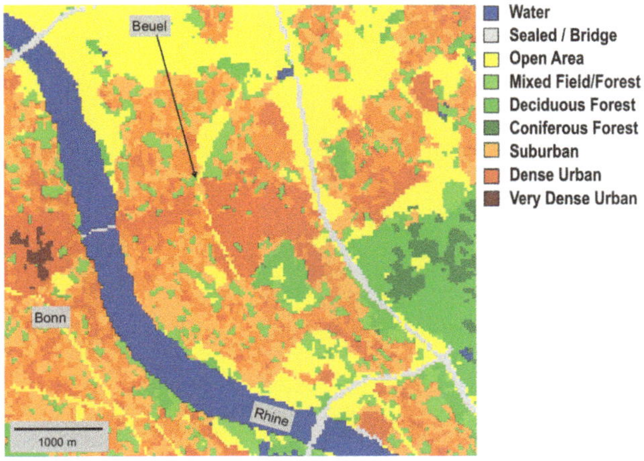

Fig. 1. Land usage of the regarded area. The land usage and 3-D building data outside the black polygon is modified in order to simulate a single standing city (here the original data is shown).

a single standing city, the 3-D building data were removed outside of the polygon and the land usage was set to "Open Area" there. The remaining built-up area can be regarded as a typical medium size middle European city. The evaluated city area has a size of around 8 km^2 with approx. 40 000 inhabitants.

Five network topologies have been examined, which are referred to as scenario 1 to 5. They can be divided into two general categories. In one, the sites are located within the urban area, and in the other category the sites are located outside the built-up area. Table 1 summarizes some characteristic information about the scenarios. For scenarios 3–5, two different antenna heights and varying distances to the built-up area are applied and analyzed in Sect. 3.1.

The base station antennas have a half power beam width of 33° horizontal and 6.7° vertical and 20.7 dBi gain, except for scenario 2, where antennas with 65° horizontal and 6.7° vertical half power beam width and 18 dBi gain were applied. The power settings were identical for all scenarios: the maximum BS power per cell is 43 dBm, the constant CPICH power is 33 dBm and the overall power for all other common channels is 36 dBm.

2.4 Service and traffic assumptions

The 3G simulations have been performed with a service mix consisting of four services as listed in Table 2 together with their main properties. The maximum allowed power to serve a single user in the downlink is 37 dBm for all services. The mobiles are modeled as stationary in the dynamic simulation. For the services WWW and High data, an antenna gain of 2 dBi (laptop device) is used, otherwise 0 dBi. The maximum active set size for the soft/softer handover algorithms is 3. Table 2 also includes the average number of users who simultaneously wish to use the corresponding service in the simulation area per service (offered traffic).

Table 1. Overview of the 5 different network topologies.

Sites are ...	Scenario	Sites	Sectors per Site	Antenna height
... inside town	1	1	6	40 m
	2	7	3	25 m
... outside town	3	1	4	30 m/60 m
	4	2	4	30 m/60 m
	5	5	3	30 m/60 m

Table 2. Overview of the services applied in the 3G simulations for the reference traffic assumption T_1.

Service	Bitrate DL kbps	Bitrate UL kbps	max. UL power dBm	active UE
Voice	16	16	21	5.0
Video Telephonie	64	64	21	39.0
WWW	128	56	24	5.9
High Data	384	56	24	6.3

In order to determine a deterioration of the service quality with increasing network load, the simulations have been performed with a varying amount of offered traffic. A reference traffic assumption called T_1 – as given in Table 2 – is scaled linearly, leading to traffic assumptions from $T_{0.25}$ ($=T_1$ multiplied with 0.25) up to T_3 (equals T_1 increased by a factor of 3).

In addition, the traffic is spatially distributed according to the land usage types, e.g. at land usage suburban, the offered traffic is half of that on the land usage dense urban. This leads to an inhomogeneous spatial traffic distribution, which yields a realistic modeling. The relative distribution of the active users over the services according to Table 2 is equal for all land usage types.

3 Simulation results

3.1 Coverage

Assuming the reference traffic assumption T_1 and the 384 kbps data service, Fig. 2 and Fig. 3 depict the coverage probability for scenario 2 and for scenario 5, respectively.

The plots show very well that the network topology that uses sites outside the city leads to a restricted coverage especially of the city centre. This is due to the fact that especially

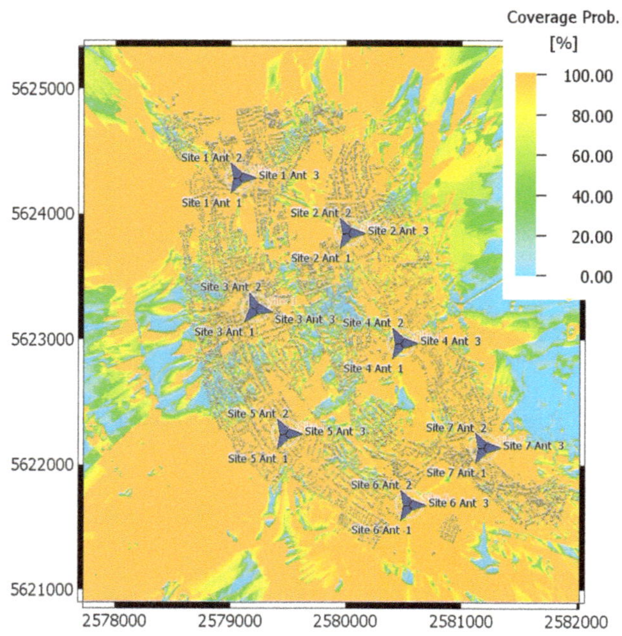

Fig. 2. Coverage probability for 384 kbps data service for scenario 2.

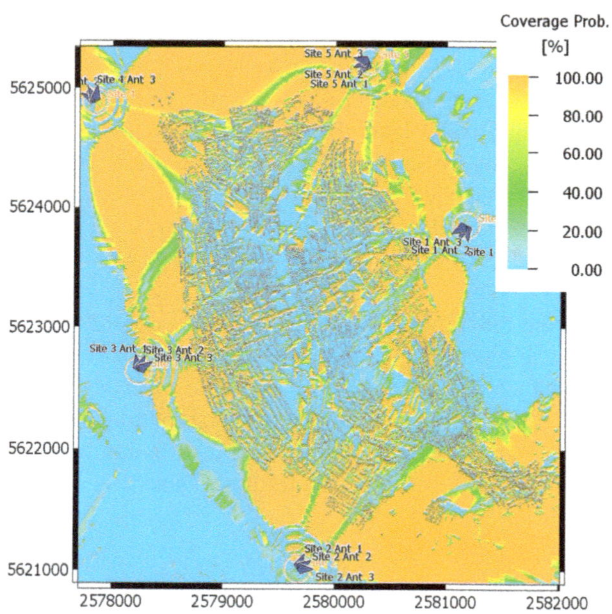

Fig. 3. Coverage probability for 384 kbps data service for scenario 5.

in the area in the middle between the base stations the cell isolation is very bad, leading to a poor interference situation.

As described in Sect. 2.2, the coverage plots represent a quasi-static evaluation. To account for shortages in the code resources of the network, the results should be interpreted in

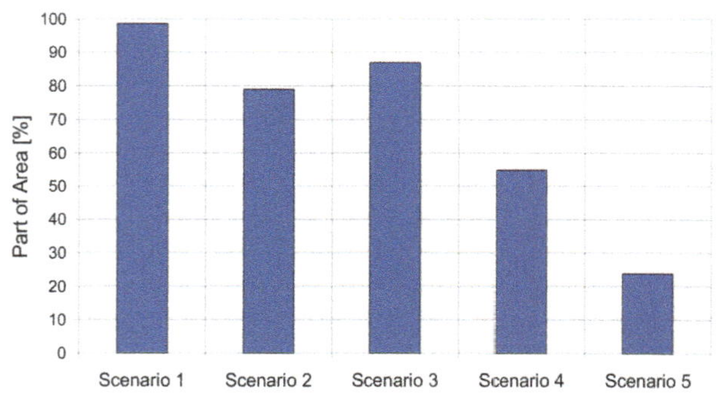

Fig. 4. Area with at least 90% coverage probability (384 kbps data service).

conjunction with the statistic results presented in Sect. 3.2 of this chapter.

Figure 4 shows the portion of the area that has a coverage probability of at least 90%. This measure allows a good comparison between the scenarios.

Scenario 1 with its central base station is very advantageous in terms of the interference situation yielding excellent results. However, due to the low number of 6 sectors, the capacity of this scenario is limited (see Sect. 3.2).

When analyzing scenarios 3, 4 and 5 with sites outside the built-up area, it is noticeable that the coverage quality diminishes with a growing number of sectors. This reflects the growing interference problems with the increasing number of sectors located outside the city. On the other hand, in scenario 2 the different cell areas are better separated and the interference impact is thus more limited to the local environment. Therefore the overall interference level is comparably low, despite the high number of 21 sectors.

It should be mentioned that the coverage differences are less distinct for the other services (see Table 2), however high data rate services are one of the major reasons for using 3G networks.

When analyzing the resulting cell layout, the scenarios 3, 4 and 5 lead to a much worse cell isolation than scenario 1 and 2. The cell layout is more inhomogeneous in scenarios 3 to 5, also leading to unfavorable handover conditions.

For the sites located outside the city, their distance to the built-up area was varied from 250 to 1000 m, and the base station antenna heights were chosen to be 30 and 60 m. There were no considerable differences in dependence of the distance; however, choosing a too short distance is disadvantageous, as the cells become more cluttered. The higher antenna heights turn out to be beneficial, as they lead to a better cell isolation when combined with higher antenna downtilt values. Therefore a distance of 500 m at 60 m antenna height is used for the detailed evaluation of scenarios 3, 4 and 5.

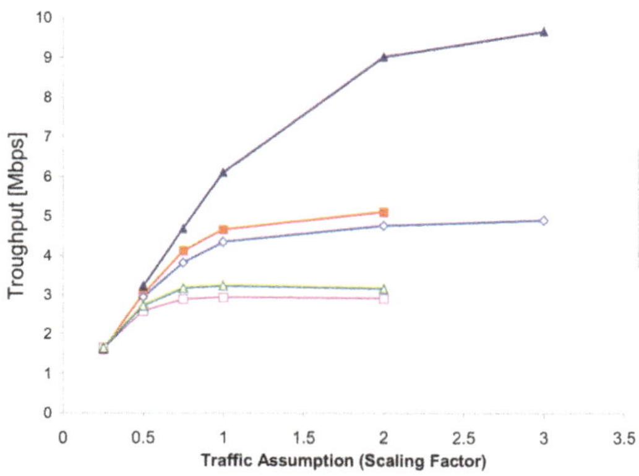

Fig. 5. Total throughput considering all services.

3.2 Network capacity

Figure 5 shows the total throughput for the whole network. As the number of sectors differs in the scenarios (see Table 1), higher total throughput values are expected if more sectors are available. The abscissa represents the scaled traffic (see Sect. 2.4). According to Fig. 5 scenario 2 yields the highest capacity, which is expected due to the high number of sectors. Scenario 5, which has 15 sectors, performs badly in terms of capacity, especially compared to scenario 1 with only 6 sectors. This reflects the poor interference situation in scenario 5.

3.3 Exposure

For the following downlink exposure analysis the transmission power is derived as mean value from the dynamic simulation. A prediction of the received power has been performed for all base station sectors under consideration of the averaged transmission powers of the individual sectors. Finally, the propagation results were incoherently summed up in the whole simulation area, leading to the plots in Fig. 6 and Fig. 7.

Due to the increased radio channel attenuation if the sites are located outside the city, the exposure in the city is lower for scenario 5 compared to scenario 2. On the other hand, the difference is less distinct as could be expected because of the fact that the required transmission powers are higher in scenario 5 (median values over all sectors: 24.4 dBm for scenario 2, 27.8 dBm for scenario 5).

Figure 8 shows a comparison between the exposure and the capacity for all five scenarios. The pink curve shows the total throughput in Mbps, the blue curve shows the 90% quantiles of the exposure values derived from results as shown in Fig. 6 and Fig. 7.

This comparison impressively shows that network topologies with base station sites outside the city yield the lowest exposure, yet at the cost of a large reduction of the network capacity. Scenario 1 yields low exposure values at a reasonable network capacity.

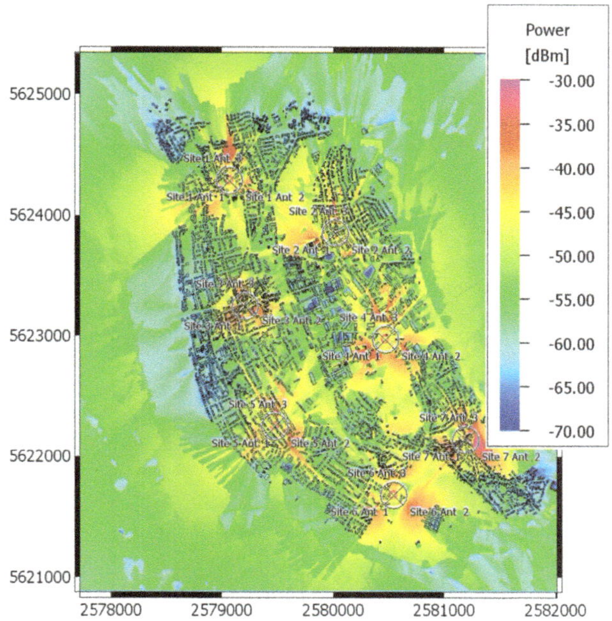

Fig. 6. Exposure for Scenario 2.

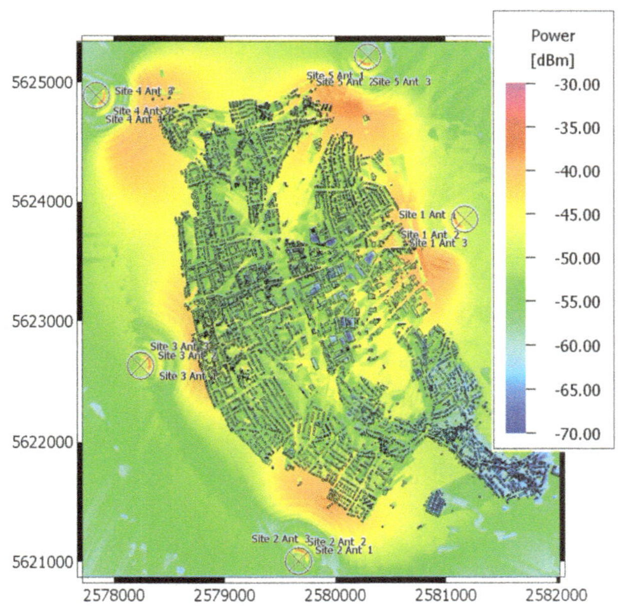

Fig. 7. Exposure for Scenario 5.

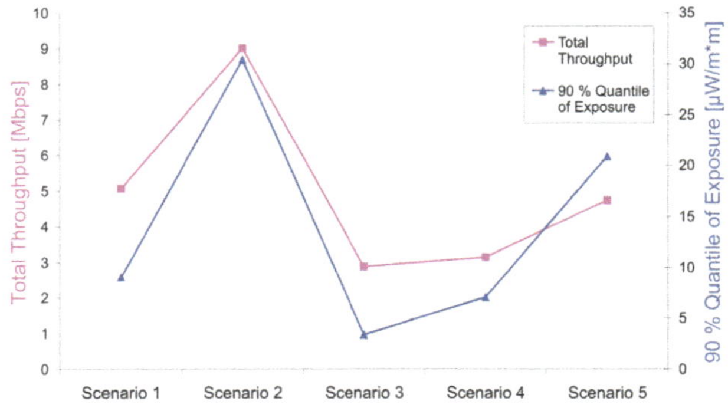

Fig. 8. Comparison of exposure and capacity for all scenarios.

3.4 HSDPA and GSM/EDGE

Furthermore simulations have been carried out investigating the impact of the different scenarios on HSDPA performance (Buddendick et al., 2004). It turned out that the mean expected data rate for a user in scenario 2 is twice as high as for a user in scenario 5.

In contrary, assuming that different carrier frequencies are assigned to all sectors, the coverage for GSM/EDGE systems obtained by scenario 5 is comparable to that of scenario 2. Thus, concerning GSM/EDGE, it might be regarded as a realistic alternative to cover a medium size city with several sites located outside the city.

4 Summary

In this paper, five different network topologies were analyzed with regard to their capability to cover a medium size city with 3G mobile service. The impact of locating the sites inside or outside the city on coverage, capacity and exposure is examined. The simulations were performed with a dynamic 3G network simulator and a ray optical propagation model applying 3-D building data.

The scenarios with several sites located outside the city provide only unsatisfactory coverage for high data rate services. A service with 384 kbps in the downlink is available in only 35% of the city area in case of five outer-city sites, but in 80% of the area in case of seven inner-city sites. However, for services with lower data rates all scenarios yield sufficient coverage in the complete city area.

A single outer-city site gives the lowest average exposure – which could be expected. However, scenarios with more than one outer-city sites yield average exposure values which exceed those of the scenario with only one site located in the city.

Furthermore, the required mobile transmission power to establish connections is lowest for the scenario with seven inner-city sites which leads to a low exposure of the active mobile users.

Regarding all three aspects (exposure, coverage and capacity), the scenarios with inner-city sites are in advantage. One outstanding site yields low exposure but insufficient capacity. Five outstanding sites yield insufficient coverage — and the exposure is comparable to those of scenarios with inner-city sites. In addition, several outstanding sites lead to an inhomogeneous cell structure, which is generally of disadvantage.

However, one result of this paper is that a single inner-city site with medium antenna height (40 m) and six sectors yields a very good coverage and acceptable capacity in the city area at a low exposure. Thus, covering a medium-size city with only one or a few inner-city sites with a higher number of sectors is a topic of further investigations. Further work will also cover more detailed investigations on HSDPA network performance for different network topologies.

Acknowledgements. The authors would like to thank T-Mobile Germany for their initiation of the presented work and especially Dr. Beyer for the valuable help.

References

Rautiainen, T., Wölfle, G., and Hoppe, R.: Verifying Path Loss and Delay Spread Predictions of a 3D Ray Tracing Propagation Model in Urban Environments, 56th IEEE Vehicular Technology Conference (VTC) 2002 – Fall, Vancouver (British Columbia, Canada), Sept. 2002.

Zhang, Z., Yun, Z., Iskander, and Iskander, M.F.: Ray Tracing method for propagation models in wireless communication systems, IEE Electronics Lett., 36(5), 464–465, March 2000.

Wahl, R., Wölfle, G., Wertz, P., and Wildbolz, P.: "Dominant Path Prediction Model for Urban Scenarios", 14th IST Mobile and Wireless Communications Summit, Dresden (Germany), June 2005.

Heyn, T., Heuberger, A., and Keip, C.: Propagation Measurements for the Characterization of a Hybrid Mobile Channel in S-band, 14th IST Mobile Wireless Communications Summit 2005 – Dresden (Germany), June 2005.

Wertz, P., Hoppe, R., Wölfle, G., Buddendick, H., and Landstorfer, F.: Dynamic Simulator for the Analysis of the Performance of WCDMA Radio Networks, 4th COST 273 MCM-Meeting in Lisbon, Portugal, COST 273 TD(02)148, Sept. 2002.

Buddendick, H., Wölfle, G., Burger, S., and Wertz P.: Simulator for Performance Analysis in UMTS FDD Networks with HSDPA, 15th IEEE International Symposium on Personal, Indoor and Mobile Radio Communications (PIMRC) 2004, Barcelona (Spain), Sept. 2004.

WinProp, Models and software for Wave Propagation and Radio Network Planning, see http://www.awe-communications.com/, 2006.

High-End Silicon PDICs

H. Zimmermann
Institute of Electrical Measurements and Circuit Design, Vienna University of Technology, Vienna, Austria

Abstract. An overview on integrated silicon photodiodes and photodiode integrated circuits (PDICs) or optoelectronic integrated circuits (OEICs) for optical storage systems (OSS) and fiber receivers is given. It is demonstrated, that by using low-cost silicon technologies high-performance OEICs being true competitors for some III/V-semiconductor OEICs can be realized. OSS-OEICs with bandwidths of up to 380 MHz and fiber receivers with maximum data rates of up to 11 Gbps are described. Low-cost data comm receivers for plastic optical fibers (POF) as well as new circuit concepts for OEICs and highly parallel optical receivers are described also in the following.

1 Introduction

There is a wide field of applicatons for Opto ASICs. Inexpensive optical receivers with top performance are in great demand. III-V semiconductors normally achieve the best peformance, while Si technologies offer low production costs. For long distance applications, with data rates of more than 40 Gbps, it is useful to accept high costs of III-V semiconductors to get the best performance. On the other hand these high costs make them unattractive for the mass markets of short distance applications where lower data rates are adequate, e.g. local-area networks, fiber to/in the home, optical interconnects on and between boards. Nevertheless it is necessary to use advanced circuit design to reach competitive results with Si technologies. Si photodetectors can detect visible and near-infrared light which is used for short distance communications, as well as for plastic-optical-fiber (POF) receivers and in optical pickup units for instance in DVD systems.

Optoelectronic integrated circuits (OEICs) are used for very compact designs and low-cost applications. They combine the photodetector and the receiver circuit on a single chip and therefore mounting and handling are much cheaper and the area consumption and weight is far below a multichip solution. In the next years there will be a demand for high volumes of low-cost highly reliable OEICs for optical buses in cars and in optical storage systems, e.g. CD-ROM, DVD, Digital Video Recording and Enhanced Video Disk (EVD). For high performance OEICs, pin photodiodes are integrated in Si processes.

Correspondence to: H. Zimmermann
(horst.zimmermann@tuwien.ac.at)

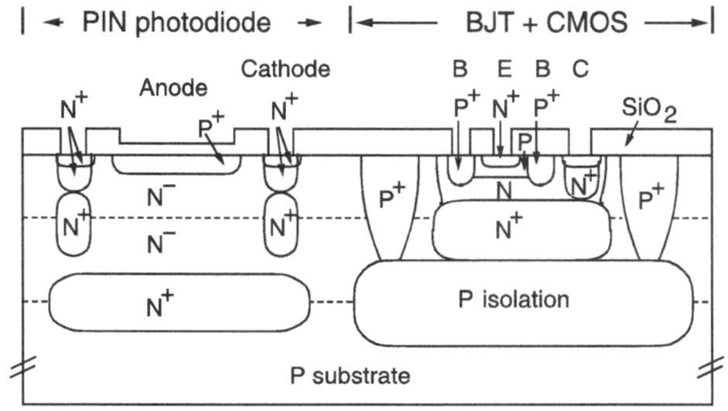

Fig. 1. PIN photodiode integration in standard buried collector (SBC) bipolar technology (Yamamoto, 1995).

2 Large–diameter integrated–photodiode OEIC

A vertical PIN photodiode was integrated in a 0.6 μm BiCMOS process. It is similar in design to the diode presented in (Yamamoto, 1995) which cross section is depicted in Fig. 1. Deep N plugs after low-doped N-epitaxial steps form the connection from the Si surface to the N+ buried–layer cathode. The electrical isolation between the npn transistors and the N-wells is formed by a P-isolation layer, which separates them. (Yamamoto, 1995) presented a bandwidth of 300 MHz at 780 nm light with a reverse bias voltage of 3 V. However, the reverse bias voltage can be larger than the chip supply

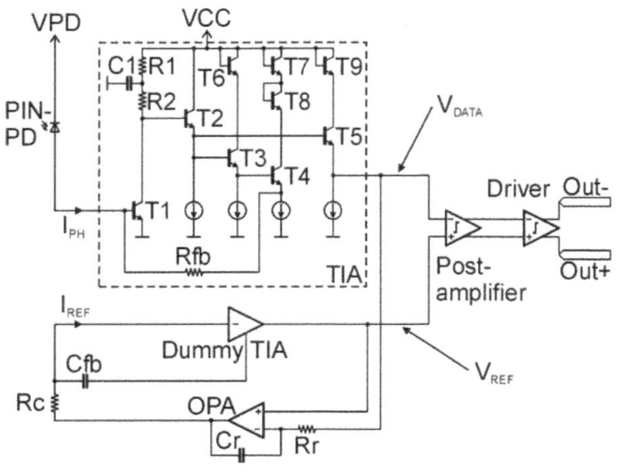

Fig. 2. Schematic of a 2.5 Gbps silicon receiver OEIC with large-diameter photodiode (Swoboda, 2004).

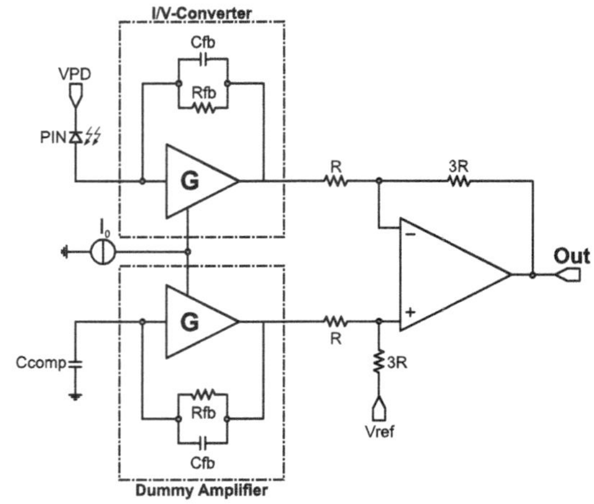

Fig. 4. Schematic of a 380 MHz two-stage OEIC for the use in DVD (Leeb, 2004).

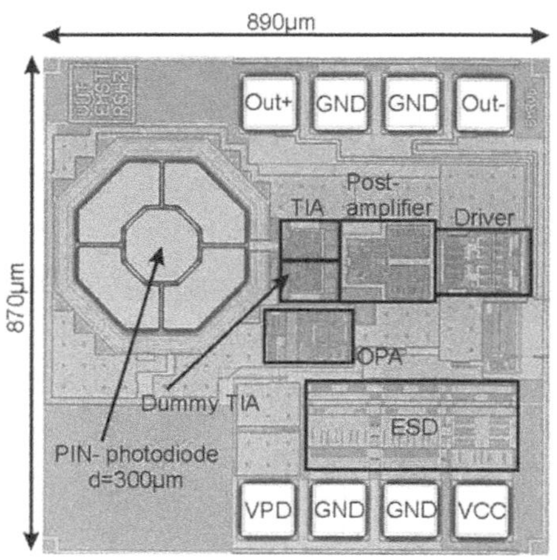

Fig. 3. Microphotograph of a 2.5 Gbps silicon receiver OEIC with large-diameter photodiode (Swoboda, 2004).

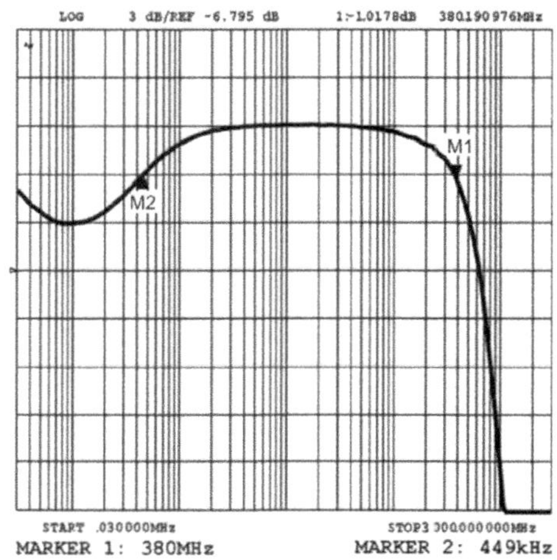

Fig. 5. Frequency response of a 380 MHz two-stage OEIC for the use in DVD (Leeb, 2004).

voltage of 5 V due to the fact that the N cathode is realized in a P-type substrate. This is important to speed up the carrier drift and to increase the bandwidth.

Run with a bias voltage of 12 V, a bandwith of more than 1.35 GHz was reached with a 660 nm laser diode (Swoboda, 2004). The simplified schematic of this 2.5 Gbps Si receiver OEIC with large-diameter photodiode is depicted in Fig. 2. A resopsivity of 0.36 A/W was achieved due to the thick intrinsic layer. In Fig. 3, a microphotograph of the OEIC receiver is depicted. The large photodiode with a diameter of 300 μm can be clearly seen. The chip dimensions are 890 μm times 870 μm. The Si OEIC achieved a sensitivity of −17.2 dBm compared to −15.7 dBm of a GaAs OEIC (Lang, 2001), whereby the die area of the Si OEIC was one fifth of that of the GaAs OEIC and the power consumption was one third.

3 Two-stage OEIC for the use in DVD

In (Leeb, 2004) an amplifier architecture that uses a differential transimpedance amplifier in the first stage and an operational amplifier in a second stage is presented.

The used amplifier structure is shown in Fig. 4. The main I/V converter and the dummy I/V converter are connected by a current source (I_0) and therefore both of them contribute gain to the input stage. The second stage is designed as subtractor amplifier and subtracts and amplifies the differential output voltage of the first–stage amplifiers. Furthermore a reference voltage (V_{ref}) is added to reach a single-ended output with low offset voltage compared to (V_{ref}). This architecture together with the capacitance (C_{COMP}) enables a doubled transimpedance with the same value of the feedback resistor (R_{fb}) compared to other concepts. This leads to a doubled differential transimpedance compared to the ar-

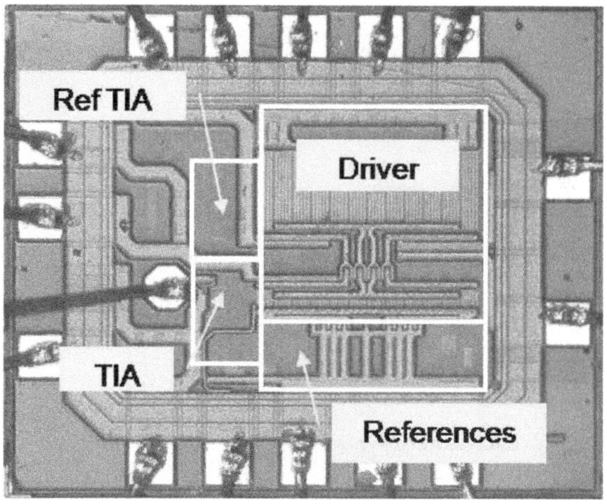

Fig. 6. Schematic of a 130 MHz-bandwidth, 120 dB-dynamic-range TIA in 0.35 μm SiGe BiCMOS (Micusik, 2007).

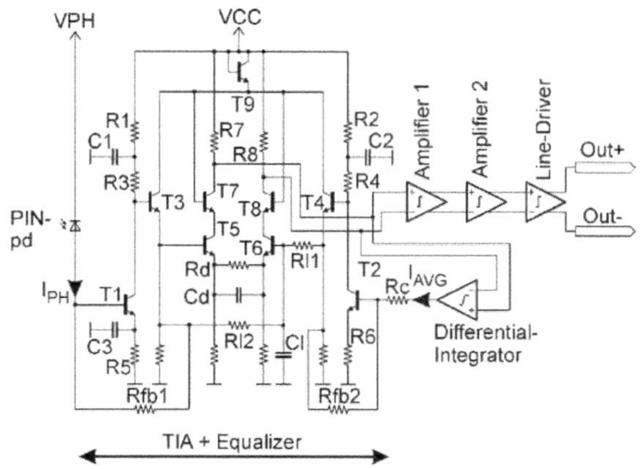

Fig. 7. Micorphotograph of a 130 MHz-bandwidth, 120 dB-dynamic-range TIA in 0.35 μm SiGe BiCMOS (Micusik, 2007).

chitecture used in (Kieschnick, 1999, Zimmermann, 2000, and Kieschnick, 2003) for the same (R_{fb}) value. The bandwidth was increased by a factor of 6.8 compared to a two-stage amplifier presented in (Takimoto, 1998) and by a factor of 2.6 compared to (Kieschnick, 1999). In addition, the photo-sensitivity was increased by a factor of about 10 compared to (Kieschnick, 1999, Zimmermann, 2000). In total, the transimpedance–bandwidth product was increased by a factor of 26 compared to (Kieschnick, 1999).

The analog modulation port of the network analyzer was used to modulate the laser source for the measurement of the frequnency response. In Fig. 5 the measured frequency response is depicted. It can be seen that the upper −3 dB cut-off frequency of the two-stage transimpedance amplifier is 380 MHz. The lower cut-off frequency is 450 kHz, for frequencies below that lower cut-off frequency the gain drops by 6 dB.

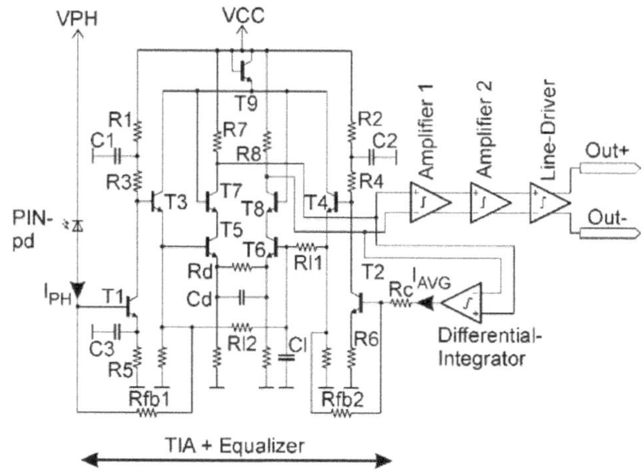

Fig. 8. Simplified schematic of a 11 Gbps optical receiver (Swoboda, 2006).

4 130 MHz-bandwidth, 120 dB-dynamic-range transimpedance amplifier (TIA) in 0.35 μm SiGe BiCMOS

The basic circuit topology of the 130 MHz-bandwidth, 120 dB-dynamic-range TIA in 0.35 μm SiGe BiCMOS (Micusik, 2007) with implemented compression is shown in Fig. 6.

The circuit consists of three transistor stages: the first in common–collector configuration (Q0) followed by a common–emitter voltage amplifier (Q1) and as third stage a common–collector voltage buffer (Q2). The feedback resistor (R_f=22 kΩ) is connected from the output of the third stage to the input of the first transistor stage. For minimum electronic noise the input common–collector stage (Q0) can be biased without any influence of the following stages. There are two types of operation. In the linear region the transistors Q5 and Q6 are off and therefore do not affect the circuit for small input currents. In the compression region the transistor Q6 is on and provides current to the input and therefore a parallel path to (R_f) which leads to a lower overall transimpedance of the TIA. Stability over the full input current range is provided by the compensation circuit consisting of Q3, Q4 and Q5. The open-loop gain of the circuit is lowered by the compensation circuit dynamically and therefore the phase margin of the loop gain is kept large enough. In the linear region the TIA itself is the limiting element regarding to the bandwidth (BW). In the compensation region the BW of the TIA increases and the limiting elements are the following stages, e.g. the differential 50 Ω driver. The full width at half maximum (FWHM) of the input current pulses used for characterization was 10 ns with a rise and fall time of 1.5 ns. The circuit proposed in (Micusik, 2007) exhibits the highest dynamic range (120 dB) for the input current compared to the literature. The power consumption of

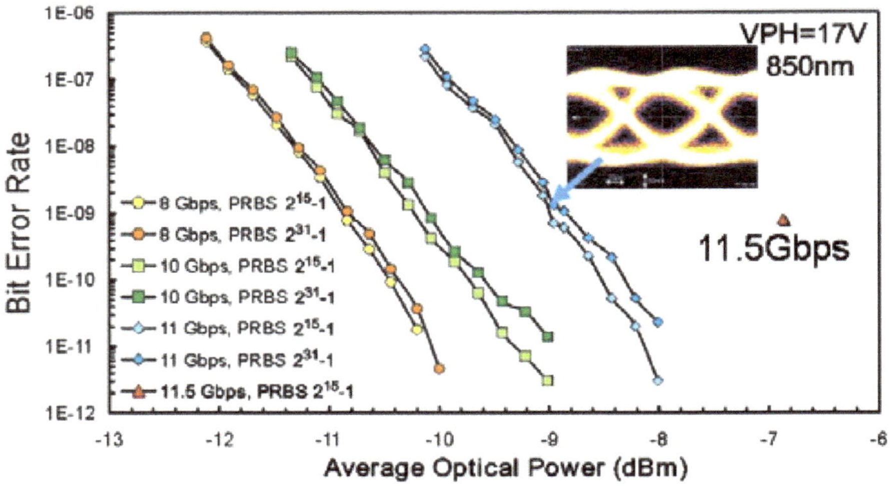

Fig. 9. Bit error rate vs. average optical power of a 11 Gbps optical receiver (Swoboda, 2006).

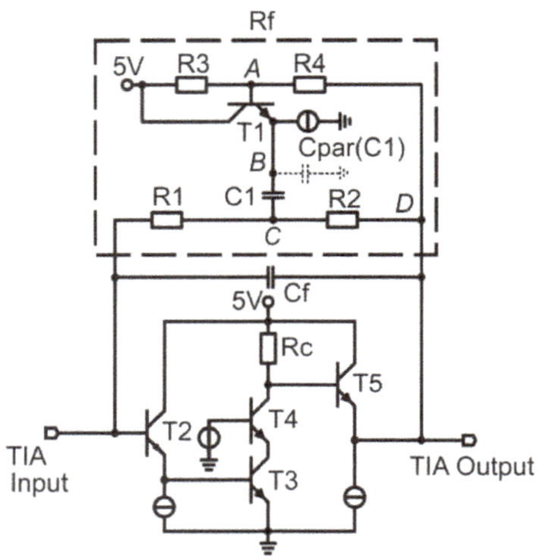

Fig. 10. Schematic of a 1-stage transimpedance amplifier with CCVD (Seidl, 2004).

the TIA alone is 2 mA from a 5V power supply. The complete receiver circuit including a 50 Ω, 1.7V peak-to-peak differential output driver consumes an area of 1.24 mm^2 in 0.35 μm SiGe BiCMOS technology and the complete power consumption is 390 mW.

A microphotograph of the 130 MHz-bandwidth, 120 dB-dynamic-range TIA in 0.35 μm SiGe BiCMOS is depicted in Fig. 7.

5 11 Gbps optical receiver

(Swoboda, 2006) presented a monolithically integrated silicon optical receiver for 850 nm light. The used photodi-

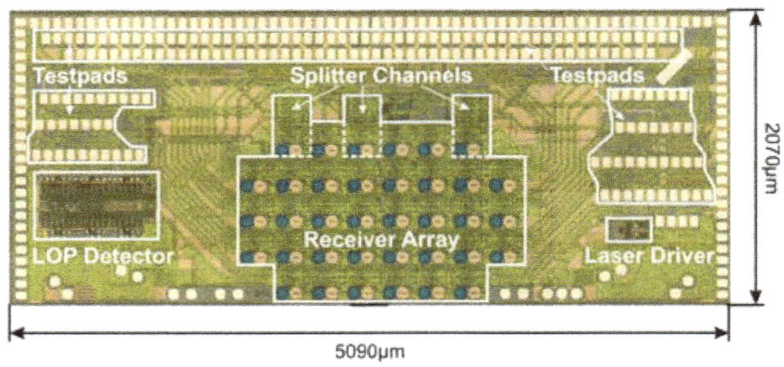

Fig. 11. Highly parallel optical receiver (Swoboda et al, 2006).

ode has a diameter of 50 μm and therefore matches exactly a 50 μm multimode fiber.

The schematic of the input circuit including a simplified schematic of the following stages is depicted in Fig. 8. A TIA consisting of T1, T3, R1, R3, R5, C1, C3, and Rfb1 is used to convert the photo current into an output voltage. The transimpedance is 500 Ω. The output voltage of the TIA subtracted by a reference voltage is amplified frequency dependent. The gain has the inverse characteristic to the photodiode frequency response. This is realized by a cascode difference amplifier consisting of T5, T6, T7, T8, R7 and R8 including frequency-dependent coupling elements. Rd and Cd compensate the photodiode drift time and the small diffusion part is compensated with C1, Rl1 and Rl2. A nearly flat frequency response with a linear phase is reached with this equalizer. TIA and equalizer are followed by three limiting amplifiers to improve the gain and drive a differential 100Ω load. The average photo current (I_{AVG}) generated by a differential integrator, a controlling circuit and a dummy TIA is provided to the dummy TIA via Rc. Therefore symmetrical clipping of the "0" s and "1" s is provided in case of

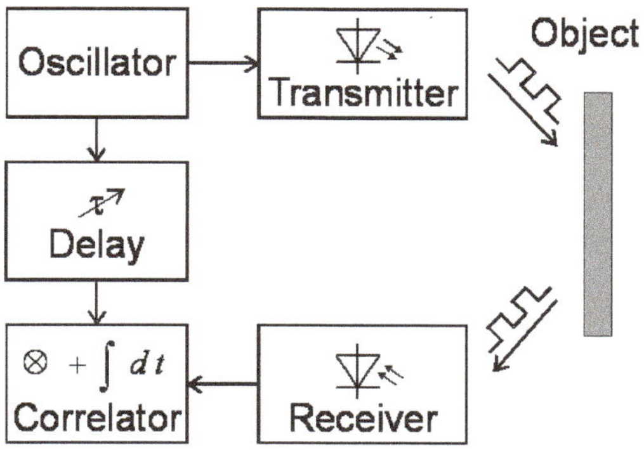

Fig. 12. Block diagram of optical distance measurement/3-D camera with continuous wave system.

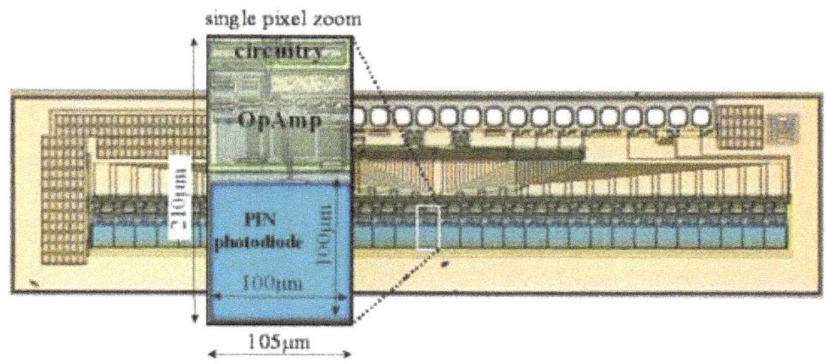

Fig. 13. Microphotograph of a line sensor for optical distance measurement with a continuous–wave system.

optical overdrive. The background light is suppressed by a lower cut-off frequency of 30 kHz introduced by the controlling circuit.

The bit–error rate (BER) was measured with a bit error analyzer for different data rates, optical power levels, and two different lengths of the pseudo random bit sequence (PRBS). The summarized results are depicted in Fig. 9. An eye diagram at 11 Gbps and a PRBS of $2^{31}-1$ at minimum optical power of -8.9 dBm for a BER of 10^{-9} is shown in the same figure.

6 TIA with CCVD feedback

In conventional TIA topologies in integrated circuits the high-ohmic feedback resistor can be the bandwidth limiting element due to parasitic capacitances to the substrate. To avoid this, a new concept was presented in (Seidl, 2004). It introduces a capacitive-coupled voltage divider (CCVD) instead of an ohmic resistor in the feedback path of the TIA, like depicted in Fig. 10. With this new topology the bandwidth was raised from 67 MHz to 387 MHz. Both topologies were realized for better comparison.

The CCVD TIA offers the opportunity to realize higher bandwidth-transimpedance products in one amplifier stage than others known from the literature did in 2- or 3-stage TIAs (Hehemann, 2002). This results in less chip area, less power consumption and less offset of TIAs.

7 Highly parallel optical receiver OEIC

In (Swoboda et al, 2006) a highly parallel optical receiver OEIC in a 0.6 μm Si BiCMOS technology was presented. It contains 36 channels with integrated PIN photodiode and a data rate of 3 Gbps each. This very compact design with a pitch of 250 μm requires a special layout of the receiver to minimize crosstalk. The measured sensitivity at 3 Gbps is -17 dBm at 850 nm in each channel. This means that the presented receiver OEIC offers an over–all data rate of more than 100 Gbps combined with a matching laser array on the sender side. Figure 11 shows a microphotograph of a test chip for the receiver chip.

The power consumption of the OEIC is 809 mW including that of a laser driver for a return channel and a loss-of-power detector. The crosstalk between the channels is below -40 dB.

8 Optical distance measurement

Sensors for capturing 3-D information of a scene in real time i.e. in a few milliseconds are still not available on the low-cost market on a grand scale. Mass products, for example neighbour-lane supervision or passenger recognition airbag control in a modern car, can use them. The interesting distance range for these sensors is form cm up to some meters. It is important for robust low-cost sensors to avoid moving mechanical parts. The integrated optical distance measurement sensors presented here work on the principle of time-of-flight. It is based on the measurement of runtime of modulated light from a laser source to the surface of a diffuse reflecting object and back to the receiver (see Fig. 12). The main difficulties of this measurement principle are the very weak amplitude of the received signal, especially compared to the high backround light, as well as the short time intervals to be measured. The optical power of the laser is limited for eye-safety reasons to the mW range. This leads to received signals per pixel in the nW range only. On the other hand, the time interval to be measured is in the ps range, e.g. the runtime of light to an object 10 m away from the sensor (c_0=2.998$\times$10^{8}m/s) is 66.7 ns which leads to time intervals of 66.7 ps for a resolution of 10 mm. To realize the measurement of these short time intervals the transmitting laser diode is 10 MHz square–wave modulated. The distance is determined out of the phase shift between sent and received signal. The correlation of the received light and the transmitted signal therefore provides the distance information. (Nemecek, 2006) describes this method in more detail.

Figure 13 shows a microphotograph of a line sensor which consists of 32 single pixels. Results of one pixel were published in (Nemecek et al, 2006). For a measurment time of 5 ms at an object distance of 3.2 m, a standard deviation of 3 cm was achieved with an optical power of 2 mW at 650 nm. The area of each pixel was $210\times105\ \mu m^2$, whereby the photodiode occupied an area of $100\times100\ \mu m^2$ in each pixel.

9 Conclusions

OEICs for different application areas were presented. The presented work of the circuit design group of the Institute of Electrical Measurements and Circuit Design of Vienna University of Technology is on top of the state of the art. E.g. the achieved data rate of 11 Gbps with a Si OEIC is the highest possible with a thick vertical Si photodiode having a quantum efficiency close to 100%. Another quantum leap was reached with the capacitive–coupled voltage driver (CCVD) feedback. It enables top performances of single–stage TIAs, without the stability problems of multi–stage TIAs. Another outstanding optical receiver is the highly parallel receiver OEIC which enables a data rate of more than 100 Gbps with a low-cost Si BiCMOS technology. A 32 pixel line sensor for distance measurement shows the potential of Si OEICs to realize a 3-D camera on a chip. All in all it could be demonstrated that high-performance low-cost Opto ASICs for a wide variety of high-volume applications are possible with BiCMOS OEICs.

Acknowledgements. The author thanks the members of the circuit design group at EMST of TU Vienna and R. Swoboda, M. Förtsch and J. Leeb from A3PICs for careful work. Financial funding in the EU project INSPIRED, the FIT-IT project HYBOR as well as the A3-projects 3-D-Sensors and SENSAFE is acknowledged. Last but not least, many thanks for the great help with preparing this manuscript go to K. Schneider–Hornstein.

References

Yamamoto, M., Kubo, M., and Nakao, K., et al.: Si-OEIC with a built-in pin-photodiode, IEEE T. Electron. Dev., 42, 58–63, 1995.

Swoboda, R. and Zimmermann H.: 2.5Gbit/s silicon receiver OEIC with large diameter photodiode, Electron. Lett., 40, 505–507, 2004.

Lang, M., Bronner, W., Benz, W., et al.: Complete monolithic integrated 2.5 Gbit/s optoelectronic receiver with large area MSM photodiode for 850 nm wavelength, Electron. Lett., 37, 20, 1247–1249, 2001.

Leeb, J., Schneider, K., and Zimmermann, H.: A 380MHz two-stage OEIC for the use in DVD pickup units, Proceedings of IEEE International Symposium on Consumer Electronics, 381–384, 2004.

Kieschnick, K., Heide, T., Ghazi, A., and Zimmermann, H.: Advanced CMOS and BICMOS Photonic Receiver ICs, Proceedings of European Solid-State Circuits Conference (ESSCIRC), 398–401, 1999.

Zimmermann, H. and Kieschnick, K.: Low-Offset BiCMOS OEIC for Optical Storage Systems, Electron. Lett., 36, 1223–1224, 2000.

Kieschnick, K. and Zimmermann, H.: High-Sensitivity BiCMOS OEIC for Optical Storage Systems, IEEE Journal of Solid-State Circuits, 38, 579–584, 2003.

Takimoto, T., Fukunaga, N., Kubo, M., and Okabayashi, N.: High Speed SI-OEIC (OPIC) for optical pickup, IEEE Transactions on Consumer Electronics, 44, 137–142, 1998.

Micusik, D. and Zimmermann, H.: Transimpedance amplifier with 120 dB dynamic range, Electron. Lett., 43, 3, 159–160, 2007.

Swoboda, R. and Zimmermann, H.: 11 Gb/s Monolithically Integrated Silicon Optical Receiver for 850nm Wavelength, IEEE International Solid-State Circuit Conference, Digest of Technical Papers, ISSCC, 49, 0193–6530, 240–241, 2006.

Seidl, C., Knorr, J., and Zimmermann, H.: Single-stage 378 MHz–178 k-transimpedance amplifier with capacitive coupled voltage divider, IEEE ISSCC, 470–471 and 540, 2004.

Hehemann, I., Brockherde, W., Bussmann A., Hoffmann, H., Kemma, A., Kokzinski, R., Richter, H., and Hosticka, B. J.: A new detector architecture for optical pickup units in DVD systems, IEEE Int. Symp. on Circuits and Systems (ISCAS), 4, 41–44, 2002.

Swoboda, R., Förtsch, M., and Zimmermann, H.: 3 Gbps-per-Channel Highly-Parallel Silicon Receiver OEIC, Proc. 33rd European Conference and Exhibition on Optical Communication (ECOC), 2, 255–256, 2007.

Nemecek, A., Oberhauser, K., and Zimmermann, H.: Distance Measurement Sensor With PIN-Photodiode and Bridge Circuit, IEEE Sensors Journal, 6, 2, 391–397, 2006

Nemecek, A., Oberhauser, K., Zach, G., and Zimmermann, H.: Distance Measurement Line Sensor with PIN Photodiodes, IEEE Sensors Conference, 275–278, 2006.

34

Analysis of communications and implementation performance of FFT based carrier synchronization of BPSK/QPSK bursts

U. Wasenmüller, T. Brack, and N. Wehn

Microelectronic System Design Research Group, University of Kaiserslautern, Erwin-Schrödinger-Straße, 67663 Kaiserslautern, Germany

Abstract. Carrier synchronization is a vital part of every inner receiver in wireless communications. Frequency offset and phase offset of the received burst must be estimated and the received samples must be corrected according to the estimation. An approximation of the maximum likelihood estimation of frequency offset is given by the Fast Fourier Transform (FFT). The accuracy of the estimation as well as the hardware complexity and throughput depend on the number of FFT points. We demonstrate three techniques to improve the accuracy of the FFT based frequency offset estimation respectively to reduce the number of needed FFT points. These three techniques as well as the reference algorithm without improvements combine phase offset estimation with frequency offset estimation. We present the communications and FPGA implementation performance of these techniques.

1 Introduction

The transmission over a wireless channel results in timing, frequency and phase offsets. A sophisticated synchronization is mandatory, to circumvent the severe losses of communications performance by these offsets. In this paper we will focus on the frequency and phase estimation and correction (carrier synchronization) of bursts with BPSK/QPSK modulation. In our targeted system these bursts are given in the complex base band with one sample per symbol after timing synchronization is properly carried out. An approximation for the maximum likelihood estimation of frequency offset is the FFT, see e.g. (Mengali et al., 1997). The implementation complexity as well as latency and throughput of the FFT depend on the number of points of the FFT. Usually zero padding with a factor of 2 or even 4 is used for the FFT to reach the needed accuracy for the communications performance.

We analyze three different techniques for reducing the number of needed points for the FFT to reduce the implementation complexity while still obtaining the required communications performance. These techniques are extensions of a FFT based algorithm for the combined estimation of frequency and phase offset with the objective to improve the accuracy of the frequency offset estimation.

- The first technique is a smart technique of sample rate reduction.
- The second technique uses an interpolation technique between the bins of the FFT result.
- The third demonstrated technique uses a decision directed approach after a FFT based coarse synchronization.

Additionally all techniques also combine the estimation of frequency and phase offset, which makes a separate module for phase offset synchronization obsolete. We present the communications performance of these three techniques for different signal to noise ratios. We make heavely use of optimizez Xilinx IP cores. Hence implementation complexities and results are given based on Xilinx devices and the trade-off between communications performance and implementation performance is analyzed.

The paper is structered as follows. In Sect. 2 our FFT based reference algorithm for the combined frequency and phase offset estimation is shortly described. In Sect. 3 the essential ideas of the three techniques for reducing the number of FFT points are presented. The additional computational effort for these techniques is derived. Sect. 4 describes the common hardware architecture and implementation details. Communications performance and implementation results are given in Sect. 5. Section 6 concludes the paper.

Correspondence to: U. Wasenmüller
(wasenmueller@eit.uni-kl.de)

2 Base carrier synchronization algorithm

We assume that timing synchronizaiton is properly carried out and the received sample sequence r is given with one sample per symbol. This symbol sequence r is given in the complex baseband according to Eq. (1):

$$r(l) := s(l) \cdot e^{j(2\pi f_o l + \Phi)} + n(l) \qquad l = 0, 1, .., L-1 \quad (1)$$

The sample sequence r with L elements is based on MPSK symbols s with one sample per symbol; i.e. for BPSK M equals 2 and for QPSK M equals 4. The sequence r is disturbed by a noise sequence n, which represents an Additive White Gaussian Noise (AWGN). The frequency offset f_o and phase offset Φ are considered fixed during the transmission of a burst. The symbol duration is denoted by T and the frequency offset f_o in Eq. 1 is given as a fraction of the symbol rate $1/T$. The problem consists in estimating the frequency offset $\tilde{f_o}$ and phase offset $\tilde{\Phi}$. The received sample sequence r has to be corrected according to the estimation of frequency and phase offset.

In the following the base algorithm is shortly described according to (Brack et al., 2005). For the estimation of frequency offset the effect of modulation by the symbol sequence s has to be removed. We restrict our investigations to the so called non data aided estimation, where the unknown parameters of frequency and phase offset are estimated with the only help of the unknown data symbols. In general, modulation removal is carried out with a power of M operation on the received sample sequence r as shown in Eq. 2:

$$\hat{r}(l) = r(l)^M \quad (2)$$

In (Wasenmüller et al., 2004) we proposed an improved modulation removal scheme for MPSK based on investigations in (Wang et al., 2003). This technique obtains much better communications performance. Modulation removal is performed as shown in Eq. 3:

$$\hat{r}(l) = |r(l)| \cdot e^{j \cdot M \arg(r(l))} \qquad l = 0, 1, .., L-1 \quad (3)$$

with M being the modulation index, and $\hat{r}$ describing the determined received sequence without modulation.

An approximation for maximum likelihood frequency estimation is given by application of the FFT algorithm to the sequence $\hat{r}$. The FFT implements the Discrete Fourier Transformation formula with N points according to Eq. 4:

$$X(k) = \sum_{n=0}^{N-1} x(n) \cdot e^{-j\left(\frac{2\pi \cdot k \cdot n}{N}\right)}$$
$$\text{with} \quad x(i) = \hat{r}(i) \quad i \leq L \quad x(i) = 0 \quad i > L \quad (4)$$

The estimated frequency offset $\tilde{f_o}$ is given by spectral analysis of the FFT output. In general the FFT bin corresponding to the estimated frequency offset is given by:

$$k_f, \ s.t.: \quad |X(k_f)| = \max_k |X(k)| \qquad k = 0, 1, .., N-1 \quad (5)$$

Equation 5 yields the index of the FFT bin with the maximum amplitude; the frequency offset $\tilde{f_o}$ is simply calculated as:

$$\tilde{f_o} = \frac{k_f}{M \cdot N} \quad \text{for} \quad k_f < N/2$$
$$\tilde{f_o} = \frac{k_f - N}{M \cdot N} \quad \text{for} \quad k_f \geq N/2 \quad (6)$$

Each bin $X(k)$ of the FFT represents a range of $(M \cdot N \cdot T)^{-1}$ Hz, not only one distinct frequency. It is obvious, that the accuracy of the FFT based frequency offset estimation depends on the number N of FFT points. The maximum frequency offset, which can be estimated with the described method, is given by:

$$|\tilde{f}_o^{max}| = \frac{1}{2 \cdot M} \quad (7)$$

This limit is given by the Nyquist criterion for the sequence $\hat{r}$. If the range of the real occurring frequency offset is known in advance, we can significantly improve the communications performance by limiting the spectral analysis with a window. This windowing technique is given by Eq. (8):

$$k_{fw} = \max_k |X(k)| \qquad k = 0, .., w_u, w_l, .., N-1 \quad (8)$$

The parameters w_u and w_l define the frequency range in which the frequency offset has to be found.

Setting w_u to $N/2 - 1$ and w_l to $N/2$ respectively describes the FFT bin k_f as given in Eq. (5).

For the estimation of the phase offset a sample sequence without frequency offset is needed and the effect of the modulation has to be removed again from the frequency corrected burst. As shown in (Brack et al., 2005) the phase offset can be estimated by the FFT bin with the index k_f

$$\tilde{\Phi} = M^{-1} \cdot \arg(X(k_f)) \quad (9)$$

Equation 9 uses the result of Eq. (4) for the bin k with $k{=}k_f$. Thus frequency and phase estimation can be efficiently combined. Note that the estimated phase offset has a M times ambiguity, which must be resolved in later processing step.

Finally the carrier synchronized sequence r_c is calculated by:

$$r_c(l) = r(l) \cdot e^{-j(2\pi \tilde{f_o} l + \tilde{\Phi})} \qquad l = 0, 1, ... L-1 \quad (10)$$

3 Improvements

As shown in the previous section the resolution of the estimated frequency offset depends on the number of FFT points. The computational complexity for the FFT operation is proportional to $N \lg_2 N$. Thus area and achievable throughput of a FFT building block are directly related to the number of FFT points. To get an efficient implementation the number of FFT points should be reduced while still obtaining a sufficient communications performance. In the following subsections three techniques are presented, which allow to reduce the number of FFT points.

3.1 Technique 1: Sample Rate Reduction (SRR)

The resolution of the estimated frequency can be obviously increased by a reduction of the sample rate. A reduction of the sample rate is equivalent to an increasing of the parameter T in Eq. (7). For the estimation there are only the L elements of the received burst sequence r available. Thus a sample rate reduction by a factor of D implies that the number of available samples for the estimation is reduced to L/D. However a reduction of the number of symbols for the estimation leads to a performance degradation. To circumvent this degradation the information of the received symbols is maintained by a averaging operation over D symbols of the modulation removed sequence $\hat{r}$ given by:

$$\hat{r}_s(k) = \sum_{i=0}^{D-1} \hat{r}(D \cdot k + i) \tag{11}$$

The operation of frequency and phase offset estimation is done with the new sequence $\hat{r}_s$ according to Eq. (4) to Eq. (9). The sample rate reduction (SRR) technique according to Equation 11 implies another interpretation for the bins of the FFT. The frequency range of a FFT bin is now given by $(M \cdot N \cdot T \cdot D)^{-1}$ Hz. The maximum frequency offset, which can be estimated is decimated by the factor D.

$$\tilde{f}_o^{\max} = \frac{1}{2 \cdot M \cdot D} \tag{12}$$

The phase estimation according to Eq. (9) yields the phase according to the sequence $\hat{r}_s$. This failure must be corrected; thus the estimated phase offset is given by:

$$\tilde{\phi} := \tilde{\phi} - \pi \tilde{f}_o / D \tag{13}$$

3.2 Technique 2: Interpolation (INT)

As mentioned in the explanation of the base algorithm for the FFT based carrier synchronisation the resolution of of the frequency estimation is limited by the number of FFT points. To improve the resolution of a given discrete fourier transformation the information of the neighbouring FFT bins of the bin k_f are taken into account too. Our technique uses a parabolic interpolation, which uses three bins of the discrete Fourier transformation. The behaviour of the Fourier transformation around the bin with the maximum energy is approximated by a parabolic function. The parabolic function is determined by the energy E_C of the bin k_f, the energy E_L of the left neighbour bin, and the energy E_R of right neighbour bin. The argument $\triangle$ of the maximum value of this parabolic function is given by:

$$\triangle = \frac{1}{2} \cdot \frac{E_R - E_L}{2E_C - E_R - E_L} \tag{14}$$

By construction, the value of $\triangle$ will be in the range $[-0.5 \cdots + 0.5]$. This implies that the improvement of the estimation of the base algorithm will be within in the range of half a bin. For the correction of the frequency offset a virtual bin $k_f + \triangle$ will be used. The estimation of the phase must also be adapted to the new virtual bin. The Fourier transformation of the new virtual bin is calculated by linear interpolation between the corresponding bins. For a positive $\triangle$ the value of virtual bin $X(k_f + \triangle)$ is calculated as:

$$X(k_f + \triangle) := X(C) + \triangle \cdot (X(R) - X(C)) \tag{15}$$

Applying the *arctan* function to X gives the estimation $\tilde{\phi}$ of the phase offset for the interpolation technique.

3.3 Technique 3: Decision Directed Approach (DD)

The idea of any decision directed approach for parameter estimation in synchronization is to use an estimation of the unknown symbols. The investigated decision directed technique of this paper derives estimation of the unknown symbols by demodulation of the carrier synchronized received samples. In the further processing the residual frequency and phase offset of the carrier synchronized received sample sequence is estimated and the sample sequence is corrected. The estimated symbols are treated as known symbols and are used to assist the estimation. The estimation of frequency offset with known symbols could be done with several well known techniques for frequency estimation, see e.g. (H. Meyer et al. , 1997). For the the base algorithm the residual frequency offset $|\tilde{f}_o - f_o|$ will be quite small. A technique for carrier synchronization can be applied, which is also used in the so called turbo synchronization (Alles, M. et al. , 2007). The estimation of the transmitted symbol sequence s will be denoted by $\tilde{s}$. The modulation removal for the further estimations is done in a data aided way for the carrier synchronized sequence r_c.

$$\hat{r}_c(l) = r_c(l) \cdot \tilde{s}^*(l); \qquad l = 0, \cdots, L-1 \tag{16}$$

The additional frequency and phase estimation is carried out with the sequence $\hat{r}_c$. By an averaging process over the first half of the sequence and over the last half of the sequence a measure for the phase offset of the first half and of the last half of the sequence $\hat{r}_c$ is calculated:

$$z_1 = \sum_{l=0}^{L/2-1} \hat{r}_c(l) \qquad z_2 = \sum_{l=L/2}^{L-1} \hat{r}_c(l) \tag{17}$$

Applying the arctan operation to z_1 and z_2 gives an estimate for the phase of the first half and for the phase of the second half of the sequence $\hat{r}_c$. The frequency offset will be estimated by the difference of the phases of the first and second half:

$$\tilde{f}_o = \frac{\arctan(z_1 \cdot z_2^*)}{\pi \cdot L/2} \tag{18}$$

The new phase estimation is given by the averaging over the whole sequence $\hat{r}_c$ according to

$$\tilde{\phi} = arctan(z_1 + z_2) \tag{19}$$

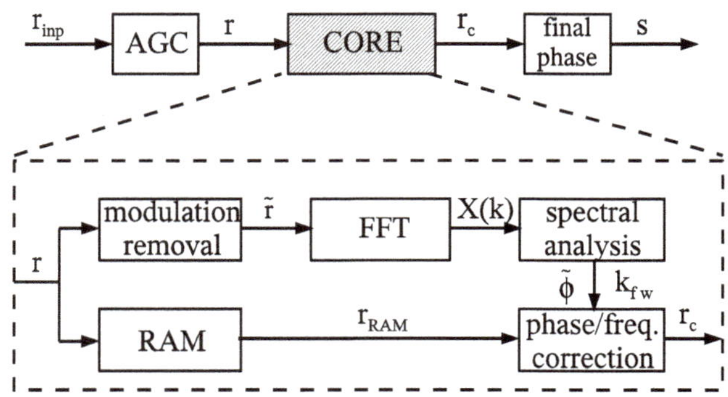

Fig. 1. Base architecture for carrier synchronization

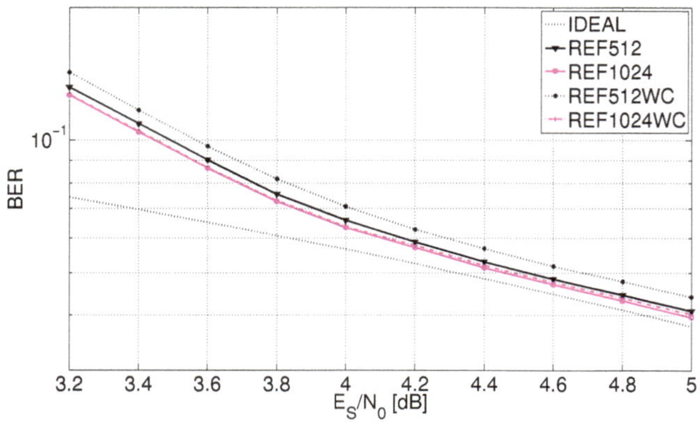

Fig. 2. Impact of different FFT length at low SNR.

The algorithm outlined in Eq. 17 to Eq. 19 needs as a prerequisite, that the residual frequency offset is quite small, i.e., the phase increase over $L/2$ elements of the sequence r_c must be less than π. If this condition is not fullfilled this algorithm produces considerable variations of the frequency estimation value.

4 Hardware architecture and implementation

4.1 Architecture

The base hardware architecture of the analyzed techniques is shown in Fig. 1. As mentioned we assume, that the processing steps of matched filtering and timing synchronization are properly carried out. As well an automatic gain control unit (AGC) is required to adapt the input sample bitwidth to the core input bit width without limiting dynamic range. The back-end must deal with the M-times phase ambiguity introduced by Eq. (9).

The building blocks of the core correspond directly to the base algorithm described in Sect. 2. The modulation removal component realizes Eq. (3) and provides the input for the FFT operation, which is given in Eq. (4). The spectral analysis component provides the estimates of frequency and phase offset according to Eq. 5 and Eq. (9). The RAM buffer is used to garantee constant throughput by hiding the FFT latency. Thus a new sample sequence r can be processed while the already analyzed sequence r_{ram} is being corrected. The phase and frequency correction component corresponds to Eq. (10). The shown techniques for improving the FFT accuracy are realized by additional computations in the building blocks of Fig. 1. The SRR technique only impacts the modulation removal component, the INT technique only impacts the spectral analysis component and the DD technique only impacts the correction component.

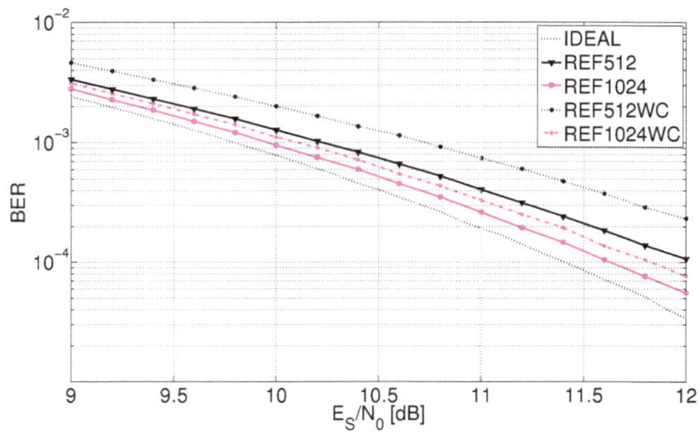

Fig. 3. Impact of different FFT length at high SNR

4.2 Implementation

We used synthetizable VHDL for implementing the architecture shown in Fig. 1 for the investigated techniques. For rapid development and to minimize debug effort we relied heavily on IP cores included in the XILINX Core Generator 9.1. We also used specific XILINX resources like the internal multipliers (MULT) and block RAM (BRAM) available on the Spartan3-E FPGA. The modulation removal in Eq. 3 uses polar coordinates. The input samples r however are given in Cartesian coordinates. The calculation of arg (r) and $|r|$ can be done with the CORDIC algorithm (Volder, J., 1959) which is available as an IP core. We utilized the fully pipelined version of this CORDIC IP core to achieve a throughput of one sample per cycle. To realize the technique of sample rate reduction (see Equation 11) additionally D samples with modulation removal must be added. This is accomplished by some control logic, two adders and additional BRAM storage. The resulting sample sequence $\tilde{r}$ without modulation has to be transformed back to Cartesian coordinates for the FFT calculation (see Eq. 4) for all techniques. This is accomplished with a sine-/cosine-look-up-table (SCL) also available as an IP core and internal multipliers to realize the e^{jx} operation.

The FFT needed for frequency and phase estimation is realized using another XILINX IP core configured to sustain the selected maximum burst length. For implementation of the FFT three XILINX IP cores are available, which differ in throughput, latency and needed resources. Because again the

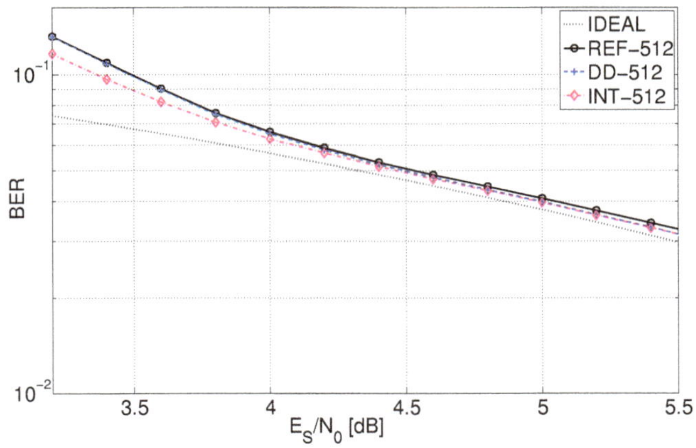

Fig. 4. Performance of DD and INT at low SNR

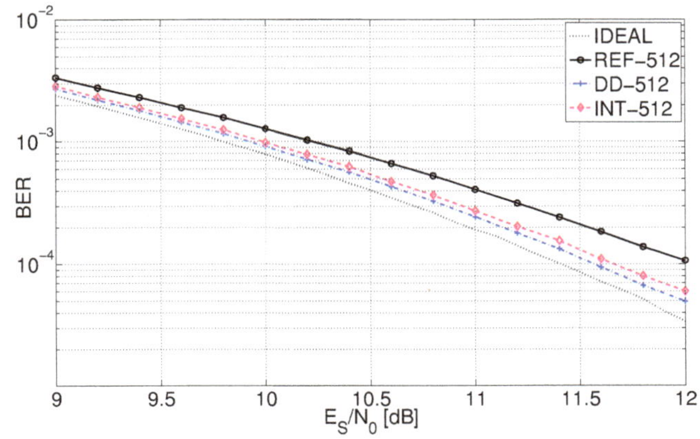

Fig. 5. Performance of DD and INT at high SNR

desired throughput was one sample per cycle, the resulting core is fully pipelined and covers most of the area.

The windowing feature proposed in Equation 8 for evaluating k_{fw} is realized in the spectral analysis block, using a limited maximum absolute value search on the determined FFT bins. This can be implemented in a very efficient way by using only counters and comparators. After determining the frequency bin k_{fw}, the spectral analysis block estimates the phase offset according to Eq.9. Again, the argument function needed for this equation is realized with a CORDIC IP core. Because only one argument calculation per phase estimation has to be performed, we can use a rather small serial CORDIC core without the ability to calculate $|r|$. The implementation of the INT technique requires some additional computations between search of the bin k_f and the argument calculation for phase estimation. Additionally the right and left neighbour bin of k_{fw} must be selected and stored. The realization of Eq. (14) requires a division, which is also available as an XILINX IP core. Furthermore some adders and control logic is required to implement Eq. (15).

In the frequency and phase correction component rotation of the samples as described in Eq. (10) is accomplished with complex number multiplications and a second SCL. The decision directed technique requires additional hardware in this

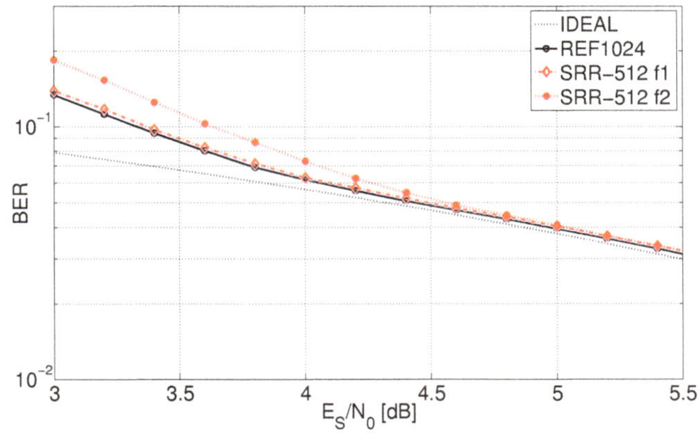

Fig. 6. Performance of SRR

component before the mentioned final correction can be performed. Calculation of the estimates $\tilde{s}$, the data aided modulation removal according to Eq. (16) and the correlation according to Eq. (17) can be carried out in a pipelined manner. This requires addditional control logic, adders and internal multipliers. Again the arg function and the divider are realized by a XILINX IP cores.

5 Results

5.1 Communications performance

In this section we present communications performance simulations based on a bit true C model. All graphs are obtained using an addiditive white Gaussian noise (AWGN) channel model. A reference graph *ideal* is provided to allow comparison of the simulation results to a perfect carrier synchronization reflecting the performance limits of the AWGN channel. Figure 2 and Fig. 3 show the bit error rate (BER) graphs for a burst with 300 QPSK modulated samples obtained by the base algorithm with a 512 point FFT and a 1024 point FFT, respectively. Two graphs result from a uniformly distributed frequency offset in the range $[0.01 \cdots 0.02]$. The other two graphs with the suffix identifier WC demonstrate the BER for a worst case szenario with frequency offset equal to a frequency exactly between two FFT bins. In high SNR region the 512 point FFT features a degradation up to 0.4 dB; in low SNR region a degradation of 0.1 dB can be observed. These simulations demonstrate the necessity to improve the results of the 512 point FFT.

Figure 4 and Fig. 5 demonstrate the performance of the DD and the INT technique. Both techniques offer with a 512 point FFT the same performance as the base algorithm with 1024 point FFT. A very slight performance degradation in comparison to the base algorithm can be observed in low SNR ranges. This behavior results from the fact, that both techniques can only improve the frequency offset estimation in a small range as explained in Sect. 3. In low and high

Table 1. Implementation resources of the analyzed techniques.

Technique	Slices	MULT	BRAM	rel. t.p.
Ref 1024	3756	21	12	1
Ref 512	3433	21	8	2
SRR 1024	3813	21	13	1
SRR 512	3491	21	9	2
INT 1024	4172	25	12	1
INT 512	3843	25	8	2
DD 1024	4685	30	13	1
DD 512	4375	30	9	2

Technique	Slices	MULT	BRAM	rel. t.p.
Ref 1024	3756	21	12	1
SRR 512	-256	-3	0	2
INT 512	+87	+4	-4	2
DD 512	+619	+9	-3	2

Table 2. Resource comparison

SNR regions the decision directed technique shows a very slight better performance than the interpolation technique.

For performance evaluation of the SRR technique the lower maximum frequency offset correction capability of this technique must be regarded. Hence we applied a window to the FFT results of the reference technique according to Eq. 8. Fig. 6 shows the performance of the SRR technique with D=2 for frequency offsets of 0.01 and 0.03. A severe degradation for the higher frequency offset can be observed. This behavior results from the fact that succesional modulation removed sample posses a higher phase difference, such that the average operation of Eq. (11) reduces the energy of the resulting sample sequence $\hat{r}_s$.

5.2 Implementation results

Target device was a low cost XILINX Spartan3-E FPGA device. Synthesis as well as place and route was carried out with Xilinx ISE 9.1 software tools. Table 1 shows the XILINX resources for the different techniques using a 512 point FFT building block and a 1024 point FFT building block respectively. The achievable clock speed of 95 MHz is identical for all techniques. The throughput of all investigated techniques depends linearely on the number of FFT points. Hence the throughput for the designs with 512 pint FFT is doubled in comparison to the 1024 point FFT based designs. In Table 1 the Xilinx resources of the reference technique with 1024 point FFT are compared with the resources for the other techniques based on a 512 point FFT.

6 Conclusions

The SRR technique exhibits the smallest implementation complexity of the analzed techniques. However the usabilty of this technique is limited by maximal occuring frequency offset and the performance requirements of the application. The other two techniques offer a comparable communications performance and implementation complexity as the reference technique. However both techniques offer a doubling of the achievable througput and thus provide clearly better efficiency. The DD technique requires more implementation ressources than the INT technique and provides only a negligible better communications performance. With the three demonstrated techniques the design space for FFT based carrier synchronization is enlarged, that a fitting of the implementation to the throughput requirements and commucation performance requirements can be obtained.

References

Alles, M., Lehnigk-Emden, T., Wasenmüller, U., and Wehn, N.: Implementation Issues of Turbo Synchronization with Duo-Binary Turbo Decoding, Invited paper, 18th Annual IEEE International Symposium on Personal, Indoor and Mobile Radio Communications (PIMRC), September, Athens, Greece, 2007.

Brack, T., Wasenmüller, U., Schmidt, D.,and Wehn, N.: Design Space Exploration for Frequency Synchronisation of BPSK/QPSK Bursts, Kleinheubacher Berichte, September 2004.

Brack, T., Wasenmüller, U., and Wehn, N.: A configurable IP Core for Combined Blind Frequency and Phase Synchronization of MPSK Bursts, Proc. 2005 IST Mobile and Wireless Communication Summit, 2005.

Mengali, U. and D'Andrea, A.: Synchronization Techniques for Digital Receivers, Plenum Publishing Corporation, 1997, New York, ISBN 0-306-45725-3, 1997.

Meyr, H., Moeneclaey, M., and Fechtel, S.: Digital Communication Receivers, John Wiley & Sons, Inc., New York, ISBN 0-471-50275-8, 1997.

Volder, J.: The CORDIC Trigonometric Computing Technique, IRE Transactions on Electronic Computing, 1959, September, 8, 330–334, 1959.

Wang, Y. and Serpedin, E., and Ciblat, P.: Optimal Blind Carrier Recovery for MPSK Burst Transmissions, IEEE Transactions on Communications, 51, 9, 1571–1581, 2003.

The influence of different PAST-based subspace trackers on DaPT parameter estimation

M. Lechtenberg and J. Götze

Information Processing Lab, TU Dortmund University, Otto-Hahn-Str. 4, 44227 Dortmund, Germany

Correspondence to: M. Lechtenberg (matthias.lechtenberg@tu-dortmund.de)

Abstract. In the context of parameter estimation, subspace-based methods like ESPRIT have become common. They require a subspace separation e.g. based on eigenvalue/-vector decomposition. In time-varying environments, this can be done by subspace trackers. One class of these is based on the PAST algorithm. Our non-linear parameter estimation algorithm DaPT builds on-top of the ESPRIT algorithm. Evaluation of the different variants of the PAST algorithm shows which variant of the PAST algorithm is worthwhile in the context of frequency estimation.

1 Introduction

Parameter estimation is a common task in signal processing. It is used in DoA-applications (*direction of arrival*) like radar and sonar or in frequency estimation applications like radio transmission or mass spectrometry – just to name a few. We are researching parameter estimation in the context of power quality applications, i.e. finding the signal components next to the fundamental frequency to extrapolate the transmission system's state and fault-conditions.

Next to transformation-based parameter estimation via Fourier methods like the DFT (*discrete Fourier transformation*) and STFT (*short-time Fourier transformation*) or wavelets, there are parameter estimators based on subspace analysis. Estimators based on the separation of subspaces have advantages that justify the higher computational complexity: They have a significantly higher resolution not bounded by the sampling frequency and the window length and do not suffer from *spectral leakage*.

Common subspace-based parameter estimators are MUSIC (*multiple signal classification*, Schmidt, 1986) and ESPRIT (*estimation parameters via rotational invariance techniques*, Paulraj et al., 1985). MUSIC identifies the frequency components excluded from the noise subspace, ESPRIT identifies the components from the signal subspace. For both algorithms, noise and signal subspace have to be separated. A wide spread way to do this is to perform a decomposition into eigenvectors and separate those in noise and signal. Apart from the expensive eigenvalue decomposition, in a time-varying environment, this can also be done by efficient subspace trackers.

A subspace tracker can be circumscribed EV (*eigenvector*) updating, but this is not exactly true. The vectors generated by subspace trackers (in general) form one possible basis of the signal's space. This basis does not have to be the EV of the correlation matrix, nor does it have to be orthogonal. The properties of the basis depend on the subspace tracker used. YAST (*yet another subspace tracker*, Badeau et al., 2008) and PROTEUS (*plane rotation-based EVD-updating scheme*, Champagne and Liu, 1998) generate an orthonormal basis, whereas PAST (*projection approximation subspace tracking*, Yang, 1995b) in its original form provides a basis that only approximates orthonormality over time.

Depending on the application, an orthonormal basis is not needed. To verify that our application does not require an orthonormal basis, we compared the influence of different subspace trackers all based on the mentioned PAST algorithm. Variants of the original PAST algorithm introduce forced orthonormality (OPAST, *orthonormal PAST*, Abed-Meraim et al., 2000) and additional steps for numerical stability (OPAST-PGS, *PAST with pairwise Gram-Schmidt-orthogonalization*, Bartelmaos et al., 2005). We also included an unrolled version of PAST reducing computational complexity (PASTd, *PAST with deflation*, Yang, 1995a) and – as reference – pure eigenvalue decomposition (EVD).

Section two describes the environment in whose context the subspace trackers are compared. Section three presents the different expansion stages of the subspace tracker. The simulation environment is depicted section four and the simulation results are presented in five.

2 Application environment

Modern energy transmission systems can be characterized by two main features. On the one hand, they were designed and constructed for a more static system state with constant centralized generation – on the other hand, nowadays, they are stressed by decentralized *green* generation (e.g. solar, wind, biomass), economic interests (e.g. stock exchange) and cross-border transmission. Decades ago, transmission management was a question of several minutes and could easily be done by humans tweaking a few parameters (like the excitation of a power plant's machine). Today's transmission grid management depends on a lot more factors and is sometimes a question of seconds. To overcome these new challenges, it is desirable to have a system of measurement units being able to characterize the system state without the need of human interpretation.

The signal model describing a power transmission is composed of *superposed sinusoids*:

$$x(n) = \sum_{i=1}^{p} c_i e^{2\pi j \cdot f_i \frac{n}{f_s}} + w_{\text{awgn}}(n) \tag{1}$$

$$\boldsymbol{x}(n) = \sum_{i=1}^{p} c_i e^{2\pi j \cdot f_i \frac{\boldsymbol{n}}{f_s}} + \boldsymbol{w}_{\text{awgn}}(n) \tag{2}$$

$$\boldsymbol{x}(n) = \mathbf{A}(\boldsymbol{n}, \boldsymbol{f})\boldsymbol{c}(n) + \boldsymbol{w}_{\text{awgn}}(n) \tag{3}$$

(scalar model (1) with rank p, complex amplitude $c_i = a_i e^{j\varphi_i}$ and noise $w_{\text{awgn}}(n)$; vector model (2) with window length N and vector $\boldsymbol{n} = [n \ldots n+N-1]$; matrix-vector model (3) with $N \times p$ matrix $\mathbf{A} = [\boldsymbol{a}(f_1) \ldots \boldsymbol{a}(f_p)]$, where vectors $\boldsymbol{a}(f) = \left[e^{2\pi j \cdot f \frac{n}{f_s}} \ldots e^{2\pi j \cdot f \frac{n+N-1}{f_s}}\right]^T$). The dominating sinusoid will naturally be the fundamental frequency (50 Hz/60 Hz, depending on the region). Next to this sinusoid, there may be interim sinusoids. These originate in oscillations from capacitances and inductances characterizing the transmission lines and routing devices triggered by changes (intended or by fault) of the net configuration. They may also be the result of not correctly synchronized/well-controlled areas of the net (*interarea modes*).

2.1 Database assisted parameter estimation

We have designed a database-assisted parameter estimation method (DaPT, Lechtenberg and Götze, 2011). This algorithm rates incoming parameter estimations over time and provides a noise-reduced parameter estimation with reliable rank estimation.

This database has entries for each signal component containing frequency f_i and (complex) amplitude c_i. Additional fields hold

1. *Frequency drift*: this field tracks the change of the frequency estimations against the database field for threshold-based frequency field updates.
2. *Phase difference*: it tracks the change of the phase gradient as second indicator for a misestimated frequency. The phase is assumed to be constant, so a linearly changing phase estimation (based on the database's frequency) indicates a mismatch.
3. *Severity of frequency drift/phase difference*: indicators for the prior mentioned threshold-based updating rules.
4. *Quality of entry*: this is the actual rating. It rates how often a frequency was part of the last estimation recursions. Based on that, components can be marked reliable (i.e. *present*), accounted for rank estimation and output.

In every recursion, the incoming frequency estimation is mapped to the most suitable database entry. Ideally, both values match neglecting the measurement noise. For entries that have been successfully mapped, the quality is increased, for the other decreased. For the drift/difference properties, the estimation is compared to the entry's features and exponentially weighted added to the field. Severity is updated like the quality mentioned before. After the drift fields have been updated, the drift's/difference's severity is checked whether the frequency component needs to be updated. Depending on the quality, the relevant (or reliable/present) entries are selected for output and counted for rank estimation.

2.2 Estimating parameters via rotational invariance techniques

Remembering the signal model in Eq. (3), the signal can also be described by subspaces like in

$$\boldsymbol{x}(n) = \mathbf{A}(\boldsymbol{n}, \boldsymbol{f})\boldsymbol{c}(n) + \boldsymbol{w}_{\text{awgn}}(n) = \mathbf{S}(n) + \sigma^2_{\text{awgn}}\mathbf{I}\,. \tag{4}$$

Within that equation, $\mathbf{S}(n)$ describes the signal subspace, which can be written as a product of the eigenvectors $\mathbf{W}(n)$ and the eigenvalues (on a diagonal matrix) $\mathbf{D}(n)$:

$$\mathbf{S}(n) = \mathbf{W}(n)\mathbf{D}(n)\mathbf{W}(n)^H\,. \tag{5}$$

Estimating frequencies (more general: *parameters*) via rotational invariance relates to the eigenvectors of the autocorrelation matrix of the input samples. In terms of subspaces described by eigenvectors, one time-step can be described by a multiplication with a diagonal matrix $\mathbf{\Phi}$ with diagonal entries $e^{2\pi j \cdot \frac{f_i}{f_s}}$:

$$\mathbf{S}(n+1) = \mathbf{W}(n)\mathbf{D}(n)\mathbf{\Phi}\mathbf{W}(n)^H\,. \tag{6}$$

In other words, the multiplication with $\mathbf{\Phi}$ is a rotation in the complex plane or a phase-shift in the time domain. This is visualized by Fig. 1.

The key idea of Esprit is to estimate the rotation $\mathbf{\Phi}$. Due to noise, $\mathbf{\Phi}$ will have off-diagonal elements. So an EVD is performed on $\mathbf{\Phi}$. The frequencies can be extracted from these

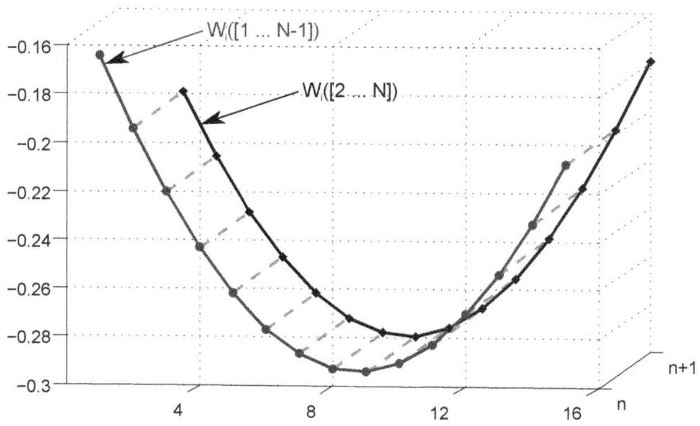

Fig. 1. Rotation of exemplary eigenvector from time-step n to $n+1$.

eigenvalues by calculating their angle. These frequencies are the input of DaPT.

Phase and amplitude estimation is obtained by a simple LS approach referring the signal model Eq. (3). Since time index and frequency are known, the matrix **A** can be built. Together with the samples the complex amplitude can be estimated.

3 Subspace tracking

The autocorrelation matrix of the input samples is defined as $\mathbf{R}_{xx} = E\left[\boldsymbol{x}\boldsymbol{x}^H\right]$. The expectation value is a statistical expression that cannot be calculated for a stream of samples, but it can be estimated e.g. via exponential filtering:

$$\mathbf{R}_{xx} \approx \tilde{\mathbf{R}}_{xx}(n) = \beta\tilde{\mathbf{R}}_{xx}(n-1) + \boldsymbol{x}\boldsymbol{x}^H \tag{7}$$

(β is called *forgetting factor*). An ordinary EVD can be applied to this approximated autocorrelation matrix to generate the eigenvectors for ESPRIT. Depending on the algorithm used, this costs $\mathcal{O}(N^2)$ to $\mathcal{O}(N^3)$, which is expensive considering that in most cases $N \gg p$. Subspace trackers realize the extraction of eigenvectors with computational costs in the range of $\mathcal{O}(p^2)$ by an recursive concept of calculating projections and deviations to a previous calculation instead of complete decompositions.

This paper will focus on one branch of subspace trackers, i.e. PAST and derivations of PAST. This is in order to be able to benchmark the influence of features like orthonormality of the subspace basis on the parameter estimation without having deviations due to different calculation approaches.

3.1 PAST

The *projection approximation subspace tracking* concept (Yang, 1995b) is based on a non-constraining cost function:

$$J(\mathbf{W}) = E\|\boldsymbol{x} - \mathbf{W}\mathbf{W}^H\boldsymbol{x}\|^2 = \mathrm{tr}(\mathbf{R}_{xx}) - \mathrm{tr}\left(\mathbf{W}^H\mathbf{R}_{xx}\mathbf{W}\right) + \mathrm{tr}\left(\mathbf{W}^H\mathbf{R}_{xx}\mathbf{W}\cdot\mathbf{W}^H\mathbf{W}\right) \tag{8}$$

(tr(...) is the trace operator). This cost function has some important properties: it is minimized *if and only if* **W** describes a basis of the signal subspace; there are no more constraints regarding **W**, especially not orthonormality. In consequence $J(\mathbf{W})$ is minimized for every basis of the signal subspace, not only the basis consisting of the eigenvectors of the autocorrelation matrix.

To find the minimum of the cost function J, a *straightforward* recursive gradient-descend method is applied (step size μ). The Gradient of J is written

$$\nabla J = \left[-2\mathbf{R}_{xx} + \mathbf{R}_{xx}\mathbf{W}\mathbf{W}^H + \mathbf{W}\mathbf{W}^H\mathbf{R}_{xx}\right]\mathbf{W} \tag{9}$$

(here, **W** is short for $\mathbf{W}(n-1)$). The resulting updating equation for a **W** approaching the minimum is given as follows:

$$\mathbf{W}(n) = \mathbf{W}(n-1) - \mu\cdot\nabla J(n-1) \tag{10}$$

By rearranging this equation with the projection $\boldsymbol{y} = \mathbf{W}^H\boldsymbol{x}$ and the approximation $\mathbf{W}^H\mathbf{W} \approx \mathbf{I}$, this becomes

$$\mathbf{W}(n) = \mathbf{W}(n-1) - \mu[\boldsymbol{x}(n) - \mathbf{W}(n-1)\boldsymbol{y}(n)]\boldsymbol{y}^H(n)\ . \tag{11}$$

With this knowledge (Eq. 11) and the approximation concept of the autocorrelation matrix (Eq. 7) the cost function can be reformulated as an exponentially weighted LS (*least squares*) problem

$$J'(\mathbf{W}(n)) = \sum_{i=1}^{n}\beta^{n-i}\|\boldsymbol{x}(i) - \mathbf{W}(n)\boldsymbol{y}(i)\|^2. \tag{12}$$

This is also known from adaptive filtering. The minimization of this cost function is expressed by the following equations

$$\mathbf{W}(n) = \mathbf{R}_{xy}(n)\mathbf{R}_{yy}^{-1}(n) \tag{13}$$

$$\mathbf{R}_{xy}(n) = \beta\mathbf{R}_{xy}(n-1) + \boldsymbol{x}\boldsymbol{y}^H \tag{14}$$

$$\mathbf{R}_{yy}(n) = \beta\mathbf{R}_{yy}(n-1) + \boldsymbol{y}\boldsymbol{y}^H\ . \tag{15}$$

The PAST algorithm by Yang is based on the RLS (*recursive LS*) algorithm described by the Eq. (13) and the inversion of $\mathbf{R}_{xy}$. Whereas the **W** in the first definition of J (Eq. 8) was not constrained to hold orthogonal columns but produced an orthogonal solution, the reformulated J' (Eq. 12) only approaches an orthogonal solution while convergence depends on SNR and β.

This section shall only give the idea of PAST. An implementation can be found in other publications like Yang (1995b) or Abed-Meraim et al. (2000). The computational complexity is $3Np + \mathcal{O}(p^2)$, see Table 1.

3.2 OPAST

As mentioned before, the original PAST algorithm does not produce orthogonal results in every run. The OPAST (*orthogonal PAST*) by Abed-Meraim et al. (2000) is based on

Table 1. Comparison of the algorithms computational complexities (N window length and p rank, note $N \gg p$).

EVD	$\mathcal{O}(N^2)$
PAST	$3Np+\mathcal{O}(p^2)$
OPAST	$4Np+\mathcal{O}(p^2)$
OPAST-PGS	$5Np+\mathcal{O}(p^2)$
PASTd	$4Np+\mathcal{O}(p^2)$

the same cost function, but introduces an additional orthogonalizing step. This step is informally written like $\mathbf{W} = \mathbf{W}(\mathbf{W}^H\mathbf{W})^{-0.5}$, with its brackets including the exponent called *inverse square root*.

This orthogonalizing step ensures an orthogonal basis (represented by $\mathbf{W}$) in every time-step. This additional calculation is incorporated in the updating algorithm of PAST. The implementation can be found in the corresponding publication. The computational complexity is $4Np+\mathcal{O}(p^2)$, see Table 1.

3.3 OPAST-PGS

Due to Bartelmaos et al. (2005), a problem of OPAST is its sensitivity to numerical rounding errors. These are originated in the inverse square root concept. The authors propose a *pairwise Gram-Schmidt orthogonalization* (OPAST-PGS) to oppose this problem.

Remembering that the recursion is indexed by n and the rank is p, two indices are defined: $k_n = \mathrm{mod}(n, p)$ and $k_{n+1} = \mathrm{mod}(n-1, p)$. With these indices, the k_{n+1}-vector of $\mathbf{W}$ is made orthogonal to the k_n-vector, before it is normalized:

$$\boldsymbol{w}_{k_{n+1}}(n) = \boldsymbol{w}_{k_{n+1}}(n) - \boldsymbol{w}_{k_n}(n)\boldsymbol{w}_{k_n}^H(n)\boldsymbol{w}_{k_{n+1}}(n) \quad (16)$$

$$\boldsymbol{w}_{k_{n+1}}(n) = \frac{\boldsymbol{w}_{k_{n+1}}(n)}{\|\boldsymbol{w}_{k_{n+1}}(n)\|} . \quad (17)$$

The computational complexity is $5Np+\mathcal{O}(p^2)$ (Table 1).

3.4 PASTd

Another derivation of the PAST algorithm was proposed by Yang (1995a). This approach focuses on a further reduction of computational costs by unrolling the PAST algorithm. This *deflation* describes the sequential estimation of the principal components (or: basis-vectors) beginning with the most dominant. The most dominant is defined by the maximal (pseudo-)eigenvalue corresponding to the principal component.

The process is the same as for the original PAST, except that it is done for each vector of $\mathbf{W}$ separately. So, PAST is performed for the most dominant vector, then a projection of the current data onto this vector is removed from the data vector. The current vector is now excluded and the process will be repeated for the new dominant vector. This method results in less matrix computations resulting in a computational complexity of $4Np+\mathcal{O}(p)$. Note that it is $\mathcal{O}(p)$ and not $\mathcal{O}(p^2)$.

4 Simulation environment

The signal will always embody a 50 Hz-component (sloppy for *fundamental system frequency ideally being 50 Hz or 60 Hz, depending on the region*), slightly varying over time, and spontaneous, temporally present components with random parameters.

The simulation is Monte-Carlo based and performs 400 iterations, each having 20 sections. The following results are consistent and bias-free.

4.1 Signal generator

The signal generator is based on a concept of random sections. A default section has a length of 5 s. The actual sections will have a length that is a small, random integer multiple of the default length. Each section varies the parameters (frequency, amplitude) of the fundamental component a little. For each section, a small random number of additional components is added and/or removed. In consequence, the additional components will endure a small random number of sections. If an additional component endures more than one section, its parameters are also slightly changed randomly.

The noise applied to the generated signal is based on the AWGN (*additional white Gaussian noise*) model and is leveled 20 dB.

4.2 Exemplary estimation

Figure 2a–b shows a typical estimation of such a signal. When viewing the total simulation time, no details of the estimation are observable. To get an insight into these algorithms, two cut-outs of typical estimations will be discussed.

Since the database concept is non-linear and relies on its own statistics, especially the first estimations of a component have a high influence. Outliers from pure incoming frequency estimations can have high impact at first. Figure 3 shows this for – in this case – faulty initial estimations of PAST(d). When the database fields become statistically reliable, the output values become consistent again.

To visualize the principle of the phase difference based frequency updating, Fig. 4 gives a cut-out of a phase estimation. The solid line is not constant but linearly in-/decreasing (with steps). The gradient of this line is a hint on the deviation between the true frequency and the current estimation. At the positions where steps occur, this gradient has been used to update the estimation.

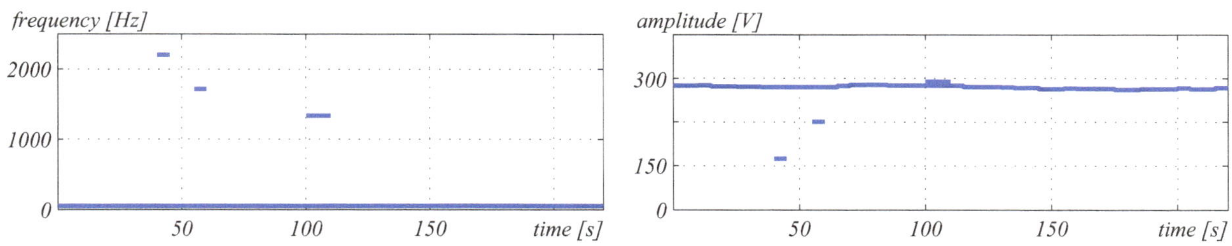

Fig. 2. A typical estimation picked from one of the simulations.

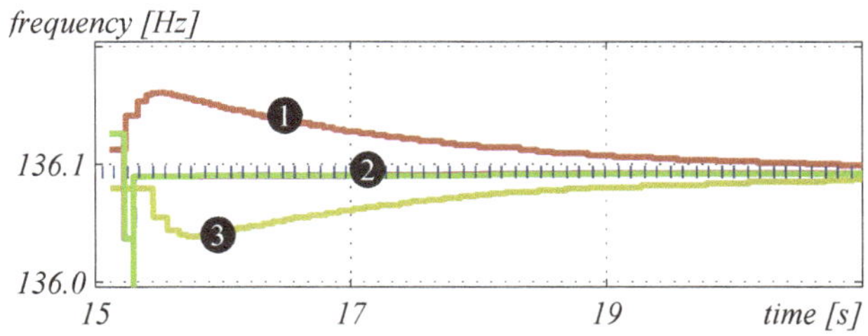

Fig. 3. A cut-out of a typical frequency estimation. (1) and (3) show aberrations due to erroneous parameter estimation and how they converge when estimations become reliable (here from PASTd and PAST); (2) points to the ideal dotted line and the EVD/OPAST-(PGS) estimation.

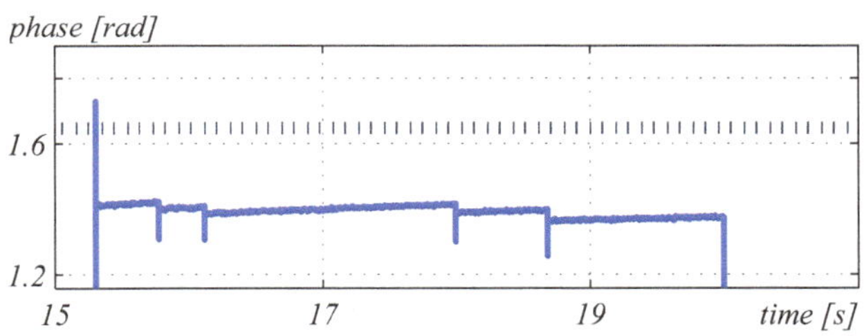

Fig. 4. A cut-out of a typical phase estimation. In the solid line, linearly increasing trends can be seen and the steps where DaPT reaches the threshold for phase difference severity and uses the gradient to correct the frequency estimation.

5 Simulation results

To summarize the results of the Monte-Carlo simulation, three cases will be presented. For the first, we measured the deviation of the frequency estimation and the theoretical values from the signal generator. We summed them for all contemporaneous components and averaged the sums over the simulation time and all simulations. Figure 5 presents the results.

First of all, it emerges that the differences between the different algorithms are small. If there are findings at all, then EVD has slight disadvantages and (O)PAST has slight advantages. EVD might suffer from the position of the exponential filtering. In EVD the autocorrelation matrix itself is exponentially filtered, which is indirect since not $\mathbf{R}_{xx}$ but the EV

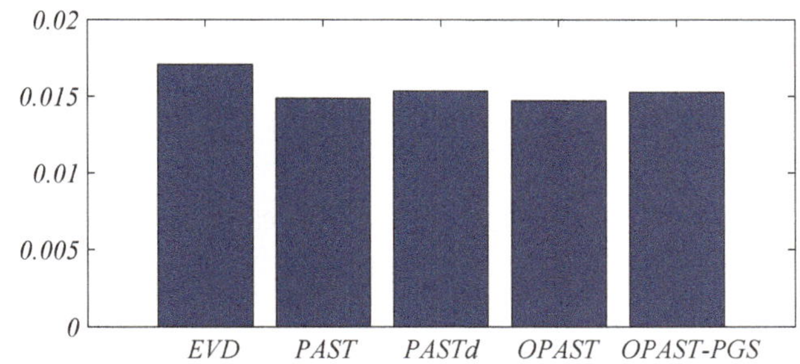

Fig. 5. Deviation of frequency estimation against true values, summed and averaged.

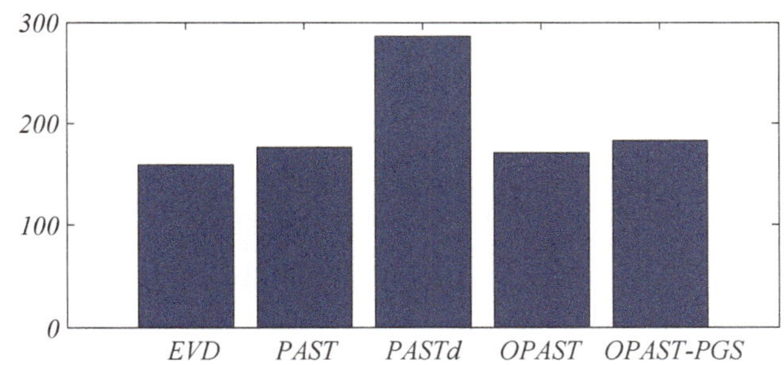

Fig. 6. Number of frequency components not found by the estimation per algorithm, summed and averaged.

are desired. The other algorithms perform that step directly on the basis of the subspace.

Also, we counted the number of frequency components that were not found by the estimator. Reasons may be (from the signal's perspecitve) a two weak amplitude or (from the statistical perspective) too high variance in the estimations. The numbers are per second and include the mandatory non-hits before DaPT regards them as present (resulting in an offset in the figure). In Fig. 6, the most significant fact is, that PASTd cannot keep up with the others. Also, EVD leaves a better mark. Both facts might be justified by the handling of estimations with high variance. A high variance in the estimations may confuse the sequential processing of PASTd. The joint cost of the serial cost functions might be higher than the cost of the (matrix-) PAST.

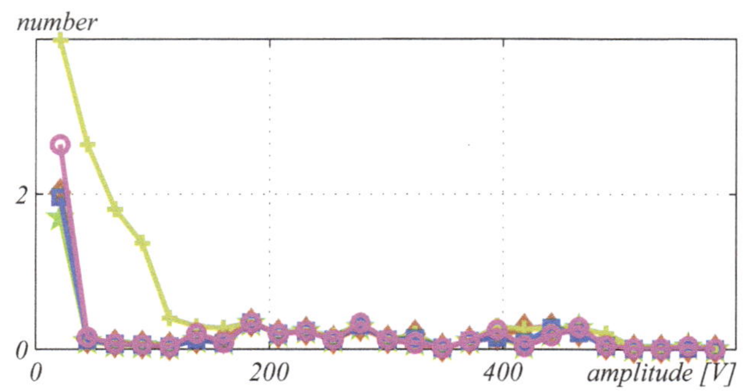

Fig. 7. Summed and averaged not-found components plotted against the amplitude: ★ EVD, ◇ PAST, □ OPAST, ○ OPAST-PGS, + PASTd.

High variances (indicating less reliable estimations) are most likely for low amplitudes (in relation to noise and dominant components). The last case surveys the unfound components against the amplitude, as (relative) amplitude is a good indicator for less reliable estimations. Figure 7 approves the previous results. PASTd is especially unsuitable for low amplitudes which may hint high variances in estimations. All the other perform equally well with slight advantages for EVD and (O)PAST. OPAST-PGS does not seem to bring advantages.

6 Conclusions

We have simulated the performance of different subspace trackers, based on the PAST theory, to evaluate the impact of the properties of the subspace basis presented to the ESPRIT algorithm. The estimated components supply the input of our DaPT non-linear parameter estimation post-processing. The subspace's basis of the signal generated from the (estimated) autocorrelation matrix of the input samples is not unique. The subspace can be described by various basis', some of them orthogonal, some even orthonormal and one of them being the signal eigenvectors of the autocorrelation matrix.

The simulations have shown, that it is not necessary for the subsequent steps (ESPRIT, DaPT) to be provided with exact eigenvectors (EVD). The basis presented to ESPRIT does not even have to be orthogonal (OPAST(-PGS)). The classic PAST algorithm, which does not guarantee an orthonormal basis, performs as well as the variations providing an orthonormal basis and as EVD providing actual eigenvectors.

A closer look reveals, that the sequential PASTd performs worse due to the lack of a joint cost function. The orthonormal OPAST is marginally better than PAST, because the approximation $\mathbf{W}^H\mathbf{W} \approx \mathbf{I}$ is more accurate for really orthonormal columns of $\mathbf{W}$. The algorithm with additional pairwise Gram-Schmidt orthogonalization does not improve the results.

In future work, the phase- and amplitude estimation shall be investigated further, i.e. other algorithms as LS relating the signal model. Also, cooperative estimations of regionally displaced measurements shall be investigated.

Acknowledgements. This work was supported by DFG (German Research Foundation).

References

Abed-Meraim, K., Chkeif, A., and Hua, Y.: Fast orthonormal PAST algorithm, IEEE Signal Proc. Let., 7, 60–62, doi:10.1109/97.823526, 2000.

Badeau, R., Richard, G., and David, B.: Fast and Stable YAST Algorithm for Principal and Minor Subspace Tracking, IEEE T. Signal Proces., 56, 3437–3446, doi:10.1109/TSP.2008.925924, 2008.

Bartelmaos, S., Abed-Meraim, K., and Attallah, S.: Mobile localization using subspace tracking, in: 2005 Asia-Pacific Conference on Communications, 1009–1013, doi:10.1109/APCC.2005.1554216, 2005.

Champagne, B. and Liu, Q.-G.: Plane rotation-based EVD updating schemes for efficient subspace tracking, IEEE T. Signal Proces., 46, 1886–1900, doi:10.1109/78.700961, 1998.

Lechtenberg, M. and Götze, J.: Database Assisted Frequency Estimation, in: 4th IEEE International Conference on Computer Science and Information Technology, 2011.

Paulraj, A., Roy, R., and Kailath, T.: Estimation Of Signal Parameters Via Rotational Invariance Techniques, in: IEEE Conference on Circuits, Systems and Computers, 83–89, 1985.

Schmidt, R.: Multiple emitter location and signal parameter estimation, IEEE T. Antenn. Propag., 34, 276–280, 1986.

Yang, B.: An extension of the PASTd algorithm to both rank and subspace tracking, 9, 179–182, doi:10.1109/97.410547, 1995a.

Yang, B.: Projection approximation subspace tracking, IEEE T. Signal Proces., 43, 95–107, doi:10.1109/78.365290, 1995b.

Matrix-Vector Based Fast Fourier Transformations on SDR Architectures

Y. He[2], K. Hueske[1], J. Götze[1], and E. Coersmeier[2]

[1]University of Dortmund, Information Processing Lab, Otto-Hahn-Str. 4, 44221 Dortmund, Germany
[2]Nokia Research Center, Bochum, Meesmannstr. 103, 44807 Bochum, Germany

Abstract. Today Discrete Fourier Transforms (DFTs) are applied in various radio standards based on OFDM (Orthogonal Frequency Division Multiplex). It is important to gain a fast computational speed for the DFT, which is usually achieved by using specialized Fast Fourier Transform (FFT) engines. However, in face of the Software Defined Radio (SDR) development, more general (parallel) processor architectures are often desirable, which are not tailored to FFT computations. Therefore, alternative approaches are required to reduce the complexity of the DFT. Starting from a matrix-vector based description of the FFT idea, we will present different factorizations of the DFT matrix, which allow a reduction of the complexity that lies between the original DFT and the minimum FFT complexity. The computational complexities of these factorizations and their suitability for implementation on different processor architectures are investigated.

1 Introduction

Recently multi-carrier modulation techniques, such as OFDM, have received great attention in high-speed data communication systems and have been selected for several communication standards (e.g. WLAN, DVB and WiMax). A central baseband signal processing task required in OFDM transceivers is the Fourier transform of the received data, which is generally realized as FFT (Cooley, 1965). To meet the real-time processing requirements, various specialized FFT processors have been proposed (Son, 2002). On the other hand Software Defined Radio (SDR) approaches for mobile terminals are gaining momentum (Dillinger, 2003). To meet the demand for high computing power and flexibility, specific reconfigurable accelerators (Heyne, 2004) or parallel processor architectures (Berthold, 2004) are used. The DFT can be implemented as FFT, which defines the optimum in terms of mathematical complexity, but as the used architectures are not necessarily optimized for butterfly structures, the question arises: Are there any efficient procedures, which reduce the computational complexity of the DFT significantly (as far as possible to the FFT complexity), but at the same time are more tailor-made to specific hardware architectures than the FFT?. In this paper a matrix-vector based formulation of the FFT algorithm (Van Loan, 1992) is derived from the respective matrix factorizations of the DFT matrix. Due to the structure of the butterfly matrices and the matrix factorizations, the matrix-vector based FFT allows several separations of the resulting matrix product. These separations have an increased computational complexity compared to the FFT, but because of their mathematical structure they allow more flexible implementations on different architectures (e.g. multi processors, vector processors, accelerated systems). The paper is organized as follows: In Sect. 2 we will clarify the definition and notation of the DFT. A short repetition of the FFT idea and the corresponding factorization is described in Sect. 3. Then, in Sect. 4 different separations of the butterfly matrices are presented, which lead to different algorithmic structures and different computational complexities. Implementation issues of these separations and their suitability for different hardware architectures are discussed. Conclusions are given in Sect. 5.

Correspondence to: K. Hueske
(klaus.hueske@uni-dortmund.de)

2 Discrete Fourier Transform (DFT)

The sequence of N (with $N{=}2^m$, $m \in \mathbb{N}$) complex numbers $x_1, \cdots, x_N$ is transformed into the sequence of N complex numbers $y_1, \cdots, y_N$ by the DFT according to

$$y_t = \sum_{n=1}^{N} x_n w_N^{(t-1)(n-1)}, \quad t = 1, \cdots, N, \tag{1}$$

where $w_N{=}e^{-j\frac{2\pi}{N}}$ is a primitive N-th root of the unity. The DFT can also be formulated as a matrix vector product (Van Loan, 1992). With $\mathbf{x}_N{=}[x_1 \cdots x_N]^T \in \mathbb{C}^N$ and $\mathbf{y}_N{=}[y_1 \cdots y_N]^T \in \mathbb{C}^N$, the DFT can be described as matrix vector product $\mathbf{y}_N = \mathbf{F}_N \cdot \mathbf{x}_N$, where $\mathbf{F}_N \in \mathbb{C}^{N \times N}$ is the DFT-Matrix with elements $f_{nt} = w_N^{(t-1)\cdot(n-1)}$ and $n, t = 1, \cdots, N$.

3 Matrix-vector based FFT

To give a repetition of the known radix-2 FFT algorithms (Cooley, 1965), this section will show the connection between the matrix $\mathbf{F}_N$ and the matrix $\mathbf{F}_{N/2}$, which enables us to compute an N-point DFT from a pair of $(N/2)$-point DFTs. To transfer the FFT idea to the DFT matrix, we introduce a so-called permutation matrix $\mathbf{P}_N$ of size $(N \times N)$, which performs an odd-even ordering of the rows (multiplied from the left with $\mathbf{P}_N$) or columns (multiplied from the right with $\mathbf{P}_N^T$) of a matrix. Furthermore $\mathbf{P}_N^T \cdot \mathbf{P}_N = \mathbf{I}_N$ holds. The following factorizations and separations are shown for the decimation in time (DIT) FFT. They can be easily transferred to the decimation in frequency (DIF) FFT.

3.1 FFT-Matrix decomposition

Applying $\mathbf{P}_N^T$ to the right of the DFT-matrix $\mathbf{F}_N$ yields

$$\mathbf{F}'_N = \mathbf{F}_N \cdot \mathbf{P}_N^T = \begin{bmatrix} \mathbf{F}_{\frac{N}{2}} & \mathbf{D}_{\frac{N}{2}} \cdot \mathbf{F}_{\frac{N}{2}} \\ \mathbf{F}_{\frac{N}{2}} & -\mathbf{D}_{\frac{N}{2}} \cdot \mathbf{F}_{\frac{N}{2}} \end{bmatrix}. \tag{2}$$

The resulting product $\mathbf{F}'_N$ has the following structure: The odd indexed columns, forming the left half of $\mathbf{F}'_N$, are composed of $\mathbf{F}_{N/2}$, while the even indexed columns, forming the right half of $\mathbf{F}'_N$, are composed of $\mathbf{D}_{N/2} \cdot \mathbf{F}_{N/2}$ and $-\mathbf{D}_{N/2} \cdot \mathbf{F}_{N/2}$. This matrix can be decomposed into the product of two matrices (as shown in Eq. 3),

$$\mathbf{F}'_N = \begin{bmatrix} \mathbf{I}_{\frac{N}{2}} & \mathbf{D}_{\frac{N}{2}} \\ \mathbf{I}_{\frac{N}{2}} & -\mathbf{D}_{\frac{N}{2}} \end{bmatrix} \cdot \begin{bmatrix} \mathbf{F}_{\frac{N}{2}} & \mathbf{0} \\ \mathbf{0} & \mathbf{F}_{\frac{N}{2}} \end{bmatrix} \tag{3}$$

where $\mathbf{D}_{N/2}$ is a diagonal matrix with elements $d_{nn} = w_N^{(n-1)}$ and $n = 1, \cdots, N/2$. The first matrix represents FFT butterfly operations, while the second matrix contains two independent DFTs of length $N/2$. Note that the second matrix is a block diagonal matrix.

3.2 Matrix-Vector based computation of the FFT

To apply the modified DFT $\mathbf{F}'_N$ to an input vector $\mathbf{x}_N$, a permutation of $\mathbf{x}_N$ is required, i.e.

$$\mathbf{P}_N \cdot \mathbf{x}_N = [\mathbf{x}_o \mathbf{x}_e]^T, \tag{4}$$

with $\mathbf{x}_o \in \mathbb{C}^{\frac{N}{2}}$ the odd and $\mathbf{x}_e \in \mathbb{C}^{\frac{N}{2}}$ the even indexed part of $\mathbf{x}_N$. Introducing $\mathbf{y}_1, \mathbf{y}_2 \in \mathbb{C}^{\frac{N}{2}}$ we can formulate the DFT as follows:

$$\mathbf{y}_N = \begin{bmatrix} \mathbf{y}_1 \\ \mathbf{y}_2 \end{bmatrix} = \mathbf{F}_N \cdot \mathbf{x}_N = \mathbf{F}_N \cdot \underbrace{\mathbf{P}_N^T \cdot \mathbf{P}_N}_{\mathbf{I}_N} \cdot \mathbf{x}_N \tag{5}$$

$$= \begin{bmatrix} \mathbf{I}_{\frac{N}{2}} & \mathbf{D}_{\frac{N}{2}} \\ \mathbf{I}_{\frac{N}{2}} & -\mathbf{D}_{\frac{N}{2}} \end{bmatrix} \begin{bmatrix} \mathbf{F}_{\frac{N}{2}} & \mathbf{0} \\ \mathbf{0} & \mathbf{F}_{\frac{N}{2}} \end{bmatrix} \cdot \begin{bmatrix} \mathbf{x}_o \\ \mathbf{x}_e \end{bmatrix}$$

Using $\mathbf{y}_o = \mathbf{F}_{\frac{N}{2}} \cdot \mathbf{x}_o$ and $\mathbf{y}_e = \mathbf{F}_{\frac{N}{2}} \cdot \mathbf{x}_e$ we obtain:

$$\mathbf{y}_1 = \mathbf{y}_o + \mathbf{D}_{\frac{N}{2}} \cdot \mathbf{y}_e = \mathbf{y}_o + \mathbf{d}_{\frac{N}{2}} \cdot * \mathbf{y}_e \tag{6}$$

$$\mathbf{y}_2 = \mathbf{y}_o - \mathbf{D}_{\frac{N}{2}} \cdot \mathbf{y}_e = \mathbf{y}_o - \mathbf{d}_{\frac{N}{2}} \cdot * \mathbf{y}_e. \tag{7}$$

With $\mathbf{d}_{\frac{N}{2}}$ the vectorized diagonal of $\mathbf{D}_{\frac{N}{2}}$, with $(\cdot *)$ the component-wise vector multiplication and taking into account that the two products are identical, Eq. (6) and Eq. (7) need altogether $\frac{N}{2}$ multiplications and $2 \cdot \frac{N}{2}$ additions. It is easy to see that a DFT of size N can be executed by two DFTs of size $\frac{N}{2}$. For $N = 2^m$ this process can be recursively repeated $m = \log_2 N$ times. The required number of operations then will be $\frac{N}{2} \log_2 N$ multiplications and $N \log_2 N$ additions, which represents the complexity of the FFT.

3.3 Factorization of the DFT matrix

To obtain alternative factorizations of the DFT matrix, the number of recursions described in the previous section can be modified. Introducing a new parameter k we can define the number of performed recursions as $m-k$ with $N = 2^m$ giving the size of the initial DFT matrix. For arbitrary m and k the resulting block submatrices in $\mathbf{F}_B$ are of size $(2^k \times 2^k)$ and can be executed as independent DFTs (or FFTs). Because the length of the input vector $\mathbf{x}_N$ is given by $N = 2^m$ the number of block submatrices equals $2^m/2^k = 2^{m-k}$. After calculating 2^{m-k} independent DFTs, $m-k$ butterfly matrices $\mathbf{B}_i$ have to be applied to finish the calculation. The entire DFT is then given as

$$\mathbf{y}_N = \mathbf{B}_{m-k} \cdot \mathbf{B}_{m-k-1} \cdots \mathbf{B}_1 \cdot \mathbf{F}_B \cdot \mathbf{x}_P, \tag{8}$$

with $\mathbf{x}_p$ the $(m-k)$-times permuted input vector $\mathbf{x}_N$. An example for $m=4$ and $k=2$ is given below.

Example 1:

$$y_N = \underbrace{\begin{bmatrix} I_8 & D_8 \\ I_8 & -D_8 \end{bmatrix}}_{B_2} \cdot \underbrace{\begin{bmatrix} I_4 & D_4 & 0 & 0 \\ I_4 & -D_4 & 0 & 0 \\ 0 & 0 & I_4 & D_4 \\ 0 & 0 & I_4 & -D_4 \end{bmatrix}}_{B_1} \cdot \underbrace{\begin{bmatrix} F_{\frac{N}{4}} & 0 & 0 & 0 \\ 0 & F_{\frac{N}{4}} & 0 & 0 \\ 0 & 0 & F_{\frac{N}{4}} & 0 \\ 0 & 0 & 0 & F_{\frac{N}{4}} \end{bmatrix}}_{F_B} \cdot x_p$$

3.4 Construction of butterfly matrices

Increasing the order of recursion $m-k$ will also increase the number of butterfly matrices $\mathbf{B}_i$. While the independent DFT submatrices in $\mathbf{F}_B$ can be processed independently, the butterfly matrices can be treated in different ways:

·As sequence of matrix multiplications

$$\prod_{i=0}^{m-k-1} \mathbf{B}_{m-k-i}$$

·As one matrix multiplication with the butterfly product matrix

$$\mathbf{B} = \prod_{i=0}^{m-k-1} \mathbf{B}_{m-k-i}$$

·As partly combined matrix multiplication with the products

$$\prod_{i=0}^{p_1} \mathbf{B}_{m-k-i} \prod_{i=p_1-1}^{p_2} \mathbf{B}_{m-k-i} \prod_{i=p_r-1}^{m-k-1} \mathbf{B}_{m-k-i}$$

with arbitrarily chosen integer numbers p fulfilling $(1<p_1<p_2<\ldots<p_r<m-k)$.

Example 2: The following example illustrates the construction of the butterfly matrices. For the given parameters m=8 and k=4 already 2^{m-k}=2^4=16 small DFTs are executed by $\mathbf{F}_B$. The structure of the remaining butterfly matrices $\mathbf{B}_i$ is depicted in Fig. 1 on the left hand side (non-zero elements are depicted as lines). The product of the butterfly matrices for increasing i is shown on the right hand side. The matrices are all of dimension $2^8 \times 2^8$, the variable nz in the figure shows the number of non-zero elements. Focusing on the butterfly product matrix in the bottom right, we can find that only nz=4096 of all $2^8 \cdot 2^8$=65536 elements are non-zero. This means, every 16-th diagonal of the butterfly product matrix is non-zero, since the other multiplications have already been performed by the block diagonal matrix $\mathbf{F}_B$. In general, every 2^k-th diagonal is non-zero, which means, that the interspace will be enlarged with increasing k.

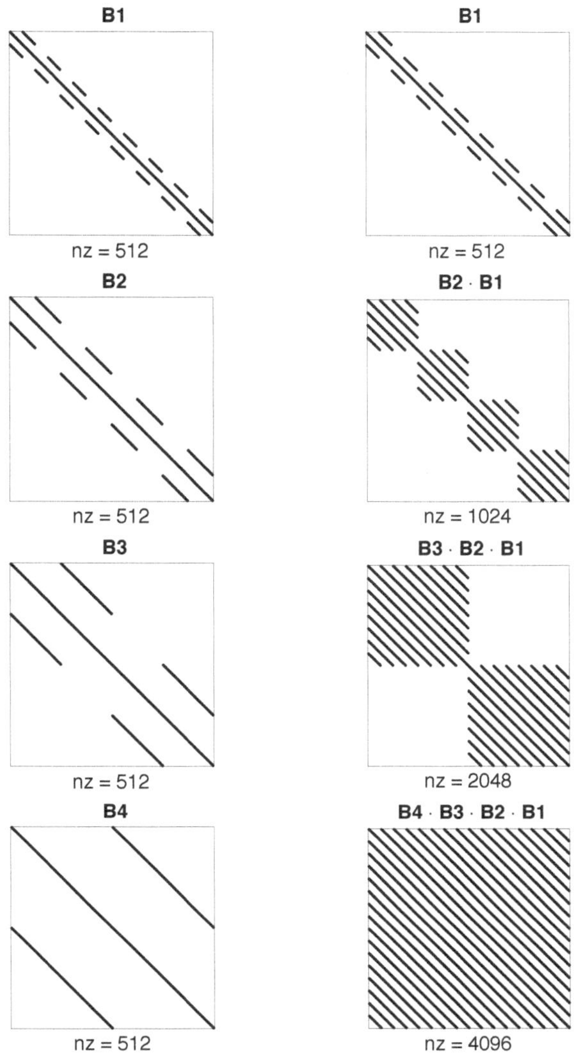

Fig. 1. The single butterfly matrices $\mathbf{B}_i$ on the left hand side; their products on the right hand side.

4 Implementation

In this section we will compare different separations of the butterfly matrix in terms of complexity and implementation issues.

4.1 Complexity

The application of the DFT matrix requires $(N-1)\cdot(N-1)$ multiplications, i.e. the product of all rows and columns, except the first row/column, which is all ones:

$$M_{\mathrm{DFT}} = N^2 - 2N + 1 = 2^m \cdot 2^m - 2 \cdot 2^m + 1.$$

The FFT is based on butterfly operations, which require $N/2$=2^{m-1} multiplications each (see Sect. 3.2). Since we have m butterfly matrices, we obtain:

$$M_{\mathrm{FFT}} = m \cdot 2^{m-1} = \frac{N}{2} \log_2 N.$$

In the following we will examine the complexity of four different separations, which we expect to lie between M_{DFT} and M_{FFT}.

Separation A. We use $(\mathbf{B}_{m-k} \cdots \mathbf{B}_1 \cdot \mathbf{F}_B)$, where the submatrices of $\mathbf{F}_B$ are executed as independent DFTs and the butterfly matrices are separately executed one by one. The number of the DFT blocks in $\mathbf{F}_B$ is 2^{m-k} and each DFT needs $2^k \cdot 2^k - 2 \cdot 2^k + 1$ multiplications. The number of butterfly matrices is $m-k$ and each butterfly needs 2^{m-1} multiplications.

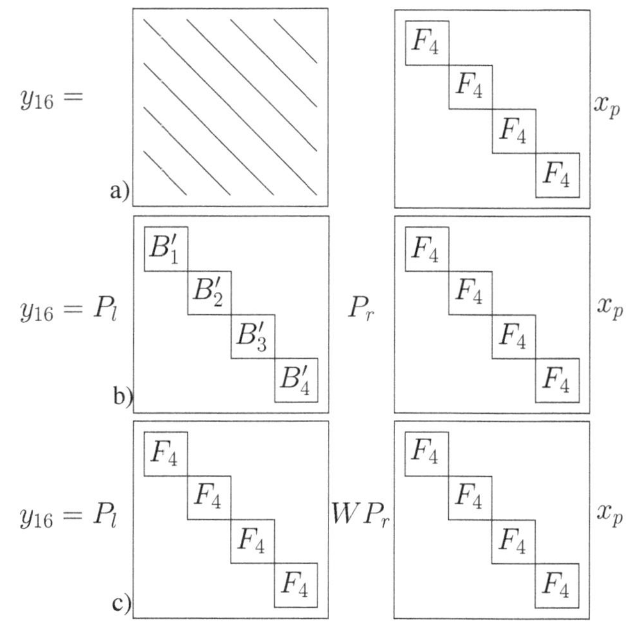

Fig. 2. Parallel FFT following Separation D.

This leads to

$$M_A = (m-k) \cdot 2^{m-1} + (2^k \cdot 2^k - 2 \cdot 2^k + 1) \cdot 2^{m-k}.$$

Separation B. We use $\mathbf{B} \cdot \mathbf{F}_B$, where the submatrices of $\mathbf{F}_B$ are still executed as independent DFTs, and $\mathbf{B}$ is the butterfly product matrix. Since only every 2^k-th diagonal is occupied, the number of required multiplications for $\mathbf{B}$ is $(2^m - 1) \cdot (\frac{2^m}{2^k} - 1)$. The number of multiplications for $\mathbf{F}_B$ is identical to separation A , such that

$$M_B = (2^m - 1) \cdot (\frac{2^m}{2^k} - 1) + (2^k \cdot 2^k - 2 \cdot 2^k + 1) \cdot 2^{m-k}.$$

Separation C. We again use $\mathbf{B} \cdot \mathbf{F}_B$, but now the submatrices of $\mathbf{F}_B$ are executed as independent FFTs. There are $k \cdot 2^{k-1}$ multiplications required for each FFT, and totally there are 2^{m-k} FFT blocks, so the number of multiplications is given as

$$M_C = (2^m - 1) \cdot (\frac{2^m}{2^k} - 1) + k \cdot 2^{k-1} \cdot 2^{m-k}.$$

Separation D. A special case of separation C is separation D, which describes a partly parallel implementation of the original FFT for the case $k = m/2$. The factorization is given as

$$\mathbf{y}_N = \mathbf{P}_l \cdot \mathbf{F}_B \cdot \mathbf{W} \cdot \mathbf{P}_r \cdot \mathbf{F}_B \cdot \mathbf{x}_p, \quad (9)$$

with $\mathbf{W}$ a twiddle factor diagonal matrix and $\mathbf{P}_l$ and $\mathbf{P}_r$ permutation matrices that sort the rows and columns of a matrix, such that $\mathbf{P}_l \cdot \mathbf{F}_B \cdot \mathbf{W} \cdot \mathbf{P}_r = \mathbf{B}$ is the butterfly product matrix. To clarify this, an example is given in Fig. 2 for $m=4$ and $k=2$.

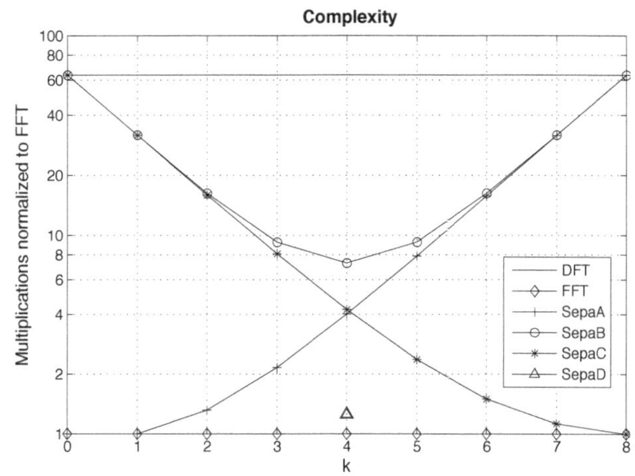

Fig. 3. Complexities for DFT, FFT and various separations.

Example 3: In a) we can see the graphical representation of example 2, with the butterfly product matrix $\mathbf{B}$ on the left and the block FFT matrix $\mathbf{F}_B$ on the right hand side. By sorting rows and columns of $\mathbf{B}$ we can transform the sparse diagonal matrix $\mathbf{B}$ to a block diagonal matrix, as shown in b). This leads to a matrix based notation of the Jeon-Reeves FFT algorithms (Jeon (1986)). As the blocks $(\mathbf{B}'_1, \cdots, \mathbf{B}'_4)$ are not identical, constant twiddle factors are extracted from the matrices, which transfers the block submatrices to DFT matrices, like shown in c). The extracted twiddle factors are collected in $\mathbf{W}$. The computational complexity of this separation is determined by the number of FFTs $(2 \cdot 2^k)$ and the additional twiddle factor multiplications $(N=2^m)$:

$$M_D = 2 \cdot 2^k \cdot k \cdot 2^{k-1} + 2^m = 2^m(1 + m/2) \quad (10)$$

The number of required multiplications for separations A–D are given in Fig. 3 for $m=8$ and varying k as multiples of M_{FFT}. The upper bound is defined by the DFT complexity, the lower bound shows the FFT complexity. Note that separation D is only defined for $k=m/2=4$.

4.2 Implementation issues

Here we will discuss possible implementations of the presented separations on different hardware architectures. The discussion is of a qualitative character, i.e. implementation details like memory access, interconnection complexity or inter processor communication time are not considered here.

Separation A. Matrix-vector processors as special kind of SIMD (Single Instruction Multiple Data) architectures are discussed as possible platforms for SDR (Schoenes (2003)). As an example we will show an implementation of separation A using a general systolic array structure for matrix multiplications, like depicted in Fig. 4. With $m=5$ and $k=2$ we have to perform 8 DFTs of length 4 in $\mathbf{F}_B$. To obtain the matrix-matrix structure we separate the input vector $\mathbf{x}_p$

into 8 parts ($\mathbf{x}_1, \mathbf{x}_2, \cdots, \mathbf{x}_8$) of length 4, which means that all $\mathbf{F}_4 \cdot \mathbf{x}_j$ with j=1, ..., 8 can be calculated independently. The multiple matrix-vector products can then be described as matrix-matrix product $\mathbf{F}_4 \cdot \mathbf{X}_p$ with $\mathbf{X}_p$=[$\mathbf{x}_1$ $\mathbf{x}_2 \cdot \mathbf{x}_8$]. The remaining butterfly operations are performed according to Eq. (6) and Eq. (7). The even numbered columns of the product $\mathbf{F}_4 \cdot \mathbf{X}_p$ are multiplied component-wise with $\mathbf{d}_4$ and then added/subtracted to the corresponding odd numbered columns. For example, the second column is multiplied by $\mathbf{d}_4$ and then added/subtracted to the first column, resulting in the intermediate vectors $\mathbf{y}_{11}$ and $\mathbf{y}_{21}$ that are stored in the registers of the first and second column of the processor array. In the next step these intermediate results are multiplied by $\mathbf{d}_8$ and added/subtracted to the other intermediate results. After $m - k$ steps the registers contain the Fourier transform of $\mathbf{x}$.
The computational complexity for this approach is only slightly increased for small values of k (compare Fig. 3), but a high speed partly parallel Fourier transformation can be realized.

Separation B. For generic matrix-vector architectures the DFT can be described as simple matrix-vector product using separation B. In this case the number of required multiplications is increased considerably compared to the FFT, but nevertheless, it is interesting to see the significant decrease of the number of multiplications (e.g. for k=3, 4, 5 in Fig. 3) compared to the DFT, just by separating the DFT matrix into a product of the two matrices $\mathbf{B}$ and $\mathbf{F}_B$.

Separation C. To match the real time requirements in SDR systems, general purpose processors are often combined with hardware accelerators, like e.g. for the Fourier transform. As the flexibility of these accelerators regarding to the DFT size might be limited, separation C can be used to map DFTs of any size onto these engines. For example in a systems with FFT accelerators of size 64 and a required FFT length of 256 (this corresponds to m=8 and k=6), the block DFTs in $\mathbf{F}_B$ can be realized on the accelerators while the remaining 2 butterfly operations are executed as a simple matrix vector multiplication on a generic host processor. This leads to a major flexibility regarding the DFT size of the used accelerators in context of the required DFT size of the specific application. Furthermore existing architectures with fixed length FFT accelerators can be easily reused.
Depending on the difference of m and k, the mathematical complexity is increased (compare Fig. 3). For the example the number of required multiplications is increased by a factor 1.5, but the 4 FFTs of size 64 can be executed in parallel, if more than one accelerator is available in the system.

Separation D. The FFT proposed by Jeon and Reeves is especially suitable for multi processor architectures (Jeon (1986)). However the extraction of the twiddle factors in separation D also enables the use of homogeneous parallel FFT cores in an SDR system, as the resulting block submatrices are all identical. This means an N-point DFT can be separated into $2\sqrt{N}$ FFTs of size $\sqrt{N}$. The first $\sqrt{N}$ FFTs are

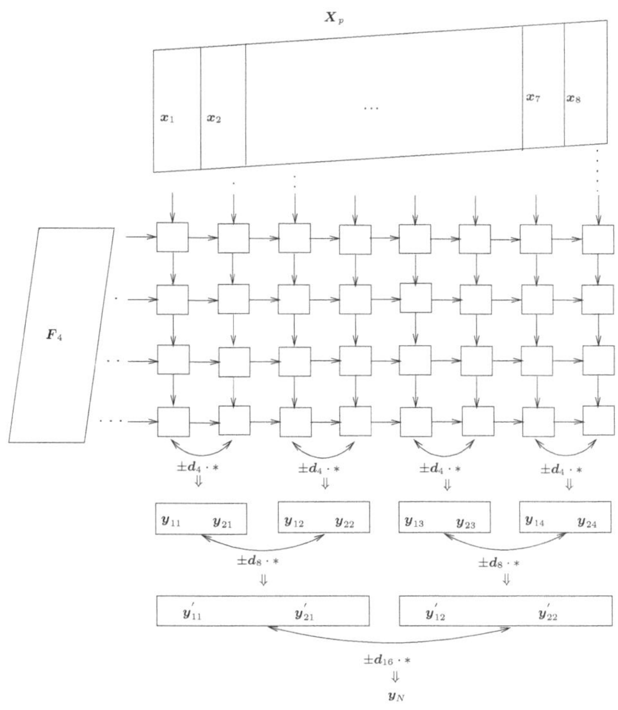

Fig. 4. Matrix-vector processor based implementation structure.

computed using the specialized processor cores, the results are then permuted and multiplied by $\mathbf{W}$. The remaining $\sqrt{N}$ blocks are again computed using the FFT cores.

5 Conclusions

The presented separations of the DFT matrix provide different algorithmic structures, which allow a flexibility in terms of the used hardware architectures for the implementation of the DFT. Depending on the architecture, the most suitable separation can be chosen, e.g. separation A for matrix processors or separation C for architectures using FFT hardware accelerators. Depending on the chosen parameters the computational complexity in terms of multiplications is only slightly increased, but some architectures may benefit from the modified algorithmic structure (e.g. addressing, bus communication). The behaviour of the presented separations on specific existing hardware architectures has yet to be examined in future work.

References

Berthold, U., Riehmeier, A.-R., and Jondral, F. K.: Spectral partitioning for modular software defined radio, 59th Vehicular Technology Conference (VTC2004), 2, 1218–1222, May 2004.

Cooley, J. W. and Tukey, J. W.: An algorithm for the machine calculation of complex Fourier series, Math. Comput., 19, 297–301, 1965.

Dillinger, M., Madani, K., and Alonistioti, N.: Software Defined Radio-Architectures, Systems and Functions, Wiley Series in Software Radio, John Wiley & Sons Ltd., 2003.

Heyne, B. and Götze, J.: A Pure Cordic Based FFT for Reconfigurable Digital Signal Processing, 12th European Signal Processing Conference (Eusipco2004), Vienna, September 2004.

Jeon, C. H. and Reeves, A. P.: The fast Fourier transform on a multicluster computer architecture, Proceedings of the 19th Annual Hawaii International Conference on System Sciences, 103–110, 1986.

Schoenes, M., Eberli, S., Burg, A., et al.: A novel SIMD DSP architecture for software defined radio, Proc. of the 46th IEEE International Midwest Symposium on Circuits and Systems (MWSCAS2003), 3, 1443–1446, December 2003.

Son, B. S., Jo, B. G., Sunwoo, M. H., and Yong, S. K.: A high-speed FFT processor for OFDM systems, Proc. of the IEEE International Symposium on Circuits and Systems (ISCAS2002), 3, 281–284, May 2002.

Van Loan, C.: Computational Frameworks for the Fast Fourier Transform, Frontiers in applied mathematics, Society for Industrial and Applied Mathematics, Philadelphia, 1992.

37

A first approach to the distortion analysis of nonlinear analog circuits utilizing X-parameters

H. Weber, C. Widemann, and W. Mathis

Institute four Theoretical Electrical Engineering, Leibniz Universität Hannover, Hanover, Germany

Correspondence to: C. Widemann (widemann@tet.uni-hannover.de)

Abstract. In this contribution a first approach to the distortion analysis of nonlinear 2-port-networks with X-parameters[1] is presented. The X-parameters introduced by Verspecht and Root (2006) offer the possibility to describe nonlinear microwave 2-port-networks under large signal conditions. On the basis of X-parameter measurements with a nonlinear network analyzer (NVNA) behavioral models can be extracted for the networks. These models can be used to consider the nonlinear behavior during the design process of microwave circuits. The idea of the present work is to extract the behavioral models in order to describe the influence of interfering signals on the output behavior of the nonlinear circuits. Hereby, a simulator is used instead of a NVNA to extract the X-parameters. Assuming that the interfering signals are relatively small compared to the nominal input signal, the output signal can be described as a superposition of the effects of each input signal. In order to determine the functional correlation between the scattering variables, a polynomial dependency is assumed. The required datasets for the approximation of the describing functions are simulated by a directional coupler model in Cadence Design Framework. The polynomial coefficients are obtained by a least-square method. The resulting describing functions can be used to predict the system's behavior under certain conditions as well as the effects of the interfering signal on the output signal.

1 Introduction

A high-frequency circuit can be designed by describing the N-port using scattering parameters (S-parameters) that allow to describe the N-port without current and voltage. Instead alternative variables are used. These variables are called the scattering variables and they are linear combination of the voltage and current of the analyzed N-port. The scattering variables are proportional to the electrical power and are measurable for high-frequencies. However, the S-parameters are restricted to linear and time-invariant N-ports. Thus, they cannot describe nonlinear circuits under large signal excitation. In the last two decades, the description with S-parameters was extended by Verspecht (1995). The linear S-parameters are advanced to the so called X-parameters introduced by Agilent Technologies (Verspecht and Root, 2006). The basic principle is called the *polyharmonic distortion* (*PHD*) which allows to describe a nonlinear circuit under large signal excitation. Moreover, it is possible to extract the X-parameters using a nonlinear vector analyzer (NVNA) (Roblin, 2011). By this means, a behavioral model for the device under test (DUT) can be developed. This model can be used for the design (Peláez Pérez, 2012) or can be added as additional information in the data sheet of an IC producer (Horn et al., 2008). The PHD assumes that only one large signal is applied to the N-port which defines the large signal operating point (LSOP). Other signals are superimposed at an integer multiple of the large signal frequency and are assumed as small signals. In this paper, a method is presented for the extraction of X-parameters using a simulator like Cadence. Therefore, the extension of scattering variables for nonlinear circuits is reviewed at first.

[1]X-parameter is a registered trademark of Agilent Technologies, Inc.

2 Scattering variables

Before describing a nonlinear N-port with X-parameters, the scattering variables are introduced. Therefore, the behavior of scattering variables for a nonlinear circuit have to be examined. For this purpose, the 1-port in Fig. 1 is given.

The nonlinear impedance in the 1-port is described by

$$i = \alpha_1 u + \varepsilon \alpha_3 u^3 . \quad (1)$$

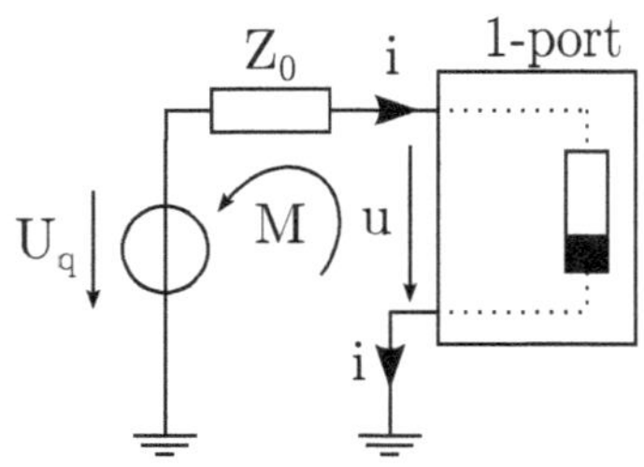

Fig. 1. Nonlinear resistor regarded as 1-port.

Table 1. Labeling of the scattering variables for a nonlinear 1-port.

Scattering Variable	
b_{11}	$A\sin(\omega t)\left(\frac{1-\alpha_1 Z_0-\frac{3}{4}A^2\varepsilon\alpha_3}{2\sqrt{Z_0}}\right)$
b_{13}	$A^3\sin(3\omega t)\frac{Z_0\varepsilon\alpha_3}{4\cdot 2\sqrt{Z_0}}$
a_{11}	$\frac{U_q}{2\sqrt{Z_0}}$

Furthermore the mesh

$$u = U_q - Z_0 \cdot i \tag{2}$$

can be set up. Assuming the port voltage to be

$$u = A\sin(\omega t), \tag{3}$$

one can insert these equations into the definitions of the scattering variables

$$a := \frac{u + Z_0 \cdot i}{2\sqrt{(Z_0)}} = \frac{U_q}{2\sqrt{Z_0}} \tag{4}$$

and with Eq. (1)

$$\begin{aligned} b := \frac{u - Z_0 \cdot i}{2\sqrt{(Z_0)}} &= \frac{(1 - Z_0\alpha_1)u - Z_0\varepsilon\alpha_3 u^3}{2\sqrt{Z_0}} \\ &= A\sin(\omega t)\left(\frac{1 - \alpha_1 Z_0 - \frac{3}{4}A^2\varepsilon\alpha_3}{2\sqrt{Z_0}}\right) \\ &+ A^3\sin(3\omega t)\frac{Z_0\varepsilon\alpha_3}{4\cdot 2\sqrt{Z_0}}. \end{aligned} \tag{5}$$

In Eq. (4), one can see that the scattering variable a is independent from the connected nonlinear 1-port. In contrast, the scattering variable b depends on the connected nonlinear 1-port. Due to the nonlinearity, harmonics can be recognized in Eq. (5). In order to describe this nonlinear effect with the scattering variables, an extended notation is introduced in Table 1. In addition to the first index, a second one is established to describe the generated harmonics separately according to the fundamental.

In Fig. 2, the scattering variables are represented in the frequency domain in order to illustrate the harmonic frequencies.

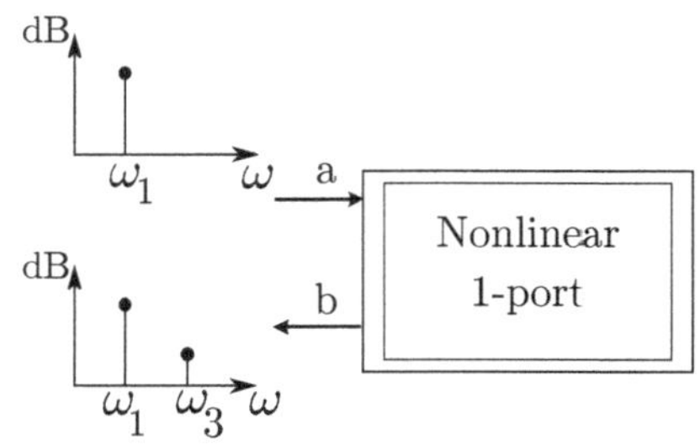

Fig. 2. Scattering variables represented in the frequency domain for a nonlinear 1-port.

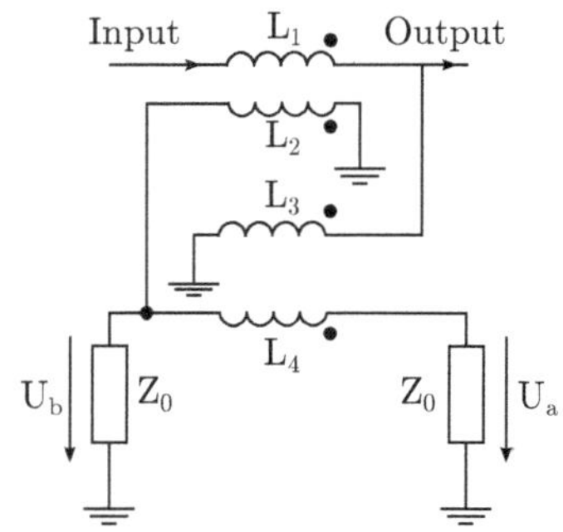

Fig. 3. Directional coupler model to obtain the scattering variables a and b.

Using the established notation for the scattering variables, it is now possible to describe all frequencies. However, it is still complicated to simulate these scattering variables because they are a linear combination of voltage and current and cannot be measured or simulated in the circuit directly. Therefore, a directional coupler is used in a real environment to determine all scattering variables. In the next section a directional coupler model is illustrated to provide all advantages of a real directional coupler in a simulator.

2.1 Directional coupler model

In the measurement technology, a directional coupler is used to obtain the scattering variables a and b. For simulation purpose, there are model approaches to the directional coupler behavior. One possibility of such a model is presented in Fig. 3 (Sierra, 2010).

This model consists of two coupled inductor pairs to transfer the current and the voltage which are used to determine the scattering variables a and b. The dimension of the inductor pairs can be determined by the coupling factor

$$C = 20\log_{10}\left(\frac{N_2}{N_1}\right). \tag{6}$$

N_2 and N_1 represent the number of windings of the coupled inductors L_1 and L_2, respectively. By the relation

$$\frac{N_2}{N_1} = \sqrt{\frac{L_2}{L_1}} \tag{7}$$

the inductors L_1 and L_2 can be dimensioned. Furthermore, $L_3 = L_2$ and $L_4 = L_1$ are assumed. The scattering variables are represented in this model by the voltages U_a and U_b. In order to determine the scattering variables, the voltages have to be multiplied with $\frac{1}{\sqrt{Z_0}}$. In the following, this model will be used to extract the scattering variables in order to describe a nonlinear N-port with X-parameters.

3 Extension to X-parameter

Certain properties of a linear N-port can be characterized by the S-parameters. Considering the scattering variables the S-parameters describe the connection between these by

$$\begin{pmatrix} b_1 \\ b_2 \\ \vdots \\ b_N \end{pmatrix} = \begin{pmatrix} S_{11} & S_{12} & \cdots & S_{1j} \\ S_{21} & S_{22} & \cdots & S_{2j} \\ \vdots & \vdots & \ddots & \vdots \\ S_{N1} & S_{N2} & \cdots & S_{NN} \end{pmatrix} \cdot \begin{pmatrix} a_1 \\ a_2 \\ \vdots \\ a_N \end{pmatrix} \tag{8}$$

Furthermore, the behavior of the nonlinear system depends on the excitation. Considering only one generated frequency, the relation between the scattering variable a_{ij} and b_{nm} can be described by

$$b_{nm} = F_{nm}(a_{11}, a_{12}, ...a_{21}, a_{22}, ...) . \tag{9}$$

The nonlinear scattering function $F_{nm}(\cdot)$ is hard to identify (Sun et al., 2010). However, under certain circumstances, simplifications can be applied. One assumption is that only one of the input signals represented by the scattering variables a_{jk} is a large signal which is typically defined as a_{11}. The rest of the signals occur at integer multiples of the large signal frequency with a small amplitude. Considering these assumptions, the system is linearized (Verspecht et al., 2005). Therefore, the superposition principle is valid for the small signals which are superimposed to the LSOP. The principle of the linearization is illustrated in Fig. 4. This allows to reduce the Eq. (9) to

$$b_{nm} = \sum_{jk} S_{nj,mk}(|a_{11}|) P^{+m-k} a_{jk} + \sum_{jk} T_{nj,mk}(|a_{11}|) P^{+m+k} a_{jk}^{*} \tag{10}$$

with $P = e^{j\varphi_{a11}}$ for the phase normalization (Verspecht and Root, 2006). The functions $S_{nj,mk}(|a_{11}|)$ and $T_{nj,mk}(|a_{11}|)$ are called X-parameters.

For linear systems, the parameters $S_{nj,mk}(|a_{11}|)$ are reduced to the linear scattering parameters in Eq. (8). Due to the two tone excitation, intermodulation occurs in nonlinear systems. This behavior is represented by the parameters $T_{nj,mk}(|a_{11}|)$ that is hence only necessary for nonlinear systems. Exploiting Eq. (10) and considering the restriction for $a_{jk} \neq a_{11}$ to be small signals, it is possible to calculate the behavior of the N-port for multiple input signals.

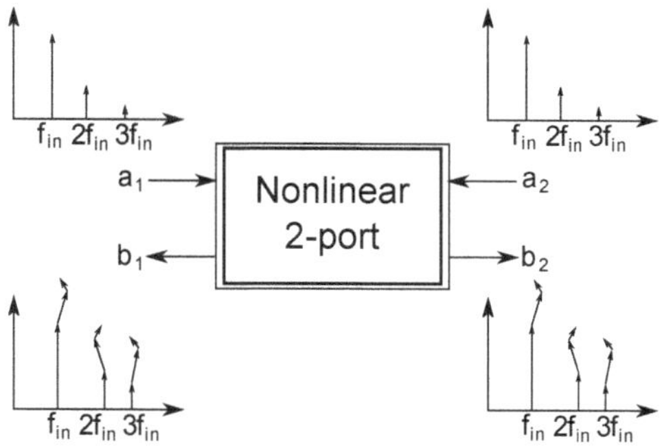

Fig. 4. Superposition for a nonlinear system (Verspecht and Root, 2006).

3.1 Determine X-parameters with a simulator

The X-parameters are extracted in the electrical measurement for a nonlinear N-port using a NVNA. In this section, a method to extract the X-parameters by using a simulator like Cadence Spectre is going to be presented. For this purpose, the network is realized in the simulator. The required scattering variables are extracted by the directional coupler model introduced in Sect. 2.1. The structure for the extraction of X-parameters is exemplarily illustrated for a 2-port network in Fig. 5.

At first, the N-port is excited only by the large signal a_{11}. Due to the phase normalization, the X-parameters $T_{n1,m1}(|a_{11}|)$ and $S_{n1,m1}(|a_{11}|)$ in Eq. (10) can be summarized to $S_{n1,m1}(|a_{11}|)$. Therefore, Eq. (10) is reduced to

$$b_{nm} = S_{n1,m1}(|a_{11}|) P^{m} |a_{11}| . \tag{11}$$

The X-parameters $S_{n1,m1}(|a_{11}|)$ can be determined by using the scattering variables b_{nm} and a_{11}. Hence, pairs of values of the scattering variable b_{nm} depending on a_{11} are necessary. For a certain range of $|a_{11}|$, the amplitude and phase of the scattering variable b_{nm} are determined. The pairs of values are interpolated by the least-square-fit (Herrmann, 2007).

After determining the X-parameters of the large signal, the X-parameters for multiply inputs ($S_{nj,mk}(|a_{11}|)$, $k > 1$) are determined. Therefore, further signals are used in addition to the large signal. A harmonic balance simulation allows to separate the intermodulation products which are represented by the X-parameters of the small signal. Exemplarily, one can assume that the excitation occurs at the frequencies $f_{a_{11}} =$ 200 MHz and $f_{a_{12}} = 400$ MHz. If all harmonics of the small signal at the frequency 400 MHz are neglected it results in three intermodulation products

$$1 \cdot f_{a_{11}} + 0 \cdot f_{a_{12}} = 200\,\text{MHz}, \tag{12}$$

$$(-1) \cdot f_{a_{11}} + 1 \cdot f_{a_{12}} = 200\,\text{MHz}, \tag{13}$$

$$3 \cdot f_{a_{11}} - 1 \cdot f_{a_{12}} = 200\,\text{MHz} \tag{14}$$

for the scattering variable b_{n1} at the frequency 200 MHz. Each of these intermodulation products is represented by a

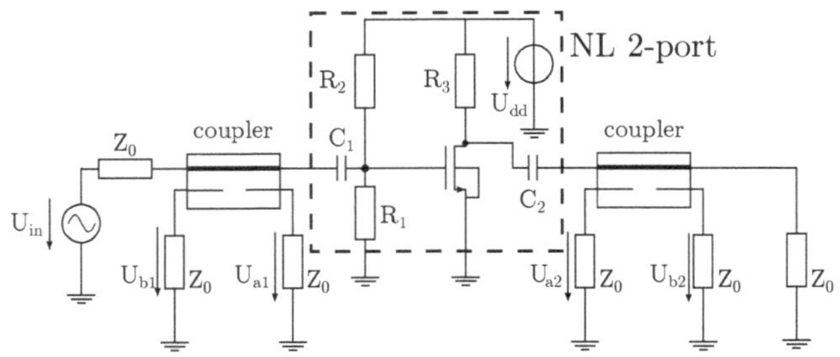

Fig. 5. Structure for the extraction of X-parameters for a 2-port.

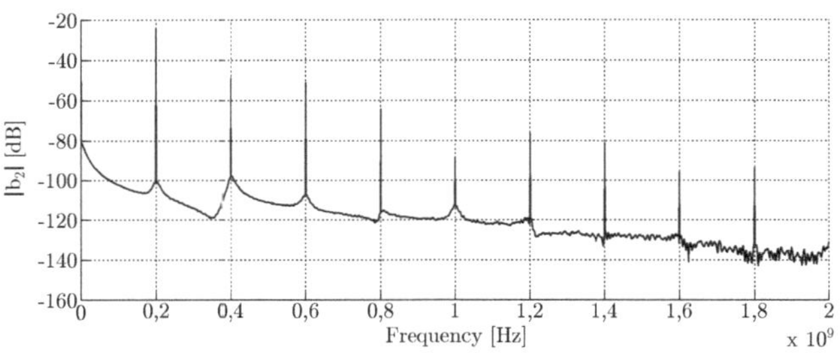

Fig. 6. Frequency domain of the scattering variable b_2.

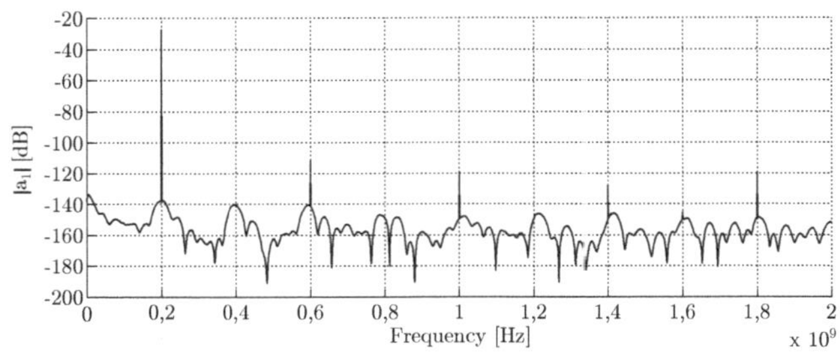

Fig. 7. Frequency domain of the scattering variable a_1.

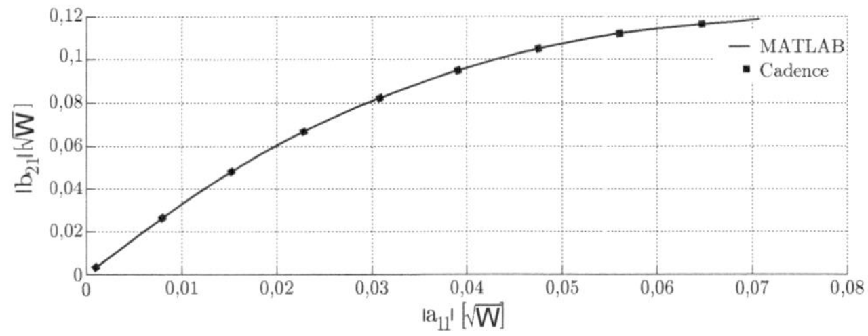

Fig. 8. Scattering variable b_{21} simulated and calculated.

X-parameter. Thus, it is possible to determine each of them separately. The intermodulation product in Eq. (13) is represented by $S_{n1,12}(|a_{11}|)$. In Eq. (14), one can see that the frequency of the small signal frequency of $f_{a_{12}}$ is multiplied by (-1). Since this factor implies the complex conjugate signal a_{12}^*, the intermodulation product Eq. (14) is represented by $T_{n1,12}(|a_{11}|)$. As functional approach for the X-parameters a polynomial approach is used for the interpolation.

3.2 Description of nonlinear 2-port with X-parameters

In the following, the nonlinear 2-port in Fig. 5 is described by using X-parameters. This example is limited to the description of the scattering variable b_{21} in order to reduce the effort.

At first, the nonlinear 2-port is excited by a_{11} at the frequency $f_{a_{11}} = 200\,\text{MHz}$. Two directional couplers are used to extract all scattered variables. Exemplarily, the spectrum of scattering variables $|a_1|$ and $|b_2|$ are illustrated in Figs. 6 and 7. In the latter, one can see that also harmonics occur in addition to the excitation at 200 MHz. However, as in Eq. (4) calculated, the scattering variable a_1 depends only on the excitation. Nevertheless, harmonics are still generated due to the imperfection of the directional coupler. In the following, these harmonics are neglected because the amplitudes are very small. In Fig. 6, harmonic frequencies can be observed in the spectrum of the scattering variable b_2 which depends on the nonlinear 2-port as stated in Sect. 2.

In the following, the extraction of the X-parameters of the scattering variable b_{21} will be executed. Therefore, the nonlinear 2-port is only excited by a_{11} at first. Pairs of values of a_{11} and b_{21} are simulated and discrete values of the connecting X-parameter is calculated by

$$S_{21,11}(|a_{11}|) = \frac{b_{21}}{a_{11}}. \tag{15}$$

These values are fitted by a polynomial approach which leads to an approximation of the X-parameter $S_{21,11}(|a_{11}|)$. The exemplary result is shown in Fig. 8 with the discrete simulation values. After determining the X-parameter $S_{21,11}(|a_{11}|)$, the small signal a_{12} is superimposed at the frequency of 400 MHz. In this case, for the scattering variable b_{21} one can say

$$\begin{aligned} b_{21} = S_{21,11}(|a_{11}|)a_{11} + S_{21,12}(|a_{11}|)P^{-1}a_{12} \\ + T_{21,12}(|a_{11}|)P^3 a_{12}^*. \end{aligned} \tag{16}$$

The X-parameters $S_{21,12}(|a_{11}|)$ and $T_{21,12}(|a_{11}|)$ are determined by a harmonic balance simulation. At first, the contribution of the intermodulation product in Eq. (13) to the scattering variable b_{21} is calculated. As for the X-parameter $S_{21,11}(|a_{11}|)$ the pairs of values are fitted by a polynomial approach which represents the X-parameter $S_{21,12}(|a_{11}|)$ as well as $S_{21,11}(|a_{11}|)$. The third intermodulation product in Eq. (14) is represented by $T_{21,12}(|a_{11}|)$. As $S_{21,12}(|a_{11}|)$ the X-parameter $T_{21,12}(|a_{11}|)$ is determined by a harmonic balance and polynomial approach.

In Fig. 9, the simulated (dots) and calculated (plane) scattering variable b_{21} is illustrated depending on a_{11} and a_{12}. For small amplitudes of a_{12} the calculated approximation coincides with the simulation whereas for higher amplitudes of a_{12} one can see a deviation. This behavior is more obvious in the two dimensional plot in Fig. 10. Therefore, the

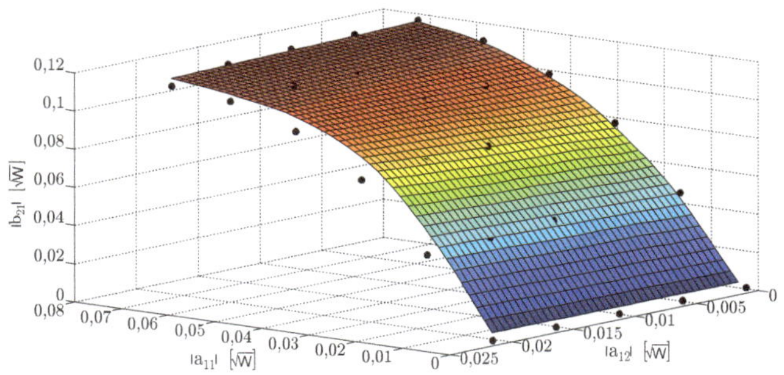

Fig. 9. Scattering variable b_{21} depending on a_{11} and a_{12}.

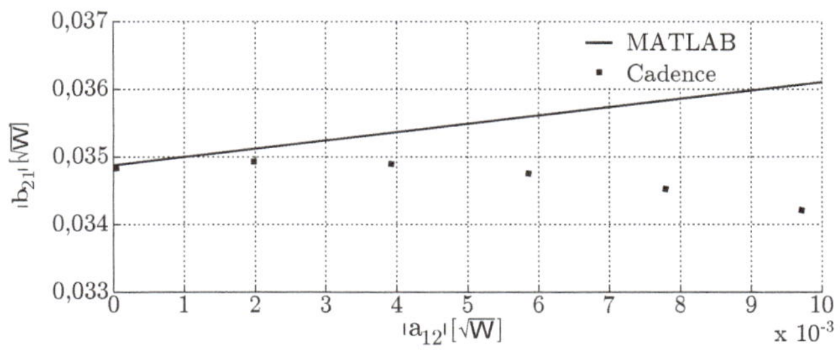

Fig. 10. Scattering variable b_{21} depending on a_{12} for fix $|a_{11}| = 0.011\sqrt{W}$.

approximation is only valid for small amplitudes of the superimposed signals as assumed for the linearization of Eq. (9). In order to demonstrate the difference between simulation and approximation the relative error is calculated by

$$\text{rel. error} = \frac{|b_{\text{sim}} - b_{\text{cal}}|}{|b_{\text{sim}}|} \cdot 100[\%]. \quad (17)$$

In Fig. 11, the relative error at discrete simulation points is illustrated for the results shown in Fig. 10. A regression line is added to illustrate the tendency. In Fig. 11, it can be seen that the smaller the amplitude of a_{12}, the better the approximation. However, the relative error raises drastically with increasing amplitude of a_{12} due to the breach of the assumption of small $a_{jk} \neq a_{11}$.

4 Conclusion and outlook

In this paper, the description of nonlinear systems with X-parameters using a circuit simulator like CADENCE Design Framework is presented. Therefore, a short introduction of the application of scattering variables on a nonlinear system was given. For the extraction of the scattering variables by a simulator a directional coupler model was used. After introducing a signal description of the nonlinear N-port with scattering variables, the X-parameters were determined. The basic idea of X-parameters allows to linearize a nonlinear system under a large signal excitation. Furthermore, a method to extract the X-parameters by a simulator was demonstrated

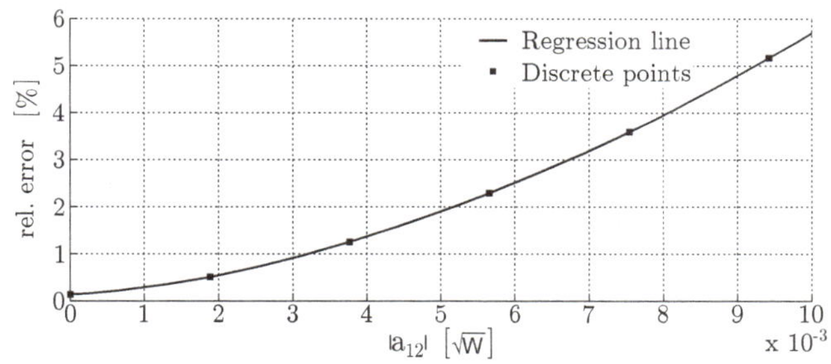

Fig. 11. Relative error of the approximation of b_{21}.

and the procedure was illustrated by an example. Future work will deal with an extension of Eq. (9) to

$$b_{nm} = \sum_{jk} S_{nj,mk}(|a_{11}|) P^{+m-k} a_{jk}^{\mu} + \sum_{jk} T_{nj,mk}(|a_{11}|) P^{+m+k} a_{jk}^{*\mu}. \quad (18)$$

It will be examined if the parameter μ can be determined by additional simulations. Considering this new parameter, the approximation probably covers a larger value range of a_{jk}.

References

Herrmann, N.: Höhere Mathematik, Oldenbourg Wissensch. Vlg., 2007.

Horn, J., Verspecht, J., Gunyan, D., Betts, L., Root, D., and Eriksson, J.: X-Parameter Measurement and Simulation of a GSM Handset Amplifier, in: Microwave Integrated Circuit Conference, 2008, EuMIC 2008. European, 135–138, 2008.

Peláez Pérez, A. M.: X-Parameters Based Analytical Design of Non-Linear Microwave Circuits: Application to Oscillator Design, Ph.D. thesis, Universidad Politécnica De Madrid, 2012.

Roblin, P.: Nonlinear RF Circuits and Nonlinear Vector Network Analyzers: Interactive Measurement and Design Techniques, Cambridge University Press, 2011.

Sierra, C. S.: Microwave directional couplers, 2010.

Sun, G., Xu, Y., and Liang, A.: The study of nonlinear scattering functions and X-parameters, in: Microwave and Millimeter Wave Technology (ICMMT), 2010 International Conference on, 1086–1089, 2010.

Verspecht, J.: Calibration of a Measurement System for High Frequency Nonlinear Devices, Ph.D. thesis, VRIJE UNIVERSITEIT BRUSSEL, 1995.

Verspecht, J. and Root, D. E.: Polyharmonic distortion modeling, Microwave Magazine IEEE, 2006.

Verspecht, J., Williams, D. F., Schreurs, D., Remley, K., and McKinley, M. D.: Linearization of large-signal scattering functions, Microwave Theory and Techniques, IEEE Transactions, 2005.

A code-aided synchronization IP core for iterative channel decoders

I. Ali, U. Wasenmüller, and N. Wehn

Microelectronic Systems Design Research Group, University of Kaiserslautern, 67663 Kaiserslautern, Germany

Correspondence to: I. Ali (imran@eit.uni-kl.de)

Abstract. Synchronization and channel decoding are integral parts of each receiver in wireless communication systems. The task of synchronization is the estimation of the general unknown parameters of phase, frequency and timing offset as well as correction of the received symbol sequence according to the estimated parameters. The synchronized symbol sequence serves as input for the channel decoder. Advanced channel decoders are able to operate at very low signal-to-noise ratios (SNR). For small values of SNR, the parameter estimation suffers from increased noise and impacts the communication performance. To improve the synchronization quality and thus decoder performance, the synchronizers are integrated into the iterative decoding structure. Intermediate results of the channel decoder after each iteration are used to improve the synchronization. This approach is referred to as code-aided (CA) synchronization or turbo synchronization.

A number of CA synchronization algorithms have already been published but there is no publication so far on a generic hardware implementation of the CA synchronization. Therefore we present an algorithm which can be implemented efficiently in hardware and demonstrate its communication performance. Furthermore we present a high throughput, flexible, area and power efficient code-aided synchronization IP core for various satellite communication standards. The core is synthesized for 65 nm low power CMOS technology. After placement and routing the core has an area of 0.194mm^2, throughput of 207 Msymbols/s and consumes 41.4 mW at 300 MHz clock frequency. The architecture is designed in such a way that it does not affect throughput of the system.

1 Introduction

Turbo and low-density parity-check (LDPC) codes are two most popular iterative channel codes nowadays. Due to their impressive communication performance near the Shannon limit they have become part of a wide range of wireless communication standards. As compared to non-iterative codes, they are capable of achieving a certain frame error rate (FER) at significantly reduced signal-to-noise ratios (SNR). However the performance of the iterative codes implicitly assumes ideal synchronization of the received signal. Small frequency and phase offsets result in a severe loss of communication performance in these decoders as stated in Mengali and D'Andrea (1997).

During the last decades, communication systems usually were operated at fairly high SNRs and synchronization was not a major source of communication performance degradation of a system as there are many well-established synchronization algorithms which can properly synchronize the systems at high SNR. Such synchronizers carry out estimation and correction of channel parameters prior to the decoding and function in either data-aided or non-data-aided modes. As mentioned earlier, Turbo and LDPC decoders can operate at low SNR. Consequently, the channel parameters estimation suffers from increased noise at low SNR which degrades the communication performance of the decoders. To achieve sufficient parameter estimations with these conventional synchronizers the amount of known data symbols i.e. pilot symbols had to be increased, which would result in a waste of bandwidth. Therefore in new family of synchronizers, the synchronization is integrated into the iterative decoding structure.

Integration of the synchronization into the iterative decoding process, commonly referred to as code-aided synchronization (CA Sync.), is a promising approach to synchronize a system at low SNR Herzet et al. (2007). The basic idea is to use the tentative a posteriori probability information of the transmitted bits which is available after each decoding iteration step to compute so-called soft symbols. These soft symbols are used as known data symbols as in data-aided synchronization to generate new parameter estimates; these new estimates are subsequently used to correct the received signal prior to performing further decoding iterations Godtmann and Meyr (2006).

Carrier phase, frequency and symbol timing of the received signal are three primary channel parameters. For parameter estimation and decoding the received signal must be down-converted to the baseband. Due to a slight difference of the oscillators frequencies in transmitter or receiver, a frequency offset will arise. Furthermore a frequency offset can appear by the Doppler effect if the transmitter and the receiver are in motion. Origin of the phase offset is given by the unknown transmission delay from the transmitter to the receiver which results in a random phase offset in the receiver. This paper concentrates on the frequency and phase offset estimation and correction (carrier synchronization) for turbo decoding.

At present iterative codes are part of various satellite communication standards like Digital Video Broadcasting – Return Channel via Satellite (DVB-RCS), Digital Video Broadcasting – Satellite services to Handhelds (DVB-SH) and GEO-Mobile Radio Interface (GMR-1 3G). These standards propose best combination of evolved data processing schemes at the transmitter to protect data from destructive channel effects under different transmission conditions and to support high data rate communication. While translating these requirements on the physical layer of a radio terminal, a flexible and high throughput hardware platform is highly desirable. The same hardware architecture should be able to be configured to the required air interface and also be able to accommodate future modifications of the standards.

1.1 State of the Art

A number of code-aided synchronization algorithms based on Expectation-Maximization (EM), the Newton-Raphson (NR), the Steepest-Descent (SD), the factor-graph representation and the sum-product for implementation of EM, NR and SD, pseudo-maximum-likelihood, and the correlation are proposed by different authors Godtmann and Meyr (2006), Herzet et al. (2007), Noels et al. (2005), Lottici and Luise (2004) and Wasenmüller et al. (2010). But no generic and flexible hardware implementation of the code-aided synchronization is published so far according to the best of our knowledge. This motivates us to introduce an algorithm which can be implemented efficiently in hardware and furthermore design a generic, high throughput, and flexible architecture for the code-aided synchronization to accommodate multiple wireless communication standards.

1.2 New contributions

This paper has three main contributions: (1) we demonstrate communication performance of a code-aided synchronization algorithm under different transmission parameters (2) we investigate the impact on throughput and communication performance by running the decoder and the code-aided synchronization algorithm in serial and parallel (3) we implement a CMOS synthesis based code-aided synchronization IP core to accomplish throughput, flexibility and energy efficiency requirements of various satellite communication standards. To the best of our knowledge we are the first to publish such an IP core.

This paper is structured as follows. Section 2 introduces turbo codes while Sect. 3 explains code-aided synchronization principle. In order to illustrate the advantages in communication performance due to the code-aided synchronization, simulation results are presented in Sect. 4. The hardware architecture and its implementation results are described in Sect. 5 and conclusions of this work are given in Sect. 6.

2 Turbo codes

With the introduction of binary turbo codes by Berrou in 1993 Berrou et al. (1993) near optimum error correction became possible. Due to these error correction capabilities, binary and duo-binary turbo codes allow for low frame error rates (FER) at a low signal-to-noise ratio (SNR). Turbo codes generally consist of a serial or parallel concatenation of two codes, so called component codes, and an interleaver.

Decoding of turbo codes is an iterative process where probabilistic information is exchanged between component decoders. A possible realization of a decoder for turbo codes is given in Fig. 1. The two component decoders that decode the two component codes are connected via interleaver and deinterleaver. They use *log likelihood ratios* (LLR) $\lambda^{s}_{d_k}$, $\lambda^{p^1}_{d_k}$ and $\lambda^{p^2}_{d_k}$ of the systematic and parity information to compute the extrinsic information $\Lambda^{e1}_{d_k}$ and $\Lambda^{e2}_{d_k}$ on the information bits. The iterative exchange of $\Lambda^{e1}_{d_k}$ and $\Lambda^{e2}_{d_k}$ between these component decoders is referred to as turbo principle. One (full) iteration is done if Decoder 1 and Decoder 2 have run once. If only one decoder has calculated new information, we call this one half iteration.

The decoding algorithm consists of a forward and a backward recursion. It computes for each possible information or parity bit d_k an a posteriori probability (APP) LLR: $\Lambda^{s}, \Lambda^{p1}, \Lambda^{p2}$.

3 Code-aided synchronization

The synchronization consists of the estimation of the unknown parameters of timing, frequency and phase offset, and the elimination of all possible negative influences introduced by these parameters. We focus on the frequency and phase synchronization in conjunction with turbo decoding. We assume, that the steps of gain control, timing synchronization, frame detection and coarse synchronization are properly carried out before. The received sample sequence r is given in the complex baseband according to Eq. (1):

$$r(l) = s(l) \cdot e^{j(2\pi f_o l + \Phi)} + n(l) \qquad l = 0, 1, \ldots, L-1 \qquad (1)$$

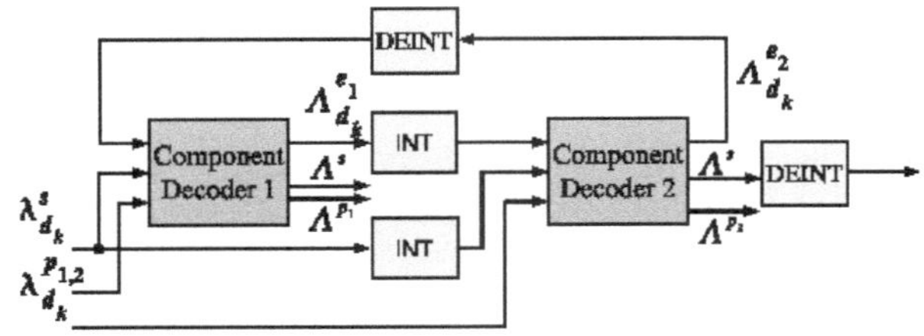

Fig. 1. Turbo Decoder.

The sample sequence r with L elements is based on modulation symbols $s(l)$ with one sample per symbol and symbol duration T, and is disturbed by a noise sequence n.

The unknown parameters of frequency offset f_o and phase offset Φ have to be estimated for every received sample sequence. These parameters are considered fixed during an estimation interval; the sample sequence has to be corrected accordingly to the estimated frequency offset $\tilde{f}_o$ and estimated phase offset $\tilde{\Phi}$. The synchronization is done in two main steps. Initially, a (coarse) carrier synchronization is performed with the help of pilot symbols. Afterwards fine synchronization is done iteratively with the additional use of tentative decoder decisions after each decoder iteration which is called here code-aided (CA) synchronization. This principle improves the estimation parameters for synchronization. The improved carrier synchronization is used to provide better input data for the decoding process. This process is done iteratively after each decoder iteration. Frequency and phase offset can be optimally estimated on an unmodulated carrier. With the assumption, that the transmitted symbol sequence s of the burst is known the effect of the modulation by each transmitted symbol $s(l)$ can be removed by:

$$\tilde{r}(l) = r(l) \cdot s^*(l) \qquad l = 0, 1, \ldots, L-1 \tag{2}$$

However it must be considered, that usually the symbols of the burst are unknown or only some symbols, used for supporting the burst detection or supporting the coarse synchronization are known. Thus we replace the transmitted symbol sequence s by an estimated symbol sequence s_e. The estimation of the transmitted symbol sequence is provided by the turbo decoder. The code-aided estimation of frequency and phase offset is based on the average phase $\tilde{Z}_0$ of the front part and on the average phase $\tilde{Z}_1$ of the rear part of the burst with a modulation removal by the estimated symbol sequence s_e. This is formally given by:

$$\tilde{Z}_k = \sum_{l=0+k\cdot L/2}^{L/2-1+k\cdot L/2} r(l) \cdot s_e^*(l) \qquad k = 0, 1 \tag{3}$$

With the two phase values of Eq. (3) the estimate of the frequency offset can be calculated with:

$$\tilde{f}_0 = \frac{\arg(\tilde{Z}_0 \cdot \tilde{Z}_1^*)}{2\pi \cdot L} \tag{4}$$

The estimate of the phase offset is calculated with the help of Eq. (4):

$$\tilde{\phi} = \arg(\tilde{Z}_0 + \tilde{Z}_1) - L \cdot \tilde{f}_0 \cdot \pi \tag{5}$$

The first decoder iteration is based on the LLR values $\lambda^s_{d_k}$, $\lambda^{p^1}_{d_k}$ and $\lambda^{p^2}_{d_k}$ calculated with the symbols of the coarse synchronized received sequence. For the code-aided synchronization an estimate of the transmitted symbols is used, which is produced by the turbo code decoder after each iteration. The estimate of the transmitted symbols is gathered by the APP LLR of the decoder. A turbo code decoder computes APP LLR values Λ^s of the systematic bits by default. In code-aided synchronization applications the decoder must additionally calculate the APP LLR values $\Lambda^{p1,2}$ for the parity bits. With the tentative soft values of systematic bits and parity bits the sequence s_e is generated and the described code-aided synchronization process can be carried out. The LLR are converted into symbols according to the Eq. (6):

$$s_e(l) = \tanh(\frac{\tau^x(2l)}{2}) + j\tanh(\frac{\tau^x(2l+1)}{2}) \qquad l = 0, 1, \ldots, L-1 \tag{6}$$

Here τ^x respresents the sequence of τ^s, τ^{p1} and τ^{p2}, which corresponds to the transmitted symbols sequence r. The received sequence r is corrected with the new estimates of frequency and phase offset. The symbols correction is performed according to the Eq. (7):

$$r(l) = r(l) \cdot e^{-j(2\pi \tilde{f}_o l + \tilde{\Phi})} \qquad l = 0, 1, \ldots, L-1 \tag{7}$$

In this way a new synchronized received sequence r^* is calculated after each full decoder iteration. The above presented code-aided synchronization algorithm is very similar to that presented in Luise and Reggiannini (1995) and Wasenmüller et al. (2010).

4 Communication performance simulations

In this section, we demonstrate the communication performance of our system. The simulations were carried out with bit true models of the hardware units to take into account the quantization losses. It is assumed that ideal timing and and frame synchronization are done in advance. Furthermore a coarse carrier synchronization is assumed, which resolves phase ambiguities and limits the residual phase and frequency offset. In current research only QPSK modulation scheme was adopted and no pilot symbols were used in the bursts. In all experiments, we used an additive white Gaussian noise (AWGN) channel. To demonstrate the performance of our CA-synchronization algorithm in challenging conditions, we chose a constant value of frequency offset. That results in linearly increasing/decreasing phase offset on

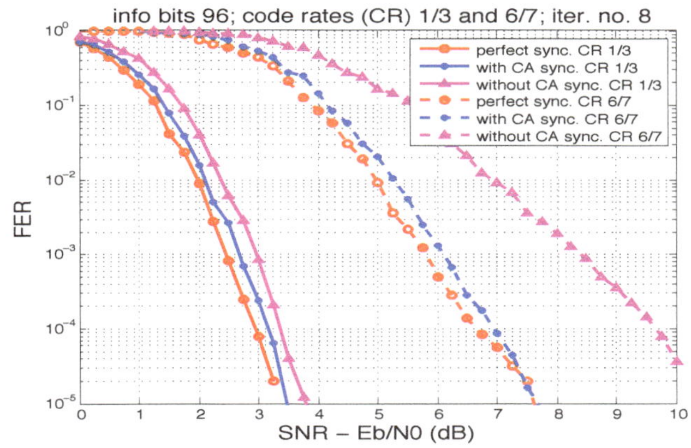

Fig. 2. Performance comparison, with and without CA Sync., short block length, in presence of residual phase and frequency offset.

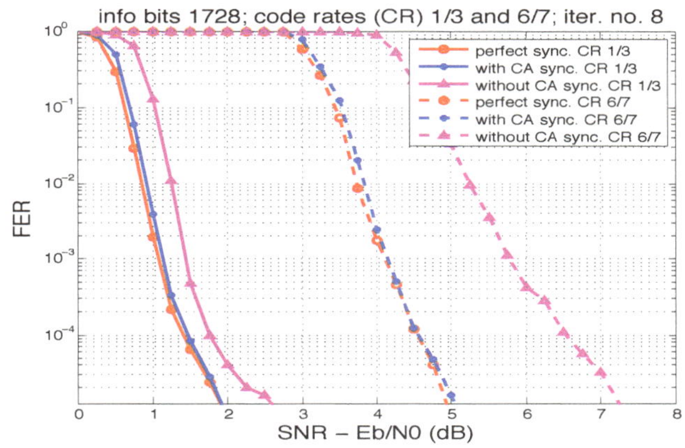

Fig. 3. Performance comparison, with and without CA Sync., long block length, in presence of residual phase and frequency offset.

each symbols of a block. So we calculated the constant frequency offset value for each block depending on its length. The first symbol of the block gets a rotation of $-22.5°$ then the offset linearly increases for each symbol in a block and the middle symbol gets $0°$ phase offset and the last symbol gets a phase offset of $22.5°$. If the input symbols are rotated with an offset of $\pm 45°$ with QPSK modulation then the decoder can not decode the data that's why in all experiments we kept the offset within the limit ($\pm 45°$). To get each result 250 000 blocks were simulated.

We used an 8-state duo-binary turbo decoder in our simulations having Max-Log-Map with ESF (extrinsic scaling factor) algorithm having value 0.75, sliding window with one trellis stage, next iteration initialization (NII) acquisition, the window length equal to the information bits, 8 iterations, 6 bit quantization for the input LLR and 7 bit for the extrinsic LLR.

We present frame error rate (FER) graphs to show the impact on communication performance of different parameters: frequency offset, phase offset, block length and code rate. Moreover, the effect on communication performance by running the synchronization in parallel to the decoder is also shown.

4.1 Phase and frequency offsets impact

Figures 2 and 3 show the performance of the decoder in presence of residual phase and frequency offsets. FER curves with ideal synchronization i.e. zero phase and frequency offset, with and without the CA synchronization are also shown in the figures. It is evident that the residual phase and frequency offsets degrade the decoder performance but the CA synchronization significantly improves the performance, i.e. the communication performance is close to the ideal synchronization by doing CA synchronization.

4.2 Block length and code rate variations

We carried out simulations with different block lenghts and code rates in presence of residual phase and frequency offsets. In this paper, we present the communication performance of the CA synchronization with the shortest and the longest block lengths and, the lowest and the highest code rates of DVB-RCS standard. The Figs. 2 and 3 show the results with 96 and 1728 bits block lengths and, 1/3 and 6/7 code rates. The gain in communication performance due to CA synchronization depends upon code rate, block length, and residual phase and frequency offset. The figures show that at FER 10^{-3} the performance gain is almost 0.25 dB with code rate 1/3 and almost 1.6–2.0 dB with code rate 6/7 depending upon the block length.

4.3 Synchronization running in parallel impact

If the synchronization and the decoder function in serial, i.e. the synchronizer waits for completion of the decoder iteration, then latency of the system is increased. As a consequence, the system throughput is heavily degraded. On the other hand, if the synchronization and the decoder function in parallel, i.e. both start their functions at the same time, but the synchronization unit uses previous iteration results of the decoder, then in this case there is no effect on the system throughput but the communication performance gain of the CA synchronization is degraded slightly. We show communication performance graphs of the both situations in Fig. 4 which shows that we get a penalty in communication performance when the code-aided synchronization is running in parallel to the decoder: 0.1 dB at iteration number 4 and 0.05 dB at iteration number 8. The insignificant loss at higher number of the decoder iterations instigates us to design an architecture running in parallel to the decoder.

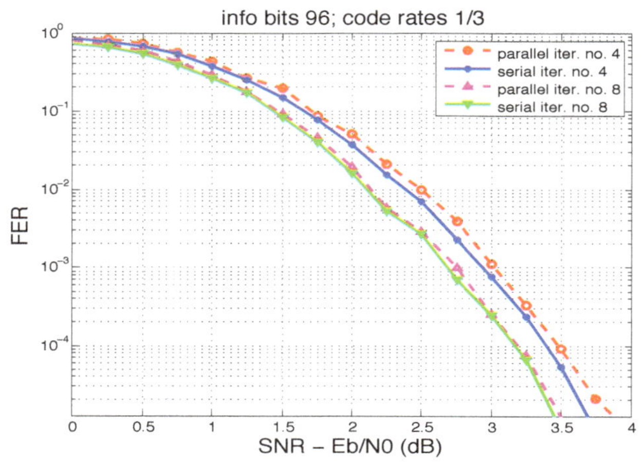

Fig. 4. Performance comparison, CA Sync. running in serial and parallel with the decoder.

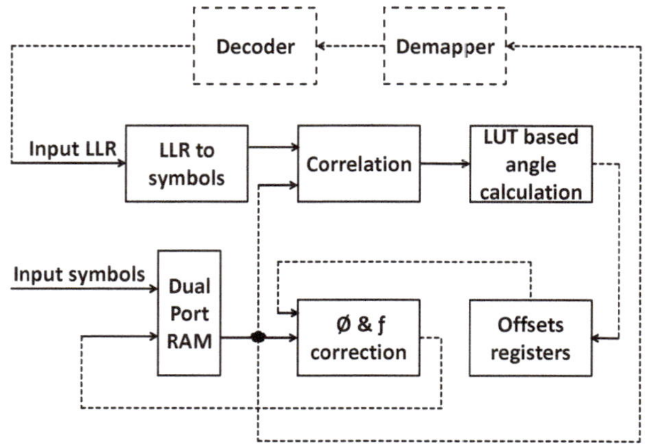

Fig. 5. Architecture of the code-aided synchronization.

5 Architecture and its implementation

The architecture of the system is shown in Fig. 5. We use synthesizable VHDL for modeling and implementing the architecture. The modules shown with dotted box in the figure, demapper and decoder, are not part of the architecture and are shown here just for architecture operation explanation. We use Lookup Tables (LUTs) to store the values of tanh, sin and cos which are required to convert the input LLR into symbols and for the symbols correction. We use STMicroelectronics ROMs for implementing all the LUTs and a DPR (dual port ram) 3072×20 to store the symbols. Two complex MAC (Multiply-Accumulate) units are implemented to perform the correlation and correction functions in parallel. Second term of Eq. (5) is also mapped on the correlation MAC unit to save an additional multiplier. Moreover a divider is implemented to get the frequency offset as described in Eq. (4).

5.1 LUT based phase calculation

We use a LUT for phase calculation (Eq. (4) and Eq. (5)) instead of using a full CORDIC (Coordinate Rotation Digital Computer). The address of the LUT is generated by the correlation function. However, the output of the correlation function is a complex number having large bitwidth. So it is required to reduce the bitwidth of the number in order to reduce the LUT size. Instead of directly truncating LSBs of a complex number which can result a number with less information carrying bits, we do scaling in a smart way. So first we truncate all possible no information carrying MSBs and then the complex number is shortened and rounded to 10 bits.

5.2 Operation

After one iteration of the turbo decoder, LLRs of the systematic and parity bits are available at the output of the decoder. Two LLR values are transferred to the "LLR to symbols" module in one clock cycle to convert them into a symbol for QPSK modulation for implementation of Eq. (6). The symbols received by the AWGN channel are stored into the DPR. One symbol from the DPR and the soft symbol at the output of the "LLR to symbols" module are transferred to the "Correlation" module. One symbol is processed in one clock cycle and in this way the whole block is processed.

The "Correlation" module implements Eq. (3) and its result is transferred to the "LUT based angle calculation" module for computation of angles (see argument function in Eq. (4) and Eq. (5)). The "Offsets registers" module saves the results of Eq. (4) and Eq. (5) for current iteration. The "Ø & f correction" module implements Eq. (7).

5.3 Synchronization running in parallel to decoder

To run the synchronizer in parallel to the decoder, the phase & frequency (Ø & f) correction module remains idle during the first decoder iteration. When the decoder executes its second iteration and the LLR are available at its output and then the correlation and correction modules start their operations in parallel. A symbol is read from the DPR and it is transferred to both the correlation and correction modules, in the same clock cycle. In this way we save a lot of accesses to the DPR. So the correlation is performed on the decoder $(n+1)$-th iteration output while the correction is performed on the decoder n-th iteration output. One symbol is corrected in one clock cycle and it is saved into the DPR again. So the DPR is used to save the corrected symbols also. After correction of a complete block, it is transferred to the demapper for next iteration of the decoder.

5.4 Implementation results

The code-aided synchronization IP Core is implemented on a 65 nm low power CMOS library. We considered the

Table 1. Post-layout area results.

Component	Case	Max. clock freq.	Area
DPR	Worst	300 MHz	$0.089\ mm^2$
Logic	Worst	300 MHz	$0.041\ mm^2$
ROMs (LUTs)	Worst	300 MHz	$0.064\ mm^2$
Sum			$0.194\ mm^2$

following process, voltage and temperature (PVT) parameters: Worst Case (WC, 1.1V, 125 °C), Nominal Case (NOM, 1.2V, 25 °C) and Best Case (BC, 1.3V, −40 °C). Synthesis was performed with Synopsys Design Compiler in topographical mode, placement and routing (P&R) with Synopses IC Compiler. The DPR can operate at maximum 300 MHz clock frequency so we synthesized the core at this frequency. Table 1 shows the results for area and the final physical design is shown in Fig. 6. We calculated the power consumption of the physical design with Synopsys PrimeTime-PX. The design consumes 51.0, 41.4 and 35.4 mW at best, nominal and worst case, respectively, at 300 MHz clock frequency.

The synthesis results show that we save a significant area by using a LUT for phase calculation. It occupies only $0.0186\ mm^2$ area while a complete CORDIC implemented on a 65 nm ASIC takes $0.86\ mm^2$ area Qi et al. (2010). However the CORDIC can also perform several other functions like conversions between polar and rectangular coordinates.

5.5 Flexibility

We analyzed requirements for utilizing the IP core for different satellite communication standards like DVB-RCS, DVB-SH, and GMR-1 3G and to support these standards we provided flexibility in the architecture at: (i) run time – block length and modulation scheme (ii) design time – bitwidths of the input symbols, the decoder LLR, and the intermediate results, and maximum number of iterations.

5.6 Throughput

The architecture has a throughput of 207 Msamples/sec for one iteration. This calculation includes both systematic and parity bits after depuncturing a block. This throughput is sufficient to meet various satellite communication standards requirements.

6 Conclusions

The code-aided synchronization improves the communication performance of the iterative decoders with a gain of 0.25 to 2.0 dB at FER 10^{-3} depending on block length, code rate, and residual phase and frequency offset. In this paper, we presented, to the best of our knowledge, the first published IP core for the code-aided synchronization. The core functions in parallel to the decoder so it does not degrade its throughput.

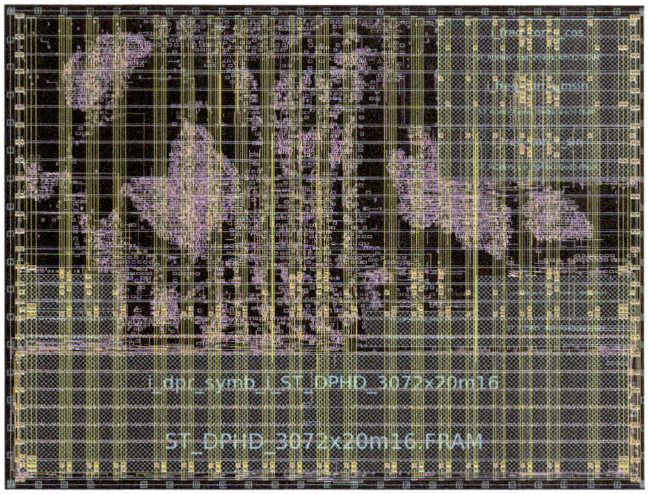

Fig. 6. Physical design including the DPR.

References

Berrou, C., Glavieux, A., and Thitimajshima, P.: Near Shannon Limit Error-Correcting Coding and Decoding: Turbo-Codes, in: Proc. 1993 International Conference on Communications (ICC '93), 1064–1070, Geneva, Switzerland, 1993.

Herzet, C., Noels, N., Lottici, V., Wymeersch, H., Luise, M., Moeneclaey, M., and Vandendorpe, L.: Code-Aided Turbo Synchronization, Proceedings of the IEEE, 95, 1255–1271, 2007.

Lottici, V. and Luise, M.: Embedding Carrier Phase Recovery Into Iterative Decoding of Turbo-Coded Linear Modulations, IEEE Transactions on Communications, 52, 661–669, 2004.

Luise, M. and Reggiannini, R.: Carrier frequency recovery in all-digital modems for burst-mode transmissions, Communications, IEEE Transactions on, 43, 1169–1178, 1995.

Mengali, U. and D'Andrea, A.: Synchronization Techniques for Digital Receivers, Plenum Publishing Corporation, New York, 1997.

Noels, N., Lottici, V., Dejonghe, A., Steendam, H., Moeneclaey, M., Luise, M., and Vandendorpe, L.: A Theoretical Framework for Soft-Information-Based Synchronization in Iterative (Turbo) Receivers, EURASIP J. Wireless Comm. and Networking, 2005, 117–129, 2005.

Qi, Z., Cabe, A., Jones, R., and Stan, M.: CORDIC implementation with parameterizable ASIC/SoC flow, in: IEEE SoutheastCon 2010 (SoutheastCon), Proceedings of the, 13–16, doi:10.1109/SECON.2010.5453930, 2010.

Godtmann, S., Hadaschik, N., Steinert, W., Pollok, A., Ascheid, G., and Meyr, H.: Coarse and Turbo Synchronization: A Case-Study for DVB-RCS, in: NEWCOM-ACoRN Workshop, Vienna, Austria, 2006.

Wasenmüller, U., Gimmler, C., and Wehn, N.: Low complexity Turbo synchronization without initial carrier synchronization, Advances in Radio Science, 8, 123–128, 2010.

A design study on complexity reduced multipath mitigation

U. Wasenmüller[1], T. Brack[1], I. Groh[2], E. Staudinger[3], S. Sand[3], and N. Wehn[1]

[1]Microelectronic Systems Design Research Group, University of Kaiserslautern, Germany
[2]Intel Mobile Communications, Germany
[3]Institute of Communications and Navigation, German Aerospace Center (DLR), Germany

Correspondence to: U. Wasenmüller (wasenmueller@eit.uni-kl.de)

Abstract. Global navigation satellite systems, e.g. the current GPS and the future European Galileo system, are frequently used in car navigation systems or smart phones to determine the position of a user. The calculation of the mobile position is based on the signal propagation times between the satellites and the mobile terminal. Further, the satellites need to be line-of-sight to the receiver for exact position calculation. However, in an urban area, the direct path may be blocked and the resulting multipath propagation causes errors in the order of tens of meters for each measurement.

In this paper an advanced algorithm for multipath mitigation known as CRMM is presented. CRMM features reduced algorithmic complexity and superior performance in comparison with other state of the art multipath mitigation algorithms. Simulation results demonstrate the significant improvements in position calculation in environments with severe multipath propagation. Nevertheless, in relation to traditional algorithms an increased effort is required for real-time signal processing due to the large amount of data, which has to be processed in parallel. Based on CRMM, we performed a comprehensive design study including a design space exploration for the tracking unit hardware part, and prototype implementation for hardware complexity estimation.

1 Introduction

Currently, personal navigation devices (PNDs) and smartphones are widely used for car and pedestrian navigation or location based services (LBS). These devices employ mass market global navigation satellite systems (GNSSs) receivers, e.g. for the current GPS and GLONASS or the future European Galileo system. A GNSS receiver determines its position uniquely by measuring and calculating the signal propagation time from at least four different satellite signals. State-of-the-art mass market GNSS receivers measure and track the time of arrival (TOA) of the satellite signals with narrow early-minus-late (NEML) delay-locked loops (DLLs). NEML DLLs result in the optimum TOA estimates when there is a direct line-of-sight (LOS) between the satellite and GNSS receiver and the signal reception is not disturbed by multipath propagation.

Often, LBS are used in cities where the satellite signals are subject to diffraction, refraction, and scattering resulting in multipath propagation. Thus, the positioning accuracy of state-of-the art mass market GNSS receivers with NEML DLLs will be degraded from a few meters to several tens of meters. In van Nee et al. (1994), maximum likelihood (ML) multipath estimation (ME) algorithms can mitigate these errors but are computationally too expensive for mass market GNSS receivers in terms of computing power and power consumption. Consequently, Selva (2004, 2005) developed complexity reduced multipath mitigation (CRMM) algorithms that employ a bank of correlators to reduce the amount of data that needs to be handled by the MLME. Several publications on CRMM discuss various implementation aspects of the algorithm, e.g. Groh and Sand (2008): Groh et al. (2011); Groh and Sand (2012). However, none provides a more detailed analysis with respect to real-time or hardware implementation issues.

In Sect. 2 we show the used system model and in Sect. 3 the considered CRMM algorithm is summarized. In Sect.4 the design study including design space exploration for the tracking unit hardware part and prototype implementation for hardware complexity estimation is presented, and in Sect. 5 the results are summarized.

2 System model and MLME

In this section, we describe briefly the underlying system model of transmitting and receiving a spread spectrum navigation signal and the MLME problem. We consider a rectangular pulse $g(t)$ with cut-off frequency f_N. The direct

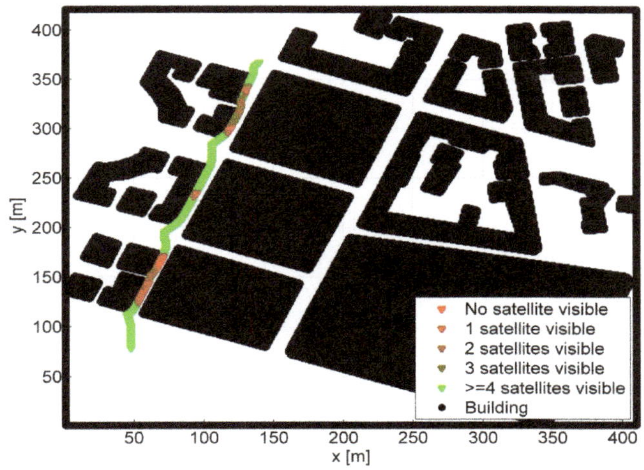

Fig. 1. GNSS satellite visibility in a Munich urban environment.

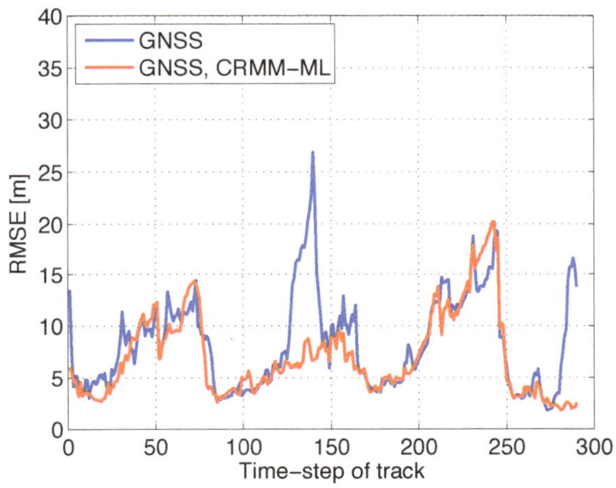

Fig. 2. RMSE vs. time-step of track for GNSS with NEML and CRMM-ML.

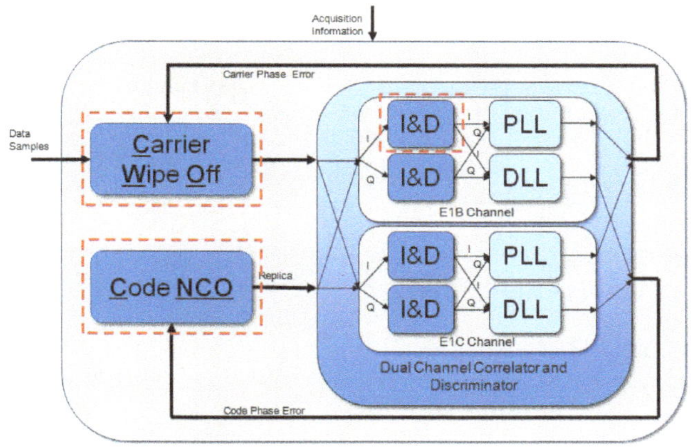

Fig. 3. Arcitecture of Tracking unit.

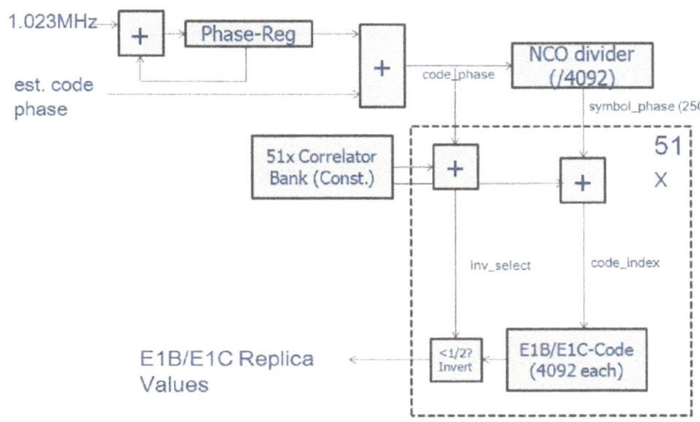

Fig. 4. Arcitecture of Code NCO and PRN unit.

sequence spread spectrum transmit signal is then

$$s(t) = \sum_{m=0}^{M-1} d_{\mathrm{P},m} \sum_{n=0}^{N-1} c_{\mathrm{P},n} g(t - mT - nT_{\mathrm{C}}),$$
$$+ \sum_{m=0}^{M-1} d_{\mathrm{D},m} \sum_{n=0}^{N-1} c_{\mathrm{D},n} g(t - mT - nT_{\mathrm{C}}) \qquad (1)$$

where $T = NT_{\mathrm{C}}$ and $T_{\mathrm{C}} = 1/R_{\mathrm{C}}$ denote the spreading code duration and chip duration with R_{C} being the chip rate. The pilot and data symbol sequences $d_{\mathrm{P},m}$ and $d_{\mathrm{D},m}$, $m = 0, \ldots, M-1$ are taken from quartenary phase shift keying modulation. M codewords form the considered time interval for MLME. The spreading code sequences $c_{\mathrm{P},n}$ and $c_{\mathrm{D},n}, n = 0, \ldots, N-1$ for the pilots and data are pseudo-noise sequences of length N as defined in EUG (2010). In order to obtain a suitable matrix vector factorization beginning from Eq. (1), we choose the sampling frequency f_{S} as integer multiple of the chip rate R_{C} yielding $Q = f_{\mathrm{S}}/R_{\mathrm{C}}$ samples per chip.

In order to demonstrate the influence of the multipath propagation on the estimation performance of the LOS path and thereby on the position estimation of the mobile station, the channel is modelled as a fixed channel of length L

$$h(\tau, t) = h(\tau) = \sum_{\ell=1}^{L} a_\ell \delta(\tau - \tau_\ell). \qquad (2)$$

where the number of channel coefficients L is assumed to be known to the receiver. a_ℓ and τ_ℓ are the amplitude and delay of the ℓ-th multipath component. Note that $l = 1$ corresponds to the LOS component. The TOA τ_1 is related to the distance between the i-th satellite and the receiver as $d_i = c\tau_1$ with c being the speed of light.

After transmission over the channel (cf. Eq. 2), we obtain the complex valued baseband-equivalent received signal as

$$y(t) = \sum_{l=1}^{L} a_l s(t - \tau_l) + n(t), \qquad (3)$$

where $s(t)$ is the navigation signal transmitted by the satellite and $n(t) \sim \mathcal{N}_{\mathrm{c}}(0, \sigma_n^2)$ describes the zero-mean additive white

Table 1. Detailed assumptions and platform properties.

Description	Parameter
Type of FPGA	Stratix II EP2S180F1020C3
ALUTs	143 520
Dedicated Logic Registers	143 520
DSP Block Resources	96
Block Memories	930 M512, 768 M4K
Processing clock speed (fclk)	100 MHz
Sample Clock (fs)	16.8 Msps (Complex Data)
ADC Output Quantisation	3 bit

Table 2. CWO complexity estimation.

Element	Parameter	Nr.	ALUT	DREG
SIN/COS-LUT	8×6 bit	1	6	6
Multiplier	3×3 bit	2	12	6
Phase Accumulator	32 bit	1	0	32
Phase Adder	32 bit	1	32	0
Phase Error Adder	32 bit	1	32	32
Overall			82	76

Table 3. CWO synthesis results.

Synthesis Alternative	ALUT	DREG	DSP
Default Options	83	51	0
Forced to use DSP	69	39	2

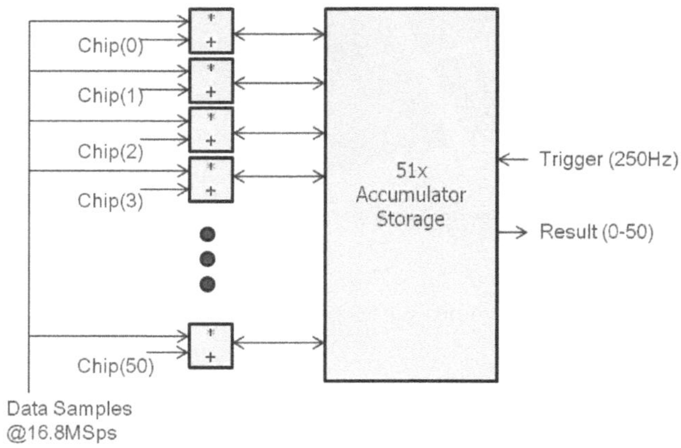

Fig. 5. Integrate and Dump unit.

Gaussian noise (AWGN) of the power $\sigma_n^2 = E\left[|n(t)^2|\right]$. Sampling $y(t)$, we obtain the receive signal vector

$$\boldsymbol{y} = \sum_{\ell=1}^{L} a_\ell \boldsymbol{s}(\tau_\ell) + \boldsymbol{n} = \mathbf{S}(\boldsymbol{\tau})\boldsymbol{a} + \boldsymbol{n} \in \mathbb{C}^{MNQ}, \tag{4}$$

where $\mathbf{S}(\boldsymbol{\tau}) = [\boldsymbol{s}(\tau_1), \ldots, \boldsymbol{s}(\tau_L)] \in \mathbb{C}^{MNQ\times L}$ and $\boldsymbol{a} = [a_1, \ldots, a_L]^{\mathrm{T}} \in \mathbb{C}^L$ form the signal matrix and the amplitude vector. The SNR γ is defined according to Eq. (4) as $\gamma = \|\boldsymbol{S}(\boldsymbol{\tau})\boldsymbol{a}\|_2^2 / \left(MNQ\sigma_n^2\right)$.

The MLME $\{\hat{\mathbf{a}}, \hat{\tau}\} \in \mathbb{C}^L$ is given according to Lentmaier and Krach (2006) by

$$\{\hat{\boldsymbol{a}}, \hat{\boldsymbol{\tau}}\} = \underset{\{\boldsymbol{a}, \boldsymbol{\tau}\}}{\operatorname{argmin}} \|\boldsymbol{y} - \mathbf{S}(\boldsymbol{\tau})\boldsymbol{a}\|_2^2, \tag{5}$$

where the first element $\hat{\tau}_1$ in $\hat{\boldsymbol{\tau}}$ denotes the TOA ML estimate. The received signal vector $\boldsymbol{y}$ may contain $M = 10$ symbols and $N = 4092$ chips per symbol with $Q = 16$ samples per chip. Thus, the optimization algorithm for Eq. (5) has to process more than 650 000 samples received within 40 ms, which is prohibitive in terms of computational complexity and power consumption for a mass market GNSS receiver.

3 CRMM

To overcome the computational complexity of Eq. (5), Selva (2004, 2005) developed the basic CRMM algorithm, which implements the following two steps:

- *Data size reduction:* the large received signal vector $\boldsymbol{y}$ is transformed by a bank of correlators into a much smaller vector $\boldsymbol{y}_\mathrm{c}$ before the ML optimization. The subspace transform results in maximum data compression and negligible performance losses. For instance, the observation vector $\boldsymbol{y}$ can be reduced from hundred thousands of samples to $\boldsymbol{y}_\mathrm{c}$ with a few tens of samples Groh and Sand (2008). $\boldsymbol{y}$ can be compressed with code matched correlators (CMCs), signal matched correlators (SMCs) and principal components (PCs) or through combining either CMCs or SMCs with subsequent PCs (cf. Lentmaier and Krach, 2006; Groh and Sand, 2008, 2012). Note, in this paper we only focus on SMCs.
- *ML optimization:* efficient and robust Newton-type optimization algorithms were developed. These algorithms employ interpolation methods to allow arbitrary delay resolution independent of the sampling rate.

Modifications of the CRMM include extension of data size reduction to time-variant signals (Groh et al., 2011), optimized correlator computation (Groh and Sand, 2012), and replacement of the Newton-type optimization with expectation maximization or space alternating general expectation maximization algorithms (Groh and Sand, 2008).

In this paper, the focus is on the data size reduction of CRMM and its realizability in hardware in the subsequent sections. Whereas the correlator input $\boldsymbol{y}$ receives samples with rate f_S, the correlator can output the samples of $\boldsymbol{y}_\mathrm{c}$ with rates between 1000 Hz to 1Hz depending on M and N. Thus, clearly subsequent complexity reduction or ML optimization will be less critical for hardware implementation.

To asses the performance of MLME with CRMM, we consider the user track depicted in Fig. 1, which shows an urban

Table 4. NCO and PRN generator complexity estimation.

Element	Parameter	Nr.	ALUT	DREG
NCO-Divider	32 bit by 12 bit	1	48	48
Correlator Banks	51 × 32 bit	1	25	32
Phase Accumulator	32 bit	1	0	32
Phase Adder	32 bit	1	32	0
Phase Error Adder	32 bit	1	32	32
Correlator Adder (large)	51 × 32 bit	1	800	0
Correlator Adder (small)	51 × 12 bit	1	36	0
E1B/E1C Code Storage	4092 × 2 bit	3	480	0
Output Registers	2 × 1 bit	51	0	102
Overall			1453	246

Table 5. NCO and PRN generator synthesis results.

Synthesis Alternative	ALUT	DREG
Default Options	1172	213

canyon scenario in Munich with average building height of 26 m. For at least four LOS satellites, the track is colored green, which is the minimum number of LOS satellites to obtain a reliable position.

Figure 2 shows the corresponding root-mean-square error (RMSE) of the position estimate in meter versus the time-step of the track in Fig. 1. The RMSE was averaged over the same track with the same satellite constellation and urban scenario for 1000 noise realizations. Comparing Figs. 1 and 2, the main error source for the time-steps 20 to 85 and 180 to 255 is due to less than four available LOS satellite signals. On the other hand, the major error source at the start, for time-steps 125 to 145, and at the end is due to multipath propagation although there should be at least four LOS satellite signals available. At these time-steps, CRMM can reduce the RMSE absolute by 7.5 m to 20 m and relative by 200 % to 650 %.

4 Prototype architecture design exploration

In the GREAT project, we investigated in detail through simulations the performance of CRMM, which is superior to most simple correlators based multipath mitigation algorithms (Hu et al., 2008). CRMM requires a large number of correlators per satellite to transform the multipath mitigation into a lower dimensional sub-space. A first complexity analysis in GREAT showed that based on the floating point operations per second, CRMM has a comparable complexity with classical correlators based multi-path mitigation algorithms (Hu et al., 2008). From GREAT, the question remained open, whether it would be feasible to implement CRMM in hardware for instance in a COTS FPGA. The following study addresses this question in more detail with the requirement of 51 integrate and dump units.

Table 6. I&D unit complexity estimation.

Element	Parameter	Nr.	ALUT	DREG
Adder	23 bit	51	1173	0
Accumulator Storage	51 × 23 bit	2	0	2346
Overall			1173	2346

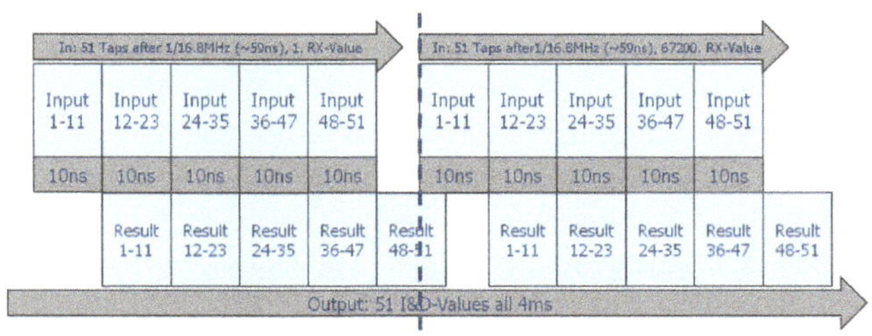

Fig. 6. Timing for alternative architecture of I&D unit.

4.1 Design drivers and top level architecture

Besides the acquisition unit, the tracking unit of GNSS receivers is by far the most complex receiver part which requires high-speed processing on dedicated hardware resources. Primary goal of the prototyping approach was to derive a preliminary architecture design for feasibility study and to acquire complexity estimation data for the tracking unit based on a given FPGA target platform. Detailed assumptions and platform properties are listed in Table 1.

The algorithms and internal quantisation used for the prototyping relies on the freely available GNSS dual-receiver MATLAB SIMULINK models and the corresponding "C"-sources. The top level architecture of the GNSS receiver tracking unit is shown in Fig. 3, consisting of the three major units "Carrier Wipe-Off" (CWO), "Code NCO and PRN

Table 7. I&D unit complexity estimation (curled-up).

Element	Parameter	Nr.	ALUT	DREG
Adder	23 bit	11	253	0
Accumulator Storage	51 × 23 bit	2	0	2346
Multiplexer	11:1	5 × 23	805	0
De-Multiplexer	11:1	5 × 23	805	0
Overall			1863	2346

generator", and the "Dual Channel Correlation and Discriminators".

Each architecture unit is considered separately in the next sections. Residual elements depicted in the top level architecture (Multipliers, Registers), glue logic, and the main FSM is considered in the final complexity estimation with an adequate lump-sum estimate.

4.2 Carrier Wipe-Off (CWO)

The CWO unit is responsible for removing the carrier of the ADC input signal by mixing this signal with a synthetically generated carrier using the estimated frequency and phase from the acquisition and tracking components. Based on the assumptions of a 3-bit ADC input at 16.8 Msps, a 6-bit output signal with the same rate has to be generated by this unit. The basic architecture derived from the available SIMULINK model uses a so called Direct Digital Synthesizer (DDS) to generate the required complex continuous wave signal, mixing is done by per-component multiplication of the complex input signal with the generated signal. The DDS uses a phase accumulator and a sine/cosine Lookup-Table (LUT) for signal generation: A 32-bit phase accumulator was employed to achieve a frequency resolution of 0.023 Hz which is adequate for the used algorithm, the LUT resolution was set to 3 bit to be compatible with the input resolution. In a first step, the hardware complexity of the individual architecture components was calculated to allow for plausibility checks after synthesis. Note that the control FSM for the unit was not considered in this calculation.

In a second step, the architecture was realized in VHDL and synthesized in the Quartus-II tool suite for the appropriate FPGA target. As an alternative, the synthesis tool was forced to apply DSP blocks for the design.

Both alternatives result in a very small core requiring $<$ 1 % of the full FPGA resources. To further reduce complexity, the architecture could be "curled up" to process only one sample every 5 clock cycles with the assumed clock speed of 100 MHz. This two-step approach with preliminary calculation based on architecture components followed by implementation and synthesis is also used for the remaining units.

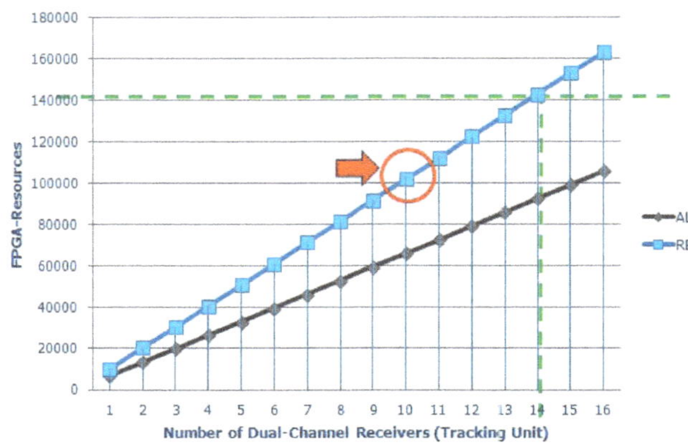

Fig. 7. Final results.

4.3 Code NCO and PRN generator unit

The Code NCO and PRN generator unit is utilized to generate a chip-sequence based on the given chip-rate for 1.023 MHz, the estimated code phase and the channel codes and correlators themselves. This unit accepts 32bit code-phase inputs in 4 ms intervals, yielding an input data rate of only 250 Hz. The outputs of the generator are 51bit code words for both E1B and E1C channels (so called replica values) at 16.8 Mcodes per second.

The architecture uses a DDS as introduced in the CWO unit to generate code phases at the chip rate of 1.023 MHz with an adjustable phase from the phase estimation. The frequency resolution is also set to 0.023 Hz, so a 32bit quantisation is derived for the phase input and accumulator.

The hardware complexity for the architecture building blocks are estimated and calculated accounting for the symmetry and actual values of the provided correlators and the resulting code index values. It is determined that instead of 51 different code index values, only three actually occur given these parameters. All these properties result in the very compact Correlators Bank, Correlation Adders, and Code Storage, cf. Table 4. The synthesis results for the VHDL implementation confirm the preliminary complexity results and even undercut the estimation, cf. Table 5.

The core resembling the Code NCO and PRN generator unit utilizes $<$ 1 % of the full FPGA resources. Same as for the CWO unit, the architecture could be "curled up" to exploit the difference between sample rate and processing speed and further reduce the complexity. It is noteworthy that for different correlators and codes, the resulting complexity could be more than 10 times higher.

4.4 Dual Channel Correlation and Discriminators unit

The Dual Channel Correlation and Discriminators unit consists mainly of two independent channel processors which in turn contain two Integrate and Dump units (I&D), a PLL, and a DLL each. The PLL and DLL are only triggered with a 4 ms (=250 Hz) interval which allows for either a highly

Table 8. I&D unit complexity estimation (RAM-based).

Element	Parameter	Nr.	ALUT	DREG	MEM
Adder	23 bit	11	253	0	0
Adder Registers	23 bit	11	0	253	0
Accumulator Storage	8×23 bit	11×2	0	0	2530 bit
Overall			253	253	2530 bit

Table 9. I&D unit synthesis results (4 units).

Synthesis Alternative	ALUT	DREG	M512	M4K
Default Options	4692	9384	0	0
Curled-Up	7452	9384	0	0
RAM-based (Auto-Fit)	1068	1068	4	60
RAM-based (Forced to use M512)	1068	1068	120	0

curled up realization or even the utilization of a general purpose processor like the NIOS-II. Using such a processor, up to 400 000 clock cycles are available to achieve one PLL or DLL result. Alternatively, multiple dual-channel GNSS receivers could share one PLL/DLL processor resource. For all these reasons, the prototyping approach only considers the four I&D units inside the Dual Channel Correlation and Discriminators unit. These units require dedicated hardware resources because of their relatively high input data rate of 16.8 MSps and large-volume data storage. Each I&D unit has to accept the 6bit receiver data coming from the CWO unit at 16.8 Msps (I or Q component) and the 51bit wide code words coming from the Code NCO and PRN generator unit at the same rate. The I&D unit basically multiply-accumulates the incoming data streams 67 200 times which is derived from the ratio between the 16.8 Msps sampling rate and the 250 symbol rate which also constitutes the output rate. Therefore, for each accumulator storage word 23bit are required ($6\text{bit} + \log 2(67\,200) > 22\text{bit}$). This task can be fulfilled by straight forward architecture shown in Fig. 5.

The complexity of the building blocks can be calculated from the two main building blocks of this approach, the adders and the accumulator storage itself (Table 6). To adapt for the very slow output data rate, a second "shadow storage" will be mandatory. This shadow storage keeps the last accumulation result until it can be transferred to the post-processing stages (PLL/DLL) and allows the continuous accumulation of newly arriving input data.

This first estimate (multiplied by 4 to get the result for the whole receiver) would already utilize $\approx 7\%$ of the complete FPGA. To reduce complexity, a "curled-up" architecture using the rate difference between processing clock (100 MHz) and data rate (16.8 Msps) was developed and analyzed. This architecture accepts 11 input samples per clock cycle instead of the 51. The timing for this "curled up" architecture is shown in Fig. 6.

Rough calculation of this alternative architecture (cf. Table 7) yielded even worse resource efficiency than the straight forward approach: the accumulator storage still consists of 2 large register banks, additionally Multiplexers and De-Multiplexers are now required to direct the 11 I/Os to the 51 storage locations.

Further investigations into the "curled-up" architecture were performed to reduce the required register resources (DREGs) and the introduced multiplexers and demultiplexers at the same time. To achieve this goal, the 51×23bit accumulator registers had to be mapped onto the dedicated RAM resources available in the Stratix-II FPGA: M512 (32×18bit) and M4K (128×36bit).

This approach requires only very limited additional logic for the adders; the multiplexers and demultiplexers are already available inside the memories addressing capability. For the accumulator storage, a memory depth of 8 was chosen to accommodate for the actually required depth of 5 words. The memories themselves were selected to be "simple dual-ported", therefore read and write accesses can occur at the same time, but not on the same word. All three alternatives have been implemented in VHDL and synthesized. For the RAM-based approach, even an option to force the synthesis to use only the M512 memory resources to increase efficiency in memory utilization was applied. Table 9 shows the results for the Dual Channel Correlation and Discriminators unit, therefore 4 I&D units were considered.

In conclusion, the Dual Channel Correlation and Discriminators unit is by far the most complex part identified in the GNSS tracking unit. Further optimizations could target an improved memory utilization by sharing memory resources at least over all 4 I&D units, even better with all I&D units if multiple receivers are deployed.

Table 10. Combined implementation results.

Unit	Alternative	ALUT	DREG
CWO	Default	82 (1 %)	76 (1 %)
Code NCO and PRN gen.	Default	1172 (18 %)	213 (2 %)
DCC and Discriminators	Default	4896 (74 %)	9384 (92 %)
Residue, Glue Logic, FSM	Estimate	500 (7 %)	500 (7 %)
Overall		6650	10 173

5 Prototype complexity estimation

The implementation results presented in detail in the previous section are now combined to yield the complexity estimation for the complete dual-channel receiver. As already explained before, residual elements, glue logic and the main FSM are introduced as a lump sum estimate only: 500 ALUTs and 500 DREGs are assumed to cover the required resources even for worst-case scenarios.

The combined implementation results for the stated alternatives (see Table 10 and Fig. 7) add up to 6650 ALUTs and 10173 DREGs which amounts to $\approx 4.6\%$ and $\approx 7.0\%$ respectively of one complete Stratix-II FPGA.

In conclusion, up to 14 dual-channel receivers could be theoretically realized on the given FPGA platform incorporating all assumptions and limitations made for this receiver prototyping approach. Because of timing and place and route (P&R) limitations, up to 10 full receivers should be feasible. By using the RAM-based approach for the realization of the Dual Channel Correlation and Discriminators unit, the results are not changed significantly: For one dual-channel receiver 7.8 % of the M4K or 12.9 % of the M512 resources of the given FPGA will be required and therefore become the bottleneck. However, if multiple receivers have to be deployed on the same FPGA, the different implementation styles can have massive effects on the scalability of the overall design, especially in terms of timing and P&R. The implementation complexity of the PLL/DLL as well as the acquisition block and all required internal and external interfaces should also be analyzed to further substantiate these findings.

References

EUG: European GNSS (Galileo) Open Service Signal In Space Interface Document (OS SIS ICD), European Union, issue 1.1, 2010.

Groh, I. and Sand, S.: Complexity Reduced Multipath Mitigation in GNSS with the Granada Bit-true Software Receiver, in: Proceedings of the IEEE/ION Position Location and Navigation Symposium (PLANS), Monterey, CA, USA, 2008.

Groh, I. and Sand, S.: Optimized Complexity Reduction for Maximum Likelihood Position Estimation in Spread Spectrum Navigation Receivers, to appear in IET Radar, Sonar & Navigation, 2012.

Groh, I., Staudinger, E., and Sand, S.: Time-Variant Maximum Likelihood Channel Estimation in Mobile Radio Navigation Systems, in: Proceedings of IEEE Vehicular Technology Conference Fall 2011, San Francisco, CA, USA, 2011.

Hu, X., Raasakka, J., Hurskainen, H., Lohan, E.-S., Groh, I., and Sand, S.: Final Report for WP1200, W2200, WP3200, GREAT Project, 2008.

Lentmaier, M. and Krach, B.: Maximum Likelihood Multipath Estimation in Comparison with Conventional Delay Locked Loops, in: Proceedings of ION GNSS, Fort Worth, Texas, USA, 2006.

Selva, J.: Efficient Multipath Mitigation Methods in Navigation Systems, Ph.D. thesis, Universitat Polytecnicia de Catalunya, Barcelona, Spain, 2004.

Selva, J.: An Efficient Newton-Type Method for the Computation of ML Estimators in a Uniform Linear Array, IEEE T. Signal Proces., 53, 2036–2045, 2005.

van Nee, R., Siereveld, J., Fenton, P., and Townsend, B.: The multipath estimating delay lock loop: approaching theoretical accuracy limits, in: Proceedings of the IEEE/ION Position Location and Navigation Symposium (PLANS), 246–251, 1994.

Application of syndrome based Turbo decoding with adaptive computational complexity in LTE downlink

J. Geldmacher[1], **K. Hueske**[1], **M. Kosakowski**[2], **and J. Götze**[1]

[1]TU Dortmund University, Information Processing Lab, Otto-Hahn-Str. 4, 44227 Dortmund, Germany
[2]Research In Motion GmbH, Universitätsstr. 140, 44799 Bochum, Germany

Correspondence to: J. Geldmacher (jan.geldmacher@ieee.org)

Abstract. This paper describes the application of an adaptive complexity decoder for the Long Term Evolution (LTE) downlink Turbo code. The proposed decoding approach is based on the block syndrome decoding principle and enables adaptive reduction of decoding effort depending on current SNR and iteration number with negligible influence on decoding performance. Numerical results in context of LTE downlink using typical mobile channels are used to demonstrate the efficiency of the approach.

1 Introduction

Mobile communication usage has shifted from mainly speech-centered use to highly bandwidth demanding data communication: A mobile phone is no longer just a *phone*, but it has become a personal communication and multimedia device, which is also used for internet browsing, video and audio streaming, online-based navigation etc. The resulting increasing bandwidth demands are satisfied in one part by using more sophisticated signal processing in the physical layer, like MIMO transmission or advanced forward error correction (FEC) methods.

Considering the Long Term Evolution (LTE), one of the most complex parts of these baseband processing components is the FEC, which is realised using Turbo coding (Berrou et al., 1993). Turbo Codes are used in several applications, because of their excellent BER performance, compared to classical codes like convolutional or block codes. In case of ideal uncorrelated Gaussian inputs and very large code block lengths, Turbo Codes can achieve close-to-capacity BERs. However, good performance has to be payed for with high computational complexity: Turbo codes are decoded iteratively by using constituent decoders, which exchange their beliefs about the transmitted information bits in the form of Log Likelihood Ratios (LLRs). Typically the LogMAP or the MaxLog algorithm (Robertson et al., 1995) are used in the constituent decoders to generate these LLRs by doing forward and backward recursions on a trellis representation of the code. Traditionally this trellis represents the encoder $\mathbf{G}$ of the code.

In this paper, a different decoding approach is considered which is based on the so called syndrome former $\mathbf{H}^T$ of the underlying code. While this decoder is equivalent to the conventional decoder in terms of trellis complexity and decoding performance, it can be modified to achieve adaptive decoding complexity based on the so called Block Syndrome Decoding (BSD) approach. The BSD concept has been described earlier in context of Viterbi decoding (Geldmacher et al., 2009) and Turbo equalization (Geldmacher et al., 2010; Hueske et al., 2010). Applying it to Turbo decoding is, however, not straightforward as will be shown in this paper, but requires the extension of the Turbo decoder by means of a pre-correction of the constituent decoders inputs.

In the following Sect. 2 the BSD concept is summarized. The syndrome based, adaptive complexity Turbo decoder is explained in Sect. 3 and numerical results for typical LTE scenarios are shown in Sect. 4. Conclusions are drawn in Sect. 5.

2 Block syndrome decoding

In syndrome decoding, instead of the trellis of the encoder, the trellis of the syndrome former $\mathbf{H}^T$ is used, where each path represents an admissible error sequence $\boldsymbol{\epsilon}$. Given the hard decision $\mathbf{r}$ of a received distorted sequence, an admissible error sequence $\boldsymbol{\epsilon}$ is defined as any sequence that results in a valid code sequence, when applied to $\mathbf{r}$. This trellis can be constructed based on the syndrome former $\mathbf{H}^T$ of the code, and the syndrome sequence $\mathbf{b} = \mathbf{r}\mathbf{H}^T$ of $\mathbf{r}$ (Forney Jr., 1970; Schalkwijk and Vinck, 1976). Consequently, in this case a MAP decoder estimates the LLRs of the error in code or information bits, and this approach is referred to as syndrome decoding. The resulting absolute LLR is identical for both approaches, and therefor the choice of the trellis has no influence on the decoding performance. Also trellis complexities of syndrome former and encoder are known to be identical in their minimal realizations (Forney Jr., 1970).

However, there is an interesting property of syndrome based decoding that can be exploited for reducing the decoding complexity. As the syndrome based decoder outputs LLRs of the error bits and not of the code bits, the probability of states and transitions depends on the number of errors in the decoders input sequence **r**: Less errors relate to a higher probability of the all-zero state in the trellis. This also means that if a subblock in the decoder's input sequence is error-free, then the zero-state is the most likely state in this subblock, and consequently that an estimation of error-free blocks can be interpreted as a state estimation.

Decoding complexity can then be reduced by identifying error-free subblocks before the decoding starts, and switching off the decoder for these subblocks. Only the remaining erroneous subblocks are actually processed by the decoder. In context of syndrome decoding, a very simple yet efficient criterion for estimating error-free subblocks is the syndrome **b** of the decoder's input sequence. Because of the orthogonality of the syndrome former $\mathbf{H}^T$ to the code, the syndrome only depends on the channel error $\boldsymbol{\epsilon}_c$,

$$\mathbf{b} = \mathbf{r}\mathbf{H}^T = (\mathbf{v}+\boldsymbol{\epsilon}_c)\mathbf{H}^T = \mathbf{u}\mathbf{G}\mathbf{H}^T + \boldsymbol{\epsilon}_c\mathbf{H}^T = \boldsymbol{\epsilon}_c\mathbf{H}^T, \quad (1)$$

so that subsequences of a certain number of consecutive zeros in **b** indicate error-free subblocks in **r** with certain probability. In (1) $\mathbf{v} = \mathbf{u}\mathbf{G}$ denotes the code sequence corresponding to the information sequence **u**.

In summary, the basic idea of BSD consists of two steps:

1. *Preprocessing* Compute the syndrome **b** of the decoders input sequence and identify subblocks of a predefined minimum number of consecutive zeros. These subblocks are considered to be error-free.

2. *Decoding* Apply the syndrome-based decoder only to the remaining erroneous subblocks.

The required minimum number of consecutive zeros is a design parameter, which is called $\ell_{\min}$ in the following. It influences complexity reduction and loss of decoding performance: A smaller $\ell_{\min}$ will result in the detection of more error-free subblocks, i.e. more reduction of decoding complexity, but also has a higher chance of identifying a subblock as error-free which actually contains errors, i.e. higher degradation of decoding performance.

It is easy to see, that the BSD concept results in a decoding complexity that adapts to the number of errors contained in the decoders input sequence. In the worst case, if the input sequence contains many errors, no error-free subblocks are identified, and the decoder has to process the whole sequence. In this case the effort is equal to a conventional decoder plus the preprocessing overhead. However, if the decoder's input contains only few errors, more error-free subblocks can be identified and the decoding effort is reduced.

Considering a Turbo decoder, it is expected that the number of remaining errors in the a priori estimate of the systematic part decreases during Turbo iterations. Additionally, if an estimate of the parity part is employed to correct errors in the parity part, then the overall number of errors in the constituent decoder's input decreases with ongoing iterations and increasing SNR. Thus, employing the BSD concept yields complexity reduction with increasing SNR and ongoing iterations.

The necessity of correcting the constituent decoder's input sequence requires a modified transition metric, which accounts for the precorrection. Therefore the syndrome decoding approach for Turbo Codes described in this work differs from other syndrome based decoding algorithms (Schalkwijk and Vinck, 1976; Yamada et al., 1983; Tajima et al., 2002; Minowa, 2003; Geldmacher et al., 2010).

3 Adaptive complexity Turbo decoder

First the operation of the syndrome based constituent decoder is described (Sect. 3.1), then the Turbo decoding framework with precorrection and the BSD extension are explained (Sect. 3.2 and Sect. 3.3).

3.1 Syndrome based Turbo decoder

It is assumed, that a half rate systematic terminated encoder is employed, and that on receiver side depuncturing is realized by replacing punctured positions with zero reliability bits. One constituent decoder takes the received soft decision sequence $\tilde{\mathbf{r}}$, a precorrection sequence **x** and a priori LLRs of the errors $L_A(\boldsymbol{\epsilon})$ in the systematic part as input. For the systematic part, it generates an *a posteriori* hard decision estimate $\hat{\boldsymbol{\epsilon}}$ of the error in the harddecision **r** of $\tilde{\mathbf{r}}$ and extrinsic LLRs $L_E(\boldsymbol{\epsilon})$. The decoder performs the following steps:

1. *Precorrection and syndrome computation* The precorrection sequence **x** is applied to **r**, and the syndrome **b** is computed as

$$\mathbf{b} = (\mathbf{r} \oplus \mathbf{x})\mathbf{H}^T. \quad (2)$$

2. *Trellis operation* Given the trellis of $\mathbf{H}^T$ subject to **b**, the MaxLog algorithm is applied to generate an *a posteriori* estimate $L(\boldsymbol{\epsilon})$. The intial state for the forward recursion is set to the zero state, while the intial state for the backward recursion is selected according to the final state of the syndrome former from (2). The transition metric at time instant t is given as

$$\tilde{x}_t^s \tilde{e}_{(p,q)}^s \left(|\tilde{r}_t^s| - L_A(\epsilon_t^s)/2\right) + \tilde{x}_t^p \tilde{e}_{(p,q)}^p |\tilde{r}_t^p|, \quad (3)$$

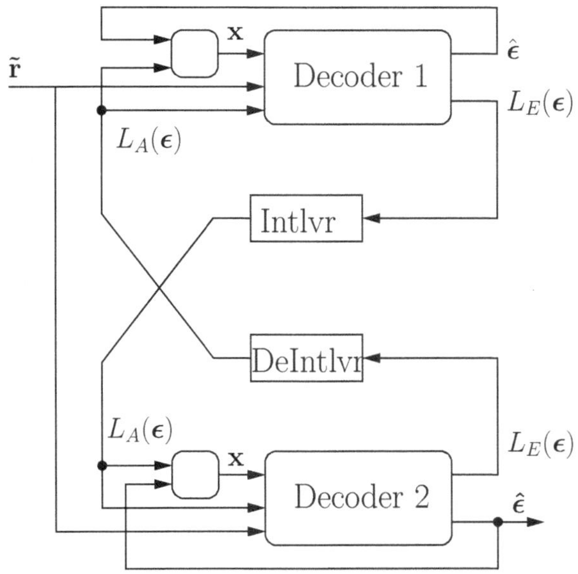

Fig. 1. Syndrome based Turbo decoder with precorrection.

where $\tilde{x}_t^s$, $\tilde{x}_t^p$ and $\tilde{e}_{(p,q)}^s$, $\tilde{e}_{(p,q)}^p$ represent BPSK modulated systematic and parity bits of precorrection sequence and error correponding to the trellis transition from states p to q. This yields the a posteriori LLR $L(\mathbf{e})$ of the incremental error $\mathbf{e}$ w.r.t. $(\mathbf{r}\oplus\mathbf{x})$. Note that for LogMAP decoding proper normalization using the channel reliability would be required.

In order to generate the LLR $L(\epsilon)$ of the absolute error in $\mathbf{r}$, the sign of $L(\mathbf{e})$, which is an estimate of the error in $\mathbf{r}\oplus\mathbf{x}$, has to be flipped according to $\mathbf{x}$,

$$L(\epsilon_t^s) = -\tilde{x}_t^s L(e_t^s). \quad (4)$$

The harddecision $\hat{\epsilon}$ of $L(\epsilon)$ delivers the *a posteriori* estimate of the channel error ϵ_c. Applying $\hat{\epsilon}$ to the systematic part $\mathbf{r}^s$ of $\mathbf{r}$ results in the estimate $\hat{\mathbf{u}}$ of the original information bits $\mathbf{u}$.

3. *Generation of extrinsic LLR* The extrinsic LLRs $L_E(\epsilon)$, that are passed to the other constituent decoder, are generated by removing the a priori information $L_A(\epsilon)$ and the received systematic softbits from $L(\epsilon_t^s)$,

$$\begin{aligned} L_E(\epsilon_t^s) &= L(e_t^s) - (-\tilde{x}_t^s)\left(\tilde{x}_t^s|\tilde{r}_t^s| - \tilde{x}_t^s L_A(\epsilon_t^s)\right) \\ &= L(\epsilon_t^s) + |\tilde{r}_t^s| - L_A(\epsilon_t^s). \end{aligned} \quad (5)$$

3.2 Turbo decoder with precorrection

An overview of the syndrome based Turbo decoder with two constituent decoders is shown in Fig. 1. Like in the conventional Turbo decoder, both constituent decoders use the systematic and their according parity part of the received softbits, and, as a priori values, the interleaved extrinsic informations $L_E(\epsilon_t^s)$ generated by the other decoder. Additionally, the precorrection sequence $\mathbf{x}$ is used.

Generally an arbitrary sequence can be selected for $\mathbf{x}$, without changing the absolute values of the generated LLRs. However, for the BSD concept it is crucial, that the input $(\mathbf{r}\oplus\mathbf{x})$ of the syndrome former contains as few errors as possible, such that the hamming weight of $\mathbf{b}$ becomes small. This can be achieved by selecting $\mathbf{x}$ as an estimate of the channel error ϵ_c, which in context of the Turbo decoding framework can be done as follows:

- The systematic part $\mathbf{x}^s$ is set according to the hard decision of the a priori values $L_A(\epsilon)$.
- For the parity part $\mathbf{x}^p$, the output $\hat{\mathbf{u}}$ from the previous iteration is reencoded, which yields an estimated code sequence $\hat{\mathbf{v}}$. The comparison of the parity part $\hat{\mathbf{v}}^p$ of $\hat{\mathbf{v}}$ with $\mathbf{r}$, yields $\mathbf{x}^p = \mathbf{r}^p \oplus \hat{\mathbf{v}}^p$.

Thus the sequence $\mathbf{x}$ depends on both constituent decoders, such that it can be used as a measure for the decoding progress. In case of convergence of the decoder, it leads to a decreasing hamming weight of the syndrome sequence $\mathbf{b}$.

The described syndrome based Turbo decoder is identical to the conventional decoder in terms of decoding performance and trellis complexity. Additional effort is required to compute the sequence $\mathbf{x}$ and the syndrome $\mathbf{b}$. However, both are binary operations and can be considered of negligible complexity.

3.3 Block syndrome decoding of Turbo codes

Based on the described syndrome based Turbo decoder, the BSD concept (Geldmacher et al., 2009, 2010) as described in Sect. 2 can be applied to achieve a reduction of decoding effort. Because of the precorrection, the syndrome sequence $\mathbf{b}$ of a constituent decoder shows subsequences of consecutive zeros, that increase with ongoing iterations in case of convergence. This can be exploited to separate the input sequence into subblocks that are considered to be erroneous and subblocks that are considered to be error-free. Consequently, a reduction of decoding effort can be achieved by only processing the erroneous subblocks and neglecting the supposely error-free subblocks.

More precisely the syndrome based Turbo decoding algorithm from Sect. 3.1 is extended as follows:

1. *Preprocessing of syndrome* Identify subsequences of $\geq \ell_{\min}$ zeros in $\mathbf{b}$. Consider the corresponding subblocks in $\tilde{\mathbf{r}}$, except a padding of $\lfloor \ell_{\min}/2 \rfloor$ at the beginning and end of each subblock, as error-free, and the remaining subblocks as erroneous.

2. *Processing of subblocks*

   (a) *Erroneous blocks* The erroneous subblocks are processed by the syndrome based MAP decoder, which generates extrinsic values $L_E(\epsilon_t^s)$ and the estimated error ϵ_t^s for all t in these subblocks. Note that these blocks can be considered to be terminated in the zero state, because the zero state is the most likely state in the preceding and succeeding error-free blocks.

   (b) *Error-free blocks* No processing is required for the supposedly error-free blocks. Instead, the extrinsic LLR is set to a sufficiently large value, $L_E(\epsilon_t^s) = \tilde{x}_t^s c$, with $c > 0$, and the estimated error is set to the systematic part of the pre-correction sequence, $\epsilon_t^s = x_t^s$. A reasonable choice for c is the largest value in the quantization range of the LLR values.

The choice of the design parameter $\ell_{\min}$ affects the achievable reduction of decoding effort and possible loss in decoding performance. Given an acceptable loss in block error rate (BLER), it may be selected heuristically.

4 Simulation results

To show the efficiency of the proposed BSD Turbo decoder, it is analyzed in context of an LTE downlink scenario (Mehlführer et al., 2009). For the simulations presented in this section, the modulation and coding scheme (MCS) is set according to the Channel Quality Indicator (CQI). A single-user, single antenna scenario is considered, where the carrier frequency is set to $f_C = 2.1$ GHz and Doppler frequency is $f_{D,\max} = 0$ Hz. For all simulations independent channel realizations are used. The bandwidth is set to 5 MHz for all simulations, and no control channel information is transmitted by the eNodeB, meaning all resources are used for payload data only. The decoder employs the Max Log MAP algorithm and is set to a maximum iteration number of $i_{\max} = 8$. CRC24 is employed for High SNR Early Termination (ET).

The computational effort will be measured as *equivalent iterations* of the conventional decoder until termination. Thus for the conventional decoder this measure equals the average number of full iterations until ET. For the BSD approach, this corresponds to the averaged number of iterations, where each iteration is weighted with the percentage of erroneous blocks. For example, if for the BSD approach the decoder would terminate after 3 full iterations, but each constituent decoder would have only processed 50 % of the code block during each full iteration, then this would correspond to 1.5 equivalent iterations.

There are two factors which have major influence on the performance of the BSD approach:

– *Code rate* The code rate determines how many positions are punctured in the transmitted block. The higher the code rate, the more punctured positions and consequently the more depunctured, zero-reliability bits are input to the decoder. However, the reliability of the syndrome-based criterion for identification of error-free blocks depends on a good estimate of the parity sequence, because of the involved precorrection. Thus, if there are more zero-reliability parity positions this criterion is less reliable and a larger value has to be selected for $\ell_{\min}$ to avoid a loss of BLER.

– *Channel type* The type of the channel influences the convergence behavior of the Turbo decoder. Naturally the BSD approach is more effective if the decoder executes more iterations before termination, because in this case the BSD has more chances to identify error-free parts.

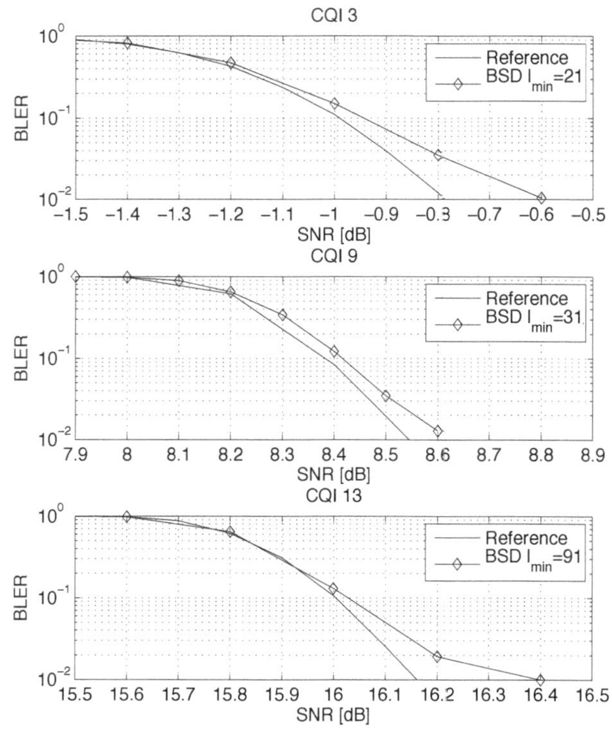

Fig. 2. Comparison of BLER for different CQI values (AWGN).

The following Sect. 4.1 compares BLER and equivalent iterations for different CQI values, i.e. different code rates. The influence of the transmission channel is analyzed in Sect. 4.2. A conventional Turbo decoder is used to provide reference results.

4.1 Code rates

To illustrate the performance of the BSD approach for different code rates, the CQI values 3, 9 and 13 are used under the assumption of AWGN transmission. The parameter $\ell_{\min}$ is adjusted such that the BLER degradation is smaller than

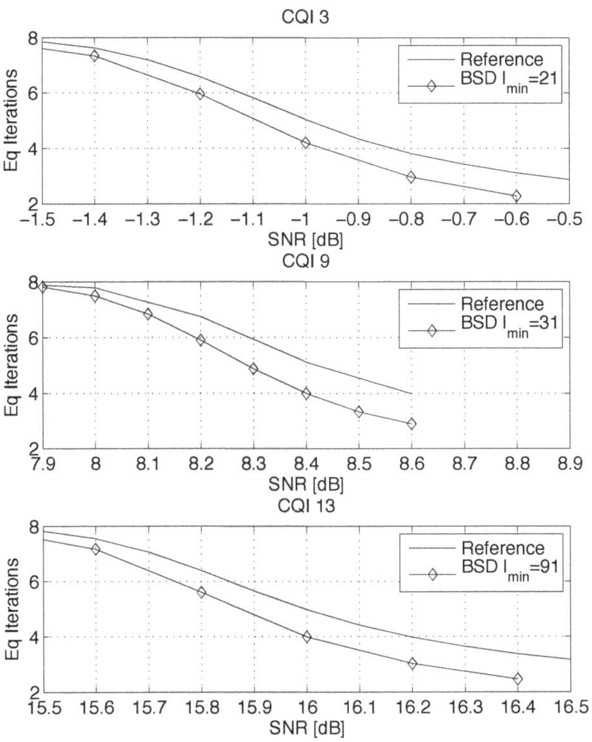

Fig. 3. Equivalent iterations for different CQI values (AWGN).

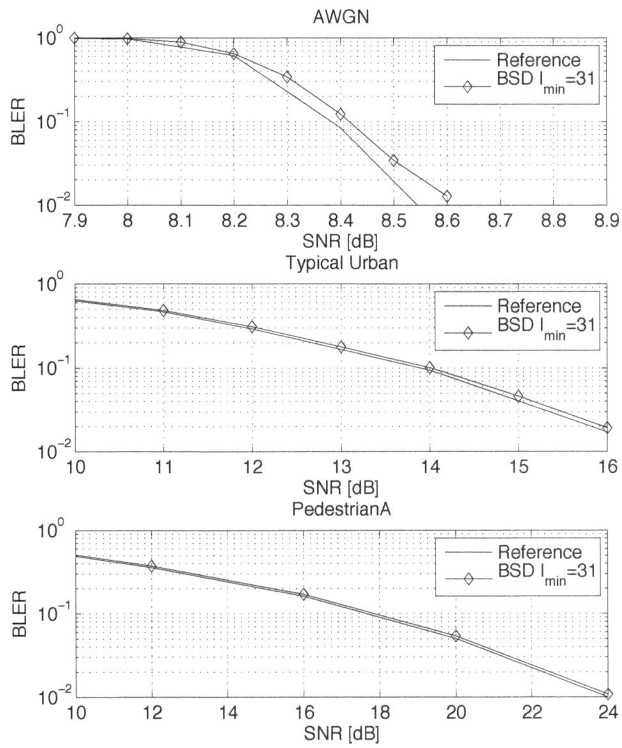

Fig. 4. Comparison of BLER for different channels (CQI9).

0.1 dB compared to the reference; it is set to $\ell_{\min} = 21$ for CQI3, $\ell_{\min} = 31$ for CQI9 and $\ell_{\min} = 91$ for CQI13. As already mentioned, a larger value is required for a higher CQI (higher code rate). The resulting BLER for the three cases is shown on Fig. 2. It can be noted, that there is a slight performance decrease, which however is still smaller than the required 0.1dB at a working point of 10 % BLER.

The equivalent number of iterations that are executed by the decoder are shown in Fig. 3. Again, the BSD approach is compared to the conventional Turbo decoder for the considered SNR range. For the conventional decoder the equivalent number of iterations only depends on the CRC24 based ET. For the BSD, it is additionally influenced by the amount of error-free subblocks that are identified during the iterations.

Naturally, the number of iterations that are required to decode a block decreases with increasing SNR for both decoders, because the quality of the received values improves. For the BSD decoder the number of equivalent iterations is always smaller than that of the conventional decoder. For example, considering a working point of 10 % BLER, the computational effort is reduced by about 0.7, 1.1 and 1 equivalent iterations at CQI3, CQI9 and CQI13. This corresponds to a reduction of decoding operations by about 20%.

4.2 Channels

In order to analyze the impact of different channel conditions, Figs. 4 and 5 compare the performance of BSD over AWGN channel to the performance over the fading channels Typical Urban (TU) and PedestrianA (PedA). The BSD is shown along with the conventional decoder for a fixed CQI of 9. The BLER is shown in Fig. 4 for the three cases. As in the previous simulation, the parameter $\ell_{\min}$ has been selected such that the impact on BLER performance is $< 0.1dB$ at the BLER working point of 10 %. For all channels the design parameter is set to $\ell_{\min} = 31$.

The reduction of decoding effort in terms of equivalent iterations is shown in Fig. 5. It can be noticed, that for the fading channels TU and PedA the average number of iterations of the conventional Turbo decoder is much smaller than in the AWGN case: While for the AWGN channel the decoder executes about 5.1 iterations in average at the working point of 10% BLER, it only executes about 2.2 and 1.9 iterations for the TU and PedA channel, respectively. This clearly affects the efficiency of the BSD approach, because if less iterations are executed until ET, there are less options for the BSD to identify error-free blocks, and thus less reduction of computational effort. Consequently the reduction in terms of equivalent iterations is about 0.5 and 0.2 for TU channel and PedA channel and thus smaller compared to the AWGN case.

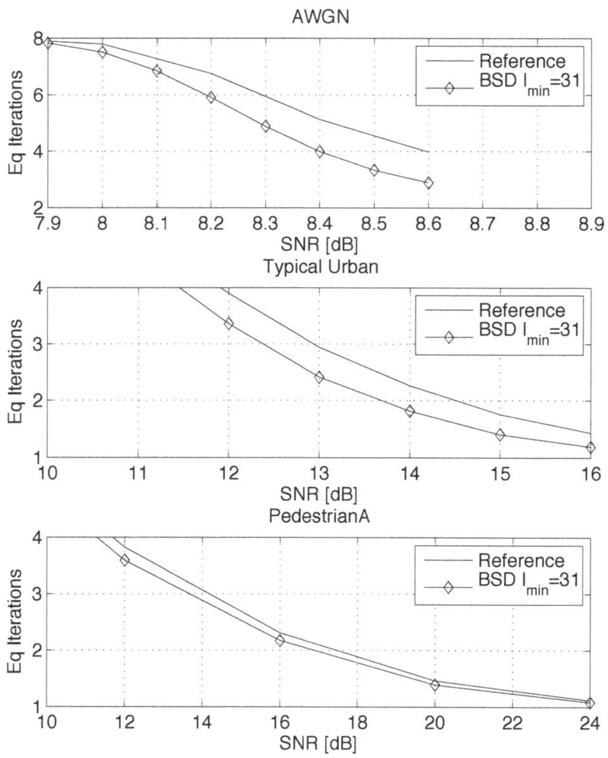

Fig. 5. Equivalent iterations for different channels (CQI9).

Table 1 summarizes the results presented in this section. For a BLER working point of 10 % and for the different CQIs and channel types, the equivalent iterations of conventional Turbo decoder and BSD are shown along with the used parameter $\ell_{\min}$. Additionally the last column shows the relative reduction of equivalent iterations. It can be seen, that in case of AWGN and TU channel the reduction of computational effort is about 20%, while in case of the PedA channel it is only 6%. It can further be noticed, that the choice of the parameter $\ell_{\min}$ depends on the CQI, i.e. the code rate, and not on the channel type. Further simulations have also shown that the variation of other system parameters has no significant influence on the choice of $\ell_{\min}$ either. Thus, the setting of $\ell_{\min}$ can be done by the decoder based on the current MCS reporting.

5 Conclusions

In this paper a decoding algorithm for Turbo Codes with reduced computational complexity has been described. The approach is based on a syndrome based identification of supposedly error-free blocks in the input sequence of a constituent decoder. Decoding only the remaining erroneous blocks results in a reduction of decoding effort while keeping the impact on the BLER performance insignificant. Depending on the channel type, the computational effort measured in equivalent iterations of the conventional decoder can be reduced by up to 20%. The approach is especially effective if the channel requires a relatively high number of iterations until termination, like e.g. in the AWGN case. On the other hand, if the channel only requires a small number of iterations (like 2–3 iterations for the PedA and TU channels), the reduction in terms of absolute iterations will be smaller. For these cases, a low SNR ET scheme may be considered, which may also be implemented based on the hamming weight of the syndrome sequence **b** (Geldmacher et al., 2011).

The realization of a syndrome based Turbo decoder is of the same complexity as a conventional Turbo decoder, because only the underlying trellis is changed. Additional effort is required for syndrome computation, the generation of an estimate of the parity errors and precorrection. These operations are based only on bits, such that they are easy to realize using XOR operations. Additionally, the identification of error-free blocks has to be done. This may be implemented by a simple counter.

Table 1. Summarized results for different CQIs and channels.

	Code Rate	$\ell_{\min}$	Eq. It. Conv. Dec.	Eq. It. BSD	Saving
AWGN, CQI3	0.19	21	3.8	3.0	21 %
AWGN, CQI9	0.60	31	5.1	3.9	22 %
TU, CQI9	0.60	31	2.3	1.8	20 %
PedA, CQI9	0.60	31	2.3	2.1	6 %
AWGN, CQI13	0.75	91	5.0	3.8	24 %

References

Evolved Universal Terrestrial Radio Access (E-UTRA): 3GPP TS 36.213 V8.8.0, 2009.

Berrou, C., Glavieux, A., and Thitimajshima, P.: Near Shannon limit error-correcting coding and decoding: Turbo-codes, in: IEEE ICC, 1064–1070, Geneva, Switzerland, 1993.

Forney Jr., G.: Convolutional codes I: Algebraic structure, IEEE Transactions on Information Theory, 16, 720–738, 1970.

Geldmacher, J., Hueske, K., and Götze, J.: An Adaptive and Complexity Reduced Decoding Algorithm for Convolutional Codes and its Application to Digital Broadcasting Systems, in: International Conference on Ultra Modern Telecommunications (ICUMT2009), St. Petersburg, Russia, doi:10.1109/ICUMT.2009.5345627, 2009.

Geldmacher, J., Hueske, K., Bialas, S., and Götze, J.: Adaptive Low Complexity MAP Decoding for Turbo Equalization, in: 6th International Symposium on Turbo Codes & Iterative Inf. Proc. (ISTC2010), Brest, France, doi:10.1109/ISTC.2010.5613804, 2010.

Geldmacher, J., Hueske, K., Götze, J., and Kosakowski, M.: Hard Decision Based Low SNR Early Termination for LTE Turbo Decoding, in: 8th International Symposium on Wireless Communication Systems (ISWCS2011), Aachen, Germany, 2011.

Hueske, K., Geldmacher, J., and Götze, J.: Syndrome Based Adaptive Complexity Channel Decoding and Turbo Equalization for ATSC DTV, in: 44th Asilomar Conference on Signals, Systems and Computers, Pacific Grove, CA, USA, 2010.

Mehlführer, C., Wrulich, M., Ikuno, J. C., Bosanska, D., and Rupp, M.: Simulating the Long Term Evolution Physical Layer, in: Proceedings of the 17th European Signal Processing Conference (EUSIPCO 2009), Glasgow, Scotland, 2009.

Minowa, T.: A syndrome decoding of high-rate turbo codes, in: IEEE International Symposium on Information Theory, 384–384, doi:10.1109/ISIT.2003.1228400, 2003.

Robertson, P., Villebrun, E., and Hoeher, P.: A comparison of optimal and sub-optimal MAP decoding algorithms operating in the log domain, IEEE ICC, 1009–1013, doi:10.1109/ICC.1995.524253, Seattle, 1995.

Schalkwijk, J. and Vinck, A.: Syndrome Decoding of Binary Rate-1/2 Convolutional Codes, IEEE Transactions on Communications, 24, 977–985, doi:10.1109/TCOM.1976.1093420, 1976.

Tajima, M., Shibata, K., and Kawasaki, Z.: Modification of Syndrome Trellises for High-Rate Convolutional Codes, Tech. rep., The Institute of Electronics, Information and Communication Engineers (IEICE), 2002.

Yamada, T., Harashima, H., and Miyakawa, H.: A new maximum likelihood decoding of high rate convolutional codes using a trellis, IEICE Transactions on Electronics and Communications in Japan, J66-A, 611–616, doi:10.1002/ecja.4400660703, 1983.

An identification procedure of multi-input Wiener models for the distortion analysis of nonlinear circuits

T. Weng[1], **S. Stegemann**[1], **W. John**[2], **and W. Mathis**[1]

[1]Institute for Theoretical Electrical Engineering, Leibniz University of Hannover, Hannover, Germany
[2]System Integration Laboratory, Paderborn, Germany

Correspondence to: S. Stegemann (stegemann@tet.uni-hannover.de)

Abstract. In this contribution, a system identification procedure of a two-input Wiener model suitable for the analysis of the disturbance behavior of integrated nonlinear circuits is presented. The identified block model is comprised of two linear dynamic and one static nonlinear block, which are determined using an parameterized approach. In order to characterize the linear blocks, an correlation analysis using a white noise input in combination with a model reduction scheme is adopted. After having characterized the linear blocks, from the output spectrum under single tone excitation at each input a linear set of equations will be set up, whose solution gives the coefficients of the nonlinear block. By this data based black box approach, the distortion behavior of a nonlinear circuit under the influence of an interfering signal at an arbitrary input port can be determined. Such an interfering signal can be, for example, an electromagnetic interference signal which conductively couples into the port of consideration.

1 Introduction

In order to ensure the reliability of electronic systems even in harsh electromagnetic environments, it is necessary to analyze the effects of EMI signals on integrated circuits at an early design stage. Generally, a disturbance signal can couple into a nonlinear circuit at an arbitrary port. The output signal will possess interference components, that can significantly distort the circuit's functionality. These interference phenomena arise due to intermodulations between the nominal input signal and the disturbing signal as an effect of the nonlinear device characteristics (Stegemann et al., 2012, 2011). To estimate those interfering products with respect to both inputs, we model the nonlinear circuits under consideration by the two input Wiener model shown in Fig. 1. In this work, the black box approach is used to identify the considered model, as in this case a rapid solution is more important than a detailed consideration of the circuit structure itself. The model structure is comprised of two linear blocks and a nonlinear static block. Both linear blocks are described by their linear transfer function and independently model the frequency characteristics of the input signal x_{in} and the distortion signal x_{dis}, respectively. The static nonlinear block is given by a multivariate polynomial that models the nonlinear behavior of the circuit. In the proposed system identification procedure for this block oriented model, the parameters of the linear blocks are obtained by adopting a correlation analysis using a white noise input. The artificial and not observable intermediate signals $\tilde{x}_{\text{in}}(t)$ and $\tilde{x}_{\text{dis}}(t)$ are avoided with the assumption of small input signals so that the whole system can be seen as linear. Then, by applying the determined linear transfer functions, a linear system of equations is set up to calculate the parameters of the nonlinear block from frequency domain data. The identified model can be used to calculate characteristic figures of merit for the distortion analysis of nonlinear circuits, such as intermodulation distortion components:

$$\mathrm{IM}_2 = \frac{A_{\mathrm{IM}_2}}{A_{\text{fundamental}}} \tag{1}$$

$$\mathrm{IM}_3 = \frac{A_{\mathrm{IM}_3}}{A_{\text{fundamental}}}. \tag{2}$$

To obtain these values, the output y is analyzed in the frequency domain. Here $A_{\text{fundamental}}$ is the magnitude at the frequency of x_{in} and $A_{\mathrm{IM}_{2,3}}$ are the magnitudes of the intermodulation products of second and third order.

Besides the premise that for small input signals the nonlinearity can be seen as linear, the proposed identification procedure does not need other specific assumptions on the structure of the analyzed circuit. E.g. there is no need to assume an

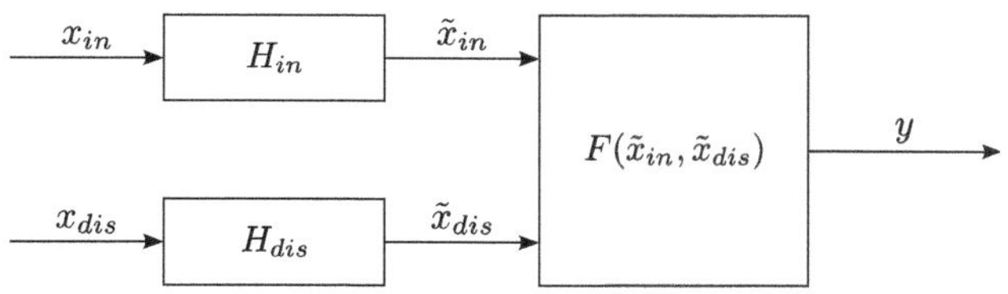

Fig. 1. Block-oriented model structure for identification procedure (cf. Stegemann et al., 2012, 2011).

invertible nonlinearity (cf. Greblicki, 1992) to calculate the intermediate signals for the identification of the linear blocks. Furthermore, each block is characterized with only one test signal, so there is no need for extensive calculations with different or complex input signals (Pintelon and Schoukens, 2001). In comparison with the classical distortion theory described in (Weiner and Spina, 1980) or (Wambacq and Sansen, 1998), the model can be used for EMI induced distortion analysis at nonsignal inputs, e.g. at the supply voltage. The applicability of the procedure for multi input distortion analysis using a simulative approach, is shown in detail for an operational amplifier in the presence of an interference signal at the power supply. In (Stegemann et al., 2012) an analytical approach for the characterization of each block is described. The remainder of the paper is organized as follows. In section 2 the predefined structure of the linear transfer functions and the multivariate polynomial are described, as well as the assumptions needed to identify their parameters. The identification procedure is then described in detail in the Sect. 2.1 for the linear dynamical and Sect. 2.2 for the nonlinear static block. In Sect. 3 the identification procedure is carried out for a two-stage operational amplifier. The results for the characterized blocks are compared with simulations from Cadence Spectre.

2 Identification procedure for distortion analysis

In this section the identification procedure for both, the linear dynamic and the nonlinear static block of the multi-input Wiener model is described. The goal is to characterize each block of the model shown in Fig. 1 by using either simulated or measured input and output data. For this purpose, the transfer characteristics of each block are given as predefined functions, for which we determine a parametrization that approximates the input/output behavior of the circuit. Since we are interested in a frequency domain analysis, we characterize each linear dynamic block by a rational function of the form:

$$H(s) = \frac{\sum_{m=0}^{M} a_m s^m}{\sum_{n=0}^{N} b_n s^n}, \tag{3}$$

i.e. the linear blocks H_{in} and H_{dis} are assumed to be completely described by linear transfer functions. The frequency domain transforms $\tilde{X}_{\text{in}}(f_{\text{in}})$ and $\tilde{X}_{\text{dis}}(f_{\text{dis}})$ of the artificial intermediate signals $\tilde{x}_{\text{in}}(t)$ and $\tilde{x}_{\text{dis}}(t)$ are then given by:

$$\tilde{X}_{\text{in}}(f_{\text{in}}) = H_{\text{in}}(f_{\text{in}}) \cdot X_{\text{in}}(f_{\text{in}}) \tag{4}$$

$$\tilde{X}_{\text{dis}}(f_{\text{dis}}) = H_{\text{in}}(f_{\text{dis}}) \cdot X_{\text{dis}}(f_{\text{dis}}) \tag{5}$$

This way, the dynamic behavior of a nonlinear circuit is modelled by a linear transfer function for each input, that filters the input signals of the circuit. Note, that from the knowledge of H_{dis}, i.e. the frequency response of the distortion signal, it is possible to estimate the frequency range in which a circuit is most susceptible to EMI as the frequency range in which H_{dis} exhibits its highest gain (Stegemann et al., 2012). The intermodulation products between $x_{\text{in}}(t)$ and $x_{\text{dis}}(t)$ at the output arise due to a nonlinear transformation, which is described by a static nonlinear function $F(\tilde{x}_{\text{in}}, \tilde{x}_{\text{dis}})$. This static nonlinearity is characterized by the K-th degree multivariate polynomial:

$$F(\tilde{x}_{\text{in}}, \tilde{x}_{\text{dis}}) = f(x, y) = \sum_{k=0}^{K} \sum_{l=0}^{K-k} \alpha_{k,l} x^k y^l. \tag{6}$$

The coefficients $\alpha_{k,l}$ are referred to as *nonlinearity coefficients*. For the black box characterization of every block, a relationship between the input of each block to the overall output of the system has to be determined. This can only be done by using suitable test signals (Bai et al., 2007). In Sect. 2.1, the determination of the parameters a_m and b_n of the linear blocks using a Gaussian white noise input and a correlation analysis is described. In order to characterize the linear blocks H_{in} and H_{dis} independently of the nonlinear block $F(\tilde{x}_{\text{in}}, \tilde{x}_{\text{dis}})$, the amplitude of the white noise signal has to be small. When determining the linear blocks, we assume that the output is at most weakly distorted by the nonlinearity. That is, we consider $y(t)$ to be

$$y(t) \approx \mathcal{L}^{-1}\{H_{\text{in}}(s)\} * x_{\text{in}}(t) + \mathcal{L}^{-1}\{H_{\text{dis}}(s)\} * x_{\text{dis}}(t), \tag{7}$$

where $\mathcal{L}^{-1}$ is the inverse Laplace transform and $*$ denotes the convolution operation. Once the linear blocks are identified, the artificial intermediate signals $\tilde{x}_{\text{in}}(t)$ and $\tilde{x}_{\text{dis}}(t)$ can be calculated by Eqs. (4) and (5). Using sinusoid input functions for the nominal input and the distortion signal, the nonlinearity coefficients of the multivariate polynomial in Eq. (6) can be determined using the spectral approach described in Sect. 2.2.

2.1 Identification of the linear blocks

In the context of linear and time invariant systems, a suitable way to characterize transfer characteristics is by means of white noise (Meyer, 2011). The identification procedure for the linear blocks of the multi input Wiener model is based on the definitions of the autocorrelation $r_{xx}(t_1, t_2)$ and the crosscorrelation $r_{xy}(t_1, t_2)$ of a white noise process, which can be represented by their mean value and only depends on the time differences (Meyer, 2011):

$$r_{xx}(\tau) = \lim_{T\to\infty} \frac{1}{2T} \cdot \int_{-T}^{+T} x(t)x(t+\tau)\,\mathrm{d}t, \tag{8}$$

$$r_{xy}(\tau) = \lim_{T\to\infty} \frac{1}{2T} \cdot \int_{-T}^{+T} x(t)y(t+\tau)\,\mathrm{d}t. \tag{9}$$

Here, $x(t)$ is the white noise input, $y(t)$ is the output signal and $\tau = t_2 - t_1$ is the time difference. Using Eq. (8), Eq. (9) can be written as follows (Meyer, 2011):

$$r_{xy}(\tau) = r_{xx}(\tau) \cdot h(\tau). \tag{10}$$

By transforming Eq. (10) into the frequency domain, we obtain the expression for determining the transfer function:

$$S_{xy}(s) = S_{xx}(s) \cdot H_{corr}(s), \tag{11}$$

with $s = j \cdot 2\pi f$. From Eq. (11), the characterization of the transfer functions is possible by measuring the response of a linear system to a white noise input and calculating the autocorrelation of the input, as well as the crosscorrelation between the output and input signal. By means of this correlation analysis and by the assumption Eq. (7), we obtain a data set for H_{in} and H_{dis} with respect to frequency, by recording the output wile stimulating the respective input x_{in} or x_{dis} with a white noise signal and zeroing the other input. Note, that we can only expect to obtain approximative results for the transfer functions, since no infinitely long signals can be considered. In order to characterize the parameters of the rational transfer function given in Eq. (3), we adopt a complex curve fitting method to determine the coefficients. A first guess of the degree of the nominator and the denominator polynomial $Z_{\text{est}}(s)$ and $N_{\text{est}}(s)$, respectively, has to be chosen with the help of the data obtained from the correlation analyis described above. Relying on the approximated characteristics in the frequency domain, we first chose a relatively high degree to overestimate, but to completely describe the system's dynamics (Pintelon and Schoukens, 2001). The coefficients a_m and b_n of (11) will then be determined by a nonlinear fitting problem for which we need a good initial guess. By means of the Gauss-Newton algorithm we obtain the linear transfer functions for both H_{dis} and H_{in}. As a last step, we reduce the order of the transfer function by an iterative approach. Eq. (3) is nonlinearly dependent on the coefficients a_m and b_n of it's nominator and denominator polynomials. That is, the error defined as the difference of the estimated transfer function[1] $H_{\text{est}}(s)$ and the characteristics resulting from the correlation analysis $H_{\text{corr}}(s)$ at the specific frequencies,

$$e(s) = H_{\text{corr}}(s) - H_{\text{est}}(s), \tag{12}$$

[1]Either being H_{dis} or H_{in} in the general form of Eq. (3).

is nonlinearly dependent on a_m and b_n, too. The goal is to find a set of coefficients that minimizes the error Eq. (12). An approximated solution of this minimization problem can be found using an iterative scheme, such as the Gauss-Newton method, which has been adopted in this work. However, the method requires the knowledge of an initial guess as a starting point for the iteration and, due to the nonlinear dependency on a_m and b_n, it can not be guaranteed that a global minimum will be found (Boyd, 2004) (Dahmen, 2008). In order to determine such an initial solution, we convert the problem into a similar but linear one, which we obtain by multiplying the error function in Eq. (12) by the yet unknown denominator $N_{\text{est}}(s)$:

$$\tilde{e}(s) = H_{\text{corr}}(s) \cdot N_{\text{est}}(s) - Z_{\text{est}}(s). \tag{13}$$

From this definition it is now possible to formulate a linear least squares problem (Tao and Zunde, 1981), which is solved by minimizing the squared error

$$E = \sum_{m=1}^{M} |\tilde{e}(s_m)|^2, \tag{14}$$

over every frequency $s_m \cdot j\omega_m$. For this linear optimization problem there exists a unique solution, since the problem is convex (Boyd, 2004). Taking this solution as a starting point, the Gauss–Newton iteration is applied to successively improve the solution of Eq. (12). After having found a possible solution for a transfer function, one should try to reduce the order of this function, which is possible for the numerator and the denominator polynomial separately. That is, we wish to find a solution for the linear blocks providing as less poles and zeros as possible with no loss in accuracy. For this purpose, an overall iteration scheme is applied, which successively reduces the degree of the nominator and the denominator, until a predefined error limit is reached. If a rapid increase of the error is observed during order reduction, the solution found up to this point will be used (Pintelon and Schoukens, 2001). A flow-chart of the applied reduction algorithm is depicted in Fig. 2.

2.2 Identification of the nonlinear block

Having successively identified the linear blocks, both for the input and for the distortion signal, the static nonlinear block can be determined, since the artificial intermediate signals $\tilde{x}_{\text{in}}$ and $\tilde{x}_{\text{dis}}$ can now be calculated by Eqs. (4) and (5). We calculate the coefficients of the multivariate polynomial in Eq. (6) by simulating or measuring the quasi periodic output for a single frequency excitation at both inputs, as:

$$x_{\text{in}}(t) = A_{\text{in}} \cos(2\pi f_{\text{in}} t + \phi_{\text{in}}) \tag{15}$$

$$x_{\text{dis}}(t) = A_{\text{dis}} \cos(2\pi f_{\text{dis}} t + \phi_{\text{dis}}). \tag{16}$$

The output $y(t)$ will then be composed of various harmonics and intermodulation products of $x_{\text{in}}(t)$ and $x_{\text{dis}}(t)$. Given

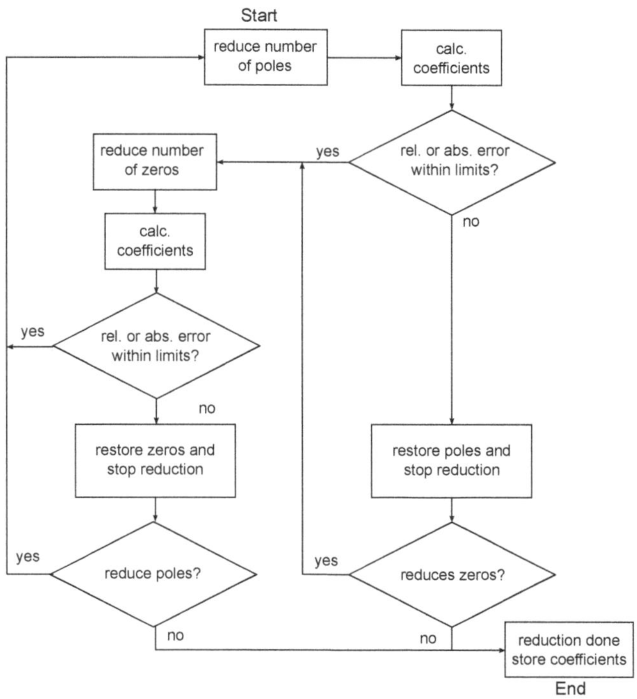

Fig. 2. Flow-chart of the algortihm for order reduction of the identified linear transfer functions.

a predefined order of the polynomial output nonlinearity $F(\tilde{x}_{\text{in}}(t), \tilde{x}_{\text{dis}}(t))$, we construct a linear system of equations in the complex domain from the output spectrum of $y(t)$. The solution of this system of equations yields the coefficients $\alpha_{k,l}$ for $k, l = 1, \ldots, K$ in (6). We fix the order K by omitting those components in the output spectrum, whose amplitudes do not exceed a predefined threshold. Given the input frequencies f_{in} and f_{dis}, the output frequencies generated by a polynomial nonlinearity are calculated using the concept of module vectors (Weiner and Spina, 1980). That is, the i-th mixing frequency of order $k+l$ is given by:

$$f_{i,k+l} = \boldsymbol{m}_{i,k+l} \cdot \boldsymbol{f}^T_{\text{exc}}. \tag{17}$$

Here $\boldsymbol{f}_{\text{exc}} = [-f_{\text{in}}, -f_{\text{dis}}, f_{\text{dis}}, f_{\text{in}}]$ is the row vector containing the input and the distortion frequency (i.e. the excitation frequencies). The row vector $\boldsymbol{m}_{i,k+l} = [m_{-f_{\text{in}}}, m_{-f_{\text{dis}}}, m_{f_{\text{in}}}, m_{f_{\text{dis}}}]$ is the i-th module vector of order $k+l$, whose entries are positive natural numbers and which obey:

$$m_{-f_{\text{in}}} + m_{-f_{\text{dis}}} + m_{f_{\text{in}}} + m_{f_{\text{dis}}} = k + l. \tag{18}$$

For a given order $k+l$ there exist[2]

$$i = 1, .., \binom{4+k+l-1}{k+l}$$

[2]Chua and Ng (1979), p. 5, for which $K = 2$ (no. of input frequencies) and $n = k+l$ (order of mixing product).

module vectors and by Eq. (17) the output frequencies are related to the input frequencies. Using this concept, we can construct the following linear system of equations with the coefficients of (6) as unknowns:

$$\mathbf{C}\tilde{\boldsymbol{\alpha}} = \boldsymbol{b}. \tag{19}$$

Here, the vector $\boldsymbol{b}$ on the right hand side contains the complex amplitudes (absolute values and phases) of the simulated or measured output spectrum of $y(t)$ at the frequencies calculated by Eq. (17) which are listed in ascending order. The entries of the matrix $\mathbf{C}$ are calculated as follows: For each frequency from Eq. (17) the resulting signal amplitude and phase is calculated by:

$$q_{k,l,i} = \frac{1}{2^{k+l}} |\tilde{x}_{\text{in}}|^k |\tilde{x}_{\text{dis}}|^l \exp[j(k \cdot \arg(\tilde{x}_{\text{in}}) + l \cdot \arg(\tilde{x}_{\text{dis}})]. \tag{20}$$

Since different module vectors can result in the same frequency, there are less unique frequencies than module vectors. Because the vector $\boldsymbol{b}$ contains the complex amplitdues for these unique frequencies only, the values Eq. (20) are connected with the rows i of $\mathbf{C}$ by their frequency, hence the index. The corresponding column j in the matrix $\mathbf{C}$ for each complex value given by Eq. (20) is then calculated by its indices k and l:

$$j = \frac{(k+l) \cdot (k+l+1)}{2} + 1 + l. \tag{21}$$

The fraction is the sum from 1 to $k+l$ which equals the number of all nonlinear coefficients up to the order $k+l-1$. Thus, it can be seen as the starting index for the coefficients of order $k+l$. As the index l (or k) is unique for each order, since the sum of k and l stays the same, it can be used to obtain the column index. Consequently, the rows of the matrix $\mathbf{C}$ contain the complex amplitudes from Eq. (20) for a specific frequency equal to the frequency in the vector $\boldsymbol{b}$, whereas the complex amplitudes are linked with the columns of $\mathbf{C}$ by their indices k and l according to Eq. (21). Due to the number of frequencies generated by the nonlinearity, Eq. (14) is an overdetermined system of equations, for which a least squares solution can be obtained. After having solved Eq. (19) to get $\tilde{\boldsymbol{\alpha}}$, we obtain the nonlinearity coefficients by division with the corresponding binomial coefficient (see e.g. Weiner and Spina, 1980 or Maas, 2003):

$$\alpha_{k,l} = \frac{k! \cdot l!}{(k+l)!} \tilde{\alpha}_{k,l} \tag{22}$$

3 Modelling a two-stage operational amplifier

The described procedure is now used to model the operational amplifier given in Fig. 3 within a simulation approach. Comparative simulations were made with Cadence Spectre.

The input signal equals the differential signal between the inputs v_{out1} and v_{out2}, where the interfering signal $v_{\text{dis}}(t)$

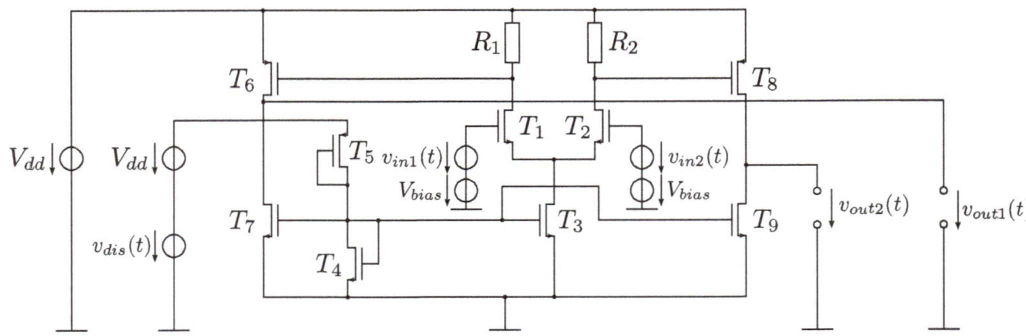

Fig. 3. Operational Amplifier with interferring signal $v_{\mathrm{dis}}(t)$ at the supply voltage

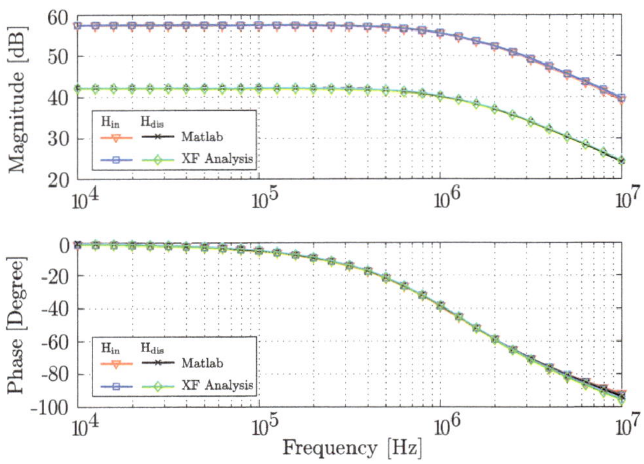

Fig. 4. Transfer functions $H_{\mathrm{in},v_{\mathrm{out1}}}$ and $H_{\mathrm{dis},v_{\mathrm{out1}}}$

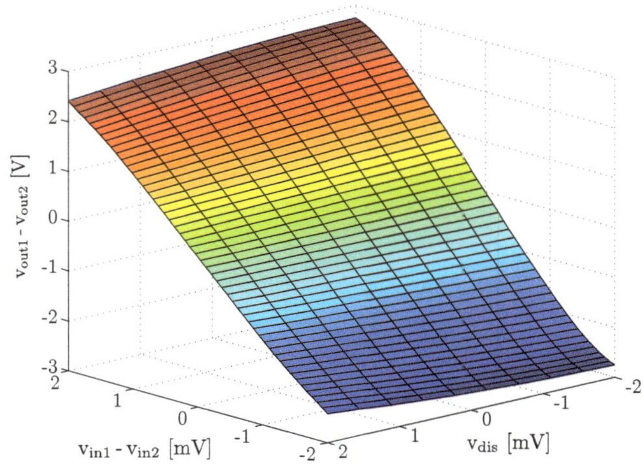

Fig. 5. Nonlinear characteristic in dependency of both input signals.

varies the reference current of the differential stage which leads to a mixing between both inputs. In this case the interfering signal does not directly influence the differential output $v_{\mathrm{out1}} - v_{\mathrm{out2}}$ because of the common mode rejection of the input stage. Thus, two transfer functions are identified for both inputs $H_{\mathrm{in},v_{\mathrm{out1}}}$, $H_{\mathrm{in},v_{\mathrm{out2}}}$, $H_{\mathrm{dis},v_{\mathrm{out1}}}$ and $H_{\mathrm{dis},v_{\mathrm{out2}}}$ according to the outputs v_{out1} and v_{out2}. The resulting bode plots of the transfer functions $H_{\mathrm{in},v_{\mathrm{out1}}}$ and $H_{\mathrm{dis},v_{\mathrm{out1}}}$ are shown in Fig. 4. The procedure yields the same results for the output v_{out2}, with just a 180° phase shift for $H_{\mathrm{in},v_{\mathrm{out2}}}$. These results were achieved by a transient simulation with Gaussian white noise input with a maximum level of $45\,\mu$V. Based on the identified frequency response, a frequency range of 10 kHz up to 10 MHz was chosen to conduct the order reduction scheme of 2.1.

For comparison purposes a counter simulation was done, using the XF-analysis in Cadence Spectre. As it can be seen from Fig. 4, there is a very good agreement between the calculated and the simulated transfer functions. The resulting polynomial from Eq. (6) is shown in Fig. 5, which was calculated for a magnitude of 2 mV for both inputs at frequencies of 200 kHz for the nominal input and 18 kHz for the interfering signal, respectively. In this case, the order of the polynomial was set to $K = 9$. The obtained model is now used for the distortion analysis of the given circuit. As the intermodulation products of second and third order and their figures of merit are of particular interest, these intermodulation components can be calculated from the amplitudes at the frequencies $f_{\mathrm{in}} \pm f_{\mathrm{dis}}$ for the second order and $f_{\mathrm{in}} \pm 2f_{\mathrm{dis}}$ for the third order components as:

$$\mathrm{IM}_2 = \frac{\alpha_{11} \cdot |H_{f\mathrm{dis}}(f_{\mathrm{dis}})X_{\mathrm{dis}}(f_{\mathrm{dis}})|}{\alpha_{10}} \tag{23}$$

$$\mathrm{IM}_3 = \frac{\alpha_{12} \cdot |H_{f\mathrm{dis}}(f_{\mathrm{dis}})X_{\mathrm{dis}}(f_{\mathrm{dis}})|^2}{\alpha_{10}}. \tag{24}$$

The estimated characteristics of IM_2 and IM_3, in dependency of the distortion amplitude v_{dis}, are depicted in Fig. 6. In this case the magnitude of the nominal input signal was set to 2 mV. The counter simulation was done by a Harmonic Balance simulation of the whole circuit in Cadence Spectre with swept v_{dis}.

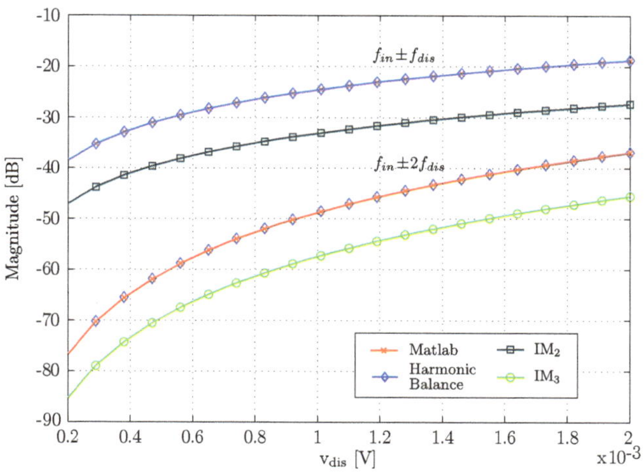

Fig. 6. Second and third order intermodulation distortion in dependency of the distortion amplitude v_{dis}.

4 Conclusion and outlook

In order to derive expressions for analyzing the intermodulation products of an input signal and an interfering signal of a nonlinear circuit, an identification procedure to obtain parametrized transfer functions for a multi input Wiener model from input and output data has been described. The linear transfer functions of the linear blocks for both inputs were derived by a correlation analysis with white noise excitation signals at both inputs and deriving the output by a transient simulation. In this context, an order reduction scheme was proposed to calculate a solution according to specific needs of accuracy or simplicity, regarding the maximum number of poles and zeros of the estimated linear transfer functions. From the simulated output spectrum with single tone stimulation at both inputs, a set of linear equations has been set up to determine the nonlinearity coefficients of the nonlinear block. The proposed identification procedure is suitable to examine the frequency range in which a circuit is most susceptible to a distortion signal, such as an electromagnetic interference signal. Having identified each block of the multi input Wiener model, expressions for the emerging intermodulation products can be obtained. Although the results in this paper were obtained by conducting simulation, the proposed identification procedure can be used in a measurement setup as well. Future work will deal with this measurement set up and consideration of the input signals used for the identification process.

Acknowledgements. The project EM4EM upon which this contribution is based is funded by the BMBF (Bundesministerium fuer Bildung und Forschung) under the project number 01M3092K. The responsibility for this publication is held by the authors only.

References

Bai, E. W., Cerone, V., and Regruto, D.: Separable inputs for the identification of block-oriented nonlinear systems, in: American Control Conference, 2007, ACC '07, 1548–1553, 2007.

Boyd, S.: Convex optimization, Cambridge University Press, Cambridge, UK New York, 2004.

Chua, L. and Ng, C.-Y.: Frequency domain analysis of nonlinear systems: general theory, IEE J. Electron. Circuits and Syst., 3, 165–185, 1979.

Dahmen, W. and Reusken, A.: Numerik fuer Ingenieure und Naturwissenschaftler, Springer, 2008.

Greblicki, W.: Nonparametric identification of Wiener systems, IEEE T. Inform. Theory, 38, 1487–1493, 1992.

Maas, S. A.: Nonlinear Microwave and RF Circuits, Artech House, 2nd Edn., 2003.

Meyer, M.: Signalverarbeitung: Analoge und digitale Signale, Systeme und Filter (German Edition), Vieweg+Teubner Verlag, 2011.

Pintelon, R. and Schoukens, J.: System Identification: A Frequency Domain Approach, IEEE Press, 2001.

Stegemann, S., John, W., and Mathis, W.: Blockorientierte Analyse der Empfindlichkeit nichtlinearer Schaltungen gegenüber elektromagnetischen Störungen an Nicht-Signaleingängen, 12. GMM/ITG-Fachtagung ANALOG 11, 2011.

Stegemann, S., John, W., Widemann, C., and Mathis, W.: A Block-Oriented Approach to EMI-induced Distortion Analysis for Nonlinear Analog Circuits, EMC Europe, Rom, 2012.

Tao, K. M. and Zunde, P.: On Levy's identification, its generalization and applications, Int. J. Syst. Sci., 12, 1317–1334, 1981.

Wambacq, P. and Sansen, W.: Distortion Analysis of analog integrated circuits, Kluwer Academic Publishers, 1998.

Weiner, D. D. and Spina, J. F.: Sinusoidal analysis and modeling of weakly nonlinear circuits : with application to nonlinear interference effects, Van Nostrand Reinhold, New York, 1980.

Design of RF MEMS based switch matrix for space applications

S. Di Nardo[1], P. Farinelli[2], T. Kim[3], R. Marcelli[4], B. Margesin[5], E. Paola[1], D. Pochesci[6], L. Vietzorreck[3], and F. Vitulli[6]

[1]ThalesAlenia Space, L'Aquila, Italy
[2]University of Perugia, Perugia, Italy
[3]Technische Universität München, Munich, Germany
[4]CNR – Institute for Microelectronics and Microsystems, Roma, Italy
[5]FONDAZIONE BRUNO KESSLER, Trento, Italy
[6]ThalesAlenia Space, Roma, Italy

Correspondence to: T. Kim (kim@tum.de)

Abstract. RF MEMS based switch matrices have several advantages compared to the mechanical or solid-state switch based ones for space applications. They are compact, light and less lossy with a high linearity up to high frequency. In this work, a 12 × 12 switch matrix with RF MEMS and LTCC technologies is presented based on the planar Beneš network. The simulated performance of the 12 × 12 switch matrix is below −12 dB IL (Insertion Loss) up to C band and −15 dB RL (Return Loss) up to Ku band. Moreover, it has a good isolation better than −50 dB. A 4 × 4 switch matrix with the same design process and technologies is fabricated and measured to verify the 12 × 12 switch matrix design process. The measured performance agrees very well to the simulations.

1 Introduction

In a communication satellite, the received signals are simultaneously to be delivered to the assigned transmission outputs with high flexibility, which is normally achieved by high-order switch matrices. Microwave or millimeter-wave switch matrices provide a full and high connectivity to the satellite broadcasting system. The switch matrices which are used up to now are based on the mechanical switches. The mechanical switches are reliable, less lossy and can drive a high power with high linearity. However, they are slow, bulky, heavy and require motors for switching, which are normally unsuitable for the cost-effective high speed system. Switch matrices using solid state devices, such as PIN diodes, MESFETs (MEtal Semiconductor Field Effect Transistors) or HEMTs (High Electron Mobility Transistor), have been also discussed due to their high speed and compact size with cheaper prices. However, they consume a DC power and may generate intermodulation spurious modes. Moreover, their applicable frequency bands are limited. Recently, Radio Frequency Micro Electro Mechanical System (RF-MEMS)s are referred to as a good candidate to get over the reported drawbacks of the mechanical or solid state based switches. They are small and virtually consume no power. RF MEMS can be operated up to high frequency with high linearity. Moreover, on maturing microfabrication technologies, the cost for RF MEMS fabrication is going down (De Angelis et al., 2012).

In the present paper, the design of the RF MEMS switch matrix is discussed. The aimed switch matrix has 12 inputs and 12 outputs. Starting from the network topology and the control algorithm, the design process for the 12 × 12 switch matrix is described including physical assembly with the control boards and the housing. Based on the results from full wave and circuit simulators, the RF performance is presented as well. A 4 × 4 switch matrix with same design process and technologies is fabricated and the measured performance is compared to simulations.

2 Discussion of network topology

Network topologies have been investigated mainly in the field of computer architecture. The representative network topologies are Crossbar, Omega, Beneš or Clos network, etc. The proper choice of the network topology for the RF MEMS networks is important because the number of required switching elements and stages makes a strong effect to the construction feasibility together with the signal transition performance and the network functionallity. Considering the required switch numbers and control algorithm implementations, Beneš and Clos networks are good candidates to construct a high order switch matrix using RF MEMS technologies.

A Beneš network is a powerful rearrangeable network with simple control algorithms. Rearrangeable networks mean that any permutation can be statically assigned by re-arranging network paths if necessary. The Beneš network is one of the oldest and best-known rearrangeable networks in use for around 40 years. It finds widespread use in shared memory-multiprocessor systems, telecommunication networks, TDMA (Time Division Multiple Access) systems for satellite communications and newer applications such as switching fabrics in internet routers. To realize a RF MEMS switch matrix, the Beneš network can be implemented in a planar way. Figure 1a shows the network block diagram for the 12 × 12 planar Beneš network using DPDT (Double-Pole Double-Throw) switches. A DPDT switch has two inputs and two outputs. The 12 × 12 Beneš network is constructed based on the 16 × 16 Beneš network (Chang and Melhem, 1997).

A Clos network is also a subclass of the rearrangeable networks. Such a network consists of two symmetric outside stages of rectangular switches with an inner stage of square switches. Those three-stage-arranged switch modules are interconnected with every module in the adjacent stage by a unique link (Yeh and Feng, 1992). More practical implementation to the RF MEMS switch matrix is 3D Clos-Beneš network, which is to construct small sub-matrices using SPnT (Single-Pole n-Throw) or DPDT RF MEMS switches and assemble them to the complicated Clos network as described in Fig. 1b. In the figure, it is depicted that 8 SP4T switches are used to construct a 4 × 4 switch sub-matrix. For an easy connection, the inner stage is 90° rotated and assembled to the outer stages in a three-dimensional way.

Physically, planar Beneš and 3D Clos Beneš networks can be realized as Fig. 2. The figure also shows the estimated sizes and weights. The planar Beneš networks consist of several planar circuit boards. They are simple and easy to assemble. They are also robust with compact size. The drawback of the planar Beneš network is that the isolation can be poor because all circuits in the network are installed in one metal housing. However, the circuit coupling can be suppressed by inserting several metal vias and frames with covers. Compared to the planar Beneš network, 3D Clos Beneš networks have better isolation. They are easier to be extended to the higher order networks. The design of a single sub-matrix is also simple. However, their assembly is mechanically complex. As described in the figure, the middle stage is rotated by 90° and connected to the other stages by several coaxial cables, which requires high costs and a large footprint together with complex DC control input port connects. The 3D Clos Beneš network is expected to be bigger and heavier than the planar Beneš network's. In this work, a planar Beneš network is decided to be fabricated due to its simplicity, smaller footprint and lower weight.

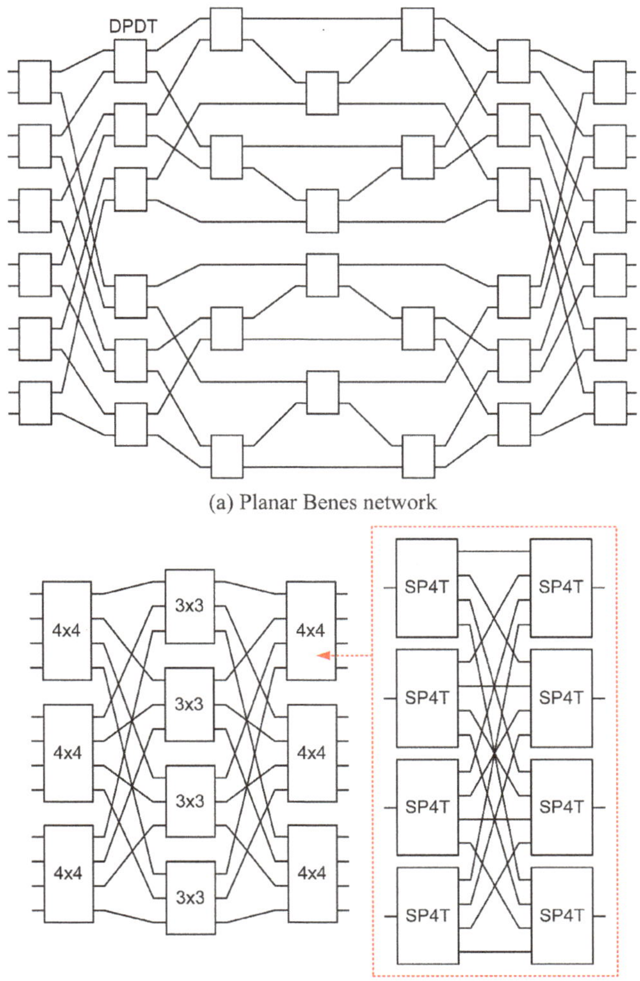

(a) Planar Benes network

(b) 3D Clos Benes network

Fig. 1. 12 × 12 Planar Beneš and 12 × 12 3-D Clos Beneš network.

3 12 x 12 Switching matrix

3.1 Switching matrix design

In the present work, the network physically consists of 6 RF boards and 6 DC boards which are actually 3 kinds of RF boards and two types of DC boards. RF boards consist of 4 LTCC (Low Temperature Co-fired Ceramic) and 2 alumina boards. The LTCC is an attractive solution to integrate electric components for high frequency applications. Compared to the printed resin board, it has good high frequency characteristics, thermal stability and is also suitable for passive components integrations (Imanaka, 2005). The RF signal which comes to the alumina board input is transmitted to the circuit on the LTCC board. After being routed to the assigned output port of the LTCC board stages (Stage 1, Stage 2 in Fig. 3), the RF signal is transmitted to the final network output through circuits on alumina boards. The DC signals to control RF MEMS switches are applied to the network by 6 DC boards.

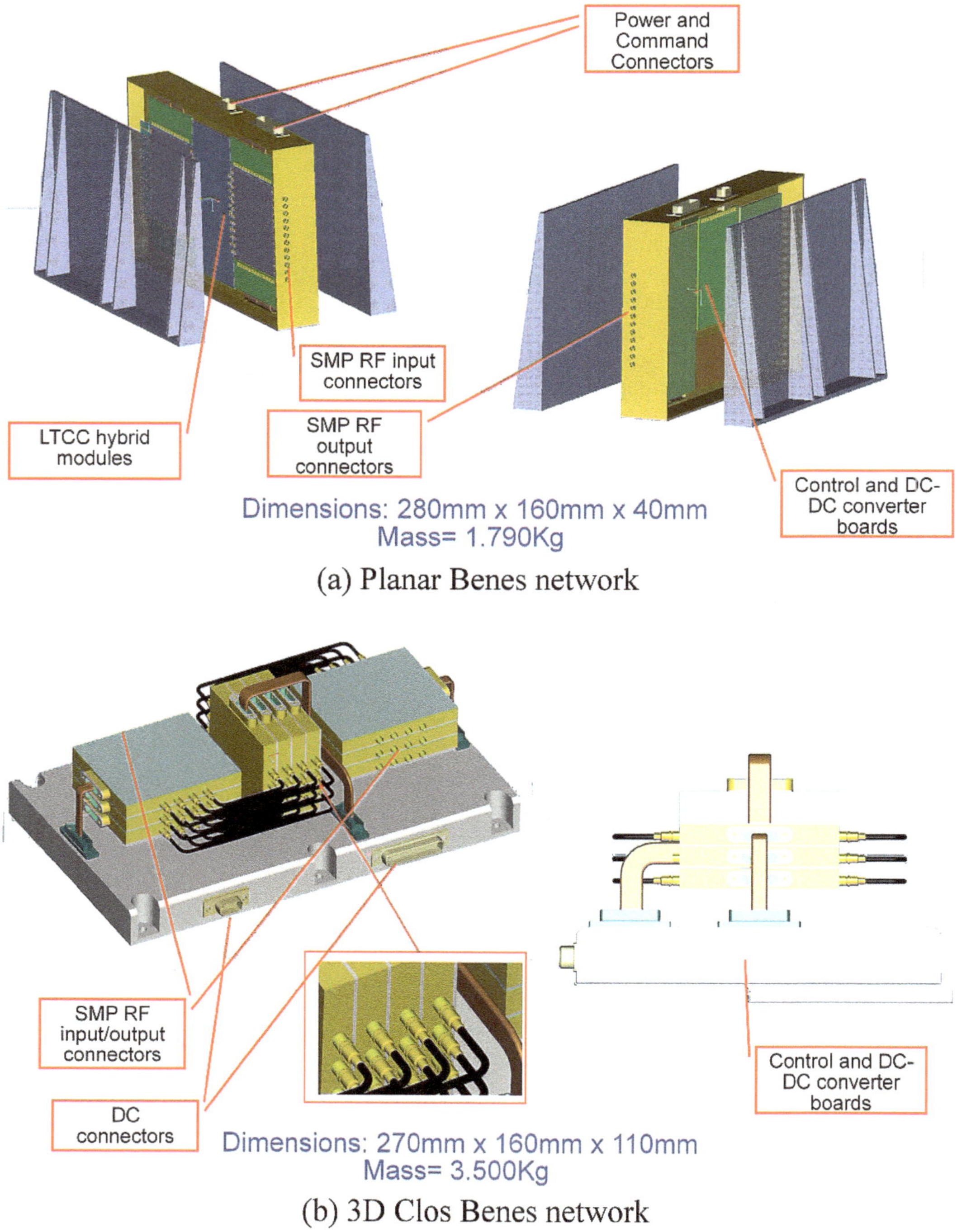

Fig. 2. Physical realization of planar Beneš network and 3-D Beneš network.

3.1.1 RF connection

The present 12 × 12 switch matrix adopts LTCC technology to construct multi-level circuits which enable the RF lines to be crossed without interfering each other. Figure 3 shows the LTCC board layouts which are designed for the 12 × 12 planar Beneš network. For the maximally feasible size of the LTCC board is 8.4cm by 8.4cm, the entire network cannot be realized in the single board. Therefore, the full 12 × 12 network is splitted in 4 sections and each section is constructed on a different LTCC board of feasible size. Between LTCC boards, bond ribbons connect the RF circuits. Generally, bond ribbon connections are robust and immune against shocks and vibrations, which is critical for the space applications because the device should survive under the launching condition. Figure 4 shows how the DPDT switch is mounted

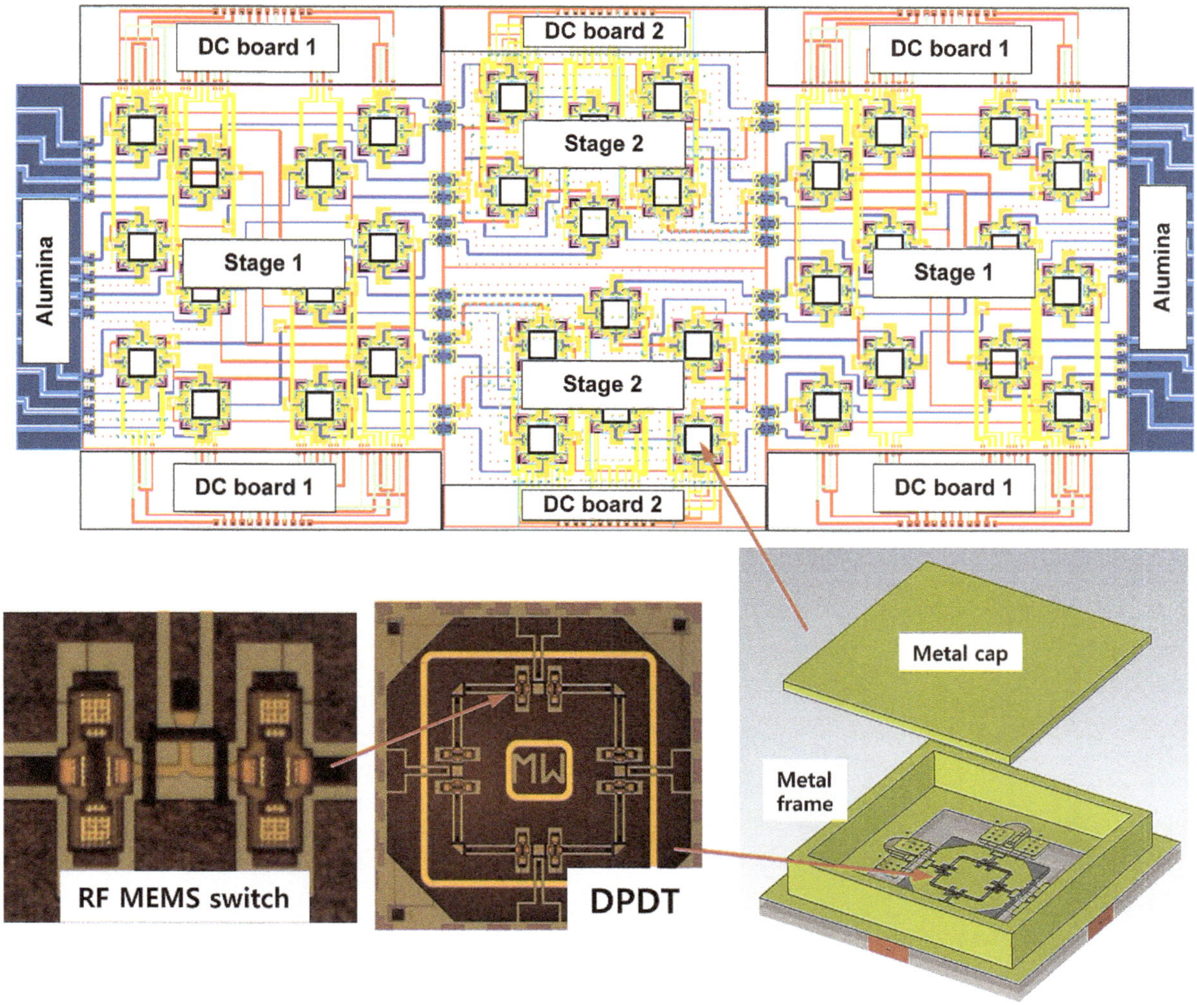

Fig. 3. 12 × 12 planar Beneš switch matrix layout and DPDT switch mount.

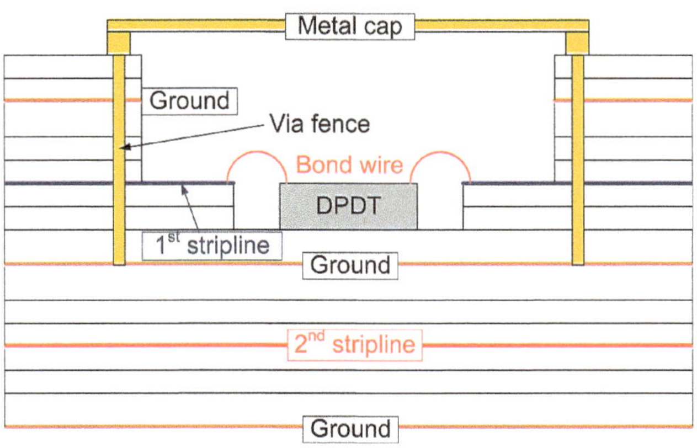

Fig. 4. DPDT switch mount on LTCC cavity.

on the LTCC board and electrically isolated from the environment. The DPDT switch is mounted in the deep LTCC cavity, where the switch circuit is located deep further to the level of the 1st stripline (SL) level on LTCC. The top and bottom layers of the switching element are grounded and connected with several vias. Consequently, each DPDT switch is capsulated by grounds and via fences, which suppress coupling between DPDTs. In the RF line design, every horizontal and vertical transition is optimized to have minimum return loss (RL) and insertion loss (IL). Additionally, metal vias are inserted between all SLs to suppress line couplings. On the input/output sections of the LTCC boards, the alumina boards are located. The reason of using alumina boards is to avoid physical collisions between LTCC boards and RF pin connectors on the housing during LTCC board installation.

3.1.2 DC connection

On the side (top and bottom side of the layout in Fig. 3) of the RF boards, 6 DC boards are connected. These DC boards are to decrease the number of DC connects. The single DPDT switch needs 4 DC contacts to control its 8 RF

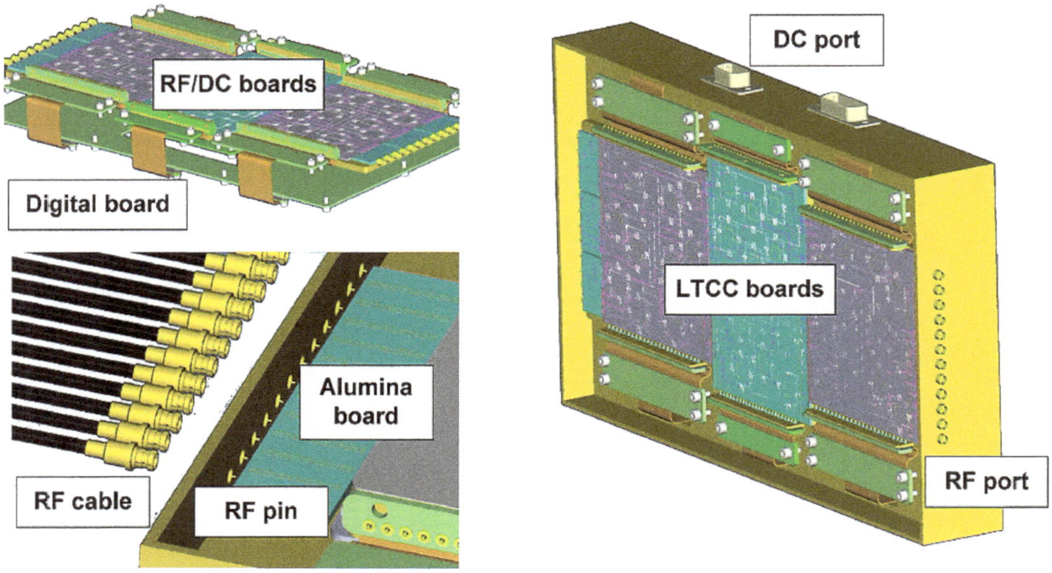

Fig. 5. Physical assembly of 12 × 12 network.

Fig. 6. Commercially available 12 × 12 Electromechanical RF Switching System (DowKey Microwave, 2012).

MEMS switches. And 36 DPDT switches are used for the entire network construction. This means, 144 DC contacts should be designed and simultaneously controlled. In this work, the number of DC contacts is decreased by binding DC contact pairs which are commonly activated during operation. This can also be done in the DPDT switch level. But the simpler approach is to bind common DC connects in the board level, which is done in the DC board. As a result, the number of DC contacts is reduced to 72.

3.2 Physical assembly

Figure 5 shows the entire physical assembly of the 12 × 12 switching matrix. RF and DC boards are installed on the metal housing. Under the RF and DC boards, a digital board to generate switch matrix control signals is assembled and connected to the DC ports. The circuits on the RF boards are connected to the RF cables through RF pins. The connection between RF board circuits and RF pins are achieved by bond ribbons for high shock and vibration immunity. As described in Fig. 1, the present switch matrix assembly has a 28cm × 12cm × 2cm size and a 1.79 kg weight. Figure 6 shows the commercially provided 12 × 12 mechanical switching system (DowKey Microwave, 2012). Its size is 48.3cm × 63.5cm × 17.8cm and the weight is 38.56 kg. Even after considering it has local control systems and a redundant power supply, it is much bigger and heavier than the proposed 12 × 12 RF MEMS switch matrix.

3.3 Network control

A high-order switch matrix needs a specific control algorithm. To control a 12 × 12 switch network, the 144 switch control inputs should be simultaneously assigned. Moreover, the number of feasible switch state configurations is 479 001 600 (=12!), which cannot be easily assigned by manual way or look-up tables. Therefore a switch matrix need the specific algorithm which can easily calculate the switch state and generate corresponding control inputs according to the given port assignments. Many network control algorithms have been reported, such as Looping algorithm, m-Matching algorithm, Euler partition algorithm, Karol's algorithm or Frame-based matching algorithm, etc. In this work, a Looping algorithm is adopted to control the planar Beneš network. A Looping algorithm is a simple sequential optimal matching algorithm. It is simple and has a low complexity. The constraint of this algorithm is that the network should have two center modules and input/output modules have to be 2 × 2. In the present work, DPDT (Double-Pole Double-Throw) RF MEMS switches, which have 2 × 2 size, are used and the center stage consists of two identical modules, which fulfills the constraint of the Looping algorithm. (Chao and Liu, 2007). When the port assignments are given,

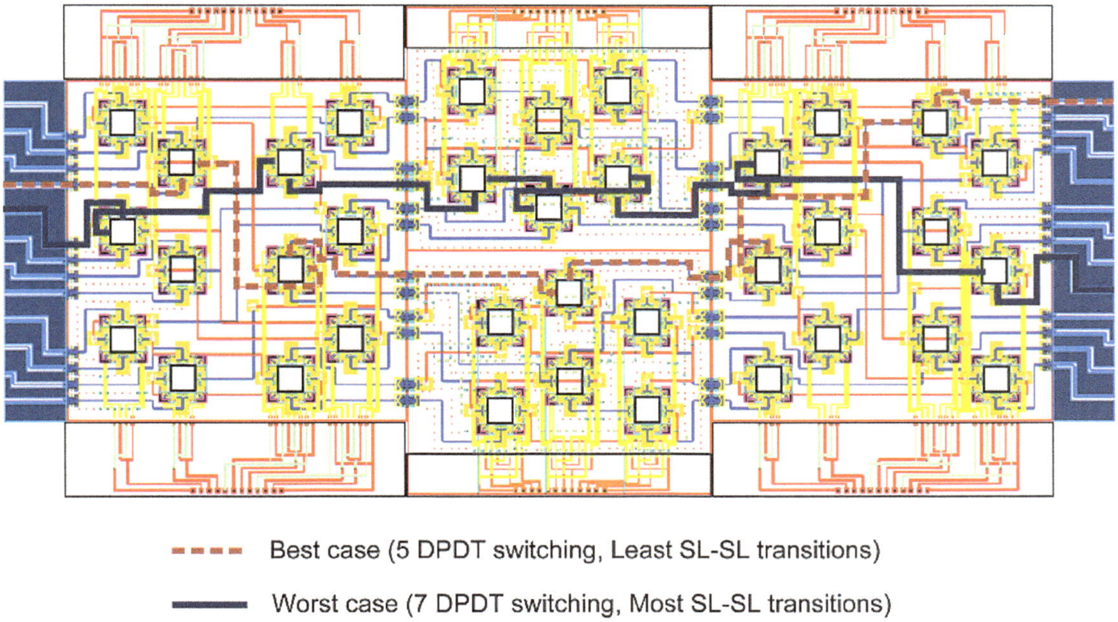

Fig. 7. Test line select.

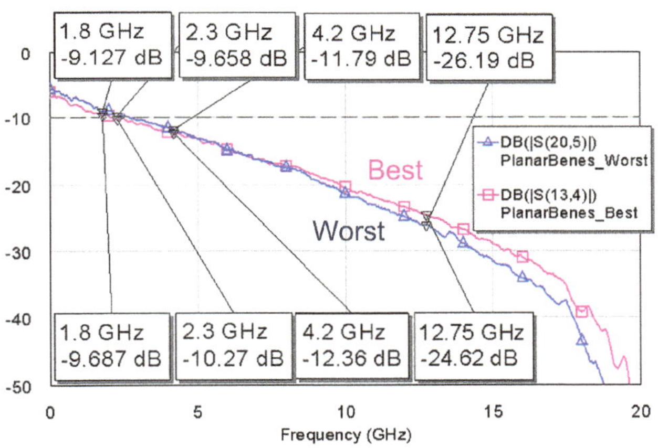

Fig. 8. Simulated IL of 12 × 12 network.

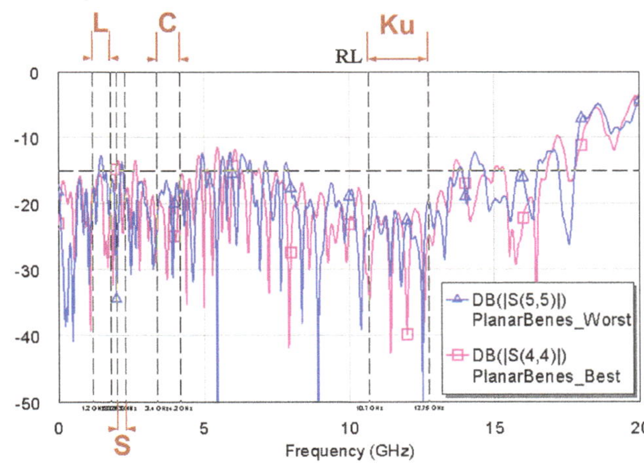

Fig. 9. Simulated RL of 12 × 12 network.

the corresponding switch states are calculated and switch control inputs are generated to be delivered to the digital board in the switch matrix (see Fig. 5). The digital board receives calculated DC inputs and controls high-voltage (typ. 60 V) drivers which practically activate RF MEMS switches.

3.4 RF signal transition performance

Unlike the 16 × 16 network, the 12 × 12 network is not symmetric, which means the number of DPDT switches is different according to the path assignment. To simulate the RF signal transition performance, two paths are selected as the best and worst cases in terms of the DPDT switching, layer-to-layer transition number and line lengths. Figure 7 shows the selected paths. Because the network is large and includes several complicated transitions, it is difficult to simulate the whole structure by full wave simulators. Therefore the detailed transition parts are full wave simulated and their results are combined with circuit simulators. The frequency band definitions are based on the commercial broadcasting satellite up/down link standards, which are L band (1.2-1.8 GHz), S band (2.025-2.3 GHz), C band (3.4-4.2 GHz) and Ku band (10.7-12.75 GHz).

3.4.1 Insertion loss (IL) and Return loss (RL)

As the 12 × 12 network is not symmetric, the IL is different according to the path assignment. This is not preferred in satellite application because the switch matrix should be

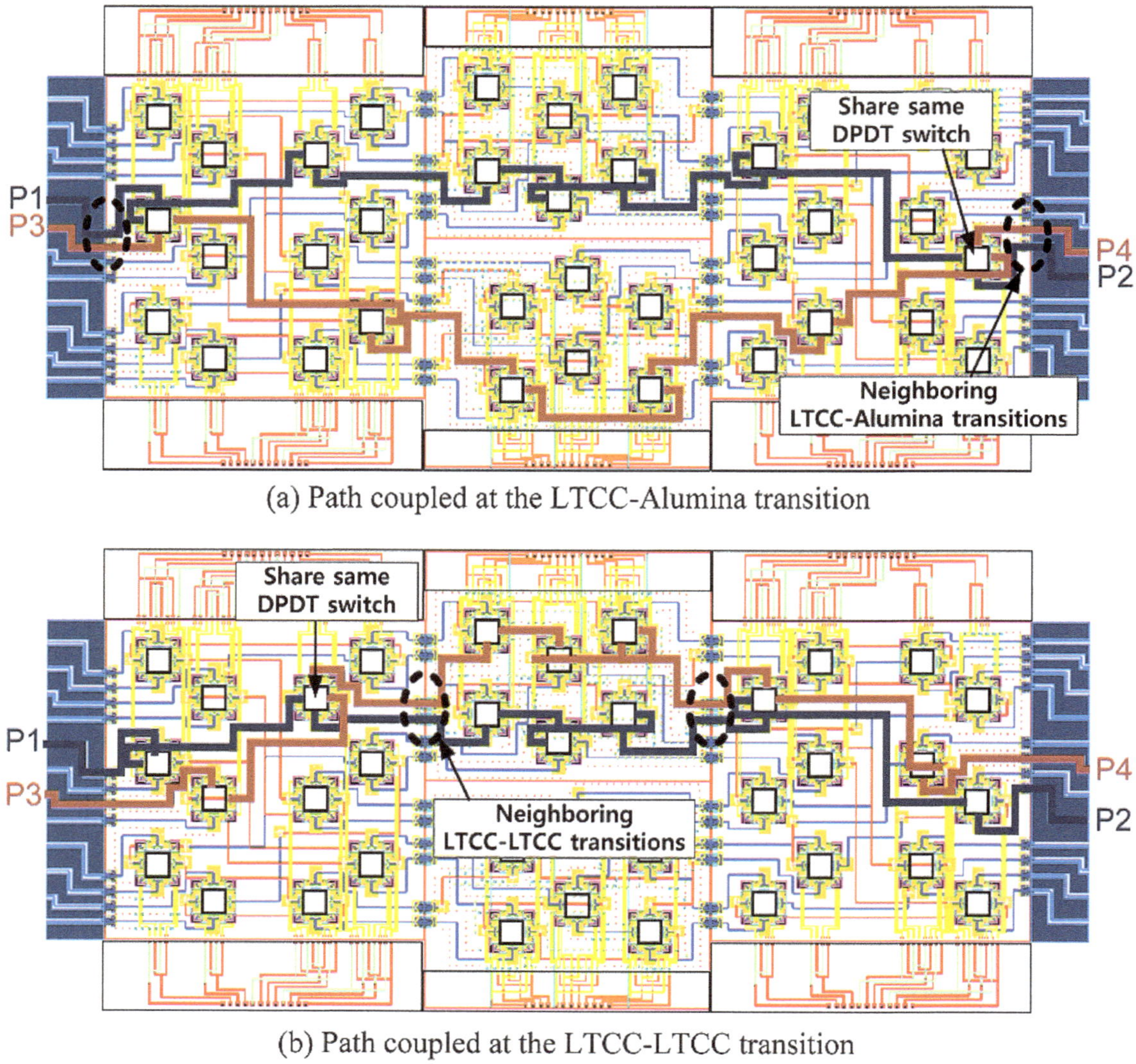

Fig. 10. Path coupling cases in the network.

transparent in terms of the RF signal transition. This means ILs should be the same for all connections regardless of the input/output port assignment. In the work, those differences are balanced by inserting resistive materials to the path which has less switching stages compared to the other paths. The resistivity and the size are carefully selected not to make severe RL increases. Figure 8 shows that the ILs are well-balanced and below −12 dB up to C band. The main IL contributions are long transmission lines and the high number of switches, which cannot be easily decreased. However, the IL can be compensated by putting amplifiers in the output stage. Generally, high order switch matrices have high linearity amplifiers to compensate system IL as the commercially available switch matrix presented in Fig. 6. Concerning RL, the present matrix has some temporal peaks, but in general RL is below −15 dB up to Ku band (see Fig. 9). This is better than the commercially available 12×12 mechanical switch matrix which has about −7 dB RL at Ku band (DowKey Microwave, 2012).

3.4.2 Isolation

Circuit simulators cannot effectively simulate the isolation of the network as they only consider the cascaded sub-circuit parameters without considering interferences between surrounding components. Therefore proper test structure definitions and approximations are to be involved to estimate the isolation. Generally, couplings mainly occur in the DPDT switch. In the planar Beneš network, the paths share an identical DPDT switch by two times at the maximum, which is the worst case for the isolation. As the worst case, the two paths in Fig. 10 are selected, which are coupled two times at the DPDT switch. Line couplings are accounted by inserting coupled lines at every switching stage. The coupled lines have the feasible maximum length in the network and

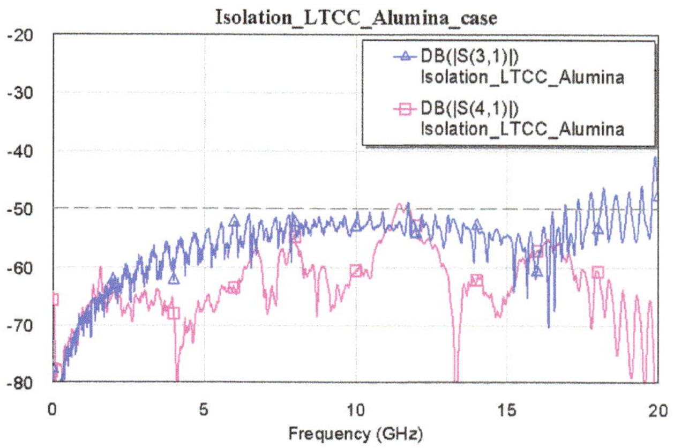

(a) Coupled in LTCC-Alumina transition

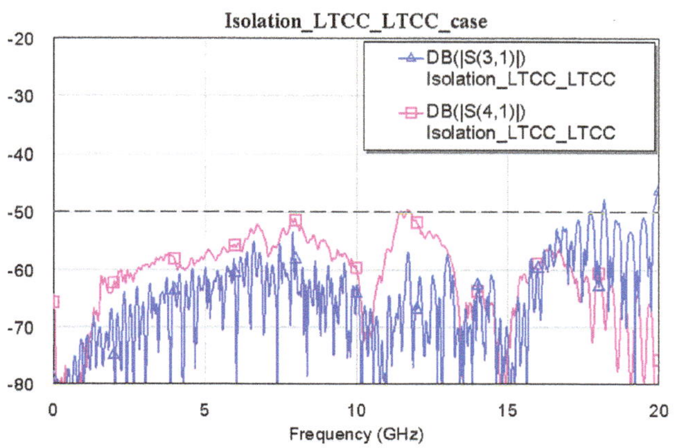

(b) Coupled in LTCC-LTCC transition

Fig. 11. Simulated isolation of 12 × 12 network.

via fences are inserted between lines. A coupling also occurs at the board-to-board transitions, LTCC-LTCC and LTCC-Alumina. In terms of the worst isolation, the couplings in the board-to-board transitions can be separated into two cases. One case is when the LTCC-Alumina transition ports are two times neighbored (see Fig. 10a). Another case is that LTCC-LTCC transition paths are neighbored and two times coupled (see Fig. 10b). Due to the nature of the Beneš network, those two cases can not happen at the same time. Figure 11 shows the isolation of the 12 × 12 network. For all frequency bands of interest, the isolation is better than −50 dB. The commercially available 12 × 12 mechanical switch matrix provided in DowKey Microwave (2012) has a −50 dB input/output isolation, which is comparable to the present matrix isolation.

3.4.3 4 x 4 switch matrix fabrication and measurement

The 4 × 4 switch matrix in Fig. 12 is to test and verify the design process of the 12 × 12 switch matrix. The 4 × 4 matrix is the simplified form of the 12 × 12 matrix and fabricated by the same design process and technologies. There are some differences from the 12 × 12 matrix, which are mainly

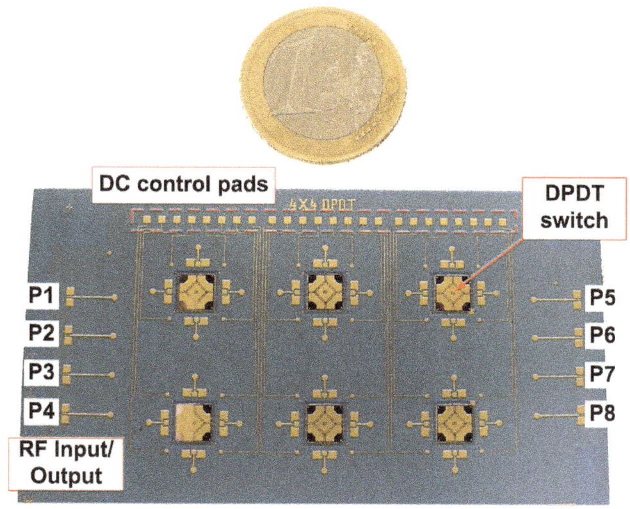

Fig. 12. 4 × 4 switch matrix using RF MEMS switches.

from the design developments after the 4 × 4 fabrication has started. The main difference is the mounting level of DPDT switches. In the 4 × 4 matrix, the DPDT switch is mounted on the higher level, where the DPDT circuit is the level of microstip line (MSL) of the LTCC board. Therefore the matrix has not only SL-SL transitions but also microstrip line (MSL) to SL transitions which are not required in the 12 × 12 matrix. Additionally, in the 4 × 4 matrix, no bondwire compensation structures which suppress the high bondwire inductance and high RL are made on the DPDT switch. Moreover, the DPDT switch and LTCC circuits are connected by a single bondwire. In the 12 × 12 matrix, a double bondwire connection is adopted and bondwire compensation structures are designed on the DPDT switch side as well. Another difference is that the transmission lines of the 4 × 4 matrix are closely distanced compared to the 12 × 12 matrix and no via fences are inserted. However, the performance and comparison to the simulations is still of interest because the same design methods and technologies are used. Figure 13 shows the measured performance of the 4 × 4 matrix and the comparisons to the simulations. The left three graphs show the IL, RL and Isolation when the port 1 and 5 are connected, and the right three are from the connection of port 4 and 5. To measure the isolation, the neighboring port 6 is measured. The measurements agree well to the simulations even though the transition performance is not as good as 12 × 12 matrix's due to the design differences as already noted. Isolation deviations at the high frequency are mainly due to the nature of the circuit simulator which does not account for the component coupling.

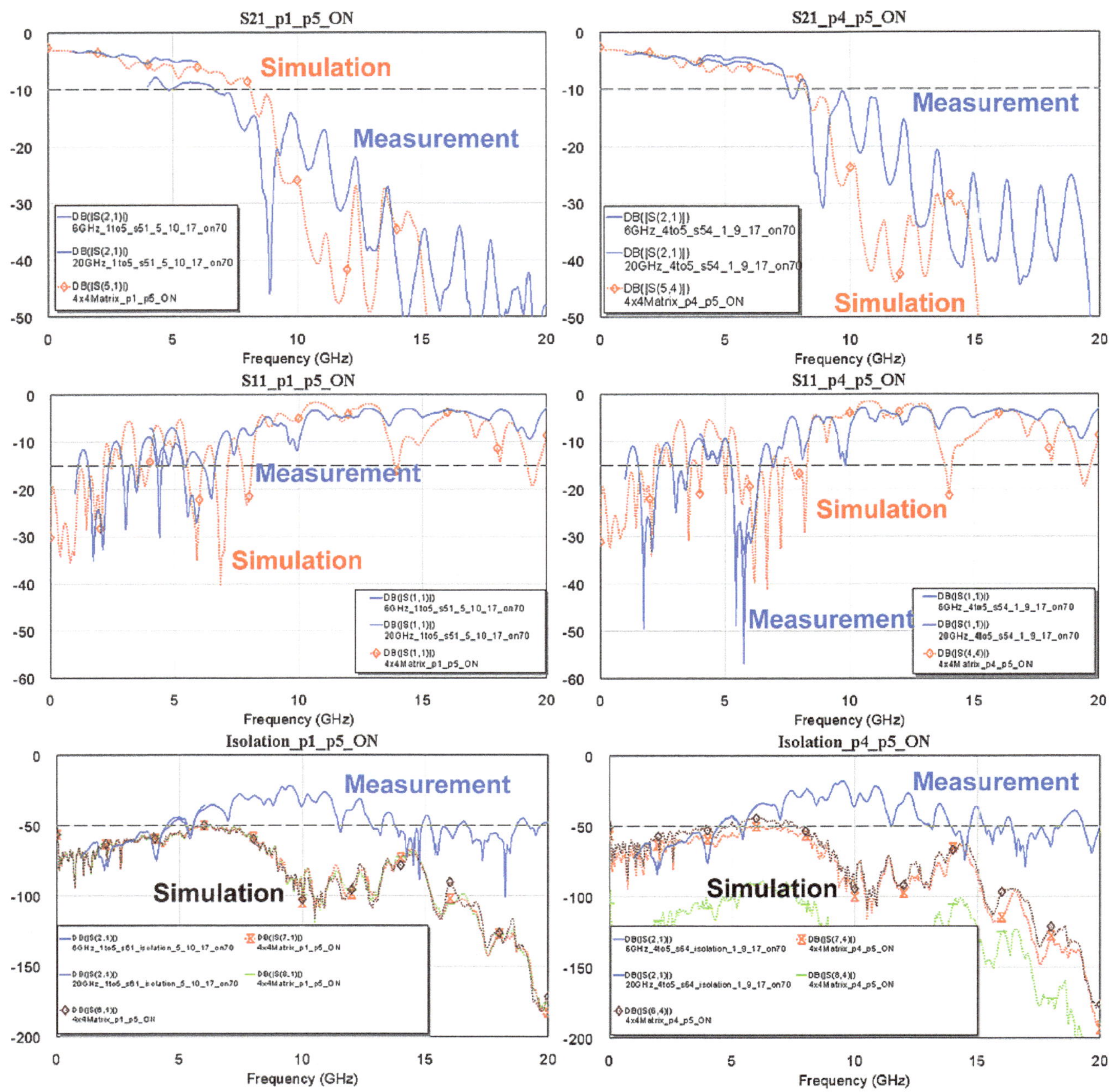

Fig. 13. Measured performance of 4 × 4 matrix and comparison to the simulations.

4 Conclusions

In this work, a 12 × 12 switch matrix using RF MEMS and LTCC technologies is designed based on the planar Beneš network controlled by Looping algorithms. For the switching components, RF MEMS DPDT switches are used and they are connected by multi-layer transmission lines fabricated by LTCC technology. The matrix shows −12 dB IL up to C band and below −15 dB RL up to Ku band. Isolation better than −50 dB is also expected at the frequency bands of interest. Compared to the 12 × 12 mechanical switch matrix which is commercially available, the designed 12 × 12 switch matrix has less than 10 % of size and volume and better RL with comparable isolation. For the verification of the matrix performance, a 4 × 4 switch matrix with the same design process and technology is fabricated and measured. The measurements and the simulations agree well each other, which makes the 12 × 12 switch matrix design process and technology more confident.

Acknowledgements. The work has been carried out in the frame of the ESA Project No AO/1-5136/06/NL/GLC. The authors will thank Francois Deborgies from ESA/ESTEC for his support and suggestions.

References

Chang, C. and Melhem, R.: Arbitrary Size Benes Networks, J. Parallel Process. Lett., 7, 279–284, 1997.

Chao, H. J. and Liu, B.: High performance switches and routers, John Wiley & Sons, Inc., Hoboken, New Jersey, USA, 2007.

De Angelis, G., Lucibello, A., Proietti, E., Marcelli, R., Bartolucci, G., Casini, F., Farinelli, P., Mannocchi, G., Di Nardo, S., Pochesci, D., Margesin, B., Giacomozzi, F., Vendier, O., Kim, T., and Vietzorreck, L.: RF MEMS ohmic switches for matrix configurations, Int. J. Microw. Wirel. Technol., 4, 421–433, doi:10.1017/S1759078712000074, 2012.

Electromechanical – RF Switching System: http://www.dowkey.com, last access: 7 December 2012.

Imanaka, Y.: Multilayered Low Temperature Cofired Ceramics (LTCC) Technology, Springer Science + Business Media, Inc., Boston, USA, 2005.

Yeh, Y.-M. and Feng, T.: On a class of rearrangeable networks, IEEE T. Comput., 41, 1361–1379, doi:10.1109/12.177307, 1992.

Permissions

All chapters in this book were first published in ARS, by Copernicus Publications; hereby published with permission under the Creative Commons Attribution License or equivalent. Every chapter published in this book has been scrutinized by our experts. Their significance has been extensively debated. The topics covered herein carry significant findings which will fuel the growth of the discipline. They may even be implemented as practical applications or may be referred to as a beginning point for another development.

The contributors of this book come from diverse backgrounds, making this book a truly international effort. This book will bring forth new frontiers with its revolutionizing research information and detailed analysis of the nascent developments around the world.

We would like to thank all the contributing authors for lending their expertise to make the book truly unique. They have played a crucial role in the development of this book. Without their invaluable contributions this book wouldn't have been possible. They have made vital efforts to compile up to date information on the varied aspects of this subject to make this book a valuable addition to the collection of many professionals and students.

This book was conceptualized with the vision of imparting up-to-date information and advanced data in this field. To ensure the same, a matchless editorial board was set up. Every individual on the board went through rigorous rounds of assessment to prove their worth. After which they invested a large part of their time researching and compiling the most relevant data for our readers.

The editorial board has been involved in producing this book since its inception. They have spent rigorous hours researching and exploring the diverse topics which have resulted in the successful publishing of this book. They have passed on their knowledge of decades through this book. To expedite this challenging task, the publisher supported the team at every step. A small team of assistant editors was also appointed to further simplify the editing procedure and attain best results for the readers.

Apart from the editorial board, the designing team has also invested a significant amount of their time in understanding the subject and creating the most relevant covers. They scrutinized every image to scout for the most suitable representation of the subject and create an appropriate cover for the book.

The publishing team has been an ardent support to the editorial, designing and production team. Their endless efforts to recruit the best for this project, has resulted in the accomplishment of this book. They are a veteran in the field of academics and their pool of knowledge is as vast as their experience in printing. Their expertise and guidance has proved useful at every step. Their uncompromising quality standards have made this book an exceptional effort. Their encouragement from time to time has been an inspiration for everyone.

The publisher and the editorial board hope that this book will prove to be a valuable piece of knowledge for researchers, students, practitioners and scholars across the globe.

List of Contributors

I. Gaspard
T-Systems, BU Technologiezentrum, Am Kavalleriesand 3, 64295 Darmstadt, Germany

G. Zimmermann
T-Systems, BU Technologiezentrum, Am Kavalleriesand 3, 64295 Darmstadt, Germany

B. Heyne
University of Dortmund, Information Processing Lab, Otto-Hahn-Str. 4, 44227 Dortmund, Germany

J. Götze
University of Dortmund, Information Processing Lab, Otto-Hahn-Str. 4, 44227 Dortmund, Germany

C. Buchacher
Infineon Technologies, Am Campeon 1–12, 85579 Neubiberg, Germany

M. Zimmermann
Infineon Technologies, Am Campeon 1–12, 85579 Neubiberg, Germany

S. Paul
Institut für Theoretische Elektrotechnik und Mikroelektronik, Univ. of Bremen, Otto-Hahn Allee, 28359 Bremen, Germany

M. Huemer
Embedded Systems and Signal Processing Group, University of Klagenfurt, Lakeside B02b, 9020 Klagenfurt, Austria

D. S. Waldhauser
Institute for Circuit Theory and Signal Processing, Munich University of Technology, Germany

J. A. Nossek
Institute for Circuit Theory and Signal Processing, Munich University of Technology, Germany

T. Arafa
University of Kaiserslautern, Communications Engineering, Gottlieb-Daimler-Str., 67653 Kaiserslautern, Germany

W. Sauer-Greff
University of Kaiserslautern, Communications Engineering, Gottlieb-Daimler-Str., 67653 Kaiserslautern, Germany

R. Urbansky
University of Kaiserslautern, Communications Engineering, Gottlieb-Daimler-Str., 67653 Kaiserslautern, Germany

C. de Schryver
Microelectronic Systems Design Research Group, University of Kaiserslautern, Erwin-Schrödinger-Str., Germany

S. Weithoffer
Microelectronic Systems Design Research Group, University of Kaiserslautern, Erwin-Schrödinger-Str., Germany

U. Wasenmüller
Microelectronic Systems Design Research Group, University of Kaiserslautern, Erwin-Schrödinger-Str., Germany

N. Wehn
Microelectronic Systems Design Research Group, University of Kaiserslautern, Erwin-Schrödinger-Str., Germany

C. Brugger
Microelectronic Systems Design Research Group, University of Kaiserslautern, 67663 Kaiserslautern, Germany

S. Weithoffer
Microelectronic Systems Design Research Group, University of Kaiserslautern, 67663 Kaiserslautern, Germany

C. de Schryver
Microelectronic Systems Design Research Group, University of Kaiserslautern, 67663 Kaiserslautern, Germany

U. Wasenmüller
Microelectronic Systems Design Research Group, University of Kaiserslautern, 67663 Kaiserslautern, Germany

N. Wehn
Microelectronic Systems Design Research Group, University of Kaiserslautern, 67663 Kaiserslautern, Germany

M. Kühn
AUDI AG, 85045 Ingolstadt, Germany

W. John
SIL System Integration Laboratory GmbH, Technologiepark 32, 33100 Paderborn, Germany

R. Weigel
Lehrstuhl für Technische Elektronik, Friedrich-Alexander-Universität Erlangen-Nürnberg, Cauerstrasse 9, 91058 Erlangen, Germany

S. Kiefhaber
Technical University of Kaiserslautern, Chair of Communications Engineering, Germany

M. Rosenbaum
Technical University of Kaiserslautern, Chair of Communications Engineering, Germany

W. Sauer-Greff
Technical University of Kaiserslautern, Chair of Communications Engineering, Germany

R. Urbansky
Technical University of Kaiserslautern, Chair of Communications Engineering, Germany

R. Streiter
Chemnitz University of Technology, Professorship of Communications Engineering , Reichenhainer Str. 70, 09126 Chemnitz, Germany

G. Wanielik
Chemnitz University of Technology, Professorship of Communications Engineering , Reichenhainer Str. 70, 09126 Chemnitz, Germany

O. Weikert
Helmut Schmidt University - University of the Federal Armed Forces Hamburg, Department of Signal Processing and Communications, Holstenhofweg 85, 22043 Hamburg, Germany

U. Zölzer
Helmut Schmidt University - University of the Federal Armed Forces Hamburg, Department of Signal Processing and Communications, Holstenhofweg 85, 22043 Hamburg, Germany

M. Kühn
AUDI AG, 85045 Ingolstadt Germany

W. John
SIL System Integration Laboratory GMBH, Technologiepark 32, 33100 Paderborn Germany

R. Weigel
Lehrstuhl für Technische Elektronik, Friedrich-Alexander-Universität Erlangen-Nürnberg, Cauerstrasse 9, 91058 Erlangen Germany

S. Miropolsky
TU Dortmund, Dortmund, Germany

S. Frei
TU Dortmund, Dortmund, Germany

C.Widemann
Institut für theoretische Elektrotechnik, Leibniz Universität Hannover, Appelstraße 9A, 30167 Hannover, Germany

S. Stegemann
Institut für theoretische Elektrotechnik, Leibniz Universität Hannover, Appelstraße 9A, 30167 Hannover, Germany

W. John
Institut für theoretische Elektrotechnik, Leibniz Universität Hannover, Appelstraße 9A, 30167 Hannover, Germany

W. Mathis
Institut für theoretische Elektrotechnik, Leibniz Universität Hannover, Appelstraße 9A, 30167 Hannover, Germany

D. Rinas
Dortmund University of Technology, Dortmund, Germany

J. Jia
Dortmund University of Technology, Dortmund, Germany

A. Zeichner
Dortmund University of Technology, Dortmund, Germany

S. Frei
Dortmund University of Technology, Dortmund, Germany

S. Edinger
University of Mannheim, Chair of Electrical Engineering, Germany

C. Bauer
University of Mannheim, Chair of Electrical Engineering, Germany

N. J. Fliege
University of Mannheim, Chair of Electrical Engineering, Germany

K. Hueske
Information Processing Lab - AG DT, University of Dortmund, Germany

J. Geldmacher
Information Processing Lab - AG DT, University of Dortmund, Germany

J. Götze
Information Processing Lab - AG DT, University of Dortmund, Germany

P. Weitkemper
Department of Communications Engineering, University of Bremen, Otto-Hahn-Allee, 28359 Bremen, Germany

K. Zielinski
Institute for Electromagnetic Theory and Microelectronics, University of Bremen, Otto-Hahn-Allee, 28359 Bremen, Germany

K.-D. Kammeyer
Department of Communications Engineering, University of Bremen, Otto-Hahn-Allee, 28359 Bremen, Germany

R. Laur
Institute for Electromagnetic Theory and Microelectronics, University of Bremen, Otto-Hahn-Allee, 28359 Bremen, Germany

R. Müller
Institute of Traffic Communications Engineering, Dresden University of Technology, Germany

H.-J. Jentschel
Institute of Traffic Communications Engineering, Dresden University of Technology, Germany

O. Schmitz
Leibniz Universität Hannover, Institut für Hochfrequenztechnik und Funksysteme, Appelstr. 9A, 30167 Hannover, Germany

S.K. Hampel
Leibniz Universität Hannover, Institut für Hochfrequenztechnik und Funksysteme, Appelstr. 9A, 30167 Hannover, Germany

H. Eul
Leibniz Universität Hannover, Institut für Hochfrequenztechnik und Funksysteme, Appelstr. 9A, 30167 Hannover, Germany

D. Schwingshackl
Infineon Austria AG, Siemensstr. 2, 9500 Villach, Austria

M. C. Neuenhahn
Chair of Electrical Engineering and Computer Systems, RWTH Aachen University, 52062 Aachen, Germany

H. Blume
Chair of Electrical Engineering and Computer Systems, RWTH Aachen University, 52062 Aachen, Germany

T. G. Noll
Chair of Electrical Engineering and Computer Systems, RWTH Aachen University, 52062 Aachen, Germany

T. Thiessen
Institute of Theoretical Electrical Engineering, Leibniz University of Hannover, Germany

W. Mathis
Institute of Theoretical Electrical Engineering, Leibniz University of Hannover, Germany

J. Dörr
TU Kaiserslautern, Institute of Communications Engineering, 67653 Kaiserslautern, Germany

M. Rosenbaum
TU Kaiserslautern, Institute of Communications Engineering, 67653 Kaiserslautern, Germany

W. Sauer-Greff
TU Kaiserslautern, Institute of Communications Engineering, 67653 Kaiserslautern, Germany

R. Urbansky
TU Kaiserslautern, Institute of Communications Engineering, 67653 Kaiserslautern, Germany

A. Reibiger
Dresden University of Technology, Department of Electrical Engineering, Mommsenstr. 13, 01062 Dresden, Germany

C. Gimmler-Dumont
University of Kaiserslautern, Germany

N. Wehn
University of Kaiserslautern, Germany

T. Maier
Robert Bosch GmbH, Reutlingen, Germany

D. Droste
Robert Bosch GmbH, Reutlingen, Germany

M. Siegel
Karlsruher Institut für Technologie (KIT), Karlsruhe, Germany

O. A. Pfänder
Microelectronics Department, University of Ulm, Germany

R. Nopper
Microelectronics Department, University of Ulm, Germany

H.-J. Pfleiderer
Microelectronics Department, University of Ulm, Germany

S. Zhou
Electrical and Computer Engineering Department, The Hong Kong University of Science and Technology, Hong Kong S.A.R.

A. Bermak
Electrical and Computer Engineering Department, The Hong Kong University of Science and Technology, Hong Kong S.A.R.

S. Scholl
Microelectronic Systems Design Research Group, University of Kaiserslautern, 67663 Kaiserslautern, Germany

E. Leonardi
Institute of Informatics, Federal University of Rio Grande do Sul, 91501970 Porto Alegre, Brazil

N. Wehn
Microelectronic Systems Design Research Group, University of Kaiserslautern, 67663 Kaiserslautern, Germany

E. Hintzen
Institute for Theoretical Electrical Engineering, Leibniz Universität Hannover, Hannover, Germany

T. Vennemann
Institute for Theoretical Electrical Engineering, Leibniz Universität Hannover, Hannover, Germany

W. Mathis
Institute for Theoretical Electrical Engineering, Leibniz Universität Hannover, Hannover, Germany

M. Lechtenberg
Information Processing Lab, TU Dortmund, Otto-Hahn-Str. 4, 44227 Dortmund, Germany

J. Götze
Information Processing Lab, TU Dortmund, Otto-Hahn-Str. 4, 44227 Dortmund, Germany

K. Görner
Institute of Energy Systems, Energy Efficiency and Energy Economics, TU Dortmund, Emil-Figge-Str. 70, 44227 Dortmund, Germany

C. Rehtanz
Institute of Energy Systems, Energy Efficiency and Energy Economics, TU Dortmund, Emil-Figge-Str. 70, 44227 Dortmund, Germany

S. Wunderlich
Hochschule Darmstadt, Department of Electrical and Computer Engineering, Darmstadt, Germany

M.Welpot
Hochschule Darmstadt, Department of Electrical and Computer Engineering, Darmstadt, Germany

I. Gaspard
Hochschule Darmstadt, Department of Electrical and Computer Engineering, Darmstadt, Germany

H. Buddendick
Institute of Radio Frequency Technology, Universität Stuttgart, 70550 Stuttgart, Germany

P. Wertz
Institute of Radio Frequency Technology, Universität Stuttgart, 70550 Stuttgart, Germany

T. F. Eibert
Institute of Radio Frequency Technology, Universität Stuttgart, 70550 Stuttgart, Germany

F. M. Landstorfer
Institute of Radio Frequency Technology, Universität Stuttgart, 70550 Stuttgart, Germany

G. Wölfle
AWE Communications GmbH, Otto-Lilienthal-Str. 36, 71034 Böblingen, Germany

H. Zimmermann
Institute of Electrical Measurements and Circuit Design, Vienna University of Technology, Vienna, Austria

U. Wasenmüller
Microelectronic System Design Research Group, University of Kaiserslautern, Erwin-Schrödinger-Straße, 67663 Kaiserslautern, Germany

T. Brack
Microelectronic System Design Research Group, University of Kaiserslautern, Erwin-Schrödinger-Straße, 67663 Kaiserslautern, Germany

N. Wehn
Microelectronic System Design Research Group, University of Kaiserslautern, Erwin-Schrödinger-Straße, 67663 Kaiserslautern, Germany

M. Lechtenberg
Information Processing Lab, TU Dortmund University, Otto-Hahn-Str. 4, 44227 Dortmund, Germany

J. Götze
Information Processing Lab, TU Dortmund University, Otto-Hahn-Str. 4, 44227 Dortmund, Germany

Y. He
Nokia Research Center, Bochum, Meesmannstr. 103, 44807 Bochum, Germany

K. Hueske
University of Dortmund, Information Processing Lab, Otto-Hahn-Str. 4, 44221 Dortmund, Germany

J. Götze
University of Dortmund, Information Processing Lab, Otto-Hahn-Str. 4, 44221 Dortmund, Germany

E. Coersmeier
Nokia Research Center, Bochum, Meesmannstr. 103, 44807 Bochum, Germany

H. Weber
Institute four Theoretical Electrical Engineering, Leibniz Universität Hannover, Hanover, Germany

C. Widemann
Institute four Theoretical Electrical Engineering, Leibniz Universität Hannover, Hanover, Germany

W. Mathis
Institute four Theoretical Electrical Engineering, Leibniz Universität Hannover, Hanover, Germany

I. Ali
Microelectronic Systems Design Research Group, University of Kaiserslautern, 67663 Kaiserslautern, Germany

U. Wasenmüller
Microelectronic Systems Design Research Group, University of Kaiserslautern, 67663 Kaiserslautern, Germany

N. Wehn
Microelectronic Systems Design Research Group, University of Kaiserslautern, 67663 Kaiserslautern, Germany

U. Wasenmüller
Microelectronic Systems Design Research Group, University of Kaiserslautern, Germany

T. Brack
Microelectronic Systems Design Research Group, University of Kaiserslautern, Germany

I. Groh
Intel Mobile Communications, Germany

E. Staudinger
Institute of Communications and Navigation, German Aerospace Center (DLR), Germany

S. Sand
Institute of Communications and Navigation, German Aerospace Center (DLR), Germany

N. Wehn
Microelectronic Systems Design Research Group, University of Kaiserslautern, Germany

J. Geldmacher
TU Dortmund University, Information Processing Lab, Otto-Hahn-Str. 4, 44227 Dortmund, Germany

K. Hueske
TU Dortmund University, Information Processing Lab, Otto-Hahn-Str. 4, 44227 Dortmund, Germany

M. Kosakowski
Research In Motion GmbH, Universitätsstr. 140, 44799 Bochum, Germany

J. Götze
TU Dortmund University, Information Processing Lab, Otto-Hahn-Str. 4, 44227 Dortmund, Germany

T. Weng
Institute for Theoretical Electrical Engineering, Leibniz University of Hannover, Hannover, Germany

S. Stegemann
Institute for Theoretical Electrical Engineering, Leibniz University of Hannover, Hannover, Germany

W. John
System Integration Laboratory, Paderborn, Germany

W. Mathis
Institute for Theoretical Electrical Engineering, Leibniz University of Hannover, Hannover, Germany

S. Di Nardo
ThalesAlenia Space, L'Aquila, Italy

P. Farinelli
University of Perugia, Perugia, Italy

T. Kim
Technische Universität München, Munich, Germany

R. Marcelli
CNR - Institute for Microelectronics and Microsystems, Roma, Italy

B. Margesin
FONDAZIONE BRUNO KESSLER, Trento, Italy

E. Paola
ThalesAlenia Space, L'Aquila, Italy

D. Pochesci
ThalesAlenia Space, Roma, Italy

L. Vietzorreck
Technische Universität München, Munich, Germany

F. Vitulli
ThalesAlenia Space, Roma, Italy